Oxidative Stress in Plants

Oxidative Stress in Plants

Edited by

Dirk Inzé

and

Marc Van Montagu

Vakgroep Moleculaire Genetica
Universiteit Gent
Belgium

CRC Press
Taylor & Francis Group
Boca Raton London New York

CRC Press is an imprint of the
Taylor & Francis Group, an **informa** business

A TAYLOR & FRANCIS BOOK

First published 2002 by Taylor & Francis

Published 2019 by CRC Press
Taylor & Francis Group
6000 Broken Sound Parkway NW, Suite 300
Boca Raton, FL 33487-2742

© 2002 by Taylor & Francis Group, LLC
CRC Press is an imprint of Taylor & Francis Group, an Informa business

First issued in paperback 2019

No claim to original U.S. Government works

ISBN 13: 978-0-367-44717-5 (pbk)
ISBN 13: 978-0-415-27214-8 (hbk)

Visit the Taylor & Francis Web site at
http://www.taylorandfrancis.com

and the CRC Press Web site at
http://www.crcpress.com

Typeset by EXPO, Malaysia

British Library Cataloguing in Publication Data
A catalogue record for this book is available from the British Library

Library of Congress Cataloging in Publication Data
A catalog record has been requested

CONTENTS

PREFACE

All oxygen-consuming organisms have to deal with the highly reactive nature of oxygen derivatives, such as superoxide radicals, hydrogen peroxide, hydroxyl and lipid radicals. To this end, aerobic organisms have developed sophisticated defense systems that are essential for their survival. Plants have an additional source of oxygen intermediates through photosynthesis, thus rendering them more prone to oxidative damage. Therefore, it comes as no surprise that higher plants possess an array of antioxidant molecules, such as vitamin C, vitamin E, carotenoids, and flavonoids. All these compounds serve as important antioxidants in human nutrition. In plants exposed to environmental stress conditions, the fine balance between the unavoidable production of reactive oxygen species and the antioxidant systems is disturbed, leading to cellular deterioration. The importance of oxidative stress during environmental adversity has been documented widely and has opened new strategies to improve the tolerance of crop plants.

In this book, world-leading scientists have summarized our current knowledge on the various aspects of research on oxidative stress in plants. Chapter 1 reviews the cellular production sites of reactive oxygen species and the most commonly used enzymatic and nonenzymatic strategies to inactivate these toxic oxygen derivatives. In Chapter 2, Christine Foyer highlights the pivotal role of photosynthetic oxygen metabolism to oxidative stress in plants; an important area, which is further detailed in Chapter 9.

Three chapters deal with oxidative stress, generated during abiotic stress, such as low temperature stress (Chapter 3), air pollutants and UV-B radiation (Chapter 4), and metals (Chapter 7).

Reactive oxygen species also play a crucial role in plants' defense against invading pathogens. Pathogen infection can lead to excess production of reactive oxygen species, thereby stopping the invading pathogen and triggering a complex defense system (reviewed in Chapter 5). Chapter 6 discusses the remarkable resemblance between tetrapyrrole-dependent oxidative stress and oxidative stress caused by both abiotic and biotic stress conditions. Furthermore, a thorough understanding of the antioxidant defense systems has opened perspectives to improve the tolerance of plants to environmental stress. Different chapters address this important area of research. In Chapter 8, attempts are discussed to improve the antioxidant systems in maize. In the last three chapters, novel insights are given into the synthesis and role of ascorbic acid and glutathione, two of the most important antioxidants in plants.

It is our hope that this book will further stimulate both new and established scientists to do research on oxidative stress in plants. Although oxygen metabolism and antioxidant defense systems are highly complex, we are now at a stage where clear strategies to improve tolerance to oxidative damage are emerging.

Finally, we would like to thank all who made this book possible: in the first place, all contributing authors; the Editorial staff at Harwood Academic Publishers who convinced us to take on the challenge to edit the book; and, last but not least, Dr Martine De Cock, who continuously helped in getting this volume to the publishers.

Dirk Inzé
Marc Van Montagu

CONTRIBUTORS

Jean-François BRIAT
Biochimie et Physiologie Moléculaire des
  Plantes
Centre National de la Recherche
  Scientifique
Institut National de la Recherche
  Agronomique et Ecole Nationale
  Supérieure d'Agronomie
Place Viala
F-34060 Montpellier Cédex 1
France

Gary P. CREISSEN
John Innes Centre
Norwich Research Park
Colney
Norwich NR4 7UH
United Kingdom

Mark W. DAVEY
Vakgroep Moleculaire Genetica and
  Departement Plantengenetica
Vlaams Interuniversitair Instituut voor
  Biotechnologie
Universiteit Gent
K.L. Ledeganckstraat 35
B-9000 Gent
Belgium

Dieter ERNST
GSF – National Research Center for
  Environment and Health
Institute of Biochemical Plant Pathology
Ingolstädter Landstrasse 1
D-85764 Neuherberg
Germany

Christine H. FOYER
Department of Biochemistry and
  Physiology
IACR-Rothamsted
Harpenden Hertfordshire AL5 2JQ
United Kingdom

Bernhard GRIMM
Institut für Pflanzengenetik und
  Kulturpflanzenforschung
Correnstrasse 3
D-06466 Gatersleben
Germany

Jan-Erik HÄLLGREN
Department of Forest Genetics and Plant
  Physiology
Swedish University of Agricultural
  Sciences
S-90183 Umeå
Sweden

Werner HELLER
GSF – National Research Center for
  Environment and Health
Institute of Biochemical Plant Pathology
Ingolstädter Landstrasse 1
D-85764 Neuherberg
Germany

Dirk INZÉ
Vakgroep Moleculaire Genetica &
  Departement Plantengenetica
Vlaams Interuniversitair Instituut voor
  Biotechnologie
Universiteit Gent
K.L. Ledeganckstraat 35
B-9000 Gent
Belgium

Barbara KARPINSKA
Department of Forest Genetics and Plant
  Physiology
Swedish University of Agricultural
  Sciences
S-90183 Umeå
Sweden

Stanislaw KARPINSKI
Department of Botany
Stockholm University
SE-10691 Stockholm
Sweden

Ulrich KEETMAN
Institut für Pflanzengenetik und
  Kulturpflanzenforschung
Correnstrasse 3
D-06466 Gatersleben
Germany

Christian LANGEBARTELS
GSF – National Research Center for
  Environment and Health
Institute of Biochemical Plant Pathology
Ingolstädter Landstrasse 1
D-85764 Neuherberg
Germany

Jun'ichi MANO
Radioisotope Laboratory
Yamaguchi University
Yoshida 1677-1
Yamaguchi 753-8515
Japan

Mike MAY
Avestha Gengraine Technologies
Plant Genome Biology Laboratory
"Discoverer", 9th floor, Unit 3
International Technology Park
Whitefield Road
Bangalore 560 066
India

Hans-Peter MOCK
Institut für Pflanzengenetik und
  Kulturpflanzenforschung
Correnstrasse 3
D-06466 Gatersleben
Germany

Philip M. MULLINEAUX
John Innes Centre
Norwich Research Park
Colney
Norwich NR4 7UH
United Kingdom

Avihai PERL
Department of Fruit Tree Breeding and
  Molecular Genetics
Institute of Horticulture
Agricultural Research Organization
The Volcani Center
P.O. Box 6
Bet-Dagan 50250
Israel

Rafael PERL-TREVES
Faculty of Life Sciences
Bar-Ilan University
Ramat-Gan 52900
Israel

Rocío SÁNCHEZ-FERNÁNDEZ
Plant Science Institute
Department of Biology
University of Pennsylvania
Philadelphia, PA 19104-6018
USA

Heinrich SANDERMANN Jr.
GSF – National Research Center for
   Environment and Health
Institute of Biochemical Plant Pathology
Ingolstädter Landstrasse 1
D-85764 Neuherberg
Germany

Dierk SCHEEL
Institut für Pflanzenbiochemie
Stiftung des öffentlichen Rechts
Weinberg 3
D-06120 Halle
Germany

Martina SCHRAUDNER
GSF – National Research Center for
   Environment and Health
Institute of Biochemical Plant Pathology
Ingolstädter Landstrasse 1
D-85764 Neuherberg
Germany

Frank VAN BREUSEGEM
Vakgroep Moleculaire Genetica &
   Departement Plantengenetica
Vlaams Interuniversitair Instituut voor
   Biotechnologie
Universiteit Gent
K.L. Ledeganckstraat 35
B-9000 Gent
Belgium

Marc VAN MONTAGU
Vakgroep Moleculaire Genetica &
   Departement Plantengenetica
Vlaams Interuniversitair Instituut voor
   Biotechnologie
Universiteit Gent
K.L. Ledeganckstraat 35
B-9000 Gent
Belgium

Teva VERNOUX
Laboratoire de Biologie Cellulaire
Institut National de la Recherche
   Agronomique
Route de Saint-Cyr
F-78026 Versailles
France

Gunnar WINGSLE
Department of Forest Genetics and Plant
   Physiology
Swedish University of Agricultural
   Sciences
S-90183 Umeå
Sweden

ABBREVIATIONS

ABA	abscisic acid
ACC	1-aminocyclopropane-1-carboxylate
AFP	anti-freeze protein
AO	ascorbate oxidase
AOS	active oxygen species
APX	ascorbate peroxidase
CAP	cold acclimation protein
cAPX	cytosolic APX
CAT	catalase
CDPK	Ca^{2+}-dependent protein kinase
Ci	intracellular CO_2 concentrations
COR	cold-regulated
DHA	dehydroascorbic acid
DHAR	dehydroascorbate reductase
EEE	excess excitation energy
EL	excess light
FNR	ferredoxin-NADP reductase
γECS	γ-glutamylcysteine synthetase
GLDH	L-GL dehydrogenase
GLO	L-gulono-1,4-lactone oxidase
GPX	glutathione peroxidase
GR	glutathione reductase
GSH	glutathione, γ-glutamyl-cysteinylglycine
GSHS	glutathione synthetase
GSSG	glutathione disulphide
GST	glutathione S-transferase
GS-X	GSH conjugate
H_2O_2	hydrogen peroxide
HR	hypersensitive response
L-AA	L-ascorbic acid, L-ascorbate, vitamin C
L-GL	L-galactono-1,4-lactone, L-galactono-γ-lactone, L-galactonic acid lactone
L-GuL	L-gulono-1,4-lactone, L-gulono-γ-lactone
LL	low light

LOOH	lipid hydroperoxide
MDHA	monodehydroascorbic acid, ascorbate-free radical
MDHAR	monodehydroascorbate reductase
MRP	multidrug resistance protein
$O_2^{\bullet-}$	superoxide
$OH^{\bullet}$	hydroxyl radical
PAL	phenylalanine ammonia-lyase
PAR	photosynthetically active radiation
PCD	programmed cell death
POX	predominant peroxidase
PR	pathogenesis-related
PS	photosystem
Q_A	the primary electron acceptor to PSII
q_p	non-photochemical quenching of variable chlorophyll *a* fluorescence
ROI	reactive oxygen intermediate
ROS	reactive oxygen species, active oxygen species
Rubisco	ribulose-1,5-bisphosphate carboxylase/oxygenase
RuBP	ribulose-1,5-bisphosphate
SAA	systemic acquired acclimation
sAPX	stromal APX
SAR	systemic acquired resistance
SOD	superoxide dismutase
tAPX	thylakoid APX
VDE	violaxanthin de-epoxidase

1 Oxidative Stress: An Introduction

Rafael Perl-Treves and Avihai Perl

MOLECULAR OXYGEN AND ITS REACTIVE DERIVATIVES

Atomic oxygen is the most abundant element in the earth's crust; molecular oxygen in the atmosphere and water is required to support all forms of aerobic life. The present oxygen reservoir (37 Emol, 1 Emol $= 10^{18}$ moles) has built up as a result of photosynthesis, a process that liberates dioxygen from water. It is kept approximately constant by respiration, in which O_2 is used as the ultimate electron acceptor. In addition, oxygen atoms are "fixed" into various organic molecules by a variety of enzymes (e.g. oxygenases) and non-enzymatic processes (Gilbert, 1981; Elstner, 1982, 1987). Aerobic organisms must, however, cope with the adverse effects of oxygen. At higher-than-atmospheric concentrations, dioxygen may inhibit or inactivate certain enzymes and it also competes with photosynthetic CO_2 fixation by ribulose-1,5-bisphosphate carboxylase/oxygenase, increasing the energetic cost of photosynthesis. Still, the toxic effect of oxygen is mainly exerted by its reactive derivatives, whereas ground-state dioxygen is rather unreactive and can peacefully co-exist with organic matter. This characteristic is explained by the parallel spins of two unpaired electrons of dioxygen, imposing an energetic barrier on its reaction with non-radical compounds (the "spin restriction"). In order to become chemically reactive, dioxygen must be physically or chemically activated (Table 1).

Physical activation occurs mainly by transfer of excitation energy from a photo-activated pigment such as an excited chlorophyll molecule to dioxygen. The latter absorbs sufficient energy and, as a result, the spin of one electron is inverted. The first *singlet state of oxygen* (designated $^1O_2$ or $^1\Delta gO_2$) is a prevalent reactive species. It is highly diffusible and capable of reacting with organic molecules (whose electrons are usually paired), and damaging photosynthetic membranes.

Chemical activation is the other mechanism to circumvent spin restriction. It occurs by univalent reduction of dioxygen, i.e. addition of electrons one by one. Four electrons (and four protons) are required for the full reduction of dioxygen to water; all three intermediates of univalent reduction, namely superoxide ($O_2^{\bullet-}$), hydrogen peroxide (H_2O_2), and the hydroxyl radical ($OH^{\bullet}$), are chemically reactive and biologically toxic (Elstner, 1987; Hamilton, 1991; McKersie and Leshem, 1994; Yu, 1994). This toxicity is reflected by their short half-lives before reacting with cellular components, as compared to that of dioxygen (>100 sec; Table 1). Reactive oxygen species colliding with an organic molecule may extract an electron from it, rendering it a radical capable of propagating a chain reaction, e.g. the peroxyl ($ROO^{\bullet}$) and alkoxyl ($RO^{\bullet}$) radicals.

Superoxide is the first reduction product of ground state-oxygen, capable of both oxidation and reduction. It may react to produce several other reactive species, and may undergo spontaneous or enzymatic dismutation to H_2O_2.

Hydrogen peroxide is not a free radical, but participates as oxidant or reductant in many cellular reactions. Unlike superoxide, H_2O_2 is highly diffusible through membranes and aqueous compartments and it may directly inactivate sensitive enzymes at a low

Table 1. Formation and characteristics (compared to molecular oxygen) of major reactive oxygen species

Reaction	ΔG kcal/mole	Reactive species	Half-life (37°), sec
Ground state oxygen		dioxygen biradical (•O-O•)	>100
Physical activation			
Chlorophyll* + O_2 → chlorophyll + $^1O_2$	+ 22.0	singlet oxygen (O-O:)	1×10^{-6}
Chemical activation			
$O_2 + e^-$ → $O_2^{•-}$	+7.6	superoxide radical (•O-O:)	1×10^{-6}
$O_2^{•-} + e^- + 2H^+$ → H_2O_2	−21.7	hydrogen peroxide (H:O-:H)	
$H_2O_2 + e^- + H^+$ → $HO^• + H_2O$	−8.8	hydroxyl radical (H:O•)	1×10^{-9}

concentration. Much like superoxide, H_2O_2 is rather stable and therefore less toxic than other reactive oxygen species; the main threat imposed by both superoxide and H_2O_2 lies in their ability to generate highly reactive hydroxyl radicals.

The *hydroxyl radical* is the most powerful oxidizing species in biological systems. It will react non-specifically with any biological molecule, and this will limit its diffusion within the cell to a distance of two molecular diameters from its site of production. No specific scavengers of OH• are known, although several metabolites, such as urea or glucose, were proposed as hydroxyl scavengers in animal systems. Recently, a role for OH• in cell wall polysaccharide metabolism has been proposed (Fry, 1998).

The different reactive species described above will cause, to varying extents, (i) inhibition of sensitive enzymes (some specific examples are discussed below), (ii) chlorophyll degradation or "bleaching", (iii) lipid peroxidation; free radicals, H_2O_2, and singlet oxygen readily attack unsaturated fatty acids, yielding lipid hydroperoxides, and, in the presence of metal catalysts, alkoxyl and peroxyl radicals that propagate chain reactions in the membranes, changing and disrupting lipid structure and membrane organization and integrity (Yu, 1994). In addition, some aldehydes and hydrocarbons produced by lipid peroxidation exert cytotoxic effects in animal systems (Esterbauer *et al.*, 1990). (iv) Indiscriminate attack by hydroxyl radicals of organic molecules, including DNA. A variety of oxidatively altered DNA species can be identified following OH• attack, including base alterations and strand breaks that may be difficult to repair or tolerate (Kasai *et al.*, 1986). Proteins exposed to OH• undergo typical modifications, including specific amino acid alterations, polypeptide fragmentation, aggregation, denaturation, and susceptibility to proteolysis (Wolff *et al.*, 1986).

BIOLOGICAL SOURCES OF REACTIVE OXYGEN

It is well-established that the formation of reactive oxygen species (ROS) accompanies normal metabolic processes in all aerobic organisms. We will describe in some detail the source of different reactive species in plant cells. Figure 1 illustrates many of the physiological pathways that are discussed along the chapter. However, many pioneer discoveries on oxygen radicals were made with a facultative aerobe, *Escherichia coli* (Fridovich, 1991). In aerobically grown *E. coli*, $O_2^{•-}$ is produced by reduction of dioxygen during membrane-associated electron transport. Only 0.04% of the electrons "leak", but a mechanism to prevent superoxide accumulation is required. Superoxide levels were measured using

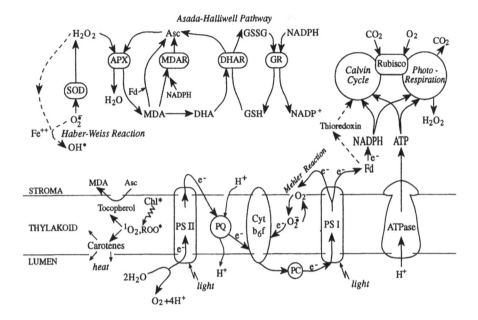

Figure 1. Biochemical chart of the main physiological processes involving reactive oxygen species in the chloroplast.

E. coli membrane preparations of mutant strains devoid of superoxide dismutase (*sodA, sodB*) (Carlioz and Touati, 1986; Imlay and Fridovich, 1991). Such mutants exhibit slow growth and several auxotrophies, demonstrating the vital role of SOD in decreasing steady state levels of $O_2^{\bullet-}$ down to 2×10^{-10} M. It is interesting to note that one of the most superoxide-sensitive enzymes in this system is *E. coli* aconitase of the citric acid cycle (Gardner and Fridovich, 1991). The reversible oxidative inactivation of aconitase by $O_2^{\bullet-}$ has been suggested to fulfill a defensive role of "circuit breaking" to cut off NADPH production, thus avoiding further build-up of superoxide by respiration. Examples of metabolic inactivation/deviation pathways used to avoid reducing conditions in the photosynthetic apparatus are discussed below.

Reactive Oxygen Formation in Plant Chloroplasts

Chloroplasts are the major source of reactive oxygen forms in plants: they harvest light energy at high efficiency, produce reducing equivalents, such as NADPH, and generate fluxes of dioxygen: indeed the most "radical-prone" conditions one could imagine. Several independent pathways, or sites of oxygen activation, have been described in chloroplasts, leading to the production of all of the above reactive species (Elstner, 1982; 1991; Asada, 1994; Foyer and Harbinson, 1994). The most important is the reducing side of photosystem I (PSI), where an electron may be passed from a membrane-bound carrier to O_2 (the "Mehler reaction"), instead of flowing to downstream carriers that finally reduce $NADP^+$ (Figure 1). Under conditions that limit the availability of electron acceptors from PSI, i.e., when the Calvin cycle does not consume NADPH rapidly enough, *superoxide* will form within the membrane. At low pH, superoxide may spontaneously dismutate to the more diffusible H_2O_2; otherwise it may interact with plastocyanine or cytochrome f and reduce them, resulting in a superoxide-mediated cyclic electron flow around PSI (Hormann *et al.*, 1993). This mechanism actually suggests a regulatory role for superoxide production, namely to divert, or cycle, excessive flow of electrons and, at the same time, prevent the diffusion of radicals away from the membrane.

Hydrogen peroxide is mostly produced in chloroplasts by disproportionation of superoxide by SOD, which is much faster than spontaneous dismutation. Another source of H_2O_2 is photorespiration that is initiated by the oxygenase activity of ribulose-1,5-bisphosphate carboxylase in the chloroplast stroma, followed by the production of H_2O_2 in the peroxisomes (Figure 1). Photorespiration may be interpreted as a protective mechanism that recycles electron acceptors and allows photosynthetic electron flow to continue under conditions of low carbon fixation. Compared to electron-cycling around PSI (discussed above as a possible protective pathway), the photorespiratory cycle would dissipate both ATP and NADPH (Wu *et al.*, 1991). An important mechanism of regulation that couples carbon fixation in the stroma to photosynthetic electron flow, is the reversible inactivation of Calvin cycle enzymes when electron carriers of the light reaction are oxidized, and their re-activation when the carriers are reduced. Such regulation is mediated by thioredoxin, stromal pH, and other factors (Figure 1; Foyer *et al.*, 1992). Hydrogen peroxide will disrupt this delicate mechanism by oxidizing thiol groups and inactivating the Calvin cycle enzymes irreversibly; it must therefore be kept below micromolar concentrations in chloroplasts (Foyer and Harbinson, 1994). It will also inactivate copper/zinc (Cu/Zn) SOD.

Hydroxyl radicals may be formed in all living cells in a reaction catalyzed by the transition metal ions, iron and copper, when both superoxide and H_2O_2 are present (Halliwell and Gutteridge, 1992):

(1) the Fenton reaction, resulting in the production of $OH^\bullet$ from H_2O_2:

$$H_2O_2 + Fe^{+2} \rightarrow OH^\bullet + OH^- + Fe^{+3}$$

(2) recycling of ferrous ion by superoxide, which acts as a reductant, allowing reaction (1) to continue:

$$O_2^{\bullet -} + Fe^{+3} \rightarrow O_2 + Fe^{+2}$$

(3) the net sum of reactions (1) and (2) is the so-called Haber-Weiss reaction:

$$H_2O_2 + O_2^{\bullet -} \rightarrow O_2 + OH^\bullet + OH^-$$

Because a common source of H_2O_2 is dismutation of $O_2^{\bullet -}$, SOD is both a scavenger and a source of reactive species. The main danger of superoxide and hydrogen peroxide is indirect, and occurs when they are allowed to accumulate in the same cellular site, and iron and copper metabolism are intimately connected with oxidative stress (see Chapter 7, this volume). According to a somewhat challenging report by Yim *et al.* (1990), hydroxyl radicals are liberated to the solution by the Cu/ZnSOD enzyme itself that reacts with its hydrogen peroxide product; the $OH^\bullet$ radical may then be the direct cause underlying the well known phenomenon of Cu/ZnSOD inactivation by H_2O_2.

Singlet oxygen is formed in chloroplasts when photo-excited chlorophyll in the triplet state reacts with dioxygen. Again, rates are higher when ATP and NADPH utilization by the Calvin cycle reactions is low. Many stress factors that limit CO_2 assimilation (e.g. those that cause stomatal closure) may enhance the process. Surplus excitation energy must therefore be dissipated, for instance via fluorescence or quenching by carotenoid pigments. In addition, singlet oxygen may be produced by certain plant peroxidases.

Reactive Oxygen Produced in Other Cellular Compartments

Mitochondria

Mitochondria consume oxygen during respiratory electron transport. Different sites of electron leakage and release of superoxide and hydrogen peroxide in respiration have been proposed. One of the sites is specific to plant mitochondria, namely the cyanide-insensitive alternative oxidase (Rich and Bonner, 1978; Elstner, 1991; McKersie and Leshem, 1994).

Endoplasmic reticulum, peroxisomes, and glyoxysomes

The smooth endoplasmic reticulum and the microsomes derived from it harbour various oxidative processes. Mixed-function oxygenases, such as cytochrome P450, perform important hydroxylation reactions in the mevalonic acid pathway, adding oxygen atoms to substrate molecules. NAD(P)H is the electron donor and superoxide may be released by such reactions. Peroxisomes and glyoxysomes are single membrane organelles that compartmentalize enzymes involved in the β-oxidation of fatty acids, and the C2 photorespiratory cycle, where glycolate oxidase transfers electrons from glycolate to oxygen and produces H_2O_2 (Lindqvist *et al.*, 1991). Xanthine oxidase, urate oxidase, and NADH oxidase generate superoxide (Elster, 1991; McKersie and Leshem, 1994).

Plasma membrane and the apoplast compartment

NAD(P)H oxidases are ubiquitous components of plasma membranes and may produce superoxide and H_2O_2 (Vianello and Macri, 1991). Oxygen activation occurs also in the apoplast: being the first site of pathogen invasion, it contains a first line of plant defense reactions and these involve reactive oxygen (see below). The most common biosynthetic pathway in the apoplast is lignin biosynthesis, where phenypropanoid precursors of lignin are cross-linked by H_2O_2 in reactions initiated by peroxidases (Gross, 1980). The required NADH is generated by a cell wall malate dehydrogenase, and then used to form H_2O_2, possibly by an NADH oxidase. Amine oxidases produce activated oxygen in the cell wall by using diamines or polyamines to reduce a quinone with concomitant formation of peroxides (Vianello and Macri, 1991; Elstner, 1991; McKersie and Leshem, 1994).

DEFENSE AGAINST OXIDATIVE STRESS

Before discussing specific components that protect plants from reactive oxygen, we should note that the exquisitely sophisticated mechanisms that regulate electron transport and photosynthesis in general, are, in fact, the primary defense against oxidative stress. These mechanisms are responsible for the fine coupling of the light and dark reactions of photosynthesis and for the adaptation of the light-harvesting apparatus to changing conditions. Overexcitation and overreduction of the photosynthetic apparatus are thus avoided, and the formation of reactive oxygen is reduced to a minimum.

Nevertheless, we know that a trickle of active oxygen species is constantly produced in all cellular compartments as a byproduct of normal cellular metabolism and that cell survival will depend upon adequate protection. All aerobic forms of life have evolved multiple defense lines, that include both scavenging enzymes and non-enzymatic antioxidants. Such multiplicity is required because reactive oxygen species (ROS) are produced in different cellular and extracellular compartments, and because reactive species differ in properties such as diffusibility, solubility, and propensity to react with various biological molecules. We thus need a correspondingly diverse set of defense molecules to act in both aqueous and membranal phases, in all cellular compartments, to promptly inactivate radicals as soon as they are formed.

A sequence of detoxification steps is often required to avoid the conversion of one reactive species into a second, more harmful one. The most notable example is the conversion of $O_2^{\bullet -}$ to H_2O_2 by SOD; an insufficiency in the next step, the H_2O_2 detoxification, would lead to H_2O_2 accumulation, inactivation of SOD, and formation of $OH^{\bullet}$ radicals. This also implies that multistep defense systems, if tilted out of balance, may collapse and get out of control, e.g. under extreme stress situations, as well as in aging, cancer, and degenerative syndromes. The coordination of the multiple defense components into one integrated and efficient network is not well understood, and constitutes a challenge to both plant and animal oxidative stress researchers. An outline of the defense systems found in plants (with some reference to non-plant systems) is given below.

Superoxide Dismutase

Superoxide dismutases (SOD) are metalloenzymes first discovered by McCord and Fridovich (1969) that convert $O_2^{\bullet -}$ to H_2O_2 in all aerobic organisms as well as some anaerobes (Hassan, 1989) in the following reaction:

$$2\, O_2^{\bullet -} + 2H^+ \xrightarrow{\text{SOD}} O_2 + H_2O_2$$

Because superoxide is the first product of univalent reduction of oxygen and also the first species to form in many biological systems, SOD is considered as the "primary defense" against oxygen radicals (Bannister *et al.*, 1987). SOD is the fastest enzyme known and its three-dimensional structure has been intensively studied (Kitagawa *et al.*, 1991; Getzoff *et al.*, 1992). The dismutation is catalyzed by the metal ion (Cu, manganese, or iron) at the active site. Superoxide is attracted to such a site by appropriately positioned, positively charged amino acid residues. SODs are classified as MnSODs, FeSODs—both types are phylogenetically related—, and a third unrelated class of Cu/ZnSODs. Plants contain all three types, and distinct SOD isozymes have been identified in the cytosol, mitochondria, and chloroplasts (e.g. Kwiatowski *et al.*, 1985; Palma *et al.*, 1986; Kanematsu and Asada, 1990; for a review, see Bowler *et al.*, 1992). SOD also exists in peroxisomes (Sandalio and del Rio, 1988), glyoxisomes (Bueno and del Rio, 1992), and in the extracellular space

(Castillo *et al.*, 1987; Schinkel *et al.*, 1998). Genes and cDNA-encoding SODs have been cloned from many plant species (Perl-Treves *et al.*, 1988; Bowler *et al.*, 1992; Kliebenstein *et al.*, 1998). All plant SODs are encoded by the nuclear genome (Perl-Treves *et al.*, 1990) and organellar isozymes are transported post-translationally to the appropriate compartment.

SODs are differentially regulated and respond to a variety of stress conditions, such as paraquat application, drought (Perl-Treves and Galun, 1991; Mittler and Zilinskas, 1994), and chilling (Karpinski *et al.*, 1994; see also Chapters 3 and 8, this volume). Several experiments on SOD overproduction in transgenic plants (but also in animal systems) have been reported, implicating SOD in stress tolerance (e.g., Perl *et al.*, 1992; Allen *et al.*, 1997; Van Camp *et al.*, 1997).

Ascorbate Peroxidase, Glutathione Reductase, and Monodehydroxyascorbate Reductase

The product of SOD, hydrogen peroxide, requires further detoxification. This is achieved by other enzymes and non-enzymatic antioxidants that may differ among the various cellular compartments. In the chloroplasts, the so-called "Halliwell-Asada pathway" of detoxification has been studied in detail (Figure 1) (Foyer and Halliwell, 1976; Nakano and Asada, 1981; for a review, see Creissen *et al.*, 1994). H_2O_2 is reduced to water by ascorbate peroxidase (APX). This heme-containing enzyme uses a large pool of 10 mM ascorbate present in the chloroplast and oxidizes it to monodehydroxyascorbate (MDA; Figure 2) in the following reaction:

$$2 \text{ ascorbate } + H_2O_2 \xrightarrow{\text{APX}} 2 \text{ MDA } + 2H_2O$$

In the chloroplasts, both stromal and thylakoid-bound forms of APX were found (Miyake and Asada, 1992), and cytosolic isozymes were described and cloned (Mittler and Zilinskas, 1992; for a review, see Creissen *et al.*, 1994). Although glutathione (GSH) is present in a similarly large pool in plastids, its utilization for direct reduction of H_2O_2 is not an important process in plants. Animals differ in this respect: their glutathione peroxidases (GSH-PRX) are very important both in mitochondria and cytosol and include selenium-containing and selenium-independent enzymes (Yu, 1994). In the plastids, MDA may give rise to dehydroxyascorbate (DHA). Both must be reduced to regenerate the ascorbate pool, which can be achieved by several reactions:

(1) non-enzymatic reduction by ferredoxin:

$$MDA + Fd_{red} \rightarrow \text{ascorbate} + Fd_{ox}$$

(2) reduction of MDA by MDA reductase (MDAR) in the stroma, using NADPH:

$$2 \text{ MDA } + \text{ NADPH } \xrightarrow{\text{MDAR}} 2 \text{ ascorbate } + \text{ NADP}^+$$

(3) reduction of DHA to ascorbate by DHA reductase (DHAR) with GSH as the reducing substrate:

$$DHA + 2 \text{ GSH } \xrightarrow{\text{DHAR}} \text{ascorbate} + GSSH$$

Scavenging is thought to occur at the thylakoid surface, near PSI, minimizing the risk of escape and reaction of ROS with each other. A second line of defense operates in the stroma, to protect the sensitive enzymes of the Calvin cycle. An APX-based cycle for removal may also operate in the plant cytoplasm (Dalton *et al.*, 1987).

Glutathione Reductase

Glutathione reductase (GR) completes the "Asada-Halliwell pathway" by regenerating the glutathione pool with NADPH as electron donor (Foyer and Halliwell, 1976). It should be noted how the repair or prevention of oxidative damage finally consumes reducing equivalents from the light reaction.

$$GSSG + NADPH \xrightarrow{GR} 2\,GSH + NADP^+$$

GR is a flavoprotein, of which homologous enzymes were studied in humans, animals, and microorganisms (Karplus *et al.*, 1989). In plants, most GR activity is found in chloroplasts. Mitochondrial and cytosolic isozymes were described as well, and pea GR was cloned (Creissen *et al.*, 1994). GR and APX activities increase in response to ethylene, ozone, SO_2, and NO_2 (Creissen *et al.*, 1994). Pea GR and *E. coli* GR were overproduced in transgenic tobacco. Results were rather variable: some of the lines with elevated GR activity exhibited increased paraquat or ozone tolerance, whereas others did not (Aono *et al.*, 1993; Broadbent *et al.*, 1995; Creissen *et al.*, 1996).

Catalase

Catalases (CATs) efficiently scavenge H_2O_2 and do not require a reducing substrate to perform the task:

$$2\,H_2O_2 \xrightarrow{CAT} 2\,H_2O + O_2$$

In animal cells, peroxisomal and cytosolic catalases are the primary scavengers of H_2O_2. In leaf tissue, catalase is localized in peroxisomes, to scavenge the H_2O_2 produced by glycolate oxidase in the C2 photorespiratory cycle. Other oxidases are also present in the peroxisomes and produce H_2O_2 as part of ureide, and fatty acid metabolism (Willekens *et al.*, 1995a). In chloroplasts, little or no catalase was found (Asada, 1994), but a mitochondrial isozyme was identified in maize (Scandalios *et al.*, 1980). The peroxisome is linked to the photosynthetic metabolism via the photorespiratory process, and, according to recent findings, may take part in oxidative stress tolerance. Catalase cDNAs were cloned from several plants (Scandalios, 1994; Willekens *et al.*, 1995a). Catalase isozymes differ in biochemical properties, as well as in developmental specificity: some seem related to germination, their principal role involving probably fatty acids conversion, whereas others are related to lignification, photorespiration, or aging processes. A *CAT-2*-deficient maize mutant had no phenotypic lesion (possibly because in C_4 plants photorespiration is less important), but a low-catalase mutant of barley was injured under photorespiratory conditions (Kendall *et al.*, 1983). Catalase undergoes photoinactivation and requires continuous *de novo* synthesis. Stress factors that affect protein synthesis (e.g. heat, chilling) may lead to catalase inactivation. This effect may be compensated for by decreased H_2O_2 production through PSII photoinhibition (see below), as well as compensatory increases in ascorbate, glutathione, and APX (Volk and Feierabend, 1989). Acclimation to chilling by pre-exposure treatment probably includes induction of catalase (Prasad *et al.*, 1994). The *CAT-2* gene of *Nicotiana* is induced by UV-B, ozone, and SO_2 (Willekens *et al.*, 1995b). Recently, a role for H_2O_2 and catalase in the hypersensitive reaction (HR) and systemic-acquired resistance (SAR) responses to plant-pathogen infection has been proposed. The model proposed by Durner and Klessig (1995) (see below) suggests that salicylic acid inhibits catalase and that the resulting H_2O_2 burst is part of the pathogenesis signal transduction chain. Interestingly, the APX enzyme is also inhibited by salicylic acid (Durner and Klessig, 1995).

Chamnongpol *et al.* (1996) described transgenic tobacco plants deficient in specific CAT isozymes. Under dim illumination, the plants looked like wild type, but upon exposure of CAT-1- deficient lines to 500 $\mu E \ m^{-2} \ sec^{-1}$ light, necrotic lesions appeared. Under high light, CAT seemed necessary for good protection against photooxidation. The relationship between H_2O_2 overproduction and defense activation of pathogenesis related proteins via salicylic acid was investigated with such plants (Chamnongpol *et al.*, 1998).

Additional Proteins and Enzymes

Thioredoxin is a small ubiquitous protein that plays a redox-regulatory role in plants (see above). It has also been suggested to protect organisms by scavenging reactive oxygen, as well as regenerating oxidized proteins (Fernando *et al.*, 1992; Takemoto *et al.*, 1998).

Extracellular scavenging systems (for instance in the plasma) are of great importance in animals. Considering the role of transition metal ions in oxidative stress, proteins involved in the homeostasis of Cu and Fe are sometimes regarded as part of the defense against, or regulatory aspects of, oxidative stress. Transferrin, a plasma Fe carrier protein, ferritin, an intracellular Fe-storing complex, as well as ceruloplasm, a Cu-binding glycoprotein, have all been suggested to have antioxidant functions *in vivo*. Nevertheless, when fully loaded with metal, these proteins may actually enhance oxidative stress, and their physiological role in this regard remains unclear (Gutteridge and Halliwell, 1992). Investigations of the role played by the metabolism of metal ions in plant stress (see Chapter 7, this volume) may unravel additional proteins that control the availability of Fe and Cu to the Haber-Weiss reaction.

In the conceptual framework of defense mechanisms, some authors have included additional cellular processes that are activated after severe oxidative damage, such as phospholipases and proteases that degrade and repair biological macromolecules, whose activities are typically induced in animal and bacterial cells by oxidative stress (Davies, 1988; Yu, 1994). According to these authors, such mechanisms allow the cell to regain its homeostasis and should be regarded as secondary defenses. The large and variable class of plant glutathione-*S*-transferases (GST) may qualify as "secondary defense enzymes" (Marrs, 1996). These enzymes catalyze the conjugation of a GSH molecule to a variety of chemical compounds, for example, in the detoxification of herbicides. The conjugate is marked for secretion to the apoplast or vacuole through glutathione pumps. GST conjugations are also important in the synthesis of secondary metabolites. In our context, GSTs detoxify toxic breakdown products of lipid peroxidation or oxidative DNA degradation; they may also function as peroxidases and scavenge radicals. Interestingly, GSTs are induced, among others, by ROS, ozone, wounding, ethylene, heavy metals, and pathogen attack.

Attempts to isolate additional, yet unknown components of oxidative stress defenses have been reported. For example, Kushnir *et al.* (1995) used an *Arabidopsis* cDNA library to transform yeast and identified clones that imparted oxidative stress tolerance to the yeast host. Three clones that encoded previously unknown plant proteins have been selected, one of which is involved in glutathione metabolism. Lin and Culotta (1995) looked for yeast genes that would complement the growth deficiency of SOD-depleted yeast strains and isolated a novel gene (*Atx1*) encoding a small protein similar to bacterial metal transporters. *Atx1* protects against H_2O_2, $O_2^{\bullet-}$, and $OH^{\bullet}$, and is induced by oxygen. Homologous sequences apparently exist in higher plants.

Non-Enzymatic Antioxidants

Ascorbate

Ascorbate (Figure 2), also known as vitamin C, is an important antioxidant in animal systems, where it was shown to react not only with hydrogen peroxide, but also with $O_2^{\bullet-}$, $OH^{\bullet}$, and lipid hydroperoxides (Yu, 1994). Its role as the APX substrate that scavenges H_2O_2 in the chloroplast stroma has been discussed above. Ascorbate is water soluble, but has an additional role on the thylakoid surface in protecting or regenerating oxidized carotenes and tocopherols. The ascorbate pool is also important in the cytosol (Foyer and Harbinson, 1994, and refs. therein). Tappel (1977) proposed that the antioxidant synergism between vitamins C and E in animal tissues (the former present in 100-fold higher concentrations than the latter) is due to the reduction of tocopherol radicals by ascorbate. On the other hand, a too high (>1 mM in animals) concentration of ascorbate may reduce Fe^{+3} to Fe^{+2} and enhance the Haber-Weiss reaction (Girotti, 1985). Once again, we realize how in oxidative defense supra-optimal levels of one scavenger may worsen the situation.

Tocopherol (vitamin E)

The major isomer of vitamin E is α-tocopherol (Figure 2), a phenolic antioxidant present in both plants and animals. Being a lipid-soluble molecule, it is very important as a chain terminator of free-radical reactions that cause lipid peroxidation (Burton *et al.*, 1982). The high degree of lipid unsaturation in chloroplast membranes requires large amounts of α-tocopherol. Plants synthesize tocopherol by enzymes localized in the inner chloroplast membrane (Soll *et al.*, 1984). Animals, though, must acquire it through their diet (a particularly rich source being vegetative oils). The numerous health benefits of vitamin E have been documented (Nesarentam *et al.*, 1992).

Carotenoids

Carotenoids are lipid-soluble molecules that protect both plants and animals against oxidative damage; β-carotene (Figure 2) is the main precursor of vitamin A. Plant

ascorbate (Asc) 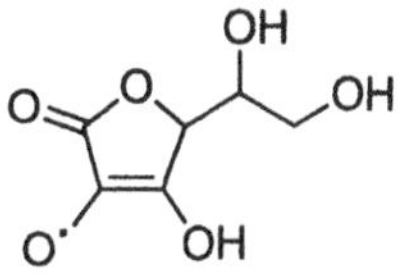　　　　monodehydroascorbate (MDA) 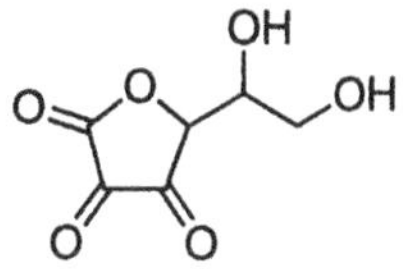　　　　dehydroascorbate (DHA)

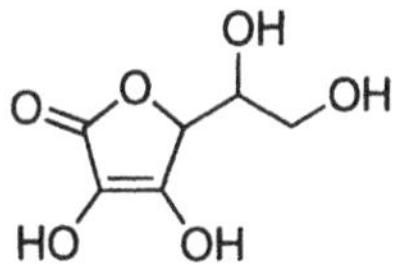

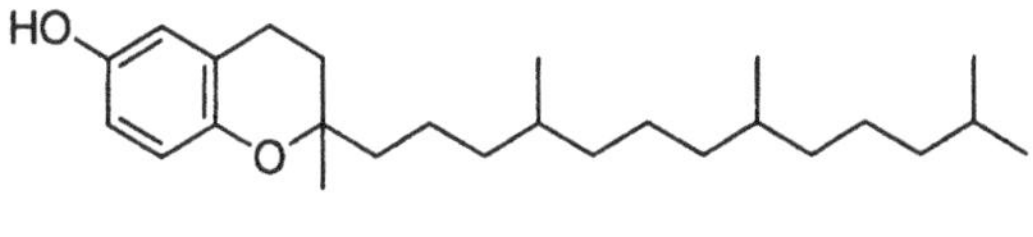

tocopherol

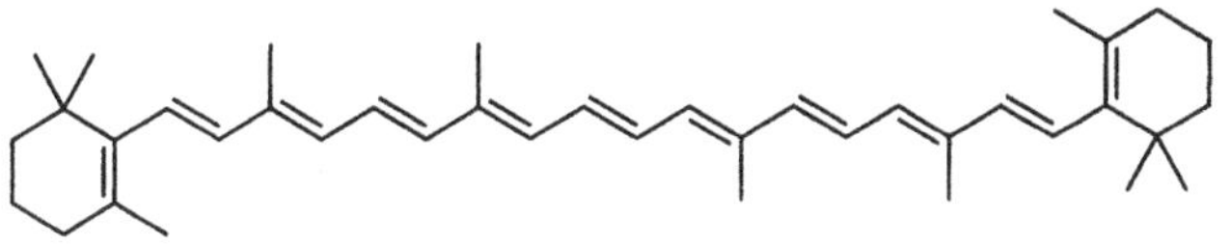

β carotene

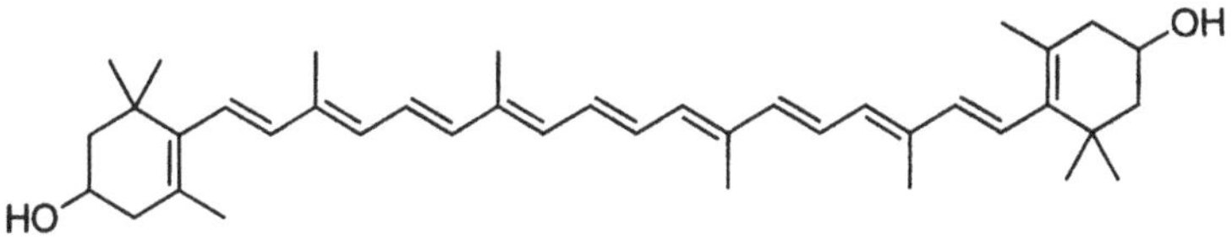

zeaxanthin

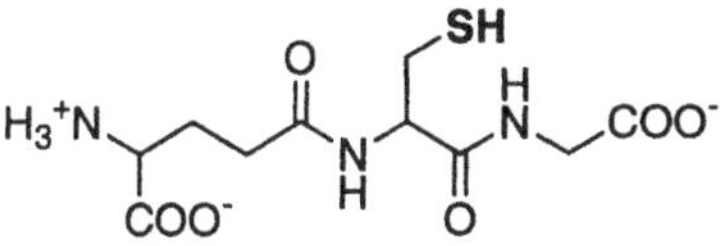

glutathione (GSH)

Figure 2.　Chemical structure of important cellular antioxidants.

carotenoids are formed from isopentenyl diphosphate in the chloroplasts and chromoplasts (Beyer, 1989). In the photosynthetic apparatus, β-carotene quenches both excited triplet-state chlorophyll and singlet oxygen, preventing them from initiating lipid peroxidation (Parker and Joyce, 1967). An excited carotene molecule can return to the ground state either by energy transfer to other pigments in the antenna (acting as a secondary photosynthetic pigment), but also by heat dissipation (acting as a quencher of superfluous excitation energy).

A particular class of carotenoids, the xantophylls, constitute a pool that undergoes prominent changes in response to strong light. The size as well as the composition of the pool change, allows better photoprotection. Zeaxanthin, the de-epoxidized form (Figure 2), better dissipates excitation energy and is derived from the epoxidized pigment violaxanthin. The rapid changes between the two forms, brought about by special enzymes, constitute the "xantophyll cycle" (Demming-Adams and Adams, 1996).

Glutathione

Glutathione (GSH), the ubiquitous γ-tripeptide (Glu-Cys-Ala) (Figure 2) effectively reduces and detoxifies many oxidant species. Enzymatic reactions involving GSH have been discussed above. GSH is important for recycling all the above vitamins, while its own regeneration depends on NAD(P)H consumption. A decrease in reduced glutathione pools accompanies many stress and disease conditions, but it is not always clear whether such a decrease is the cause of the disease or stress situation, or rather reflects a greater demand for GSH (Yu, 1994).

Miscellaneous compounds

Flavonoids are a group of phenolic compounds that may have antioxidant activity (Yuting *et al.*, 1990). Their synthesis in plants increases under strong or UV-enriched light, and they may protect the cell against lipid peroxidation (Torel *et al.*, 1986). Uric acid may play an antioxidant role in animals, both in intra- and extracellular compartments (Davies *et al.*, 1986). Sugar alcohols, such as mannitol, are produced by many plants undergoing osmotic stress. They may serve as compatible solutes and osmoprotectants, but may also function as antioxidants (Smirnoff and Cumbes, 1989; Stoop *et al.*, 1996). Mannitol is a more efficient carbon sink for light reaction products (such as NADPH), and may therefore alleviate photooxidative stress under some circumstances. The potential of plant antioxidant compounds, such as resveratrol from grapes and polyphenols from tea, in human therapeutics has recently drawn much attention (Waffo *et al.*, 1998; Katiyar and Mukhtar, 1997).

OXIDATIVE STRESS AS RELATED TO OTHER PLANT STRESSES

Plants are exposed to abrupt daily and seasonal changes in the environment and they display a wide spectrum of developmental responses and biochemical adaptations to stress conditions. The idea that tolerance mechanisms to several kinds of stress are interconnected and partially overlapping is certainly interesting from a basic, as well as an applied, perspective and seems, today, more plausible than ever. The new and exciting data on the intricacies of signal transduction pathways in plants, and the multiple roles that oxygen radicals play in plant metabolism may indeed point to shared, rather than separate, protective pathways in the plant (Leshem and Kuiper, 1996; Smirnoff, 1998).

Along these lines, a plethora of physiological studies found correlations between levels of antioxidants and the level of stress tolerance among plant species, varieties, and biotypes. Seasonal and developmental variation in detoxifying agents, as well as their induction by many different stresses has been reported. Some of the data linking oxidative stress to other stresses is discussed below. It appears that, in many cases, radical-induced damage underlies stress situations such as heat, cold, UV, air pollutants, and drought. Another important facet of ROS is their newly discovered involvement in signalling, which may imply that tolerance could be selected (or genetically engineered) against one stress, resulting in cotolerance against other stresses. Alternatively, we could elicit defense responses and render plants physiologically tolerant by proper pretreatments, or by affecting the signal transduction chain. A genetic approach to test the physiological relevance of different protectants and to investigate cotolerance phenomena have seldom been taken (Gressel and Galun, 1994). More recently, however, molecular genetic studies are playing an increasingly important role in addressing such questions.

Photoinhibition

Photoinhibition is defined as a decrease in photosynthetic activity after exposure to strong light. In other words, when the supply of reducing power generated by the light reactions exceeds the demands by the dark reactions, the chloroplast may experience oxidative stress and undergo photooxidation, including pigment bleaching and lipid peroxidation. The term "photoinhibition" refers, however, to the faster response of inhibiting photosynthetic electron flow, mainly in PSII, while avoiding further damage (Krause, 1994). Reversible photoinhibition can therefore be regarded as a protective measure to prevent further formation of reactive oxygen and more extreme photooxidative injury. The physiological condition of a given plant will specifically determine whether it will experience photoinhibition at a given light regime. Examples of photoinhibitory conditions include strong light accompanied by low temperatures, or exposure to stronger light without previous gradual adaptation. Other processes that alleviate excess reduction of the photosystems have already been mentioned, e.g. photorespiration and cyclic electron flow. Dioxygen and reactive oxygen species might play a role in photoinhibition, but the mechanism of initial PSII inhibition is still a matter of debate. Does O_2 directly react with reduced plastoquinone (Q_A)? Or does superoxide, generated by the Mehler reaction, inhibit the PSII reaction center? Other possibilities implicate reactions of $O_2^{\bullet-}$, H_2O_2, $OH^{\bullet}$, or singlet oxygen with the D1 reaction center protein, or the P680 chlorophyll (Durrant *et al.*, 1990; Krause, 1994).

Once PSII is inactivated, the D1 reaction center protein becomes altered, undergoes degradation, and must be replaced for the recovery of PSII from photoinhibition. Marking inactivated D1 for degradation depends on oxygen and may be done by singlet oxygen (Sopory *et al.*, 1990).

Recently, attention has been drawn to photoinhibition of PSI as well because of new evidence for its occurrence in higher plants (Sonoike, 1996). In the chilling-sensitive plants, *Cucumis* and *Phaseolus*, PSI was selectively inactivated *in vivo* under chilling conditions; the *psaB* gene product, a PSI reaction center subunit, was degraded following chlorophyll P700 destruction and PSI inactivation. Oxygen was required for this effect, and the addition of active-oxygen scavengers prevented PSI inactivation and PSA-B degradation. Restoration of PSI activity requires *de novo* PSA-B synthesis and would last a few days; the latter protein has a slower turnover than does the D1 protein of PSII. PSI inhibition is therefore relevant when PSII is still active, namely at chilling temperatures and under dim light.

Herbicides, Toxins, and Triazole Compounds

Several herbicides generate reactive oxygen. Paraquat and diquat are bipyridinium herbicides, known to acquire electrons from PSI and generate superoxide in the light (Foyer *et al.*, 1994). Natural or synthetic photosensitizers induce oxidative damage in the light. The most studied natural photosensitizer is a fungal toxin, cercosporin produced by the pathogen *Cercospora* (Daub and Ehrenshaft, 1993). Cercosporin is activated by light and reacts with oxygen to form $^1O_2$, causing severe lipid peroxidation (Daub and Briggs, 1983). Several herbicides, such as aciflurfen, act as photosensitizing compounds and promote the accumulation of metabolic intermediates of chlorophyll. Upon excitation by light such tetrapyrrole intermediates produce singlet oxygen that kills the plant.

Triazole compounds such as paclobutrazol, uniconazole, and triamimenol are plant growth regulators that inhibit a cytochrome P450-dependent oxidation in the gibberellin biosynthesis pathway. Surprisingly, they have recently been shown to confer tolerance to active oxygen generated by paraquat. Paclobutrazol stimulated an increase in antioxidant enzyme activities in wheat plants, increasing their oxidative stress tolerance (Kauss and Jeblick, 1995). Triazole compounds had been previously shown to enhance tolerance of plants to several environmental stresses (Fletcher and Hofstra, 1988).

Metal Toxicity

Accumulation of phytotoxic metals in the environment results from industrial and agricultural practices. Zn, Cu, Fe, and Cd are widespread pollutants and damage plants in two different modes: by (i) direct inhibition of plant growth and biosynthetic pathways, and (ii) involvement in radical production (Foyer *et al.*, 1994). Plant exposed to elevated levels of copper ions were reported to exhibit lipid peroxidation and pigment bleaching (Sandmann and Gonzales, 1989). Prolonged exposure to $CuSO_4$ resulted also in chlorophyll degradation and in a decline in the endogenous level of catalase. Cu and Fe ions are redox active and catalyze the Fenton reaction (see above). Lipid peroxides also originate from the induction of lipoxygenase in the presence of Cu (Foyer *et al.*, 1994). Treatments with cadmium decreased the chlorophyll and heme levels of germinating mung bean seedlings by inducing lipoxygenase activity with the simultaneous inhibition of the antioxidative enzymes (Somashekaraiah *et al.*, 1992; Van Assche and Clijsters, 1990; Gallego *et al.*, 1996). Recent work demonstrated a link between metal toxicity, oxidative stress, and defense responses in *Arabidopsis* and *Nicotiana* (Xiang and Oliver, 1998; Kampfenkel *et al.*, 1995). In yeast, genes coding for both Fe uptake and oxidative stress response are regulated by the same transcription factor (Dancis *et al.*, 1992).

Air Pollution

Atmospheric pollutants, such as ozone and sulfur dioxide, have been implicated in the formation of free radicals (Cross *et al.*, 1998). Mehlhorn (1990) suggested that the phytotoxicity of ozone is due to its oxidizing potential and the consequent formation of radicals. Ozone seems to be a greater threat to plants than sulfur dioxide (Heagle, 1989). Plants treated with ozone exhibited lipid peroxidation, pigment bleaching, degradation of the PSII D1 protein, and a decrease in the activity and quantity of ribulose-1,5-bisphosphate carboxylase (Godde and Buchhold, 1992; Landry and Pell, 1993; Foyer *et al.*, 1994). Exposure to sulfur dioxide resulted in tissue damage and release of stress ethylene from both photosynthetic and non-photosynthetic tissues (Peiser and Yang, 1985). When cells are exposed to sulfur dioxide, an appreciable acidification of the cytoplasm occurs because this gas reacts with water to form sulfurous acid that may then be converted to sulfuric acid (Veljovic-Jovanovic *et al.*, 1993).

Ultraviolet Radiation

Increasing fluxes of UV-B (290–320 nm) radiation are reaching the earth's surface as a consequence of stratospheric ozone depletion (Kerr and McElroy, 1993). The deleterious effects of UV-B on plants have been extensively studied (Teramura and Sullivan, 1994; Bornman and Sundby-Emanuelsson, 1995). Damage by UV-B to PSII involves impairment of electron transport (Hideg *et al.*, 1993) and structural damage of the reaction center proteins, primarily D1 (Greenberg *et al.*, 1989). UV-B irradiation induces the accumulation of free radical-scavenging enzymes, such as SOD (Foyer *et al.*, 1994). Illumination of isolated thylakoid membranes by UV-B produced free radicals, mainly hydroxyl and carbon-centered ones, but did not result in singlet oxygen formation. Besides the immediate free-radical production, UV-B irradiation initiated radical-yielding reactions that can be detected in leaves even minutes after the cessation of the treatment (Hideg and Vass, 1996).

Landry *et al.* (1995) utilized *Arabidopsis thaliana* mutants to characterize physiological processes that are critical for protecting plants from UV-B stress. Mutants defective in their ability to synthesize UV-B-absorbing compounds (flavonoids and sinapate esters) were found to be more sensitive to UV-B than the wild type and exhibited the highest levels of lipid and protein oxidation. APX activity increased in response to the UV-B treatment, indicating that the plant responded to UV-B by expressing an oxidative stress response, and that sunscreen compounds reduce oxidative damage caused by UV-irradiation.

Salt Stress

Salinity affects important metabolic processes located in chloroplasts and mitochondria (Cheeseman, 1988), but little is known about its effect on activated oxygen metabolism of these organelles. Hernandez *et al.* (1995) hypothesized that the decrease in CO_2 concentration in chloroplasts brought about by stomatal closure results in $NADP^+$ shortage and O_2 reduction. Experiments with leaf mitochondria and peroxisomes from NaCl-treated pea plants, have demonstrated a salinity-induced enhancement in $O_2^{\bullet-}$ production, as well as a strong decrease in mitochondrial MnSOD (Hernandez *et al.*, 1993). The possible involvement of activated oxygen species in the mechanism of damage by NaCl stress was further studied in chloroplasts from two pea cultivars with differential sensitivity to 70 mM NaCl. In the tolerant plants, NaCl stress increased Cu/ZnSOD and ascorbate peroxidase activities, as well as the ascorbate pool. In the sensitive plants, the H_2O_2 content and lipid peroxidation products increased without change in the enzymatic activities (Hernandez *et al.*, 1995). These results support the idea proposed by Singha and Choudhuri (1990) that H_2O_2 played a role in the mechanism of salt injury. Similarly, in radish plants exposed to 100 mM NaCl, the activity of APX increased two-fold, but the levels of the respective transcripts remained unchanged, suggesting that the response was mediated by post-transcriptional events (Lopez *et al.*, 1996). In a recent study in *Citrus*, Cu/ZnSOD and APX were induced by salt stress (Gueta-Dahan *et al.*, 1997), as did SOD isozymes of the halophyte *Mesembryanthemum* (Miszalski *et al.*, 1998).

Drought and Heat Stress

During drought stress, an abscisic acid (ABA) signal causes stomatal closure and the light-exposed, over-reduced photosynthetic apparatus may experience oxidative stress. According to Price and Hendry (1991) who studied the role of oxygen radicals in different grasses exposed to drought, water deficit stress causes an overall inhibition of protein synthesis, inactivation of several chloroplast enzymes, impairment of electron transport, increased membrane permeability, and increased activity of enzymes of the H_2O_2 scavenger system. SOD mRNA were found to be induced by ABA (Guan and Scandalios, 1998). In the resurrection plant *Sorobolus stapfianus*, the levels of glutathione reductase and dehydroascorbate reductase increased upon drought (Sgherri and Navari-Izzo, 1995). Perl-Treves and Galun (1991) observed a drought-induced increase in cytosolic Cu/ZnSOD transcript, and Burke *et al.* (1985) reported an increase in GR activity in cotton upon drought. Malan *et al.* (1990) selected heat- and drought-tolerant maize inbreds and found that they had improved coping with oxidative stress. Correlations between antioxidant defense enzymes and heat stress in tomato and *Vicia* were described as well (Rainwater *et al.*, 1996; Filek *et al.*, 1997). The recovery from drought appears also to be a delicate stage that may be accompanied by oxidative stress, requiring the induction of defense systems (Mittler and Zilinskas, 1994). This observation is reminiscent of the recovery from anoxia in flooded plants, where tolerant plants exhibit increased SOD levels (Monk *et al.*, 1987; Biemelt *et al.*, 1998), as well as the phenomenon of oxidative injury upon reperfusion that threatens ischemic patients.

Chilling and Freezing

Combinations of low temperatures and strong light impart photooxidative stress on plants: Wise and Naylor (1987) demonstrated that reactive oxygen and lipid peroxidation are involved in chilling injury of cucumbers. Evergreen forests are seasonally exposed to such a stress, which is further exacerbated by air pollutants (Karpinski *et al.*, 1994; see Chapter 4, this volume). Michalski and Kaniuga (1982), who compared the effect of chilling in tomato, a sensitive species, and spinach, that is cold tolerant, reported that during a cold and dark treatment, sensitive leaves become depleted of Cu ions and of Cu/ZnSOD activity, and experience oxidative injury upon illumination. This observation implies that chilling stress also weakens protection against photooxidation.

Transgenic alfalfa with elevated Cu/ZnSOD showed improved tolerance to freezing stress (McKersie *et al.*, 1993), whereas transgenic tobacco with elevated SOD had an increased chilling stress tolerance (Sen Gupta *et al.*, 1993). Acclimation of maize seedlings to otherwise lethal chilling temperatures by a milder cold pretreatment was accompanied by catalase and peroxidase transcript accumulation (Prasad *et al.*, 1994). Other correlations between cold acclimation and antioxidant defense have recently been reported (O'Kane *et al.*, 1996; Tao *et al.*, 1998). An H_2O_2 treatment could induce defense enzymes and confer chilling tolerance, but, if delivered at cold temperatures, the induction would not work. It remains to be seen whether H_2O_2 induction operates *in vivo* during acclimation to cold stress.

THE ROLE OF REACTIVE OXYGEN IN DISEASE RESISTANCE: REACTIVE SPECIES AS SIGNALS?

The involvement of reactive oxygen in plant-pathogen and plant-pest interactions is one of the most exciting developments in the field of oxidative stress responses. Reactive oxygen species apparently play a multiplicity of roles as cell suicide agents, antimicrobial compounds, lignification substrates, and, most interestingly, as signal molecules (Mehdy, 1994).

The Respiratory Burst

An important resistance response of a plant against an invading pathogen is the hypersensitive response (HR), which induces localized cell death around the invasion site. This response often involves a respiratory burst and the rapid production of oxygen radicals: the accumulation of H_2O_2 and oxygen radicals is one of the earliest events following host-pathogen recognition. In addition to its oxidative potential in killing or inhibiting the growth of pathogens, H_2O_2 participates in a number of responses, such as phytoalexins biosynthesis (Degousee *et al.*, 1994), lignification, cross-linking of cell wall glycoproteins (Brisson *et al.*, 1994), and transcription of defense proteins (Levine *et al.*, 1994). Alvarez *et al.*, (1998) demonstrated that, after the primary burst around the inoculation site, secondary ''micro bursts'' occur that are required for systemic resistance. By using transgenic plants, Harding and Roberts (1998) evidenced the involvement of calmodulin signalling in the plant oxidative burst. Transgenic potato plants overexpressing the glucose oxidase gene from *Aspergillus niger* provided a direct link between H_2O_2 generation and disease resistance (Wu *et al.*, 1995): the increased H_2O_2 levels conferred resistance against *Erwinia carotovora* and *Phytophthora infestans*. The first plant gene encoding a catalytic subunit of a respiratory burst related-oxidase has only recently been cloned (Groom *et al.*, 1996) and it shares homology with mammalian NADPH oxidase, that produces, in neutrophils, an outwards-directed burst of superoxide during inflammation.

Role of Hydrogen Peroxide and Salicylic Acid

In some cases the pathogen elicits a systemic response in the plant, and resistance of distant tissues to subsequent infections (systemic acquired resistance or SAR) is induced by a yet unknown mobile signal. Considerable evidence supports the involvement of salicylic acid (SA) in the induction of SAR (Yalpani *et al.*, 1991). Chen *et al.* (1993)

demonstrated that SA binds to a catalase isoform in tobacco leaves and inhibit its activity, causing increased levels of H_2O_2 and concluded that SA could act via H_2O_2 in inducing SAR. Pretreatments of parsley cell cultures with SA, or methyl jasmonate, greatly enhanced the elicited H_2O_2 burst (Kauss *et al.*, 1994; Kauss and Jeblick, 1995). Vernooij *et al.* (1994) asked whether SA is the mobile signal in SAR. Grafting experiments were performed, using transgenic plants that express a bacterial SA-degrading enzyme. Transgenic root stocks, although unable to accumulate SA, were capable of delivering a signal that rendered the non-transformed scions resistant to pathogen infection (Vernooij *et al.*, 1994). This result indicated that the translocated signal is not SA, although the latter's presence is required in the distant tissue to induce SAR.

Role of Extracellular Superoxide in HR

Certain plant mutations induce dead cell-lesions in the absence of pathogens. Such a mutant, of which lesions spread beyond the well-localized boundaries of the wild-type response was studied. Appearance of the lesions could be triggered by supplying superoxide to extracellular spaces by xanthine and xanthine oxidase, but not by H_2O_2. SOD, but not CAT or APX, attenuated the response (Jabs *et al.*, 1996). This study significantly demonstrates that extracellular $O_2^{\bullet-}$ is a component of the cascade leading to programmed cell death during HR. The *lsd*1 mutation probably impairs the plant's ability to confine the response, or lowers the threshold for its initiation.

Plant Tumor Formation

Another interesting plant-pathogen interaction is tumor formation by *Agrobacterium*. From a new study by Jia *et al.* (1996), this process appears also to involve an oxidative burst! Nononcogenic strains or inoculation with *E. coli* did not provoke the oxidative burst, whereas SOD-overproducing plants had a smaller burst, and a smaller tumor. The relationship between such a burst and the well-established hormone-induced proliferation in the gall is yet unexplained. Is there a possible parallel with the role of radicals in animal tumors?

Nitric Oxide and Plant Stress?

Recently the endogenous production of the nitric oxide radical by plants has been reported (Leshem and Haramaty, 1996), which raised the possibility that this important gaseous signal molecule, intensively studied in animal systems (Moncada *et al.*, 1991), may operate also as a signal in plants and may be related to stress. First proofs of its involvement in the HR response was published recently (Delledonne *et al.*, 1998).

Plant-Pest Interactions and the Oxidative Burst

Besides its involvement in responses to viral, bacterial, and fungal pathogens, reactive oxygen appears to participate also in plant-pest interactions. Several genes whose expression is up-regulated after nematode infection have been isolated from potatoes

(Niebel *et al.*, 1995). One of them encoded a catalase isoform, *Cat2St*, whose mRNA levels increased throughout the infected root. Bi and Felton (1995) have demonstrated a shift in the oxidative status of soybeans attacked by the insect *Helicoverpa zea*: feeding caused significant increases in lipid peroxidation, and hydroxyl radical formation. The activity of several enzymes, including lipoxygenase, peroxidases, ascorbate oxidase, and NADH oxidase increased. Interestingly, the oxidative changes in the host plant correlated with increased oxidative damage in the midgut of insects feeding on previously wounded plants. Spider mite feeding on soybeans increased plant lipid peroxidation, lipogenase, and peroxidase levels, reduced carotenoid levels, but did not affect catalase and SOD (Hildebrand *et al.*, 1986). Phloem-feeding by aphids increased the glutathione reductase levels in wheat and barley (Argandona, 1994).

REACTIVE OXYGEN, DISEASE AND STRESS: HUMAN AND ANIMAL SYSTEMS

The involvement of reactive oxygen with human disease and aging is a "hot" field of study, and the relationships of oxidative stress with a great number of diseases, including diabetes, cataract, and AIDS have been studied. Only a few important examples dealing with cancer and aging will be discussed here.

Oxygen–Free Radicals in Inflammation and Cancer

The involvement of free radicals, such as $O_2^{\bullet -}$, OH$^\bullet$, and carbon-centered alkyl and peroxyl radicals, in inflammation or cancer, has been extensively studied. An important endogenous cause of chronic oxidative stress in animals and humans is the inflammatory response (Cerutti and Trump, 1991). Activated leukocytes generate $O_2^{\bullet -}$ and hypochlorous acid, which represents an important source of oxygen-free radicals *in situ* (Weiss, 1989). The radicals mediate the killing of target cells, but also induce oxidative stress in adjacent tissues. Activated neutrophils stimulate mutagenesis *in vitro* (Weitzman and Gordon, 1990), and oxidative stress from chronic inflammation promotes cancer development in many organs, which may underlie as much as one third of the world cases of cancer (Ames *et al.*, 1993). Examples of cancer induction by chronic inflammation are ulcerative colitis (Collins *et al.*, 1987), mesothelioma caused by asbestos deposits (Mossman *et al.*, 1990), urinary bladder cancer induced by *Schistosoma haematobium* infections (Rosin *et al.*, 1994), and the induction of hepatocellular carcinoma by viral hepatitis (Shimoda *et al.*, 1994).

Other studies suggest that oxygen free radicals may also directly contribute to cancer development (not via chronic inflammation) (Dreher and Junod, 1996; Comstock *et al.*, 1997; Olinski *et al.*, 1998). Radical-related lesions in proteins and DNA accumulate, and reactive oxygen is believed to stimulate the development of cancer at all three stages of the disease: initiation, promotion, and progression.

Initiation, or the first step of carcinogenesis, requires a permanent modification of the DNA in one cell. The number of oxidative hits to DNA is estimated to be approximately 10,000 per day in humans (Ames *et al.*, 1993). Many modifications result in replicative blocks, whereas others may induce point mutations that escape repair mechanisms, and accumulate with age (Lindahl, 1993). The frequent 8-OH-Gua modification produces GC to TA transversions that are frequently detected in the *RAS* oncogene (Bos, 1988), or in modified p53 genes in lung and liver tumors (Hsu *et al.* 1991).

Oxidative stress can also stimulate the proliferation of mutated cell clones, by modulating the expression of specific genes. Oxygen free radicals can induce large changes in cytosolic Ca^{2+}, brought about by the mobilization of intracellular Ca^{2+} stores, or influx of extracellular Ca^{2+}. This mechanism may regulate the transcription of genes involved in cell growth and proliferation (Larson and Cerutti, 1989; Maki *et al.*, 1992; Werlen *et al.*, 1993).

The final stage in cancer development is the acquisition of the malignant properties by the tumor. The accelerated growth of tumors requires additional DNA alterations. It has been hypothesized that an elevated generation of oxygen free radicals in tumor cells increased genomic instability (Toyokuni *et al.*, 1995) and lowered the activity of antioxidant enzymes (Punnonen *et al.*, 1994).

Oxygen–Free Radicals and Aging

Oxidative stress has been postulated, years ago, to be a causal factor in the aging process (Harman, 1956). The basic tenet of this hypothesis is that the age-associated decline in the functional capacity of biological systems is primarily due to the accumulation of irreparable oxidative molecular damage (Sohal *et al.*, 1995). The extent of oxidative damage to DNA, proteins, and lipids has been found to increase with age, providing support for this hypothesis (Agarwal and Sohal, 1994). A possible cause for such age-related accumulation of molecular damage could be a corresponding decline in the efficiency of antioxidative defenses. Extensive studies have yielded, however, a large body of confusing, often contradictory, data. A fairly common finding is that the level of some antioxidative defenses increases, while others decline or remains unchanged with age.

Utilizing X-ray irradiation as a source of reactive oxygen species, Agarwal and Sohal (1996) tried to determine whether the susceptibility of tissues to protein oxidative damage increases with age, and whether tissues of longer-lived species are less susceptible. Brain homogenates from 22-month-old rats were, in fact, more susceptible to oxidative stress than those from 3-month-old rats, and a comparison of five different species (mouse, rat, rabbit, pig, and pigeon) indicated that the maximum life span potential of the species was inversely related to their susceptibility to acute oxidative stress. An earlier study of the same group utilized *Drosophila melanogaster* that overproduced Cu/ZuSOD and catalase (Sohal *et al.*, 1995). The transgenic flies were less susceptible to protein and DNA oxidation, and lived 34% longer than control flies. Laboratory-selected long-lived *Drosophila* strains exhibited improved oxidative stress resistance (Arking, 1998) and interestingly, an analogy with late-flowering *Arabidopsis* mutants exhibiting paraquat tolerance was reported (Kurepa *et al.*, 1998).

Age-related loss of different cognitive and motoric abilities in mice was positively correlated with oxidative molecular damage in the cerebral cortex and the cerebellum, respectively (Forster *et al.*, 1996). These results support the view that oxidative stress is a causal factor in brain senescence, and suggest that age-related decline of cognitive and motoric performance progress independently, involving different regions of the brain.

Although longevity is strongly influenced by the environment, its genetic component is also being investigated (Shmookler and Ebert, 1996). Ebert *et al.* (1996) have identified five chromosomal regions that help specify life span in *Caenorhabditis elegans*, by comparing the genotypes of short-lived and long-lived progenies. This mapping study suggested that longevity is positively correlated with the levels of superoxide dismutase and catalase late in life.

REGULATION OF THE OXIDATIVE STRESS RESPONSE

The Regulatory Problem

The different defense proteins and antioxidant molecules must be present in the cell at some baseline level, to provide a constitutive defense against reactive species that form under normal metabolic conditions, or that appear as a result of a sudden stress. The *E. coli* FeSOD is an example of a constitutive defense protein, present even when the cell grows unaerobically, whereas a second isozyme, MnSOD, is induced by oxygen and oxidative stress. Many of the genes that encode defense components in bacteria, animals, and plants, can, in fact, increase their transcription in response to oxidative stress. Other levels of regulation, such as activation of existing proteins, are known as well. Some of the most interesting open questions involve the regulation of the oxidative stress response. How does the cell sense oxidative stress in its different compartments? How is such information transduced to the photosynthetic and respiratory apparati, to quickly elicit all those subtle physiological adaptations that were discussed above? What are the transcriptional regulators of such genes that encode SOD, APX, and GR? No one has yet identified such components in plants; it may be useful therefore, to have a closer look at some non-plant models, because parallel mechanisms may operate in the plant.

Oxidative-Stress Regulons in Bacteria

Escherichia coli and *Salmonella typhimurium* are enterobacteria that experience oxidative stress when leaving the host's body or when attacked by macrophages. Bacteria have very efficient induction-repression systems; they respond to environmental threats by inducing "global responses", i.e. by expressing groups of unlinked genes (named regulons) that are regulated in concert. A few regulons are related to stress. Hypersensitive mutants that are unable to mount the response, as well as mutants that "respond" constitutively, have allowed the genetic identification and cloning of important regulatory genes.

Hydrogen peroxide at micromolar amounts induces approximately 30 bacterial polypeptides. This treatment renders the microorganism resistant to otherwise lethal millimolar concentrations of H_2O_2. These polypeptides were found to include CAT and GR, but not SOD. Some of them are also part of the heat-shock and/or superoxide responses (Demple, 1991). The *OxyR* locus was found to encode a positive regulator of this response and to belong to a family of DNA-binding proteins. It promotes the transcription of most of its target genes, but downregulates its own. OXY-R is a redox-sensitive protein that undergoes a reversible, subtle conformational change as a result of oxidation. It will bind to its target gene promoters both at 1, and 100 mM dithiotreitol

(DTT) concentrations, respectively, but its DNAase footprinting patterns are different under such conditions, and only the oxidized OXY-R will promote transcription of defense genes (Demple, 1991; Storz *et al.*, 1990).

Superoxide and paraquat induce an additional set of 40 polypeptides, including *SodA*, glucose-6-phosphate dehydrogenase (required to generate NAD(P)H), endonucleases involved in DNA repair, and heat shock chaperonins. Such a diverse list of "regulon members" supports the above-mentioned inclusive view of cellular defenses that considers the various repair functions as secondary defenses against oxidative stress. A regulatory locus that encodes two proteins, SOX-R and SOX-S, positively regulates nine of the superoxide-inducible genes, and negatively regulates three others. It appears that SOX-S directly activates transcription, whereas SOX-R is a redox-sensor that activates the former. The molecular details of the OXY-R and SOX-R/S functions are still actively pursued (Hidalgo and Demple, 1996; Hidalgo *et al.*, 1998). Two additional regulons, and their respective regulators *katF* and *soxQ*, control responses to carbon starvation, antibiotic resistance, and to oxidative stress.

Individual defense genes are often members of several regulons: *sodA*, for example, is regulated by four different ones! This is a genuinely complex regulatory network, of which the coordination and integration aspects are not yet understood. When turning to eukaryotes, with their multiple organelles, and more diverse array of antioxidant molecules, we should indeed expect very sophisticated regulation.

Yeast and Animal Cells

Understanding the regulation of oxidative response in eukaryotes, and its exploitation for therapeutic purposes has become a major research target (Sen, 1998). In yeast, a transcription factor, YAP-1, was identified, the overproduction of which conferred tolerance to toxic compounds, including H_2O_2 and thiol-oxidants. YAP-1 becomes activated and promotes transcription of target genes under oxidative conditions (Kuge and Jones, 1994). This effect results from an increase in YAP-1 binding to its DNA targets, as shown by gel retardation assays. Genes that are transcriptionally induced by *YAP*-1 and confer oxidative stress tolerance were also identified (Kuge and Jones 1994). One of these encodes thioredoxin, a thiol-protein whose defensive role was discussed above. Other targets of YAP-1 include GSH-1, the rate-limiting enzyme in yeast glutathione synthesis (Wu and Moye-Rowley, 1994). The catalase promoter has been recently dissected and regions responsible for its oxidative stress inducibility have been mapped (Nakagawa *et al.*, 1998).

In mammalian tissues, several examples of induction of defense responses have been documented, for example after exposure of lung tissue to hyperoxic conditions (Harris, 1992). The transcription factor AP-1 (homologous to yeast YAP-1) has been studied in detail. It is a heterodimer made of the JUN and FOS polypeptide products. Transcription of the respective gene is induced by H_2O_2. Its DNA-binding activity responds only poorly to hydrogen peroxide, but is strongly increased by reducing conditions (Meyer *et al.*, 1993). Another transcription factor, NFκB, is transcriptionally induced, and also post-transcriptionally activated to bind DNA, by a variety of oxidants and biological inducers, including viral proteins or inflammation cytokines. All these inductory effects can be abolished by antioxidant treatments. Target genes of NFκB include *TNFα* and β-interferon.

Activation of NFκB is a complicated process: it is present in the cell as an inactive complex, and an inhibitory subunit, IκB, must be released to enable the rest of the complex to enter the nucleus and bind target DNA (Ginn-Pease and Whisler, 1998).

Regulation of Defense Systems in Plants

The existence of genetic mechanisms that regulate and coordinate the oxidative defense can be inferred from physiological and genetic studies in which increased expression of a few defense genes in concert has been reported. A thoroughly analyzed example is a paraquat-tolerant genotype of the weed *Conyza bonariensis*, that exhibits increase in the activities of three enzymes, chloroplast SOD, APX, and GR. The phenotype is inherited as a single locus, probably encoding a regulatory gene (Shaaltiel *et al.*, 1988). An alternative explanation would be that the locus regulates only SOD, but the resulting increase in H_2O_2 induces the downstream genes. Indeed, a transgenic plant overproducing Cu/ZnSOD in its chloroplasts had elevated APX as well (Sen Gupta *et al.*, 1993). Therefore, a regulatory role for SOD might be claimed. In another study, however, increased MnSOD targeted to the chloroplasts suppressed endogenous SODs and did not elevate downstream enzyme activities (Slooten *et al.*, 1995).

We have discussed the multiplicity of defenses and the need to regulate their expression according to changes in plant metabolism and in the environment. By examining the induction of individual defense genes at the molecular level, such genes have been found to be regulated both developmentally and in response to stresses of various origin (Kliebenstein *et al.*, 1998). For example, transcript levels of Cu/ZnSODs of tomato respond to light, paraquat application, ethylene, wounding, and drought, and vary according to leaf age (Perl-Treves and Galun, 1991). The cellular mechanism that mediates gene activation in response to all these factors is still unknown; for instance, whether SOD induction during drought occurs as a result of reactive oxygen production or precedes it, and whether its induction by ethylene implies an endogenous role for ethylene in the system. Is ethylene upstream or downstream a putative superoxide signal? Isolation of promoters of oxidative response genes (Hérouart *et al.*, 1993; Kardish *et al.*, 1994; Van Camp *et al.*, 1996) provides important tools to answer such questions, allowing us to "work our way up" the signal transduction chain and identify DNA-binding factors, redox sensing-proteins, etc. The first indications for redox sensing in the transcriptional regulation of plant antioxidant genes were provided by Wingsle and Karpinski (1996), who treated spruce needles with either reduced (GSH) or oxidized (GSSG) glutathione and followed the changes in enzymatic defense levels. GSSG increased GR activity by 60%, probably as a result of a posttranslational modification, because the respective protein and transcript levels remained unaffected. Chloroplast Cu/ZnSOD did not respond, while cytosolic Cu/ZnSOD transcript increased by GSSG and decreased by GSH. Quite different results have been reported by Hérouart *et al.* (1993), who followed the expression of a reporter gene fused to the cytosolic Cu/ZnSOD promoter from tobacco. Here, GSH induced transcription in tobacco protoplasts, while GSSG did not affect it. Molecular identification of putative redox-sensitive factors would allow a closer look at these systems.

Post-translational activation of plant enzymes by the cell redox potential is an established phenomenon. The best known example involves activation of Calvin cycle enzymes by thioredoxin (see above). Another indication for redox modulation involved bacterial cysteine-rich enzymes expressed in transgenic plants. Enzymatic activities of β-glucuronidase and neomycin phosphotransferase II increased 8-fold after dithiothreitol treatment, and both the enzymatic activity and the amount of protein were increased (Garcia-Olmedo *et al.*, 1994). A novel mechanism of redox sensing to mediate light regulation was discovered by Danon and Mayfield (1994), who studied chloroplast gene expression in *Chlamydomonas* and found that gene-specific translational activator proteins are imported into the chloroplast, where they bind to mRNAs, and regulate their translation: RNA binding is increased by reductants and decreased by oxidants.

A further level of physiological regulation involves the action of phytohormones that are released in response to various stresses. How do these interact with the oxidative stress response? Ethylene, ABA, jasmonate, and SA have all been implicated in stress responses such as wounding, anaerobiosis, drought, cold, and salinity (Leshem and Kuiper, 1996). Hormones could initiate defense responses before the redox situation of a cell has changed, because they can convey information about an environmental/metabolic stress that is occurring in a distant tissue. Another possibility is that some hormone-metabolic pathways are themselves sensitive to oxidative changes. As a result, the hormone will propagate the defense response to other tissues, or maintain it for extended periods, following a local change in the redox state at its primary site of production.

REFERENCES

Agarwal, S. and Sohal, R.S. (1994) DNA oxidative damage and life expectancy in houseflies. *Proc. Natl. Acad. Sci. USA*, **91**, 12332–12335.

Agarwal, S. and Sohal, R.S. (1996) Relationship between susceptibility to protein oxidation, aging, and maximum life span potential of different species. *Exp. Gerontol.*, **31**, 365–372.

Allen, R.D., Webb, R.P., and Schake, S.A. (1997) Use of transgenic plants to study antioxidant defenses. *Free Rad. Biol. Med.*, **23**, 473–479.

Alvarez, M.E., Pennell, R.I., Meijer, P.J., Ishikawa, A., Dixon, R.A., and Lamb, C. (1998) Reactive oxygen intermediates mediate a systemic signal network in the establishment of plant immunity. *Cell*, **92**, 773–784.

Ames, B.N., Shigenaga, M.K., and Hagen, T.M. (1993) Oxidants, antioxidants, and the degenerative diseases of aging. *Proc. Natl. Acad. Sci USA*, **90**, 7915–7922.

Aono, M., Kubo, A., Saji, H., Tanaka, K., and Kondo, N. (1993) Enhanced tolerance to photooxidative stress of transgenic *Nicotiana tabacum* with high chloroplastic glutathione reductase activity. *Plant Cell Physiol.*, **34**, 129–135.

Argandona, V.H. (1994) Effect of aphids infestation on the enzyme activities in barley and wheat. *Phytochemistry*, **35**, 1521–1552.

Arking, R. (1998) Molecular-basis of extended longevity in selected drophila strains. *Curr. Sci.*, **74**, 859–864.

Asada, K. (1994) Production and action of active oxygen species in photosynthetic tissues. In C.H. Foyer and P.M. Mullineaux, (eds.), *Causes of Photooxidative Stress and Amelioraton of Defense Systems in Plants*, CRC Press, Boca Raton, pp. 77–104.

Bannister, J.V., Bannister, W.H., and Rotilio, G. (1987) Aspects of the structure, function, and applications of superoxide dismutase. *CRC Crit. Rev. Biochem.*, **22**, 111–180.

Beyer, P. (1989) Carotene biosynthesis in daffodil chromoplasts: on the membrane-integral desaturation and cyclization reactions. In C.D. Bayer, J.C. Shannon, and R.C. Hardison, (eds.), *Physiology, Biochemistry and Genetics of Non-Green Plastids*, American Society of Plant Physiologists, Rockville, pp. 157–170.

Bi, J. and Felton, G. (1995) Foliar oxidative stress and insect herbivory: Primary compounds, secondary metabolites, and reactive oxygen species as components of induced resistance. *J. Chem. Ecol.*, **21**, 1511–1530.

Biemelt, S., Keetman, U., and Albrecht, G. (1998) Re-aeration following hypoxia or anoxia leads to activation of the antioxidative defense system in roots of wheat seedlings. *Plant Physiol.*, **116**, 651–658.

Bornman, J.F. and Sundby-Emanuelsson, C. (1995) Response of plants to UV-B radiation: some biochemical and physiological effects. In N. Smirnoff, (ed.), *Environment and Plant Metabolism: Flexibility and Acclimation*, Bios Scientific, Oxford, pp. 245–262.

Bos, J.L. (1988) The *ras* gene family and human carcinogenesis. *Mutat. Res.*, **195**, 255–271.

Bowler, C., Van Montagu, M., and Inzé, D. (1992) Superoxide dismutase and stress tolerance. *Annu. Rev. Plant Physiol. Plant Mol. Biol.*, **43**, 83–116.

Brisson, L.F., Tenhaken, R., and Lamb, C. (1994) Function of oxidative cross-linking of cell was structural proteins in plant disease resistance. *Plant Cell*, **6**, 1703–1712.

Broadbent, P., Creissen, G.P., Kular, B., Wellburn, A.R., and Mullineaux, P.M. (1995) Oxidative stress responses in transgenic tobacco containing altered levels of glutathione reductase activity. *Plant J.*, **8**, 247–255.

Bueno, P. and del Río, L.A. (1992) Purification and properties of glyoxisomal cuprozinc superoxide dismutase from watermelon cotyledons (*Citrullus vulgaris* Schrad.). *Plant Physiol.*, **98**, 331–336.

Burke, J.J., Gamble, P.E., Hatfield, J.L., and Quisenberry, J.E. (1985) Plant morphological and biochemical responses to field water deficit. I. Responses of glutathione reductase activity and paraquat sensitivity. *Plant Physiol.*, **79**, 415–419.

Burton, G.W., Joyce, A., and Ingold, K.U. (1982) First proof that vitamin E is major lipid-soluble, chain-breaking antioxidant in human blood plasma. *Lancet*, **2**, 327.

Carlioz, A. and Touati, D. (1986) Isolation of superoxide dismutase mutants in *Escherichia coli*: is superoxide dismutase necessary for aerobic life? *EMBO J.*, **5**, 623–630.

Castillo, F.J., Miller, P.R., and Greppin, H. (1987) Extracellular biochemical markers of photochemical oxidant air pollution damage in Norway spruce. *Experientia*, **43**, 111–115.

Cerutti, P.A. and Trump, B.F. (1991) Inflammation and cancer: role of phagocyte-generated oxidants in carcinogenesis. *Blood*, **76**, 655–663.

Chamnongpol, S., Willekens, H., Langebartels, C., Van Montagu, M., Inzé, D., and Van Camp, W. (1996) Transgenic tobacco with a reduced catalase activity develops necrotic lesions and induces pathogenesis related expression under high light. *Plant J.*, **10**, 491–503.

Chamnongpol, S., Willekens, H., Moeder, W., Langebartels, C., Sandermann, H.J., Van Montagu, M., Inzé, D., and Van Camp, W. (1998) Defense activation and enhanced pathogen tolerance induced by H_2O_2 in transgenic tobacco. *Proc. Natl. Acad. Sci USA*, **95**, 5818–5823.

Cheeseman, J.M. (1988) Mechanisms of salinity tolerance in plants. *Plant Physiol.*, **87**, 547–550.

Chen, Z., Silva, H., and Klessig, D.F. (1993) Active oxygen species in the induction of plant systemic acquired resistance by salicylic acid. *Science*, **262**, 1883–1886.

Collins, R.H.J., Feldman, M., and Fordtran, J.S. (1987) Colon cancer, dysplasia, and surveillance in patients with ulcerative colitis. A critical review. *N. Engl. J. Med.*, **316**, 1654–1658.

Comstock, G.W., Alberg, A.J., Huang, H.Y., Wu, K., Burke, A.E., Hoffman, S.C., Norkus, E.P., Gross, M., Cutler, R.G., Morris, J.S., Spate, V.L., and Helzlsouer, K.J. (1997) The risk of developing lung cancer associated with antioxidants in the blood: ascorbic acid, carotenoids, α-tocopherol, selenium, and total peroxyl radical absorbing capacity. *Cancer Epidemiol. Biomarkers Prev.*, **6**, 907–916.

Creissen, G., Broadbent, P., Stevens, R., Wellburn, A.R., and Mullineaux, P. (1996) Manipulation of glutathione metabolism in transgenic plants. *Biochem. Soc. Trans.*, **24**, 465–469.

Creissen, G., Edwards, A., and Mullineaux, P. (1994) Glutathione reductase and ascorbate peroxidase. In C.H. Foyer and P.M. Mullineaux, (eds.), *Causes of Photooxidative Stress and Amelioration of Defense Systems in Plants*, CRC Press, Boca Raton, pp. 343–364.

Cross, C.E., van der Vliet, A., Louie, S., Thiele, J.J., and Halliwell, B. (1998) Oxidative stress and antioxidants at biosurfaces: plants, skin, and respiratory tract surfaces. *Environ. Health Perspect.*, **106**, 1241–1251.

Dalton, D.A., Hanus, F.J., Russel, S.A., and Evans, H.J. (1987) Purification, properties and distribution of ascorbate peroxidase in legume root nodules. *Plant Physiol.*, **83**, 789–794.

Dancis, A., Roman, D.G., Anderson, G.J., Hinnebusch, A.G., and Klausner, R.D. (1992) Ferric reductase of *Saccharomyces cerevisiae*: molecular characterization, role in iron uptake, and transcriptional control by iron. *Proc. Natl. Acad. Sci USA*, **89**, 3869–3873.

Danon, A. and Mayfield, S.P. (1994) Light-regulated translation of chloroplast messenger RNAs through redox potential. *Science*, **266**, 1717–1719.

Daub, M.E. and Briggs, S.P. (1983) Changes in tobacco cell membrane composition and structure caused by the fungal toxin cercosporin. *Plant Physiol.*, **71**, 763–766.

Daub, M.E. and Ehrenshaft, M. (1993) The photoactivated toxin cercosporin as a tool in fungal photobiology. *Plant Physiol.*, **73**, 855–857.

Davies, K.J.A. (1988) Proteolytic systems as secondary antioxidant defenses. In C.K. Chow, (ed.), *Cellular Antioxidant Defense Mechanisms*, CRC Press, Boca Raton, pp. 25–67.

Davies, K.J.A., Sevanian, A., Muakkassah-Kelly, S.F., and Hochstein, P. (1986) Uric acid-iron ion complexes. A new aspect of the antioxidant functions of uric acid. *Biochem. J.*, **235**, 747–754.

Degousee, N., Triantaphylides, C., and Montillet, J.L. (1994) Involvement of oxidative processes in the signaling mechanisms leading to the activation of glyceollin synthesis in soybean (*Glycine max.*). *Plant Physiol.*, **104**, 845–952.

Delledonne, M., Xia, Y., Dixon, R.A., and Lamb, C. (1998) Nitric oxide functions as asignal in plant disease resistance. *Nature*, **394**, 585–588.

Demming-Adams, B. and Adams, W.W.I. (1996) The role of xantophyll cycle carotenoids in the protection of photosynthesis. *Trends Plant Sci.*, **1**, 21–26.

Demple, B. (1991) Regulation of bacterial oxidative stress genes. *Annu. Rev. Genet.*, **25**, 315–337.

Dreher, D. and Junod, A.F. (1996) Role of oxygen free radicals in cancer development. *Eur. J. Cancer*, **32A**, 30–38.

Durner, J. and Klessig, D.F. (1995) Inhibition of ascorbate peroxidase by salicylic acid and 2,6-dichloroisonicotinic acid, two inducers of plant defense responses. *Proc. Natl. Acad. Sci USA*, **92**, 11312–11316.

Durrant, J.R., Giorgi, L.B., Barber, J., Klug, D.R., and Porter, G. (1990) Characterization of triplet states in isolated photosystem II reaction centres: oxygen quenching as a mechanism for photodamage. *Biochim. Biophys. Acta*, **1017**, 167–175.

Ebert, R.H., Shammas, M.A., Sohal, B.H., Sohal, R.S., Egilmez, N.K., Ruggles, S., and Shmookler, R.R. (1996) Defining genes that govern longevity in *Caenorhabditis elegans*. *Dev. Genet.*, **18**, 131–143.

Elstner, E.F. (1982) Oxygen activation and oxygen toxicity. *Annu. Rev. Plant Physiol.*, **33**, 73–96.

Elstner, E.F. (1987) Metabolism of activated oxygen species. In D.D. Davies, (ed.), *Biochemistry of Metabolism* (The Biochemistry of Plants: a Comprehensive Treatise, Vol. 11), Academic Press, San Diego, pp. 253–315.

Elstner, E.F. (1991) Mechanisms of oxygen activation in different compartments of plant cells. In E.J. Pell and K.L. Steffen, (eds.), *Active Oxygen/Oxidative Stress and Plant Metabolism*, American Society of Plant Physiologists, Rockville, pp. 13–25.

Esterbauer, H., Zollner, H., and Schaur, R.J. (1990) Aldehydes formed by lipid peroxidation: mechanisms of formation, occurrence and determination. In C. Vigo-Pelfrey, (ed.), *Membrane Lipid Oxidation*, CRC Press, Boca Raton, pp. 240–268.

Fernando, M.R., Nanri, H., Yoshitake, S., Nagata-Kuno, K., and Minakami, S. (1992) Thioredoxin regenerates proteins inactivated by oxidative stress in endothelial cells. *Eur. J. Biochem.*, **209**, 917–922.

Filek, M., Baczek, R., Niewiadomska, E., Pilipowicz, M., and Koscielniak, J. (1997) Effect of high temperature treatment of *Vicia faba* roots on the oxidative stress enzymes in leaves. *Acta Biochim. Pol.*, **44**, 315–321.

Fletcher, R.A. and Hofstra, G. (1988) Triazoles as potential plant protectants. In D. Berg and M. Plempel, (eds.), *Sterol Biosynthesis Inhibitors: Pharmaceutical and Agrochemical Aspects*, Ellis Horwood, Chichester, pp. 321–331.

Forster, M.J., Dubey, A., Dawson, K.M., Stutts, W.A., Lal, H., and Sohal, R.S. (1996) Age-related losses of cognitive function and motor skills in mice are associated with oxidative protein damage in the brain. *Proc. Natl. Acad. Sci USA*, **93**, 4765–4769.

Foyer, C.H. and Halliwell, B. (1976) The presence of glutathione and glutathione reductase in chloroplasts: a proposed role in ascorbic acid metabolism. *Planta*, **133**, 21–25.

Foyer, C.H. and Harbinson, J. (1994) Oxygen metabolism and the regulation of photosynthetic electron flow. In C.H. Foyer and P.M. Mullineaux, (eds.), *Causes of Photooxidative Stress and Amelioration of Defense Systems in Plants*, CRC Press, Boca Raton, pp. 1–42.

Foyer, C.H., Lelandais, M., and Harbinson, J. (1992) Control of the quantum efficiencies of photosystems I and II, electron flow and enzyme activation following dark to light transitions in pea leaves. *Plant Physiol.*, **99**, 979–986.

Foyer, C.H., Lelandais, M., and Kunert, K.J. (1994) Photooxidative stress in plants. *Physiol. Plant.*, **92**, 696–717.

Fridovich, I. (1991) Molecular oxygen: friend and foe. In E.J. Pell and K.L. Steffen, (eds.), *Active Oxygen/Oxidative Stress and Plant Metabolism* (Current Topics in Plant Physiology, Vol. 6), American Society of Plant Physiologists, Rockville, pp. 1–5.

Fry, S.C. (1998) Oxidative scission of plant cell wall polysaccharides by ascorbate-induced hydroxyl radicals. *Biochem. J.* **332**, 507–515.

Gallego, S.M., Benvades, M.P., and Tomaro, M.I. (1996) Oxidative damage caused by cadmium chloride in sunflower (*Helianthus annuus* L.) plants. *J. Exp. Bot.*, **58**, 41–52.

Garcia-Olmedo, F., Pineiro, M., and Diaz, I. (1994) Dances to a redox tune. *Plant Mol. Biol.*, **26**, 11–13.

Gardner, P.R. and Fridovich, I. (1991) Superoxide sensitivity of the *Escherichia coli* aconitase. *J. Biol. Chem.*, **266**, 19328–19333.

Getzoff, E.D., Cabelli, D.E., Fisher, C.L., Parge, H.E., Viezzoli, M.S., Banci, L., and Hallewell, R.A. (1992) Faster superoxide dismutase mutants designed by enhancing electrostatic guidance. *Nature*, **358**, 347–351.

Gilbert, D.L. (1981) *Oxygen and Living Processes*, Springer-Verlag, Berlin.

Ginn-Pease, M.E. and Whisler, R.L. (1998) Redox signals and NF-kappaB activation in T cells. *Free Rad. Biol. Med.*, **25**, 346–361.

Girotti, A.W. (1985) Mechanisms of lipid peroxidation. *J. Free Rad. Biol. Med.*, **1**, 87–95.

Godde, D. and Buchhold, J. (1992) Effect of long term fumigation with ozone on the turnover of the D-1 reaction center polypeptide of photosystem II in spruce (*Picea abies*). *Physiol. Plant.*, **86**, 568–574.

Greenberg, B.M., Gaba, V., Canaani, O., Malkin, S., and Edelman, M. (1989) Separate photosensitizers mediate degradation of the 32 kDa reaction centre II protein in visible and UV spectral regions. *Proc. Natl. Acad. Sci USA*, **86**, 6616–6620.

Gressel, J. and Galun, E. (1994) Genetic control of photooxidant tolerance. In C.H. Foyer and P.M. Mullineaux, (eds.), *Causes of Photooxidative Stress and Amelioration of Defense Systems in Plants*, CRC Press, Boca Raton, pp. 237–273.

Groom, Q.J., Torres, M.A., Fordham-Skelton, A.P., Hammond-Kosack, K.E., Robinson, N.J., and Jones, J.D. (1996) *rboh*A, a rice homologue of the mammalian gp91phox respiratory burst oxidase gene. *Plant J.*, **10**, 515–522.

Gross, G.G. (1980) The biochemistry of lignification. *Adv. Bot. Res.*, **8**, 25–63.

Guan, L. and Scandalios, J.G. (1998) Two structurally similar maize cytosolic superoxide dismutase genes, Sod4 and Sod4A, respond differentially to abscisic acid and high osmoticum. *Plant Physiol.*, **117**, 217–224.

Gueta-Dahan, Y., Yaniv, Z., Zilinskas, B.A., and Ben-Hayyim, G. (1997) Salt and oxidative stress: similar and specific responses and their relation to salt tolerance in *citrus*. *Planta*, **203**, 460–469.

Gutteridge, J.M.C. and Halliwell, B. (1992) The antioxidant proteins of extracellular fluids. In C.K. Chow, (ed.), *Cellular Antioxidant Defense Mechanisms*, CRC Press, Boca Raton, pp. 1–23.

Halliwell, B. and Gutteridge, J.M. (1992) Biologically relevant metal ion-dependent hydroxyl radical generation. An update. *FEBS Lett.*, **307**, 108–112.

Hamilton, G.A. (1991) Chemical and biochemical reactivity of oxygen. In E.J. Pell and S.K. Steffen, (eds.), *Active Oxygen/Oxidative Stress and Plant Metabolism*, American Society of Plant Physiologists, Rockville, pp. 6–12.

Harding, S.A. and Roberts, D.M. (1998) Incompatible pathogen infection results in enhanced reactive oxygen and cell death responses in transgenic tobacco expressing a hyperactive mutant calmodulin. *Planta*, **206**, 253–258.

Harman, D. (1956) Aging: A theory based on free radical and radiation chemistry. *J. Gerontol.*, **11**, 298–300.

Harris, E.D. (1992) Regulation of antioxidant enzymes. *FASEB J.*, **6**, 2675–2683.

Hassan, H.M. (1989) Microbial superoxide dismutases. *Adv. Genet.*, **26**, 65–97.

Heagle, S.A. (1989) Ozone and crop yield. *Annu. Rev. Phytopathol.*, **27**, 397–423.

Hernandez, J.A., Corpas, F.J., Gomez, L.A., del Río, L.A., and Sevilla, F. (1993) Salt induced oxidative stress mediated by activated oxygen species in pea leaf mitochondria. *Plant Physiol.*, **89**, 103–110.

Hernandez, J.A., Olmos, E., Corpas, F.J., Sevilla, F. and del Río, L.A. (1995) Salt induced oxidative stress in chloroplasts of pea plants. *Plant Sci.*, **105**, 151–167.

Hérouart, D., Van Montagu, M., and Inzé, D. (1993) Redox-activated expression of the cytosolic copper/zinc superoxide dismutase gene in *Nicotiana*. *Proc. Natl. Acad. Sci USA*, **90**, 3108–3112.

Hidalgo, E. and Demple, B. (1996) Activation of SoxR-dependent transcription *in vitro* by noncatalytic or NifS-mediated assembly of [2Fe-2] clusters into apo-SoxR. *J. Biol. Chem.*, **271**, 7269–7272.

Hidalgo, E., Leautaud, V., and Demple, B. (1998) The redox-regulated SoxR protein acts from a single DNA site as a repressor and an allosteric activator. *EMBO J.*, **17**, 2629–2636.

Hideg, E. and Vass, I. (1996) UV-B induced free radical production in plant leaves and isolated thylakoid membranes. *Plant Sci.*, **115**, 251–260.

Hideg, E., Sass, L., Barbato, R., and Vass, I. (1993) Inactivation of photosynthetic oxygen evolution by UV-B irradiation: a thermoluminescence study. *Photosynthesis Res.*, **38**, 455–462.

Hildebrand, D.F., Rodrigues, J.G., Brown, G.C., Luu, K.T., and Volden, C.S. (1986) Peroxidative responses of leaves in soybeans injured by two spotted spider mites (Acari: Tetrachidea). *J. Econ. Entomol.*, **79**, 1459–1465.

Hormann, H., Neubauer, C., Asada, K., and Schreiber, U. (1993) Intact chloroplasts display pH 5 optimum of O_2 reduction in the absence of emthyl viologen: indirect evidence for a regulatory role of superoxide protonation. *Photosynthesis Res.*, **37**, 69–80.

Hsu, I.C., Metcalf, R.A., Sun, T., Welsh, J.A., Wang, N.J., and Harris, C.C. (1991) Mutational hotspot in the p53 gene in human hepatocellular carcinomas. *Nature*, **350**, 427–428.

Imlay, J.A. and Fridovich, I. (1991) Assay of metabolic superoxide production in *Escherichia coli*. *J. Biol. Chem.*, **266**, 6957–6965.

Jabs, T., Dietrich, R.A., and Dangl, J.L. (1996) Initiation of runaway cell death in an *Arabidopsis* mutant by extracellular superoxide. *Science*, **273**, 1853–1856.

Jia, S.R., Kumar, P.P., and Kush, A. (1996) Oxidative stress in *Agrobacterium*-induced tumors on *Kalanchoe* plants. *Plant J.*, **10**, 545–551.

Kampfenkel, K., Van Montagu, M., and Inzé, D. (1995) Effects of iron excess on *Nicotiana plumbaginifolia* plants. *Plant Physiol.*, **107**, 725–735.

Kanematsu, S. and Asada, K. (1990) Characteristic amino acid sequences of chloroplast and cytosol isozymes of Cu,Zn superoxide dismutase in spinach, rice and horsetail. *Plant Cell Physiol.*, **31**, 99–112.

Kardish, N., Magal, N., Aviv, D., and Galun, E. (1994) The tomato gene for the chloroplastic Cu,Zn superoxide dismutase: regulation of expression imposed in transgenic tobacco plants by a short promoter. *Plant Mol. Biol.*, **25**, 887–897.

Karpinski, S., Karpinska, B., Wingsle, G., and Hallgren, J.-E. (1994) Molecular responses to photooxidative stress in *Pinus sylvestris*. I. Differential expression of nuclear and plastid genes in relation to recovery from winter stress. *Physiol. Plant.*, **90**, 358–366.

Karplus, P.A., Pai, E.F., and Schulz, G.E. (1989) A crystallographic study of the glutathione binding site of glutathione reductase at 0.3-nm resolution. *Eur. J. Biochem.*, **178**, 693–703.

Kasai, H., Crain, P.F., Kuchino, Y., Nishimura, S., Ootsuyama, A., and Tanooka, H. (1986) Formation of 8-hydroxyguanine moiety in cellular DNA by agents producing oxygen radicals and evidence for its repair. *Carcinogenesis*, **7**, 1849–1851.

Katiyar, S.K. and Mukhtar, H. (1997) Tea antioxidants in cancer chemoprevention. *J. Cell Biochem. Suppl.*, **27**, 59–67.

Kauss, H. and Jeblick, W. (1995) Pretreatment of parsley suspension cultures with salicylic acid enhances spontaneous and elicited production of H_2O_2. *Plant Physiol.*, **108**, 1171–1178.

Kauss, H., Jeblick, W., Ziegler, J., and Krabler, W. (1994) Pretreatment of parsley (*Petroselinum crispum*) suspension cultures with methyl jasmonate enhanced elicitation of activated oxygen species. *Plant Physiol.*, **105**, 89–94.

Kendall, A.C., Keys, A.J., Turner, J.C., Lea, P.J., and Miflin, B.J. (1983) The isolation and characterization of a catalase-deficient mutant of barley (*Hordeum vulgare*). *Planta*, **159**, 505–511.

Kerr, J.B. and McElroy, C.T. (1993) Evidence for large upward trends of ultraviolet-B radiation linked to ozone depletion. *Science*, **262**, 1032–1034.

Kitagawa, Y., Tanaka, N., Hata, Y., Kusunoki, M., Lee, G.P., Katsube, Y., Asada, K., Aibara, S., and Morita, Y. (1991) Three-dimensional structure of Cu, Zn-superoxide dismutase from spinach at 2.0 A resolution. *J Biochem.* (Tokyo) **109**, 477–485.

Kliebenstein, D.J., Monde, R.A., and Last, R.L. (1998) Superoxide dismutase in Arabidopsis: An eclectic enzyme family with disparate regulation and protein localization. *Plant Physiol.*, **118**, 637–650.

Krause, G.H. (1994) The role of oxygen in photoinhibition of photosynthesis. In C.H. Foyer and P.M. Mullineaux, (eds.), *Causes of Photooxidative Stress and Amelioraton of Defense Systems in Plants*, CRC Press, Boca Raton, pp. 43–76.

Kuge, S. and Jones, N. (1994) *YAP1* dependent activation of TRX2 is essential for the response of *Saccharomyces cerevisiae* to oxidative stress by hydroperoxides. *EMBO J.*, **13**, 655–664.

Kurepa, J., Smalle, J., Montagu, M.V., and Inzé, D. (1998) Oxidative stress tolerance and longevity in *Arabidopsis*: the late-flowering mutant *gigantea* is tolerant to paraquat. *Plant J.*, **14**, 759–764.

Kushnir, S., Babiychuk, E., Kampfenkel, K., Belles-Boix, E., Van Montagu, M., and Inzé, D. (1995) Characterization of *Arabidopsis thaliana* cDNAs that render yeasts tolerant toward the thiol-oxidizing drug diamide. *Proc. Natl. Acad. Sci USA*, **92**, 10580–10584.

Kwiatowski, J., Safianowska, A., and Kaniuga, Z. (1985) Isolation and characterization of an iron-containing superoxide dismutase from tomato leaves, *Lycopersicon esculentum*. *Eur. J. Biochem.*, **146**, 459–466.

Landry, L.G. and Pell, E.J. (1993) Modification of rubisco and altered proteolytic activity in O$_3$-stressed hybrid poplar (*Populus maximowizii x trichocarpa*). *Plant Physiol.*, **101**, 1355–1362.

Landry, L.G., Chapple, C.C., and Last, R.L. (1995) Arabidopsis mutants lacking phenolic sunscreens exhibit enhanced ultraviolet-B injury and oxidative damage. *Plant Physiol.*, **109**, 1159–1166.

Larson, R. and Cerutti, P. (1989) Translocation and enhancement of phosphotransferase activity of protein kinase C following exposure in mouse epidermal cells to oxidants. *Cancer Res.*, **49**, 5627–5632.

Leshem, Y.Y. and Haramaty, E. (1996) The characterization and contrasting effects of the nitric oxide free radical in vegetative stress and senescence of *Pisum sativum* Linn. foliage. *J. Plant Physiol.*, **148**, 258–263.

Leshem, Y.Y. and Kuiper, P.J.C. (1996) Is there a GAS (general adaptation syndrome) response to various types of environmental stress? *Biol. Plant.*, **38**, 1–18.

Levine, A., Tenhaken, R., Dixon, R., and Lamb, C. (1994) H$_2$O$_2$ from the oxidative burst orchestrates the plant hypersensitive disease resistance response. *Cell*, **79**, 583–593.

Lin, S.J. and Culotta, V.C. (1995) The *ATX1* gene of *Saccharomyces cerevisiae* encodes a small metal homeostasis factor that protects cells against reactive oxygen toxicity. *Proc. Natl. Acad. Sci USA*, **92**, 3784–3788.

Lindahl, T. (1993) Instability and decay of the primary structure of DNA. *Nature*, **362**, 709–715.

Lindqvist, Y., Branden, C.I., Mathews, F.S., and Lederer, F. (1991) Spinach glycolate oxidase and yeast flavocytochrome b2 are structurally homologous and evolutionarily related enzymes with distinctly different function and flavin mononucleotide binding. *J. Biol. Chem.*, **266**, 3198–3207.

Lopez, F., Vansuyt, G., Casse-Delbart, F., and Fourcroy, P. (1996) Ascorbate peroxidase activity, not the mRNA level, is enhanced in salt-stressed *Raphamus sativa* plants. *Physiol. Plant.*, **97**, 13–20.

Maki, A., Berezesky, I.K., Fargnoli, J., Holbrook, N.J., and Trump, B.F. (1992) Role of [Ca^{2+}]i in induction of c-fos, c-jun, and c-myc mRNA in rat PTE after oxidative stress. *FASEB J.*, **6**, 919–924.

Malan, C., Greyling, M.M., and Gressel, J. (1990) Correlation between Cu/Zn superoxide dismutase and glutathione reductase, and environmental and xenobiotic stress tolerance in maize inbreds. *Plant Sci.*, **69**, 157–166.

Marrs, K.A. (1996) The function and regulation of glutathione-S-transferases in plants. *Plant Physiol. Plant Mol. Biol.*, **47**, 127–158.

McCord, J.M. and Fridovich, I. (1969) Superoxide dismutase. An enzymic function for erythrocuprein (hemocuprein). *J. Biol. Chem.*, **244**, 6049–6055.

McKersie, B.D., Chen, Y., de Beus, M., Bowley, S.R., Bowler, C., Inzé, D., D'Halluin, K., and Botterman, J. (1993) Superoxide dismutase enhances tolerance of freezing stress in transgenic alfalfa (*Medicago sativa* L.). *Plant Physiol.*, **103**, 1155–1163.

McKersie, B.D. and Leshem, Y.Y. (1994) *Stress and Stress Coping in Cultivated Plants*, Kluwer Academic Publishers, Dordrecht.

Mehdy, M.C. (1994) Active oxygen species in plant defense against pathogens. *Plant Physiol.* 105, 467–472.

Mehlhorn, H. (1990) Ethylene-promoted ascorbate peroxidase activity protects plants against hydrogen peroxide, ozone and paraquat. *Plant Cell Environ.*, **13**, 971–976.

Meyer, M., Schreck, R., and Baeuerle, P.A. (1993) H2O2 and antioxidants have opposite effects on activation of NF-κB and AP-1 in intact cells: AP-1 as secondary antioxidant-responsive factor. *EMBO J.*, **12**, 2005–2015.

Michalski, W.P. and Kaniuga, Z. (1982) Photosynthetic apparatus of chilling sensitive plants. XI. Reversibility by light of cold and dark-induced inactivation of cyanide sensitive superoxide dismutase activity in tomato leaf chloroplasts. *Biochim. Biophys. Acta*, **680**, 250–257.

Miszalski, Z., Slesak, I., Niewiadomska, E., Baczekkwinta, R., Luttge, U., and Ratajczak, R. (1998) Subcellular-localization and stress responses of superoxide-dismutase isoforms from leaves in the C-3-CAM intermediate halophyte *Mesembryanthemum crystallinum* L. *Plant Cell Environ.*, **21**, 19–179.

Mittler, R. and Zilinskas, B.A. (1992) Molecular cloning and characterization of a gene encoding pea cytosolic ascorbate peroxidase. *J. Biol. Chem.*, **267**, 21802–21807.

Mittler, R. and Zilinskas, B.A. (1994) Regulation of pea cytosolic ascorbate peroxidase and other antioxidant enzymes during the progression of drought stress and following recovery from drought. *Plant J.*, **5**, 397–405.

Miyake, C. and Asada, K. (1992) Thylakoid-bound ascorbate peroxidase in spinach chloroplasts and photoreduction of its primary oxidation product, monodehydroxyascorbate radicals, in thylakoids. *Plant Cell Physiol.*, **33**, 541–553.

Moncada, S., Palmer, R.M., and Higgs, E.A. (1991) Nitric oxide: physiology, pathophysiology, and pharmacology. *Pharmacol. Rev.*, **43**, 109–142.

Monk, L.S., Fagerstedt, K.V., and Crawford, R.M.M. (1987) Superoxide dismutase as an anaerobic polypeptide: a key factor in recovery from oxygen deprivation in *Iris pseudacorus*? *Plant Physiol.*, **85**, 1016–1020.

Mossman, B.T., Bington, J., Corn, M., Seaton, A., and Gee, J.B.L. (1990) Asbestos: scientific development and implications for public policy. *Science*, **247**, 294–301.

Nakagawa, C.W., Yamada, K., and Mutoh, N. (1998) Two distinct upstream regions are involved in expression of the catalase gene in Schizosaccharomyces pombe in response to oxidative stress. *J. Biochem.* (Tokyo), **123**, 1048–1054.

Nakano, Y. and Asada, K. (1981) Hydrogen peroxide is scavenged by ascorbate-specific peroxidase in spinach chloroplasts. *Plant Cell Physiol.*, **22**, 867–880.

Nesarentam, K., Kohr, H.T., Ganeson, J., Chong, Y.H., Sundram, K., and Gapor, A. (1992) The effect of vitamin E tocotrienols from palm oil on chemically induced mammary carcinogenesis in female rats. *Nutr. Res.*, **12**, 879–892.

Niebel, A., Heungens, K., Barthels, N., Inzé, D., Van Montagu, M., and Gheysen, G. (1995) Characterization of a pathogen-induced potato catalase and its systemic expression upon nematode and bacterial infection. *Mol. Plant-Microbe. Interact.*, **8**, 371–378.

O'Kane, D., Gill, V., Boyd, P., and Burdon, R. (1996) Chilling, oxidative stress and antioxidant responses in Arabidopsis thaliana callus. *Planta*, **198**, 371–377.

Olinski, R., Jaruga, P., and Zastawny, T.H. (1998) Oxidative DNA base modifications as factors in carcinogenesis. *Acta Biochim. Pol.*, **45**, 561–572.

Palma, J.M., Sandalio, L.M., and del Río, L.A. (1986) Manganese superoxide dismutase in higher plant chloroplasts: a reappraisal of a controverted cellular localization. *J. Plant Physiol.*, **125**, 427–439.

Parker, C.A. and Joyce, T.A. (1967) Delayed fluorescence and some properties of the chlorophyll triplets. *Photochem. Photobiol.*, **6**, 395

Peiser, G. and Yang, S.F. (1985) Biochemical and physiological effects of SO_2 on nonphotosynthetic processes in plants. In W.E. Winner, H.A. Mooney, and R.A. Goldstein, (eds.), *Sulfur Dioxide and Vegetation*, Stanford University Press, Stanford, pp. 148–161.

Perl, A., Perl-Treves, R., Galili, S., Aviv, D., Shalgi, E., Malkin, S., and Galun, E. (1992) Enhanced oxidative-stress defense in transgenic potato expressing tomato Cu, Zn superoxide dismutases. *Theor. Appl. Genet.*, **85**, 568–576.

Perl-Treves, R., Abu-Abied, M., Magal, N., Galun, E., and Zamir, D. (1990) Genetic mapping of tomato cDNA clones encoding the chloroplastic and the cytosolic isozymes of superoxide dismutase. *Biochem. Genet.*, **28**, 543–552.

Perl-Treves, R. and Galun, E. (1991) The tomato Cu,Zn superoxide dismutase genes are developmentally regulated and respond to light and stress. *Plant Mol. Biol.*, **17**, 745–760.

Perl-Treves, R., Nacmias, B., Aviv, D., Zeelon, E.P., and Galun, E. (1988) Isolation of two cDNA clones from tomato containing two different superoxide dismutase sequences. *Plant Mol. Biol.*, **11**, 609–623.

Prasad, T.K., Anderson, M.D., Martin, B.A., and Stewart, C.R. (1994) Evidence for chilling induced oxidative stress in maize seedlings and a regulatory role for hydrogen peroxide. *Plant Cell*, **6**, 65–74.

Price, A.H., and Hendry, G.A.F. (1991) Iron-catalysed oxygen radical formation and its possible contribution to drought damage in nine native grasses and three cereals. *Plant Cell Environ.*, **14**, 477–484.

Punnonen, K., Okamoto, K., Hyoto, M., Kudo, R., and Ahotupa, M. (1994) Antioxidant enzyme activities and oxidative stress in human breast cancer. *J. Cancer Res. Clin. Oncol.*, **120**, 374–377.

Rainwater, D.T., Gossett, D.R., Millhollon, E.P., Hanna, H.Y., Banks, S.W., and Lucas, M.C. (1996) The relationship between yield and the antioxidant defense system in tomatoes grown under heat stress. *Free Rad. Res.*, **25**, 421–435.

Rich, P.R. and Bonner, W.D. (1978) The sites of superoxide anion generation in higher plant mitochondria. *Arch. Biochem. Biophys.*, **188**, 206–213.

Rosin, M.P., Anwar, W.A., and Ward, A.J. (1994) Inflammation, chromosomal instability, and cancer: the schistosomiasis model. *Cancer Res.*, **54**, 1929s–1933s.

Sandalio, L.M. and del Río, L.A. (1988) Intraorganellar distribution of superoxide dismutase in plant peroxisomes. *Plant Physiol.*, **88**, 1215–1218.

Sandmann, G. and Gonzales, H.G. (1989) Peroxidative processes induced in bean leaves by fumigation with sulphur dioxide. *Environ. Pollut.*, **56**, 145–154.

Scandalios, J.G. (1994) Regulation and properties of plant catalyses. In C.H. Foyer and P.M. Mullineaux, (eds.), *Causes of Photooxidative Stress and Amelioraton of Defense Systems in Plants*, CRC Press, Boca Raton, pp. 275–315.

Scandalios, J.G., Tong, W.-F., and Roupakias, D.G. (1980) *Cat 3*, a third gene locus coding for a tissue specific catalase in maize: genetics, intracellular location, and some biochemical properties. *Mol. Gen. Genet.*, **179**, 33–41.

Schinkel, H., Streller, S., and Wingsle, G. (1998) Multiple forms of extracellular-superoxide dismutase in needles, stem tissues and seedlings of Scots pine. *J. Exp. Bot.*, **49**, 931–936.

Sen Gupta, A., Webb, R.P., Holaday, A.S., and Allen, R.D. (1993) Over-expression of superoxide dismutase protects plants from oxidative stress. *Plant Physiol.*, **103**, 1067–1073.

Sen, C.K. (1998) Redox signaling and the emerging therapeutic potential of thiol antioxidants. *Biochem. Pharmacol.*, **55**, 1747–1758.

Sgherri, C.L.M. and Navari-Izzo, F. (1995) Sunflower seedlings subjected to increasing water deficit stress: oxidative stress and defence mechanisms. *Physiol. Plant.*, **93**, 25–30.

Shaaltiel, Y., Chua, N.-H., Gepstein, S., and Gressel, J. (1988) Dominant pleiotropy controls enzymes co-segregating with paraquat resistance in *Conyza bonariensis*. *Theor. Appl. Genet.*, **75**, 850–856.

Shimoda, R., Nagashima, M., Sakamoto, M., Yamaguchi, N., Hirohashi, S., Yokota, J., and Kasai, H. (1994) Increased formation of oxidative DNA damage, 8-hydroxydeoxyguanosine, in human livers with chronic hepatitis. *Cancer Res.*, **54**, 3171–3172.

Shmookler, R.R. and Ebert, R.H. (1996) Genetics of aging: current animal models. *Exp. Gerontol.*, **31**, 69–81.

Singha, S. and Choudhuri, M.A. (1990) Effect of salinity (NaCl) on H_2O_2 mechanism in *Vigna* and *Oryza* seedlings. *Biochem. Physiol. Pflanz.*, **186**, 69–74.

Slooten, L., Capiau, K., Van Camp, W., Van Montagu, M., Sybesma, C., and Inzé, D. (1995) Factors affecting the enhancement of oxidative stress tolerance in transgenic tobacco overexpremanganese superoxide dismutase in the chloroplast. *Plant Physiol.*, **107**, 737–750.

Smirnoff, N. (1998) Plant resistance to environmental stress. *Curr. Opin. Biotechnol.*, **9**, 214–219.

Smirnoff, N. and Cumbes, Q.J. (1989) Hydroxyl radical scavenging activity of compatible solutes. *Phytochemistry*, **28**, 1057–1060.

Sohal, R.S., Agarwal, A., Agarwal, S., and Orr, W.C. (1995) Simultaneous overexpression of copper- and zinc-containing superoxide dismutase and catalase retards age-related oxidative damage and increases metabolic potential in *Drosophila melanogaster*. *J. Biol. Chem.*, **270**, 15671–15674.

Soll, J., Schultz, G., Joyard, J., Douce, R., and Block, M.A. (1984) Localization and synthesis of prenylquinones in isolated outer and inner envelope membranes from spinach chloroplasts. *Arch. Biochem. Biophys.*, **238**, 290–299.

Somashekaraiah, B.V., Padmaja, K., and Prasad, A.R.K. (1992) Phototoxicity of cadmium ions on germinating seedlings of mung bean (*Phaseolus vulgaris*): Involvement of lipid peroxidase in chlorophyll degradation. *Physiol. Plant.*, **85**, 85–89.

Sonoike, K. (1996) Photoinhibition of photosystem I: its physiological significance in the chilling sensitivity of plants. *Plant Cell Physiol.*, **37**, 239–247.

Sopory, S.K., Greenberg, B.M., Mehta, R.A., Edelman, M., and Mattoo, A.K. (1990) Free radical scavengers inhibit light dependent degradation of the 32-kDa photosystem II reaction center protein. *Z. Naturforsch. C*, **45**, 412–417.

Stoop, J.M.H., Williamson, J.D., and Mason Pharr, D. (1996) Mannitol metabolism in plants: a method for coping with stress. *Trends Plant Sci.*, **1**, 139–144.

Storz, G., Tartaglia, L.A., and Ames, B.N. (1990) Transcriptional regulator of oxidative stress-inducible genes: direct activation by oxidation. *Science*, **248**, 189–194.

Takemoto, T., Zhang, Q.M., and Yonei, S. (1998) Different mechanisms of thioredoxin in its reduced and oxidized forms in defense against hydrogen peroxide in *Escherichia coli*. *Free Rad. Biol. Med.*, **24**, 556–562.

Tao, D.L., Oquist, G., and Wingsle, G. (1998) Active oxygen scavengers during cold acclimation of Scots pine seedlings in relation to freezing tolerance. *Cryobiology*, **37**, 38–45.

Tappel, A.L. (1977) Protection against free radical lipid peroxidation reactions. In J. Roberts, R.C. Adelman, and V. Cristofalo, (eds.), *Pharmacological Intervention in the Aging Process*, Plenum Press, New York, pp. 111–131.

Teramura, A.H. and Sullivan, J.H. (1994) Effects of UV-B radiation on photosynthesis and growth of terrestrial plants. *Photosynthesis Res.*, **39**, 463–473.

Torel, J., Cillard, J., and Cillard, P. (1986) Antioxidant activity of flavonoids and reactivity with peroxy radicals. *Phytochemistry*, **25**, 383–385.

Toyokuni, S., Okamoto, K., Yodoi, J., and Hiai, H. (1995) Persistent oxidative stress in cancer. *FEBS Lett.*, **358**, 1–3.

Van Assche, F. and Clijsters, H. (1990) Effects of metals on enzyme activity in plants. *Plant Cell Environ.*, **13**, 195–206.

Van Camp, W., Hérouart, D., Willekens, H., Takahashi, H., Saito, K., Van Montagu, M., and Inzé, D. (1996) Tissue-specific activity of two manganese superoxide dismutase promoters in transgenic tobacco. *Plant Physiol.*, **112**, 525–535.

Van Camp, W., Inzé, D., and Van Montagu, M. (1997) The regulation and function of tobacco superoxide dismutases. *Free Rad. Biol. Med.*, **23**, 515–520.

Veljovic-Jovanovic, S., Bilger, W., and Heber, U. (1993) Inhibition of photosynthesis, stimulation of zeaxanthin formation and acidification in leaves by SO_2 and reversal of these effects. *Planta*, **191**, 365–376.

Vernooij, B., Friedrich, L., Morse, A., Reist, R., Kolditz-Jawhar, R., Ward, E., Uknes, S., Kessmann, H., and Ryals, J. (1994) Salicylic acid is not the translocated signal responsible for inducing systemic acquired resistance but is required in signal transduction. *Plant Cell*, **6**, 959–965.

Vianello, A. and Macri, F. (1991) Generation of superoxide anion and hydrogen peroxide at the surface of plant cells. *J. Bioenerg. Biomembr.*, **23**, 409–423.

Volk, S. and Feierabend, J. (1989) Photoinactivation of catalase at low temperature and its relevance to photosynthetic and peroxide metabolism in leaves. *Plant Cell Environ.*, **12**, 701–712.

Waffo, T.P., Fauconneau, B., Deffieux, G., Huguet, F., Vercauteren, J., and Merillon, J.M. (1998) Isolation, identification, and antioxidant activity of three stilbene glucosides newly extracted from *Vitis vinifera* cell cultures. *J. Nat. Prod.*, **61**, 655–657.

Weiss, S.J. (1989) Tissue destruction by neutrophils. *N. Engl. J. Med.*, **320**, 365–376.

Weitzman, S.A. and Gordon, L.I. (1990) Inflammation and cancer: role of phagocyte-generated oxidants in carcinogenesis. *Blood*, **76**, 655–663.

Werlen, G., Belin, D., Conne, B., Roche, E., Lew, D.P., and Prentki, M. (1993) Intracellular Ca^{2+} and the regulation of early response gene expression in HL-60 myeloid leukemia cells. *J. Biol. Chem.*, **268**, 16596–16601.

Willekens, H., Inzé, D., Van Montagu, M., and Van Camp, W. (1995a) Catalases in plants. *Mol. Breeding*, **1**, 207–228.

Willekens, H., Van Camp, W., Van Montagu, M., Inzé, D., Langebartels, C., and Sandermann, H.J. Jr (1995b) Ozone, sulphur dioxide and ultraviolet B have similar effects on mRNA accumulation of antioxidant genes in *Nicotiana plumbaginifolia* (L.). *Plant Physiol.*, **106**, 1007–1014.

Wingsle, G. and Karpinski, S. (1996) Differential redox regulation by glutathione of glutathione reductase and Cu, Zn-superoxide dismutase gene expression in *Pinus sylvestris* L. needles. *Planta*, **198**, 151–157.

Wise, R.R. and Naylor, A.W. (1987) Chilling enhanced peroxidation: evidence for the role of singlet oxygen and superoxide in the breakdown of pigments and endogenous antioxidants. *Plant Physiol.*, **83**, 278–282.

Wolff, S.P., Garner, A., and Dean, R.T. (1986) Free radicals, lipids, and protein breakdown. *Trends Biochem. Sci.*, **11**, 27–31.

Wu, A.L. and Moye-Rowley, W.S. (1994) GSH1, which encodes gamma-glutamylcysteine synthetase, is a target gene for yAP-1 transcriptional regulation. *Mol. Cell. Biol.*, **14**, 5832–5839.

Wu, G., Shortt, B.J., Lawrence, E.B., Levine, E.B., Fitzsimmons, K.C., and Shah, D.M. (1995) Disease resistance conferred by expression of a gene encoding H_2O_2-generating glucose oxidase in transgenic potato plants. *Plant Cell*, **7**, 1357–1368.

Wu, J., Weimanis, S., and Heber, U. (1991) Photorespiration is more effective than the Mehler reaction in protecting the photosynthetic apparatus against photoinhibition. *Bot. Acta*, **104**, 283

Xiang, C. and Oliver, D.J. (1998) Glutathione metabolic genes coordinately respond to heavy metals and jasmonic acid in arabidopsis. *Plant Cell*, **10**, 1539–1550.

Yalpani, N., Silverman, P., Wilson, T.M., Kleier, D.A., and Raskin, I. (1991) Salicylic acid is a systemic signal and an inducer of pathogenesis- related proteins in virus-infected tobacco. *Plant Cell*, **3**, 809–818.

Yim, M.B., Chock, P.B., and Stadtman, E.R. (1990) Copper, zinc superoxide dismutase catalyzes hydroxyl radical production from hydrogen peroxide. *Proc. Natl. Acad. Sci USA*, **87**, 5006–5010.

Yu, B.P. (1994) Cellular defenses against damage from reactive oxygen species. *Physiol. Rev.*, **74**, 139–162.

Yuting, C., Ronglian, Z., Zhongjian, J., and Yong, J. (1990) Flavonoids as superoxide scavengers and antioxidants. *Free Rad. Biol. Med.*, **9**, 19–21.

2 The Contribution of Photosynthetic Oxygen Metabolism to Oxidative Stress in Plants

Christine H. Foyer

INTRODUCTION

Under field conditions, the performance of a plant in terms of growth, development, biomass accumulation and yield depends on the ability of metabolism and physiology to adapt and acclimate to fluctuating environmental conditions (Boyer, 1982). Numerous biotic (e.g. insects, fungi, and viruses) and abiotic (e.g. temperature, availability of water, and nutrients) factors affect growth and vigour. Deficits in the availability of essential resources (light, water, carbon, and nitrogen) are detrimental to plant performance. Similarly, variations in environmental conditions away from those which are optimal may result in 'stress' that also limits vigour. Studies on the responses of plants to a range of adverse environments has led to the concept of a phenomenon called 'oxidative stress' that is characterized by the accumulation of potentially harmful active oxygen species (AOS) in plant tissues. Oxidative stress can be caused by perturbations in metabolism such as a loss of coordination between source (energy producing) and sink (energy using) processes. In photosynthesis this can occur when the rate of photon absorption exceeds the rate of photon utilization.

Photosynthesis produces a number of AOS, including superoxide, hydrogen peroxide, hydroxyl radicals, and singlet oxygen (Asada *et al.*, 1974; Allen, 1977; Macpherson *et al.*, 1993; Foyer and Noctor, 2000). Singlet oxygen and hydroxyl radicals can cause extensive membrane damage and form lipid peroxides derived from polyunsaturated fatty acids (Halliwell, 1987; Cadenas 1989). Although a series of regulatory mechanisms have evolved within the plant to limit the production of these molecules, oxidative damage remains a potential problem in photosynthesis. It is incorrect, however, to presume that oxygen activation is favoured because chloroplasts produce oxygen as a result of photosynthesis. This is not the case, because oxygen rapidly equilibrates across the thylakoid and chloroplast envelope membranes. There is no evidence to suggest that the oxygen content of the chloroplasts is higher in the light than the dark.

AOS can have both positive and negative effects on plant metabolism. In optimal conditions, AOS are produced in a controlled manner and have important functions in plants such as cell wall biosynthesis and redox signalling. AOS contribute to the redox balance in plant cells. An increase in AOS is sensed and transduced to the nucleus where the gene transcription processes are modified to compensate for fluctuations due to changes in the environment (Mayfield and Taylor, 1987; Bowler and Chua, 1994; Danon and Mayfield, 1994; Price *et al.*, 1994; Huner *et al.*, 1996). H_2O_2 accumulation in particular, appears to signal environmental change (Okuda *et al.*, 1991; Matsuda *et al.*, 1994; Prasad *et al.*, 1994; Green and Fluhr, 1995; Henkow *et al.*, 1996; Foyer *et al.*, 1997). As a diffusible signal-transducing molecule H_2O_2 may alert metabolism to the presence of both biotic and abiotic threats (for review, see Foyer *et al.*, 1997). H_2O_2 accumulation has been implicated in several stress responses in plants such as the hypersensitive response (HR) and systemic acquired resistance (Chen *et al.*, 1993; Levine *et al.*, 1994), chilling tolerance (Ornran, 1980; Okuda *et al.*, 1991; Prasad *et al.*, 1994), and in the development of cross tolerance to a variety of stresses (Bowler *et al.*, 1992; Gueta-Dahan *et al.*, 1997; Huner *et al.*, 1996; Wu *et al.*, 1997). Several genes have been shown to be induced by H_2O_2 and H_2O_2 recognition sequences have been reported (Chen *et al.*, 1996; Foyer *et al.*, 1997).

The hypothesis that chloroplast metabolism makes a significant contribution to oxidative stress when plants are exposed to suboptimal environmental conditions is widely accepted. This concept has arisen because photosynthesis produces molecular O_2 and is also capable of reducing O_2 to H_2O_2 and activating ground-state triplet O_2 to highly active singlet oxygen. However, the production of AOS is limited by efficient regulation; furthermore, the chloroplast contains an array of antioxidant defenses to destroy AOS as they are formed.

In higher plants, oxidative stress causes a variety of molecular and metabolic responses (Arnott and Murphy, 1991; Dempsey and Klessig, 1994; Doke *et al.*, 1994; Wu *et al.*, 1997) that appear to be remarkably consistent regardless of whether the oxidative stress is caused by perturbation of metabolism within plant cells, for example by discrepancies between energy producing and energy-consuming reactions (Bowler *et al.*, 1989; Gueta-Dahan *et al.*, 1997), or whether the stress is perceived outside the cell membrane, for example as a result of the oxidative burst produced by the HR (Arnott and Murphy, 1991; Chen *et al.*, 1993). Exposure to abiotic stresses, such as ozone pollution or UV-B irradiation, causes metabolic changes that resemble those observed following pathogen attack (Kangasjärvi *et al.*, 1994; Willekens *et al.*, 1994; Sandermann, 1996; Thalmair *et al.*, 1996).

Increased AOS production is a generic feature of biotic and edaphic stresses in plants (Foyer *et al.*, 1997). The chloroplast is not the only source of AOS in stress conditions. The plasmalemma-bound NADPH oxidase systems, for example, produce O_2^- and H_2O_2 in response to many forms of stress (Doke *et al.*, 1994) and large amounts of H_2O_2 are produced in the peroxisomes via photorespiration (Zelitch, 1973). Short-term increases in AOS appear to be tolerated in plant cells (Levine *et al.*, 1994). In the case of H_2O_2 generation by the plasmalemma, this tolerance may be associated with the presence of extracellular peroxidases in the cell wall (Levine *et al.*, 1994). AOS accumulation only poses a risk of causing wide-spread oxidative damage when plants suffer severe environmental or metabolic limitations because the antioxidant systems can compensate for relatively large changes in AOS production. Chlorosis is a common symptom of extreme environmental stress in plants (Giardi *et al.*, 1997; Kingston-Smith and Foyer, 2000). Destruction of photosynthetic pigments may result from photooxidative damage initiated by strong oxidants, such as singlet oxygen (Powles, 1984). A more probable explanation is that changes in pigment content are caused by redox control of the expression of genes coding for components of pigment-protein complexes in the thylakoid membrane (Pearson *et al.*, 1993; Escoubas *et al.*, 1995; Henkow *et al.*, 1996; Giardi *et al.*, 1997).

OXIDATIVE STRESS AND PHOTOSYNTHESIS

Photosynthesis and photorespiration produce substantial amounts of superoxide and/or H_2O_2 as metabolic intermediates (Zelitch, 1973, 1990; Asada, 1994; Willekens *et al.*,

1997). In chloroplasts, superoxide is converted to H_2O_2 by the action of superoxide dismutases (SOD) and H_2O_2 is rapidly destroyed by the action of ascorbate peroxidases (APX) (Asada *et al.*, 1973; Asada, 1992, 1994). Chlorophyll, in its excited triplet state, is able to transfer energy directly to ground-state oxygen that forms the highly reactive state singlet oxygen (Foyer and Harbinson, 1994, 1997). Whereas singlet oxygen can be quenched by carotenoid pigments, the first line of defense against this form of active oxygen is always avoidance. Effective regulation of energy utilization in photosynthesis serves to prevent singlet oxygen formation (Foyer and Harbinson, 1994). Regulation of the photosynthetic harvesting is geared to minimizing overreduction of the photosynthetic reaction centres, which decreases the likelihood of reduction of O_2 by photosystem II (PSII) electron acceptors, of triplet chlorophyll formation, and, hence, of singlet oxygen production (Foyer *et al.*, 1990). Similarly, control of electron flow between PSI and PSII by the transmembrane proton/electrochemical potential difference acts to regulate the reduction state of the acceptor side of PSI to prevent overreduction and minimize superoxide formation. Rapid elimination of superoxide and H_2O_2 as they are formed prevents production of both singlet O_2 and hydroxyl radical ($OH^\bullet$; Khan and Kasha, 1994). Whereas hydroxyl radicals formation was found to be relatively high in PSII of *Euglena gracilis* (Tschiersch and Ohmann, 1993), spinach chloroplasts only produced substantial amounts of this radical when the endogenous antioxidant system was impaired (Jakob and Heber, 1996). Hydroxyl radicals can be formed by interaction of superoxide and H_2O_2, particularly in the presence of transition metal ions (Elstner *et al.*, 1978; Youngman and Elstner, 1981; Cadenas, 1989). Hence, effective elimination of both of these substrates by SOD and APX within the chloroplasts largely prevents hydroxyl radical formation (Jakob and Heber, 1996). In intact chloroplasts, extensive hydroxyl radical formation was only observed when potassium cyanide was added to inhibit endogenous APX (Jakob and Heber, 1996). Nevertheless, plant cells have antioxidants that can destroy hydroxyl radicals. Ascorbate, which is present in chloroplasts at high concentrations, is able to reduce hydroxyl radicals (Halliwell, 1987). The production of mannitol in stress conditions can protect chloroplasts against hydroxyl radical-induced damage (Shen *et al.*, 1997). In the chloroplasts the antioxidants which destroy AOS are generally found in close proximity to the reactions that produce them (Halliwell, 1987). The capacity of the chloroplast antioxidant defenses is greater than that of AOS production in order to protect photosynthesis from inactivation (Kaiser, 1979).

REGULATION OF THE PHOTOSYNTHETIC ELECTRON TRANSPORT CHAIN

In photosynthesis, light energy is used as a driving force for metabolism, particularly the fixation of CO_2 into sugar phosphate. The absorption of a quantum of light by a photosynthetic pigment results in the formation of an excited state (Witt, 1979; Shipman, 1980). Excitation energy moves through the pigment bed to a reaction centre that contains a special pair of chlorophyll molecules. When the reaction centre chlorophylls become excited, they are rapidly oxidized by an adjacent electron acceptor. The chlorophyll cation radical thus formed is then reduced by an electron donor. At low light levels, photosynthesis is limited by the maximum quantum efficiency of light use. At high light, photosynthesis is limited by the capacity of photosynthetic carbon assimilation to use the assimilatory power (ATP and NADPH) generated by the photosynthetic electron transport chain. At times, when photosynthetic carbon assimilation is limited by environmental or metabolic constraints available light energy can be far in excess of that which can be effectively used in photosynthetic carbon assimilation. To avoid damage in the short term, the quantum efficiency PSII is decreased and excess excitation energy is dissipated as heat (Demmig-Adams and Adams, 1996; Ruban *et al.*, 1997) and alternative sinks for assimilatory power, such as photorespiration, have to be used (Wu *et al.*, 1991; Osmond and Grace, 1995). In the longer term, the composition of the photosynthetic apparatus is modified to correspond to prevailing environmental conditions (Anderson and Osmond, 1987). Recent evidence suggests that many proteins in the chloroplast are subject to redox control (Allen, 1992; Henkow *et al.*, 1996; Pfannschmidt *et al.*, 2000) via changes in the redox state of components such as plastoquinone, ferredoxin, or thioredoxin (Danon and Mayfield, 1994; Escoubas *et al.*, 1995; Maxwell *et al.*, 1995; Huner *et al.*, 1996; Pfannschmidt *et al.*, 2000), or via the glutathione system (Baier and Dietz, 1997; Link *et al.*, 1997) or the ascorbate system (Noctor *et al.*, 2000).

Concepts of control within photosynthetic systems are frequently discussed in terms of the independent regulation of either the electron transport processes or the carbon reduction (Benson-Calvin) cycle, occurring in the thylakoid membrane and the stroma, respectively. The electron transport system consists of two photosystems that operate sequentially to achieve light-driven reduction of $NADP^+$ with concomitant production of a proton gradient (Arnon and Chain, 1975). This system is used to generate ATP. The trans-thylakoid proton gradient also has a decisive influence on the downregulation of the quantum efficiency of PSII in excess light (Briantais *et al.*, 1979; Noctor *et al.*, 1991; Schreiber *et al.*, 1992, 1995) and on the flux of electrons between PSII and PSI (Haehnel, 1984; Lavergne *et al.*, 1992; Levings and Siedow, 1995).

NADPH and ATP produced by the electron transport processes are consumed during the assimilation and reduction of CO_2 to the level of sugar phosphate (Arnon *et al.*, 1958). The producer-consumer relationship between electron transport and CO_2 assimilation ensures their tight coupling by virtue of the cycling of intermediates (Leegood *et al.*, 1985; Siebke *et al.*, 1991; Foyer and Noctor, 2000). However, *in vivo* regulation is complicated by the necessity to reconcile the conflicting requirements of the thylakoid reactions and the stromal enzymes. High levels of ATP and NADPH are needed to drive high rates of CO_2 reduction, but high rates of electron transport are difficult to maintain if the electron acceptor, NADP, and the substrate for photophosphorylation, ADP, are not plentiful (Heber *et al.*, 1990). The hypothesis that $NADPH/NADP^+$ ratios in the chloroplasts increase at times when CO_2 assimilation is decreased is widely accepted but there is no experimental evidence to support this view. Indeed, the $NADPH:NADP^+$ ratios of the stroma are tightly controlled to ensure that sufficient $NADP^+$ is always available for electron transport (Takahama *et al.*, 1981; Gerst *et al.*, 1994). In fact, coregulation of electron transport and CO_2 assimilation serves to minimize fluctuations in the NADPH:NADP ratio (for review, see Foyer, 1984). Of central importance to this regulation is the modulation of the activities of key enzymes of the carbon reduction cycle so their activity matches the availability of assimilatory power (Leegood and Walker, 1982; Leegood *et al.*, 1985; Cseke and Buchanan, 1986; Kelly and Latzko, 1997). In many conditions the capacity of metabolism to use assimilatory power limits photosynthesis (Heber *et al.*, 1986, 1988). Electron transport is regulated to prevent excessive reduction of electron acceptors. This regulation can be transiently perturbed by rapid changes in conditions, for example a change from darkness to high light. In such situations, the $NADP^+$ pool transiently decreases (Takahama *et al.*, 1981; Siebke *et al.*,

1991). This decrease lasts only as long as it takes to activate the light-modulated enzymes of the Benson-Calvin cycle to levels commensurate with the change in light availability. Only during the induction phase of photosynthesis (which can last several minutes) does substantial reduction of the NADP pool occur (Foyer, 1984; Foyer *et al.*, 1992).

The molecular mechanisms whereby electron transport is restrained when ADP and NADP are in short supply are not fully understood, but under these conditions the quantum efficiency of PSII is downregulated and thermodynamic constraints exert a restraining control on the rate of electron flow (Horton, 1989; Foyer *et al.*, 1990; Heber *et al.*, 1996). As the light incident upon a leaf is increased, so the primary electron donor of PSI, P700, becomes progressively more oxidized (Harbinson and Hedley, 1989; Harbinson and Foyer, 1991). This occurs because the rate-limiting step of electron transport, the oxidation of plastoquinol by the cytochrome b_6/f complex, lies between PSII and PSI (Haehnel, 1984; Heber *et al.*, 1988; Hope *et al.*, 1994). Plastoquinol oxidation involves the release of protons into the thylakoid lumen, a reaction which is progressively inhibited as the pH of the lumen decreases (Joliot *et al.*, 1992; Hope *et al.*, 1994) and in the light limits the rate of electron transport between the photosystems. This phenomenon is called, 'photosynthetic control' (West and Wiskich, 1968; Weis *et al.*, 1987; Foyer *et al.*, 1990). Whether such control responds to the redox state of plastoquinone or of PSI electron acceptors, such as ferredoxin, is unclear (Hundal *et al.*, 1995), but cyclic electron transport around PSI may involve both ferredoxin and superoxide (Takahashi and Asada, 1988; Hormann *et al.*, 1993; Asada, 1994). There are other complexes in the thylakoid membrane analogous to the complex-1 of mitochondria that may be involved in redox-poising PSI electron transport. Regardless of mechanism, effective controls appear to match the supply of electrons to PSI with the rate of ferredoxin oxidation so that the $NADP^+$ pool is not depleted. The rate of electron flow to PSI only increases when the rate of oxidation of reduced electron acceptors, such as NADPH and reduced ferredoxin, by sink processes is increased. Most of the electrons from PSI are passed to $NADP^+$ from ferredoxin via ferredoxin-$NADP^+$ reductase (FNR). The ferredoxin pool acts as a feed-forward regulator of the thiol-modulated enzymes of the Benson-Calvin cycle through thioredoxin (Leegood and Walker, 1982; Cseke and Buchanan, 1986; Siebke *et al.*, 1991). Ferredoxin also supplies electrons to several other biosynthetic processes such as nitrogen assimilation. Reduced ferredoxin reacts quickly with oxygen according to the equation:

$$\text{Ferredoxin}_{\text{red}} + O_2 \rightarrow \text{Ferredoxin}_{\text{ox}} + O_2^-$$

The mechanisms that serve to decrease the quantum efficiency of PSII and facilitate the harmless conversion of light energy directly to heat to prevent overreduction of PSII (Demmig-Adams, 1990; Demmig Adams and Adams, 1992; Horton and Ruban, 1992; Horton *et al.*, 1994) are perhaps better characterized than those regulating the redox state of PSI (Figure 1). Carotenoid pigments associated with PSII and the light-harvesting antennae can quench triplet states directly and have antioxidant activity (Siefermann-Harms, 1987; Miller *et al.*, 1996). Thermal energy dissipation from excited singlet states is initiated by the proton gradient across the thylakoid membrane. This mechanism is obligatory for heat dissipation to occur but conversion of violaxanthin to zeaxanthin in the xanthophyll cycle is also involved (Gilmore and Yamamoto, 1992, 1993; Gilmore, 1997). Regulated increases in the rate of thermal energy dissipation of excess excitation energy correlate with the conversion of the carotenoid pigment violaxanthin into zeaxanthin

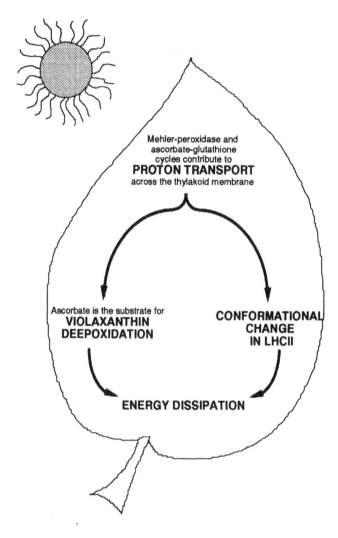

Figure 1.   Schematic representation of the major components involved in thermal energy dissipation in leaves.

(Figure 2). This reaction, which is catalyzed by the enzyme violaxanthin de-epoxidase (VDE) uses ascorbic acid as a cofactor (Hager, 1969; Hager and Holocker, 1994). When violaxanthin is present in the light-harvesting antennae complexes they are efficient light-harvesting systems (Figure 2) (Koyama, 1991; Siefermann-Harms, 1987). When violaxanthin is converted to zeaxanthin energy dissipation is favoured (Figure 2), for example by direct quenching of chlorophyll excited states via energy transfer to zeaxanthin (Frank *et al.*, 1994). Changes in the xanthophyll de-epoxidation states alone are not sufficient to bring about quenching (Noctor *et al.*, 1991), but they control the interactions between the proteins in the light-harvesting antennae (Ruban *et al.*, 1997).

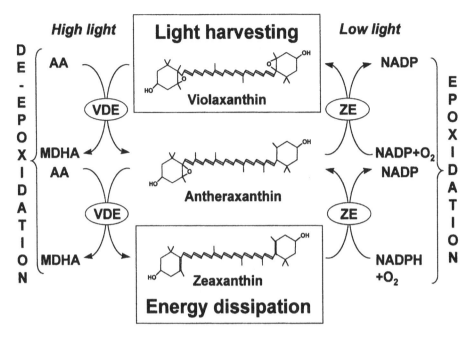

Figure 2. The xanthophyll cycle and its effects on light-harvesting and energy dissipation within the photosynthetic apparatus. The enzyme violaxanthin de-epoxidase (VDE) catalyses the sequential conversion of violaxanthin to zeaxanthin via antheraxanthin using ascorbic acid (AA) as a reductant and generating monodehydroascorbate (MDHA). The epoxidation of zeaxanthin to violaxanthin in the reverse reaction sequence is catalysed by the enzyme zeaxanthin epoxidase (ZE).

Photoinhibition

When plants, algae, and cyanobacteria are exposed to increasing light intensities, photosynthesis first becomes saturated and then begins to decline in the process termed photoinhibition (Jones and Kok, 1966; Powles, 1984; Björkman, 1987; Björkman and Demming-Adams, 1994; Long *et al.*, 1994; Osmond, 1994). This decrease in efficiency results largely from changes in PSII, although photoinhibition of PSI has also been reported in some species (Havaux and Davaud, 1994; Terashima *et al.*, 1994). The D1 reaction centre protein of PSII turns over continuously in the light, but many environmental stress situations increase the rate of turnover (Barber and Andersson, 1992; Giardi *et al.*, 1997). Accumulation of both redox and energy equivalents is known to inactivate PSII *in vitro* either via excessive reduction of the plastoquinone pool during acceptor side inhibition or by pH-induced inhibition of the water-splitting system during donor side inhibition (Theg *et al.*, 1986; Jegerschöld *et al.*, 1990; Setlik *et al.*, 1990; Eckert *et al.*, 1991; Tyystjärvi *et al.*, 1994). D1 turnover is a multistep process (Vas *et al.*, 1992) involving: (i) generation of a species within PSII that causes physical damage; (ii) irreversible modification of the D1 polypeptide leading to inactivation of PSII; (iii) cleavage of the D1 polypeptide; and (iv) migration of damaged PSII from the granal region to the stromal lamellae followed by removal of damaged D1 and its replacement by a newly synthesized D1 polypeptide (Mäenpää *et al.*, 1987; Callahan *et al.*, 1989; Barber and Andersson, 1992). In the absence of stress, the rate of damage to D1 is balanced by the rate of D1 replacement resulting in no net loss of PSII activity. Under stress conditions, however, the rate of PSII inactivation increases and can exceed the rate of D1 turnover and a net loss of PSII activity is observed. In low light, PSII activity can recover, because the rate of D1 synthesis generally exceeds the rate of PSII damage in these circumstances (Prasil *et al.*, 1992). If D1 synthesis is blocked by the addition of an inhibitor of chloroplast protein synthesis, such as lincomycin, then recovery of PSII is prevented.

Phosphorylation of thylakoid proteins, particularly the light-harvesting antenna complexes, is an important mechanism that regulates excitation energy distribution between the photosystems (Staehelin and Arntzen, 1983; Horton and Lee, 1985; Allen, 1992).

In addition, the N-threonine terminus of the D1 protein is phosphorylated in high light (Allen, 1992). Phosphorylated D1 accumulates in stress situations and may prevent substantial D1 degradation (Bracht and Trebst, 1994). Phosphorylation may also regulate the production of PSII proteins.

Singlet oxygen, superoxide, and H_2O_2 can be generated within PSII complexes (Wydrzynski *et al.*, 1989; Fine and Frisch, 1990; Macpherson *et al.*, 1993; Ananyev *et al.*, 1994). Several components associated with PSII have antioxidant activity (Ananyev *et al.*, 1994; Hundal *et al.*, 1995; Miller *et al.*, 1996). It is considered that AOS formation in PSII can initiate photoinhibitory damage (Kyle *et al.*, 1985; Kyle, 1987; Mishra *et al.*, 1993; Hideg *et al.*, 1994). Degradation of D1 protein may commence by radical attack. At least one specific protease is activated during photoinhibition (Mattoo *et al.*, 1989; Barber and Andersson, 1992; Aro *et al.*, 1993). Redox control of mRNA abundance and mRNA-binding proteins contributing to D1 synthesis and the production of light-harvesting antenna proteins has been reported (Danon and Mayfield, 1994; Escoubas *et al.*, 1995; Maxwell *et al.*, 1995). Changes in the turnover of D1 may be a general adaptive response to environmental stress mediated by the redox poise of the chloroplasts (Huner *et al.*, 1996; Giardi *et al.*, 1997). While redox signalling has been shown to regulate the expression of several genes that code for chloroplast proteins (Pearson *et al.*, 1993; Escoubas *et al.*, 1995; Henkow *et al.*, 1996; Pfannschmidt *et al.*, 1999), photoinhibition has also been shown to cause changes in the expression of genes encoding cytosolic proteins (Karpinski *et al.*, 1997). Redox equilibrium between the reduced and oxidized forms of ascorbate and glutathione may also be an important regulator of chloroplast gene expression in response to stress (Foyer *et al.*, 1997; Link *et al.*, 1997). Hence, the activities of the antioxidant enzymes determine the fate of the plant in stress situations not only by preventing excess AOS accumulation and limiting oxidative damage, but also by effective regulation of gene expression allowing appropriate signals [H_2O_2, reduced glutathione (GSH), glutathione disulfide (GSSG), ascorbic acid (L-AA), and dehydroascorbate (DHA)] to be generated in stress conditions.

Electron Transport to Oxygen: the Mehler Reaction

Whereas NADP is the preferred electron acceptor in photosynthesis, oxygen can also accept electrons from the photosynthetic electron transport chain (Mehler, 1951; Mehler and Brown, 1952; Egneus *et al.*, 1975; Allen, 1975, 1992; Marsho *et al.*, 1979). Molecular oxygen contains two unpaired electrons with parallel spins. As a consequence, O_2 is most easily reduced by single electron additions because divalent reduction of O_2 (to H_2O_2) requires a spin inversion (Cadenas, 1989). The addition of an electron to oxygen by the photosynthetic electron transport chain produces superoxide (Asada *et al.*, 1974). The reduction of molecular oxygen by the photosynthetic electron transport system is called the 'Mehler reaction' (Mehler, 1951; Mehler and Brown, 1952) and electron transport from water to molecular O_2 is called 'pseudocyclic electron flow' (Allen, 1975, 1977). Carriers within the photosynthetic electron transport chain that have electrochemical potentials commensurate with the reduction of molecular oxygen exist in both PSII and PSI (Asada *et al.*, 1974; Allen, 1977; Wydrzynski *et al.*, 1989). Oxygen reduction by the electron transport components does not appear to be a deleterious event *per se* but rather serves a useful function in preventing overreduction of the electron transport chain (Arnon and Chain, 1975; Egneus *et al.*, 1975; Heber *et al.*, 1978; Ziem-Hanck and Heber, 1980; Polle, 1996), which is considered to "poise" the electron carriers for more efficient functioning (Heber *et al.*, 1978; Ziem-Hanck and Heber, 1980; Levings and Siedow, 1995). Whereas components of PSII are capable of reducing O_2 (Wydrzynski *et al.*, 1989; Ananyev *et al.*, 1994), the major flux of electrons to oxygen occurs on the reducing side of PSI (Asada, 1994, 1996). All the electron transport components on the reducing side of PSI, from the iron-sulphur centres to reduced thioredoxin, are auto-oxidizable, i.e. they can donate electrons to oxygen and produce superoxide. Superoxide may be formed within the thylakoid membrane (Takahashi and Asada, 1988; Hormann *et al.*, 1993; Asada, 1994) or at the membrane surface (Allen, 1977; Badger, 1985). A proportion of the reduced ferredoxin pool is able to diffuse rapidly away from the thylakoid membranes and generate superoxide and H_2O_2 throughout the stroma (Misra and Fridovich, 1971; Allen, 1977). The iron-sulphur centres on the reducing side of PSI (X, A, and B) have high affinities for oxygen (Asada *et al.*, 1974; Takahashi and Asada, 1988). At the membrane surface, ferredoxin (Telfer *et al.*, 1970; Misra and Fridovich, 1971; Allen, 1975), ferredoxin-NADP$^+$ oxidoreductase (FNR) (Firl *et al.*, 1981; Goetze and Carpentier, 1994), and other members of the dehydrogenase family of flavoenzymes, such as monodehydroascorbate reductase (MDHAR) (Miyake *et al.*, 1996) and glutathione reductase (GR), can all generate superoxide at high rates when attached to thylakoid membranes, even though autoxidation rates of the charge transfer complexes of MDHAR, GR and FNR *per se* are extremely low (Hossain and Asada, 1985; Sano *et al.*, 1995). Various isoforms of these enzymes are found in chloroplasts, but only FNR is a membrane protein. Autoxidation of reduced ferredoxin produces superoxide (Telfer *et al.*, 1970; Misra and Fridovich, 1971; Allen, 1975, 1977), but the rate of this reaction is rather low (Hosein and Palmer, 1983; Asada, 1994). Flavoenzymes, such as FNR, accept electrons from PSI and produce superoxide at a much higher rate than ferredoxin (Goetze and Carpentier, 1994). Superoxide formation can be increased even further by covalent linkage of a viologen group (Bes *et al.*, 1995). The contribution of flavoenzyme-mediated superoxide production *in vivo* is unknown because the endogenous electron acceptors for these enzymes suppress superoxide formation. It has been argued earlier in this chapter that physiological electron acceptors, such as NADP$^+$ are always available for electron transport and, hence, superoxide production via components associated with PSI will be minimized.

Superoxide produced by thylakoid components can spontaneously dismutate to molecular oxygen and H_2O_2. In chloroplasts this reaction is catalyzed enzymatically via SOD (Asada *et al.*, 1973; Salin, 1988). Chloroplasts also contain large amounts of ascorbic acid that can efficiently reduce superoxide to H_2O_2 (Buettner and Jurkiewicz, 1996). The concentrations of ascorbate present in chloroplasts are frequently so high (10–50 mM) (Foyer *et al.*, 1983) that this reaction can effectively compete with the SOD for superoxide reduction.

H_2O_2 is a powerful inhibitor of photosynthetic CO_2 assimilation (Kaiser, 1976, 1979; Charles and Halliwell, 1980). In plant cells, H_2O_2 is destroyed by the action of catalases or peroxidases. Catalase is absent from chloroplasts that contain large amounts of APX activity (Anderson *et al.*, 1983a, 1983b). Chloroplasts contain APXs both in the aqueous stromal compartment and on the thylakoid membranes (Chen and Asada, 1989; Miyake *et al.*, 1991; Asada, 1992). Genes coding for glutathione peroxidase (GPX) have been described in plants and the products of these genes are located in the chloroplasts (Mullineaux *et al.*, 1997). The transcription of most *GPX* genes is found to be induced by oxidative stress, but these enzymes do not appear to catalyze glutathione-dependent reduction of H_2O_2 *per se* and have very low activity (Eshdat *et al.*, 1997). Nevertheless, constitutive expression of GPX in transformed plants led to increased stress resistance (Roxas *et al.*, 1997).

Oxygen reduction via the Mehler reaction is tightly coupled to the production and destruction of H_2O_2 (Groden and Beck, 1979; Anderson *et al.*, 1983a, 1983b) and these reactions together are therefore called the 'Mehler-peroxidase cycle' (Neubauer and Schreiber, 1989; Neubauer and Yamamoto, 1992). Intrinsic to this reaction sequence is the regeneration of the substrate for H_2O_2 reduction by APX, ascorbate (L-AA). In the reduction of H_2O_2 by APX, monovalent oxidation of L-AA produces MDHA (Miyake and Asada,1992a; Grace *et al.*, 1995), which is itself a powerful electron acceptor that is reduced by ferredoxin (Miyake and Asada, 1994). The Mehler-peroxidase cycle consists of (i) electron transfer from water through the photosynthetic electron transport chain to oxygen for superoxide at PSI, (ii) the dismutation of the superoxide radical by SOD to form H_2O_2, (iii) the reduction of H_2O_2 to water by APX, and (iv) the regeneration of ascorbate from MDHA.

The Mehler-peroxidase cycle not only performs an essential protective function in preventing oxidative stress, but also contributes to the control of photosynthetic electron transport since it is a coupled reaction sequence. Increased flux through this cycle increases the transthylakoid pH gradient and prevents excessive reduction of the PSI (Foyer *et al.*, 1990; Polle, 1996; Foyer and Harbinson, 1997). The risk of harmful back reactions within PSII that increase the frequency of direct energy exchange between the activated states of chlorophyll and ground-state molecular oxygen to produce singlet oxygen, is decreased in these circumstances (Foyer and Harbinson, 1994). In this way, the production and destruction of active oxygen species is directly involved in the regulation of photosynthetic electron transport.

The flux through the Mehler-peroxidase cycle has proven very difficult to determine (Marsho *et al.*, 1979; Steiger and Beck, 1981; Badger, 1985; Robinson, 1988). The rate of electron flow to oxygen is considered to be, at maximum, approximately 10% of the total electron flow (Egneus *et al.*, 1975; Gerbaud and Andre, 1980; Robinson, 1988; Tourneaux and Peltier, 1995). In intact leaves, the Mehler flux can only be determined when oxygen uptake by other processes has been suppressed. The most important of these is the oxygenase reaction of ribulose-1,5-bisphosphate carboxylase (Rubisco), which is the first reaction of photorespiration (Ogren, 1984). Photorespiration has a high flux and can protect the photosynthetic apparatus when CO_2 assimilation is limited by the availability of CO_2 (Osmond and Björkman, 1972; Heber *et al.*, 1996; Kozaki and Takeba, 1996). It has been suggested that the Mehler-peroxidase cycle could serve a similar photoprotective role preventing photoinhibition. Pseudocyclic electron flow would, hence, increase the transthylakoid pH gradient to allow thermal energy dissipation (Osmond and Grace, 1995; Biehler and Fock, 1996; Lovelock and Winter, 1996). Rates of oxygen reduction of 40 μmol electron $m^{-2}s^{-1}$ have been calculated in situations where photorespiration was suppressed (Guy *et al.*, 1993; Schreiber *et al.*, 1994; Osmond and Grace, 1995). Such values are much higher than calculated rates of superoxide formation by the thylakoid membranes (Asada *et al.*, 1974; Takahashi and Asada, 1982), which are, at most, approximately 30 μmol O_2 mg^{-1} Chl h^{-1}. In estimations of Mehler flux, high CO_2 concentrations have to be used to suppress photorespiration, which may not always have been sufficient (Canvin *et al.*, 1980; Osmond and Grace, 1995). Furthermore, the hypothesis that the Mehler reaction can act as an alternative sink for electrons when CO_2 assimilation is limited requires that considerations relating to photosynthetic control are taken into account. The Mehler reaction is a coupled process leading to ATP formation (Egneus *et al.*, 1975; Forti and Elli, 1995). ATP must therefore be consumed at a rate commensurate with its production to ensure continued electron flux through the pathway of pseudocyclic electron flow. If an adequate ATP sink is not present, the trans-thylakoid pH gradient will increase, leading not only to decreases in the quantum efficiency of PSII (Krieger *et al.*, 1992; Ruban *et al.*, 1992; Gilmore *et al.*, 1994) but also restriction of electron flow to PSI and, hence, decreased Mehler reaction.

The Antioxidant Enzymes of the Chloroplast

Efficient destruction of O_2^- and H_2O_2 in chloroplasts requires the concerted action of several antioxidant enzymes acting in synchrony (Figure 3). Superoxide is frequently regarded as a primary agent of oxygen toxicity because its production inevitably leads to the formation of increasingly more toxic species. Superoxide *per se* has only a limited capacity as both an oxidant and reductant (Halliwell, 1987). In contrast, to superoxide H_2O_2 is a strong oxidant and a potent inhibitor of photosynthetic CO_2 assimilation (Kaiser, 1976, 1979), because it can rapidly oxidize protein thiol groups (Charles and Halliwell, 1980). H_2O_2 cannot be allowed to accumulate in chloroplasts, since several enzymes of the Benson-Calvin cycle are only active when critical thiol groups are present (Leegood and Walker, 1982; Cseke and Buchanan, 1986). H_2O_2 rapidly oxidizes thiol-modulated enzymes of the Benson-Calvin cycle, such as fructose-1,6-bisphosphatase and sedoheptulose-1,7-bisphosphatase, causing inactivation (Kaiser, 1976, 1979; Charles and Halliwell, 1980). H_2O_2 will also inactivate other chloroplast components, such as Cu/ZnSOD (Hodgson and Fridovich, 1975; Casano *et al.*, 1997) and glutamine synthetase (Levine, 1983), and cause degradation of the D1 protein (Miyao *et al.*, 1995). Chloroplasts contain both thylakoid-bound and soluble forms of SOD and APX to destroy H_2O_2 (Hayakawa *et al.*, 1985; Miyake and Asada, 1992b, 1994; Ogawa *et al.*, 1995).

L-AA is oxidized by a successive reversible one-electron transfer process with a free radical intermediate, MDHA. The ascorbate redox system consists therefore of L-AA, MDHA, and DHA. APXs use ascorbate to reduce H_2O_2 and produce MDHA as the primary product (Sapper *et al.*, 1982; Hossain *et al.*, 1984). Rapid regeneration of ascorbate both from MDHA and DHA is essential for the continued antioxidative function of the ascorbate pool. MDHA radicals have a short lifetime and disproportionate spontaneously at neutral pH values to DHA and ascorbate at a rate similar to that observed

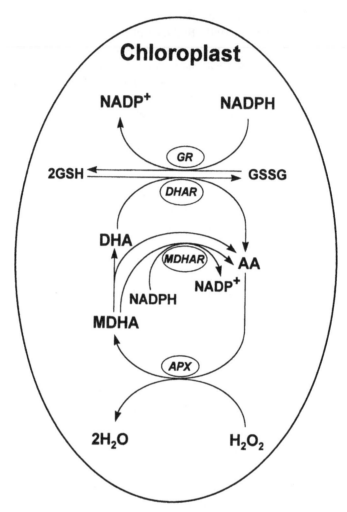

Figure 3. The ascorbate-glutathione cycle of chloroplasts.

with the disproportionation of superoxide (Bielski *et al.*, 1971; Bielski, 1982). MDHA radicals are reduced directly to ascorbate by reduced ferredoxin produced by PSI (Miyake and Asada, 1992b, 1994). The chloroplast stroma contains an NAD(P)H-dependent MDHAR (Hossain *et al.*, 1984; Hossain and Asada, 1985).

DHA is reduced to ascorbate by the action of DHA reductase which uses reduced glutathione GSH as the reducing substrate (Foyer and Halliwell, 1976). This reaction generates GSSG, which is re-reduced to GSH by the action of GR that use NADPH as a reductant. Together the enzymes that act concertedly to remove H_2O_2 form the ascorbate-glutathione cycle (Figure 3). The two reductants, ascorbate and glutathione, which are cycled in this pathway are abundant in chloroplasts (Foyer and Halliwell, 1976; Foyer *et al.*, 1983). On the thylakoid membrane SOD, APX, and ferredoxin-dependent MDHA reduction constitute a thylakoid-bound scavenging system for superoxide and H_2O_2 (Figure 4) and may form a relatively closed system acting within a

5–10 nm layer on the surface of the membrane (Miyake and Asada, 1994; Asada, 1996). O_2 and H_2O_2 escaping the thylakoid antioxidant defenses or arriving in the chloroplasts from the cytosol are effectively scavenged in the stroma by APX (sAPX), Cu/ZnSOD, MDHAR, DHAR, and GR that use NADPH produced as a result of electron transport (Figure 3). DHA is always found in leaves and other plant tissues (Foyer *et al.*, 1983; Robinson, 1997), but its existence in chloroplasts has recently been called into question (Morrell *et al.*, 1997).

The stromal ascorbate-glutathione cycle uses NADPH and regenerates NADP and, hence, contributes to the recycling of the pyridine nucleotide pools in the stroma. It is a coupled process leading to ATP synthesis. The turnover of both the ascorbate and glutathione pools when H_2O_2 is added to isolated chloroplasts suggest that both pools can contribute to H_2O_2 detoxification (Anderson *et al.*, 1983a, 1983b). Furthermore, overproduction of GR in the chloroplasts of poplar leaves led to an increase in the reduction state of the glutathione pool and an increase in L-AA (Foyer *et al.*, 1995). This observation would suggest that glutathione-mediated ascorbate regeneration contributes to the turnover of ascorbate and to the capacity to protect photosynthesis from the damaging effects of high light. This system will also destroy H_2O_2 escaping the action of catalase (Foyer *et al.*, 1994).

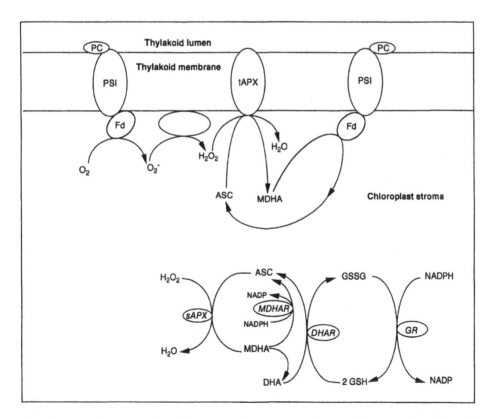

Figure 4. Schematic representation of the membrane-bound and soluble antioxidant enzymes of the chloroplast.

The Role of Glutathione in Ascorbate Recycling in Chloroplasts

Whereas DHA can be chemically reduced by GSH to ascorbate (Foyer and Halliwell, 1976), this reaction is catalyzed by several types of enzyme in plant tissues (Wells *et al.*, 1990; Trumper *et al.*, 1994; Kato *et al.*, 1997). It has recently been argued that DHA detected in chloroplast extracts is artefactual and that *in vivo* DHA levels are low or even negligible (Morrell *et al.*, 1997). This hypothesis was principally supported by the observation that DHA at concentrations thought to exist in the chloroplast could cause oxidative inactivation of two enzymes known to be regulated by the thioredoxin system (Morrell *et al.*, 1997). The conclusion that DHA accumulation must be avoided (Morrell *et al.*, 1997) did not take into account the dynamic nature of the regulation of the thiol-mediated enzymes of the chloroplasts. It is well established that the stromal enzymes regulated by the thioredoxin system require on-going reduction to remain active (e.g. Leegood and Walker, 1982; Leegood *et al.*, 1985). Data obtained by addition of oxidants to enzymes removed from a continuous system of thioredoxin reduction are not relevant to *in vivo* conditions in the chloroplast, where the activation state of thiol-regulated enzymes reflects differences between reductive and oxidative fluxes. Hence, it is important that DHA is reduced back to ascorbate. It does not, however, mean that DHA cannot exist *in vivo*. The significance of GSH as a reductant in this process has been established in relation to stress tolerance. A tropical fig mutant devoid of DHAR activity was found to be sensitive to high light (Yamasaki *et al.*, 1995). Evidence to support the critical role of DHAR, GSH, and GR in maintaining the foliar ascorbate pool has been obtained in transformed plants overproducing GR that have higher foliar ascorbate contents (Foyer *et al.*, 1995) and improved tolerance to oxidative stress (Foyer *et al.*, 1991, 1995; Aono *et al.*, 1993, 1995a), whereas transformed plants depleted of GR activity resulted in increased sensitivity to stress (Aono *et al.*, 1995b).

Environmental Stress Responses and Limitations

In general, stress enhances antioxidant capacity (Gillham and Dodge, 1987; Grace and Logan, 1996). The degree to which the activities of individual antioxidant enzymes are increased as a result of stress imposition is extremely variable. Whereas several types of peroxidase, including GPXs, have been shown to protect plants from stress-induced damage, only APXs have been implicated in the regulation of photosynthesis (Neubauer and Schreiber, 1989; Neubauer and Yamamoto, 1992; Osmond and Grace, 1995). APXs may therefore be more important in the protection of the electron transport processes against stress-induced damage than other forms of peroxidase or catalases.

The antioxidant system determines the lifetime of AOS within the cellular environment. Rather than a system designed to completely eliminate AOS, it appears to be one that permits control of the redox state of various cellular components. Many of the antioxidant enzymes consist of families of several different isoforms, for example there are at least four different forms of APX (Asada, 1992). The thylakoid-bound and stromal chloroplastic APXs are extremely unstable in the absence of ascorbate (Miyake *et al.*, 1993), whereas the cytosolic APXs are much more stable and have a broader substrate specificity (Asada, 1992). The genes coding for these different APXs respond to both metabolic and environmental signals (for example, see Karpinski *et al.*, 1997, 1999). The turnover and expression of the cytosolic *APX* genes appear to be more sensitive to changes in the environment than their chloroplastic counterparts. Overproduction of Cu/ZnSOD in tobacco chloroplasts, for example, led to increased accumulation of cytosolic APX transcripts (Sen Gupta *et al.*, 1993a). Similarly, exposure of *Arabidopsis* plants to photoinhibitory irradiances caused a rapid increase in mRNA accumulation of two cytosolic *APX* genes (*APX1* and *APX2*) (Karpinski *et al.*, 1997, 1999). Such stress-induced increases in APX or other antioxidant enzymes are considered to indicate acclimation of the plants to counterbalance increased oxidant production (Eisenstark *et al.*, 1995). Conversely, decreases in the activities of antioxidant enzymes during stress has frequently been interpreted as a failure of the plant to adapt to the prevailing stress conditions. In view of the relevance of H_2O_2 in signal transduction-regulated decreases in antioxidant activity may, however, have strategic relevance particularly in the short term. This is also perhaps the reason why overproduction of antioxidant enzymes in chloroplasts has proved to be so successful in improving protection of photosynthesis in stress situations.

Whereas many stress situations cause increases in total foliar antioxidant activity, little is known about the coordinate control of expression and activity of the different antioxidant enzyme isoforms in plant cells (Scandalios, 1983). Perhaps best characterized in terms of coordinate expression are the catalases and the SODs (Scandalios, 1994, 1997; Willekens *et al.*, 1994; Auh and Scandalios, 1997). In terms of SOD activity, all higher plant chloroplasts contain Cu/ZnSODs encoded by the nuclear *sodCp* gene. Some plant species also contain FeSODs that are coded by nuclear *sodB* genes (Bowler *et al.*, 1994). Whereas the photosynthetic membranes of cyanobacteria contain MnSODs, early reports of thylakoid-bound MnSOD in the chloroplasts of higher plants have never been substantiated (Asada *et al.*, 1975; Okuda *et al.*, 1979; Hayakawa *et al.*, 1985). FeSOD synthesis appears to be tightly linked to photosynthetic activity (Herbert *et al.*, 1992; Kurepa *et al.*, 1997). The absence of FeSOD led to PSII damage in cyanobacteria (Herbert *et al.*, 1992). In higher plants, FeSOD is a highly hydrophillic protein suggesting localization in the chloroplast stroma, but it may be able to associate with the thylakoid membranes (Van Camp *et al.*, 1996) as does the Cu/ZnSOD isoform (Van Ginkel and Brown, 1978). The presence of two forms of SOD in chloroplasts may be of physiological advantage (Droillard and Paulin, 1990).

Many studies have reported enhanced stress tolerance related to overproduction of SOD in chloroplasts (Tepperman and Dunsmuir, 1990; Bowler *et al.*, 1992; McKersie *et al.*, 1993, 1996; Perl *et al.*, 1993; Sen Gupta *et al.*, 1993a, 1993b; Foyer *et al.*, 1994; Van Camp *et al.*, 1994; Slooten *et al.*, 1995; Van Camp *et al.*, 1996; Payton *et al.*, 1997; Arisi *et al.*, 1998). Nevertheless, the precise mechanisms whereby SOD overproduction increases stress tolerance have not been clearly defined. The relevance of SOD overproduction without concomitant overproduction of other enzymes of H_2O_2 detoxification in chloroplasts is not initially apparent, because it has never been demonstrated or indeed suggested that the dismutation of superoxide is the rate-limiting step of the detoxification process. One possibility is that SOD is involved in substrate channelling ensuring that hydrogen peroxide generated from superoxide is passed directly to ascorbate peroxidase at the membrane surface. Since H_2O_2 is a signal-transducing molecule in plant cells, Prasad *et al.* (1994) suggested that increased protection by SOD overproduction may be due to enhanced H_2O_2 production. Some indications of the possible functions of excess SOD in chloroplasts come from studies on the regulation of photosynthesis in transformed plants (Slooten *et al.*, 1995; Payton *et al.*, 1997; Arisi *et al.*, 1998).

While overproduction of MnSOD in cotton did not appear to modify photosynthetic electron transport processes at low temperatures (Payton *et al.*, 1997), overproduction of

FeSOD in poplar altered the regulation of photosynthesis in two situations: (i) low CO_2 partial pressures, and (ii) in the presence of methyl viologen (Arisi *et al.*, 1998). In poplars, overproducing FeSOD in the chloroplast, which showed greatly enhanced FeSOD activity values of photochemical quenching of chlorophyll *a* fluorescence (q_P) (Krause and Weis, 1991), were similar over a range of CO_2 concentrations from 30 to 1000 $\mu l\, L^{-1}$. In marked contrast, a decline in q_P was observed in leaves from untransformed controls at intracellular CO_2 concentrations (Ci) below 200 $\mu l.L^{-1}$ (Arisi *et al.*, 1998), suggesting that in untransformed plants the primary PSII electron acceptors (such as Q_A) become progressively more reduced as CO_2 assimilation becomes limiting. In contrast, overreduction of Q_A was prevented in the leaves of the FeSOD transformants even at very low Ci. Under conditions of diminished CO_2 reduction, the consumption of electrons by pathways other than the Calvin cycle (such as the Mehler peroxidase cycle) can support noncyclic electron transport. The addition of H_2O_2 to isolated intact chloroplasts causes a rapid increase in q_P (Foyer *et al.*, 1994), supporting the conclusion that H_2O_2 generation and degradation may act as an alternative electron sink that maintains the oxidized state of Q_A. This mechanism may be responsible for the observed sustained q_P values at low CO_2 concentrations in the leaves of FeSOD transformants.

Whereas increased flux through the Mehler-peroxidase cycle may explain the q_P differences observed in poplar leaves at low Ci values, the overall rate of electron flow was similar in FeSOD transformants and in untransformed controls. It is possible that under the non-photorespiratory conditions (1% O_2) used in these experiments the reduction of oxygen limited the Mehler flux. Reported values for the K_m for O_2 of the Mehler reaction are between 2 and 60 μM (Robinson, 1988), which corresponds to 0.15 to 5% O_2 in the air at 25°C. An alternative explanation is that electron transport is restricted to such degree at low CO_2 and low O_2 that the Mehler-peroxidase cycle alone could not maintain Q_A in an oxidized state. Such observations clearly indicate that the Mehler reaction can prevent Q_A overreduction in environmental conditions that cause a decline in Ci, for example, as a result of stomatal closure during drought. Indeed, increased tolerance to water stress was observed in transformed alfalfa overproducing MnSOD in the chloroplasts (McKersie *et al.*, 1996).

When methyl viologen is added to chloroplasts, it is univalently photoreduced by PSI to its cation radical, which rapidly donates electrons to oxygen producing the superoxide anion. In this situation, it appears that superoxide production exceeds endogenous SOD capacity in untransformed plants because increased SOD activity in transformed tobacco and poplar plants improved protection of PSII from damage (Van Camp *et al.*, 1996; Arisi *et al.*, 1998). Overproduction of GR in chloroplasts (Foyer *et al.*, 1995) also results in increased protection of the photosynthetic processes against stress-induced damage. There is, therefore, considerable potential for engineering improved stress tolerance by amelioration of the antioxidant defenses of the chloroplasts.

RELATIONSHIPS BETWEEN CHLOROPLASTS AND OTHER COMPARTMENTS OF THE CELL

The antioxidant defenses of the chloroplast do not function in isolation. The chloroplasts reside in the cytosol of plant cells in close proximity to the peroxisomes and mitochondria with which substantial metabolite exchange occurs (Figure 5). The site of ascorbate biosynthesis within the plant cells is unknown and glutathione synthesis is possible in both the chloroplasts and cytosolic compartments (Noctor *et al.*, 1997a, 1997b, 1997c). None of the enzymes involved in the antioxidant defenses are coded for by genes on the chloroplast genome; all are nuclear encoded. The principal H_2O_2-scavenging enzymes in leaves are catalase, which is located in the peroxisomes, and APX which is located in the chloroplast, cytosol, and apoplast (Asada, 1992; Willekens *et al.*, 1994).

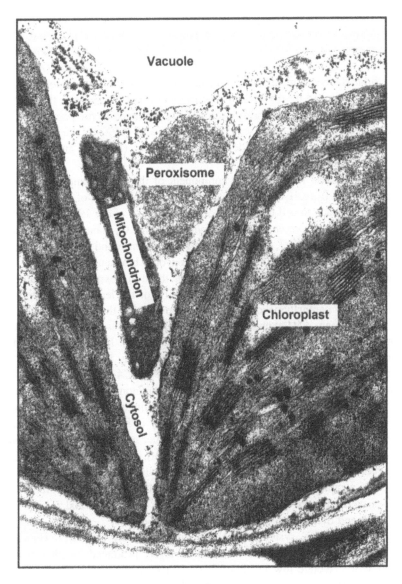

Figure 5.   Electron micrograph of a leaf palisade cell from the C_3 plant, *Trifolium dubium*, showing the close association of chloroplasts with mitochondria and peroxisomes. On one side the cytoplasm is boarded by the cell wall and leaf space, on the other by the vacuole. The magnification is 39,000. (This photograph was produced by Barry Thomas at IGER, Aberystwyth, UK).

The mode of action of catalases and peroxidases is essentially different (Willekens *et al.*, 1994). Catalases catalyze the dismutation of two molecules of H_2O_2 to water and molecular oxygen at near diffusion-controlled rates. Hence, whereas the V_{max} of the reaction is high, the affinity H_2O_2 is poor because two molecules of H_2O_2 have to impinge simultaneously at the active site. Even in the peroxisomes of photosynthetic cells which contain very high concentrations of catalase, APXs are also found presumably to prevent catalase inactivation by low concentration of H_2O_2. APX has a much higher affinity for H_2O_2 than catalase and can therefore maintain lower steady-state concentrations of H_2O_2 in peroxisomes than catalase alone. Catalase turns over rapidly in mature green leaves in a light-dependent manner similar to that of the D1 reaction centre protein of PSII (Hertwig *et al.*, 1992; Streb *et al.*, 1993; Feierabend and Dehne, 1996). APXs have a much higher affinity for H_2O_2 than catalases (Asada, 1992). The subcellular distribution of these enzymes would suggest that chloroplastic APX removes H_2O_2 produced by the Mehler reaction and other reactions in the chloroplasts whereas catalase scavenges photorespiratory H_2O_2 produced in the peroxisomes. H_2O_2 can diffuse freely through plant cell membranes and APX in the chloroplast can detoxify H_2O_2 arising from outside the chloroplasts (Nakano and Asada, 1980; Anderson *et al.*, 1983a).

In C_3 plants, catalase is an essential component of the antioxidant system protecting photosynthetic cells from H_2O_2-induced oxidation. In the leaves of C_3 plants, catalase deficiency rapidly leads to cell death and necrosis (Kendall *et al.*, 1983; Willekens *et al.*, 1997). In the C_3 plant *Nicotiana plumbaginifolia* and in the C_4 plant maize, three unlinked structural catalase genes (*Cat1*, *Cat2*, and *Cat3*) have been found (Guan and Scandalios, 1995; Willekens *et al.*, 1997). These genes encode biochemically distinct catalase isoenzymes (*CAT-1*, *CAT-2*, and *CAT-3*). Two of these are expressed in mature leaves (Willekens *et al.*, 1994). *CAT-1* comprises about 80% of the leaf catalase activity and is located in palidase parenchyma cells. *CAT-2* accounts for the remaining 20% and is localized in the phloem. The expression of *CAT-1* in C_3 leaves suggests a role in H_2O_2-scavenging during photorespiration (Willekens *et al.*, 1994). Epidermal cells, which are not photosynthetically active, did not contain detectable levels of *CAT-1* mRNA. Transformed tobacco lines deficient in either *CAT-1* or *CAT-2* (or both) had necrotic lesions on their leaves when exposed to high light (Chamnongpol *et al.*, 1996). In *CAT-1* deficient tobacco the development of leaf necrosis was linked to the production of H_2O_2 through photorespiration. Similarly, catalase deficiency in barley was lethal when plants were grown in air, but could be avoided by growth under non-photorespiratory conditions (Kendall *et al.*, 1983). In contrast, in maize (a C_4 plant), catalase deficiency produced no marked phenotype (Scandalios 1994), indicating that catalase activity is less important in the absence of a rapid flux through the photorespiratory pathway. Nevertheless, when maize leaves were exposed to low temperature stress, the CAT-1 and CAT-2 isoforms increased (Auh and Scandalios, 1997).

The catalase and APX systems act co-operatively to remove H_2O_2 (Foyer *et al.*, 1994). While catalase will remove the bulk of H_2O_2 produced by photorespiration, peroxidases will scavenge low levels of H_2O_2 not destroyed by catalase action. This assures maximal H_2O_2 destruction at a minimal cost in terms of reducing power. The ascorbate-glutathione cycle located in both the cytosol and the chloroplasts will remove H_2O_2 from the peroxisome if catalase is insufficient to cope with H_2O_2 production (Foyer *et al.*, 1994).

Photorespiration

The inhibition of photosynthesis in C_3 plants by oxygen is attributed to the competitive effect of O_2 on Rubisco activity (Cornic and Louason, 1980; Ogren, 1984; Björkman, 1996). During carbon assimilation, Rubisco uses CO_2 from the atmosphere to carboxylate ribulose-1,5-bisphosphate (RuBP). This enzyme can also use oxygen to oxygenate RuBP to initiate the reaction sequence called photorespiration (Ogren, 1984). Photorespiration can significantly decrease carbon gain and is a major source of H_2O_2 (Zelitch and Ochoa, 1953), producing two molecules of H_2O_2 for every molecule of CO_2 released. Photorespiratory CO_2 loss would be much higher if catalase were not effective in preventing chemical decarboxylation of the keto-acids glyoxylate and hydroxypyruvate in the peroxisome (Zelitch 1973, 1990).

Together carbon assimilation and photorespiration use most of the assimilatory power generated by the electron transport reactions. Photorespiration is able to use excess electron transport capacity when Benson-Calvin cycle activity is limited by the availability of CO_2 and the carboxylation activity of Rubisco is decreased in favour of oxygenation (Wu *et al.*, 1991). Photorespiration thus provides an alternative sink for photosynthetic energy and prevents excessive reduction of the electron transport chain that would inevitably lead to singlet oxygen formation and damage to PSII and other vital components. Photorespiration appears to be more effective than the Mehler reaction in protecting photosynthesis from the damaging effects of high light (Wu *et al.*, 1991; Osmond and Grace, 1995). Stimulation of photosynthesis is frequently observed in C_3 plants when the oxygen concentration is decreased to low levels (2%) or when CO_2 is increased. These conditions strongly inhibit photorespiration because they favour the carboxylation reaction of Rubisco and depress the oxygenation reaction. Several studies have shown, however, that low oxygen concentrations can inhibit photosynthesis under high light and high CO_2 conditions and that this inhibition can be reversed by normal levels of O_2 (Jolliffe and Tregunna, 1972; Cornic and Louason, 1980; McVetty and Canvin, 1981; Peterson, 1991).

Glutathione Biosynthesis

The possible adaptive significance of the apparently wasteful process of photorespiration has been largely discussed in terms of providing an alternative sink of assimilatory power at times when CO_2 availability is limited (Wu *et al.*, 1991; Osmond and Grace, 1995). Carbon entering this pathway is subsequently returned to the chloroplasts to support continued photosynthesis (Keys, 1980; Ogren, 1984). It has been argued that the supply of amino acids for biosynthetic reactions cannot be a possible function of photorespiration because it would drain the Benson-Calvin cycle of carbon and inhibits CO_2 assimilation (Ogren, 1984). Recently, however, it has become clear that the synthesis of GSH, an essential component of plant defense responses, requires photorespiratory glycine (Noctor *et al.*, 1997a, 1997b, 1998a).

Glutathione is synthesized from glutamate, cysteine, and glycine (Figure 6) in reactions catalyzed by γ-glutamylcysteine synthetase (γ-ECS) and glutathione synthetase (GS). This two-step reaction sequence can take place in both chloroplastic and non-chloroplastic compartments (Noctor *et al.*, 1998a). The biosynthesis and accumulation of GSH depends on the activity of γ-ECS, the availability of cysteine, and the light-dependent formation of

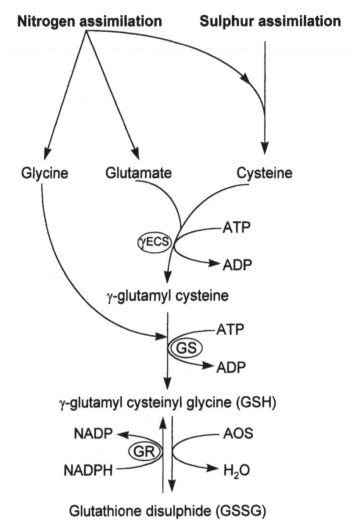

Figure 6. The pathway of GSH biosynthesis in plants showing the enzymes: γ-glutamyl/cysteine synthetase (γ-ECS), glutathione reductase (GR) and glutathione synthetase (GS).

glycine through the photorespiratory pathway (Noctor *et al.*, 1996, 1997a, 1997b, 1997c). In poplar leaves, the relationship between the foliar contents of γ-EC and GSH was strongly influenced by illumination (Noctor *et al.*, 1997a), the effect of illumination being most pronounced in leaves from transformed poplars overproducing the first enzyme of the GSH biosynthesis, γ-ECS (Figure 6), which possess markedly enhanced γ-EC and GSH contents (Arisi *et al.*, 1997). In the dark, GSH contents of these leaves were relatively low and the foliar γ-EC contents were higher than these of untransformed poplar. Overproduction of γ-ECS, therefore, elevated γ-EC from the ranks of a trace metabolite to one of major quantitative importance (Arisi *et al.*, 1997). On illumination, γ-EC content decreased and GSH increased while cysteine contents were little affected by light (Noctor *et al.*, 1997a).

Light-dependent conversion of γ-EC to GSH could not be explained in terms of light-modulation of the activities of the enzymes that catalyze the biosynthesis of GSH, which were similar in light and dark (Noctor *et al.*, 1997a). Conversely, the light-induced increase in serine and glycine contents correlated with the decrease in γ-EC and increase in GSH, suggesting that the absence of photorespiratory glycine production is the factor which limits conversion of γ-EC to GSH in the dark (Noctor *et al.*, 1997a). This hypothesis is supported by the effects of illumination of leaves under non-photorespiratory conditions (Noctor *et al.*, 1997a). In conditions of low O_2 or of CO_2 enrichment, the inverse light-induced increase in foliar glycine was prevented and the light-induced conversion of γ-EC to GSH was severely attenuated. Neither light nor O_2 content had a significant effect on foliar glutamate and cysteine pools.

The effect of photorespiration can be replaced during darkness by supplying glycine (Noctor *et al.*, 1997b). Supplying glycine through the petiole largely prevented the dark-induced increase in γ-EC and caused accumulation of GSH even in the dark (Noctor *et al.*, 1997b). Similarly, when leaf discs were incubated on glycine, the dark-induced increases in γ-EC was largely abolished, but changes in the cysteine pool were still apparent (Noctor *et al.*, 1997b). These results demonstrate that a major factor restricting GSH synthesis in situations where γ-ECS activity is increased, during darkness or the absence of photorespiration, is the availability of glycine.

The requirements of light and glycine to support maximal rates of conversion of γ-EC to GSH are evident in both untransformed poplars and in poplars strongly overproducing γ-ECS in either the chloroplast or cytosol (Noctor *et al.*, 1998a). The greater the capacity for γ-EC synthesis the more stringent the kinetic limitation by glycine availability will become. In response to environmental triggers, acclimatory responses, such as accelerated cysteine synthesis, increased *de novo* synthesis of γ-ECS or alleviated inhibition of γ-ECS by GSH may all favour increased rates of γ-EC synthesis (Noctor *et al.*, 1997c). It is pertinent to ask whether GSH biosynthesis would deplete the Benson-Calvin cycle of carbon in such circumstances. The rate of GSH biosynthesis is relatively low compared to that of RuBP oxygenation. Under optimal conditions the loss of carbon to GSH biosynthesis would be only about 0.5% of net assimilate (Noctor *et al.*, 1998b), but under stress conditions, when photosynthesis declines and γ-EC synthesis increases, the demand for photorespiratory glycine would be much greater.

COMPARTMENTATION OF ASCORBATE

In contrast to cyanobacteria, which contain only low concentrations (30–100 μm) of ascorbate (Tel-Or *et al.*, 1985), L-AA is a major metabolite of chloroplasts from higher plants (Foyer *et al.*, 1983; Foyer, 1993a; Gillham and Dodge, 1986, 1987) and represents about 10% of the soluble carbohydrate pool in leaves (Smirnoff, 1996). It is therefore remarkable that the pathway of ascorbate biosynthesis in plants has only recently been resolved (Wheeler *et al.*, 1998). An *Arabidopsis* mutant deficient in ascorbic acid biosynthesis showed increased sensitivity to ozone, UVB irradiation, and sulphur dioxide fumigation, confirming the role of ascorbate in protection against oxidative damage (Conklin *et al.*, 1996, 1997).

L-AA ionizes at the hydroxyl C-2 (pk 4.77) or C-3 (pk 11.57) and exists as a monovalent anion at physiological pH values (Rose and Bode, 1993). Carrier-mediated movement of ascorbate in the direction of the electrochemical gradient has been suggested for the chloroplast envelope (Anderson *et al.*, 1983b; Beck *et al.*, 1983). Active transport against an electrochemical gradient may occur on the plasmalemma. The plasmamembrane contains at least three different mechanisms of ascorbate transport and could facilitate ascorbate-mediated transport of reducing equivalents between the cytosol and apoplast (Horemans *et al.*, 2000). A highly specific b-type cytochrome transferring electrons from cytosolic ascorbate to extracellular acceptors, including MDHA, has been found on the plasmamembrane as well as MDHAR and other ascorbate carriers selectively transporting L-AA and DHA between the cytosol and apoplast (Asard *et al.*, 1995; Horemans *et al.*, 1997; Foyer and Lelandais, 1996).

Whereas L-AA transport into isolated intact chloroplasts across the chloroplast envelope has been observed (Anderson *et al.*, 1983b; Beck *et al.*, 1983; Foyer and Lelandais, 1996), the thylakoid membrane system appears to have no ascorbate carrier system to transport ascorbate from the stroma to the thylakoid lumen (Foyer and Lelandais, 1996). This is surprising because the enzyme VDE requires L-AA to convert violaxanthin to zeaxanthin (Figure 2). Since ascorbate is vital to both zeaxanthin-dependent energy dissipation and the Mehler-peroxidase reaction sequence, it is possible that ascorbate availability might regulate these two processes. The addition of H_2O_2 to illuminated thylakoid membrane preparations induces a transient inhibition of zeaxanthin formation (Neubauer and Yamamoto, 1992). The ascorbate pool in the chloroplast stroma is substantial and it is generally found to be largely in the reduced form (Foyer *et al.*, 1983; Foyer 1993b; Foyer and Lelandais, 1995). VDE is located on the lumenal surface of the thylakoid membrane and ascorbate must cross the thylakoid membrane to reach this enzyme (Rockholm and Yamamoto, 1996). If the permeability of the thylakoid membrane for ascorbate restricts the rate of zeaxanthin formation it could have important consequences for thermal energy dissipation. Since no transport system for ascorbate was found only the uncharged species (ascorbic acid) can diffuse across the membrane. In the dark, the concentration of ascorbate should be similar in both compartments but in the light ascorbate will tend to diffuse out of the thylakoid lumen. Upon illumination, the pH of the lumen falls and VDE binds to the lumenal side of the thylakoid membrane and becomes active (Hager and Holocker, 1994). The affinity of VDE for ascorbate is strongly dependent on pH, the acid form being the true substrate for the enzyme (Bratt *et al.*, 1995).

Differential Intercellular Partitioning of Antioxidants in Maize Leaves

In the discussion of the antioxidant defenses of the chloroplasts thus far we have assumed that the antioxidant components are equally distributed between all photosynthetic cells. Certain C_4 plants, such as maize, have two types of photosynthetic cells whose functions are very different (Furbank and Foyer, 1988). In maize leaves, GR and DHAR were almost exclusively localized in the mesophyll cells whereas the majority of the APX and SOD activities were localized in the bundle sheath tissue (Doulis *et al.*, 1997; Pastori *et al.*, 1999). Catalase and MDHAR were found to be approximately equally distributed between the two cell types. H_2O_2 was found to accumulate only in the mesophyll compartment in optimal growth conditions (Doulis *et al.*, 1997). These observations are interesting because the enzymes of the Benson-Calvin cycle, which are very sensitive to inhibition by H_2O_2, are found only in bundle sheath chloroplasts (Furbank and Foyer, 1988).

The localization of GR and DHAR in the mesophyll tissues may result from the requirement of these enzymes for reducing power (Doulis *et al.*, 1997). Bundle sheath cells are deficient in PSII and may not generate sufficient NADPH to support the reduction of GSSG and DHA. GSSG and DHA produced in the bundle sheath tissues must be transported to the mesophyll tissues to be reduced (Figure 7). Because of this requirement for cycling of reduced and oxidized forms of ascorbate and glutathione, the bundle sheath cells may be less protected against oxidative damage than the mesophyll cells. This appears to be the case when maize leaves are treated with methyl viologen because substantial oxidative damage to proteins present in the bundle sheath, but not in the mesophyll, is observed (Figure 8). Comparisons of carbonyl formation on mesophyll and bundle sheath proteins from methyl viologen-treated leaves and from plants grown at 15°C and 20°C revealed that oxidative damage was not uniformly distributed between these tissues, but was localized almost exclusively in the bundle sheath (Kingston-Smith and Foyer, 2000), suggesting that oxidative damage induced by these treatments is restricted to the bundle sheath tissue because of inadequate antioxidant protection during stress in this tissue.

CONCLUSIONS

The atmosphere of the earth contained very low concentrations of oxygen until the cyanobacteria evolved and photosynthesis became a predominant metabolic process (Berkner and Marshall, 1965). Since molecular oxygen is toxic, the antioxidant systems were of fundamental importance in allowing photosynthetic organisms to thrive and proliferate. Antioxidant defenses thus govern the balance between life and death in an oxygen atmosphere (Fridovich, 1995). In the chloroplasts these are not only essential protective agents but they have become intricately involved with the regulation of photosynthesis. The presence of photodynamic pigments in the thylakoid membranes commits the chloroplasts to light absorption. The energy of the absorbed radiation is then used to oxidize water and produce the NADPH required to reduce CO_2 to the level of sugar phosphate. Intrinsic to this process is the generation of a transmembrane proton potential difference that can be used not only to produce the ATP needed for CO_2 assimilation but also to regulate the flux of electrons. Both energy and electrons can be transferred to oxygen by the electron transport system. These processes may have beneficial effects if they are effectively coupled to efficient quenching reactions.

The network of interactions between photosynthesis and oxygen metabolism is highly complex. Oxygen is used both as an electron acceptor for photosynthesis and also as a substrate for assimilatory power in photorespiration. In turn, photorespiration not only serves as a sink for assimilatory power but it also provides essential glycine for the synthesis of glutathione. Since glutathione synthesis is greatly enhanced in stress situations, photorespiration becomes an essential feature of this process as the source of glycine that permits acclimation of this antioxidant to stress conditions.

The chloroplast is frequently considered to be a source of oxidative stress in plants. This hypothesis has largely arisen from observations of AOS production by isolated intact chloroplasts in *in vitro* experiments. In such studies the endogenous antioxidant systems are frequently impaired or the chloroplast fractions contain large amounts of cell debris and broken chloroplasts. When care is taken to retain chloroplast integrity and function,

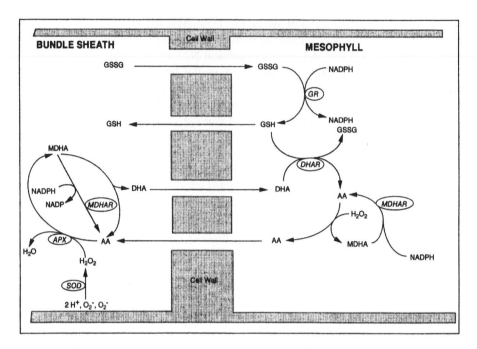

Figure 7. Differential localisation of the antioxidant enzymes between the bundle sheath and mesophyll cells of maize. Localisation of glutathione reductase (GR) and dehydroascorbate reductase (DHAR) in the mesophyll cells necessitates transport of the reduced and oxidised forms of ascorbate (AA, DHA) and glutathione (GSH, GSSG) between the cell types.

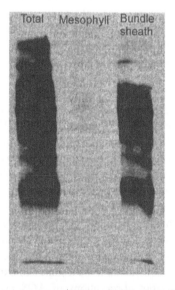

Figure 8. Oxidative damage to proteins extracted from whole leaf (total) and mesophyll and bundle sheath tissue of maize and determined by the extent of carbonyl group formation. Mesophyll and bundle sheath extracts of leaves treated with methyl viologen, were prepared as described by Kingston-Smith and Foyer (1989), loaded on an equal protein basis of 5 μg per track and separated on 15% SDS-PAGE.

isolated chloroplasts do not liberate AOS but are more than capable of destroying H_2O_2 added to the medium when illuminated. The chloroplasts are therefore equipped with sufficient antioxidant defenses to prevent oxidative stress.

The production of AOS by the thylakoid membranes is regulated and limited by efficient control of both electron transport and light-harvesting processes. This regulation limits the potential for oxidative damage and prevents uncontrolled electron transport to oxygen. Perhaps more than any other organelle within the plant cell, chloroplasts have embraced the potential for using the chemistry of oxygen for metabolism while minimizing the deleterious effects of uncontrolled interactions with oxygen (Foyer and Noctor, 2000). Hence, whereas the potential of the chloroplast as a source of oxidative stress is large, in reality this organelle offers minimal risk because of pre-emptive regulation and effective defense.

REFERENCES

Allen, J.F. (1975) Oxygen reduction and optimum production of ATP in photosynthesis. *Nature,* **256,** 599–600.

Allen, J.F. (1977) Oxygen–a physiological electron acceptor in photosynthesis? *Curr. Adv. Plant Sci.,* **29,** 459–468.

Allen, J.F. (1992) Protein phosphorylation in regulation of photosynthesis. *Biochim. Biophys. Acta,* **1098,** 275–335.

Ananyev, G., Renger, G., Wacker, U., and Klimov, V. (1994) The photoproduction of superoxide radicals and the superoxide dismutase activity of photosystem II. The possible involvement of cytochrome b559. *Photosynthesis Res.* **41,** 327–338.

Anderson, J.W., Foyer, C.H., and Walker, D.A. (1983a) Light-dependent reduction of hydrogen peroxide by intact spinach chloroplasts. *Biochim. Biophys. Acta,* **724,** 69–74.

Anderson, J.W., Foyer, C.H., and Walker, D.A. (1983b) Light-dependent reduction of dehydroascorbate and uptake of exogenous ascorbate by spinach chloroplasts. *Planta,* **158,** 442–450.

Anderson, J.M. and Osmond, C.B. (1987) Sun-shade responses: compromises between acclimation and photoinhibition. In D.J. Kyle, C.B. Osmond, and C.J. Arntzen, (eds.), *Photoinhibition* (Topics in Photosynthesis, Vol. 9), Elsevier Science Publishers, Amsterdam, pp. 1–37.

Aono, M., Kubo, A., Saji, H., Tanaka, K., and Kondo, N., (1993) Enhanced tolerance to photooxidative stress of transgenic *Nicotiana tabacum* with high chloroplastic glutathione reductase activity. *Plant Cell Physiol.,* **34,** 129–135.

Aono, M., Saji, H., Sakamoto, A., Tanaka, K., Kondfo, N., and Tanaka, K., (1995a) Paraquat tolerance of transgenic *Nicotiana tabacum* with enhanced activities of glutathione reductase and superoxide dismutase. *Plant Cell Physiol.,* **36,** 1687–1691.

Aono, M., Sahi, H., Fujiyama, K., Sugita, M., Kondo, N., and Tanaka, K., (1995b) Decrease in activity of glutathione reductase enhances paraquat sensitivity in transgenic *Nicotiana tabacum. Plant Physiol.,* **107,** 645–648.

Arisi, A.-C.M., Cornic, G., Jouanin, L., and Foyer, C.H. (1998) Overexpression of FeSOD in transformed poplar modifies the regulation of photosynthesis at low CO_2 partial pressures or following exposure to the pro-oxidant herbicide methyl viologen. *Plant Physiol.,* **117,** 565–574.

Arisi, A.-C.M., Noctor, G., Foyer, C.H., and Jouanin, L. (1997) Modification of thiol contents in poplars (*Populus tremula* x *P. alba*) over-expressing enzymes involved in glutathione biosynthesis. *Planta,* **203,** 362–372.

Arnon, D.I. and Chain, R.K. (1975) Regulation of ferredoxin-catalysed photosynthetic phosphorylations. *Proc. Natl. Acad. Sci. USA,* **72,** 4961–4965.

Arnon, D.I., Whatley, F.R., and Allen, M.B. (1958) Assimilatory power in photosynthesis. *Science,* **127,** 1026–1034.

Arnott, T. and Murphy, T.M. (1991) A comparison of the effects of a fungal elicitor and ultraviolet radiation on ion transport and hydrogen peroxide synthesis in rose cells. *Environ. Exp. Bot.* **31,** 209–216.

Aro, E.-M., McCaffery, S., and Anderson, J.M. (1993) Photoinhibition and D_1 protein degradation in peas acclimated to different growth irradiances. *Plant Physiol.,* **103,** 835–843.

Asada, K. (1992) Ascorbate peroxidase—a hydrogen peroxide scavenging enzyme in plants. *Physiol. Plant.*, **85**, 235–241.

Asada, K. (1994) Production and action of active oxygen species in photosynthetic tissues. In C.H. Foyer, and P.M. Mullineaux, (eds.), *Causes of Photooxidative Stress and Amelioration of Defence Systems in Plants*, CRC Press, Boca Raton, pp. 77–104.

Asada, K. (1996) Radical production and scavenging in the chloroplasts. In N.R. Baker, (ed.), *Photosynthesis and the Environments*, Kluwer Academic Publishers, Dordrecht, pp. 128–150.

Asada, K., Kiso, K., and Yoshikawa, K. (1974) Univalent reduction of molecular oxygen by spinach chloroplasts on illumination. *J. Biol. Chem.*, **247**, 2175–2181.

Asada, K., Urano, M., and Takahashi, M. (1973) Subcellular location of superoxide dismutase in spinach leaves and preparation and properties of crystalline spinach superoxide dismutase. *Eur. J. Biochem.*, **36**, 257–266.

Asada, K., Yoshikawa, K., Takahashi, M., Maeda, Y., and Enmanji, K. (1975) Superoxide dismutases from a blue-green algae *Plectonema boryanum*. *J. Biol. Chem.*, **250**, 2801–2807.

Asard, H., Horemans, N., and Caubergs, R.J. (1995) Involvement of ascorbic acid and a b-type cytochrome in plant membrane redox reactions. *Protoplasma*, **184**, 36–41.

Auh, C.-K. and Scandalios, J.G. (1997) Spatial and temporal responses of the maize catalases to low temperature. *Physiol. Plant.*, **101**, 149–159.

Badger, M.R. (1985) Photosynthetic oxygen exchange. *Annu. Rev. Plant Physiol.*, **36**, 27–53.

Baier, M. and Dietz, K.-J. (1997) The plant 2-Cys peroxiredoxin BAS1 is a nuclear-encoded chloroplast protein: its expressional regulation, phylogenetic origin, and implications for its specific physiological function in plants. *Plant J.*, **12**, 179–190.

Barber, J., and Andersson, B. (1992) Too much of a good thing: Light can be bad for photosynthesis. *Trends Biochem. Sci.*, **17**, 61–66.

Berkner, L.V., and Marshall, L.C. (1965) On the origin and rise of oxygen concentration in the earth's atmosphere. *J. Atmos. Sci.*, **22**, 225–261.

Beck, E., Burkert, A., and Hofmann, M. (1983) Uptake of L-ascorbate by intact chloroplasts. *Plant Physiol.*, **73**, 41–45.

Bes, M.T., De Lacey, A., Peleato, M.L., Fernandez, V.M., and Gómez-Moreno, C. (1995) The covalent linkage of a viologen to a flavoprotein reductase transforms it into an oxidase. *Eur. J. Biochem.*, **233**, 593–599.

Biehler, K. and Fock, H. (1996) Evidence for the contribution of the Mehler-peroxidase reaction in dissipating excess electrons in drought-stressed wheat. *Plant Physiol.*, **112**, 265–272.

Bielski, B.J.J. (1982) Chemistry of ascorbic acid radicals. In P.A. Seib, and B.M. Tolbert, (eds.), *Ascorbic Acid, Chemistry, Metabolism and Uses*, American Chemical Society, Washington, D.C., pp. 81–100.

Bielski, B., Comstock, D., and Bowen, A. (1971) Ascorbic acid free radicals. I. Pulse radiolysis study of optical absorption and kinetic properties. *J. Amer. Chem. Soc.*, **93**, 5624–5629.

Björkman, O. (1987) Low temperature chlorophyll fluorescence in leaves and its relationship to photon yield of photosynthesis in photoinhibition. In D.J. Kyle, C.B. Osmond, and C.J. Arntzen, (eds.), *Photoinhibition*, (Topics in Photosynthesis, Vol. 9), Elsevier, Amsterdam, pp. 123–144.

Björkman, O. (1996) The effect of oxygen on photosynthesis in higher plants. *Physiol. Plant.*, **19**, 618–633.

Björkman, O. and Demmig-Adams, B. (1994) Regulation of photosynthetic light energy capture, conversion and dissipation in leaves of higher plants. In E. Schulze, and M.M. Caldwell, (eds.), *Ecophysiology of Photosynthesis*, Springer-Verlag, Berlin, pp. 17–47.

Bowler, C., Alliotte, T., De Loose, M., Van Montagu, M., and Inzé, D. (1989) The induction of manganese superoxide dismutase in response to stress in *Nicotiana plumbaginifolia*. *EMBO J.*, **8**, 31–38.

Bowler, C., and Chua, N.-H. (1994) Emerging themes of plant signal-transduction. *Plant Cell*, **6**, 1529–1541.

Bowler, C., Van Camp, W., Van Montagu, M., and Inzé, D. (1994) Superoxide dismutase in plants. *Crit. Rev. Plant Sci.*, **13**, 199–218.

Bowler, C., Van Montagu, M., and Inzé, D. (1992) Superoxide dismutase and stress tolerance. *Annu. Rev. Plant Physiol. Plant Mol. Biol.*, **43**, 83–116.

Boyer, J.S. (1982) Plant productivity and the environment. *Science*, **218**, 443–448.

Bracht, E. and Trebst, A. (1994) Hypothesis on the control of D_1 protein turnover by nuclear coded proteins in *Chlamydomonas reinhardtii*. *Z. Naturforsch.*, **49C**, 439–446.

Bratt, C.E., Arvidsson, P.-O., Carlsson, M., and Akerlund, H.-E. (1995) Regulation of violaxanthin de-epoxidase by pH and ascorbate concentration. *Photosynthesis Res.*, **45**, 169–175.

Briantais, J.-M., Vernotte, C., Picaud, M., and Krause, G.H. (1979) A quantitative study of the slow decline of chlorophyll *a* fluorescence in isolated chloroplasts. *Biochim. Biophys. Acta*, **548**, 128–138.

Buettner, G.R. and Jurkiewicz, B.A. (1996) Chemistry and biochemistry of ascorbic acid. In E. Cadenas, and L. Packer, (eds.), *Handbook of Antioxidants*, Marcel Dekker, New York, pp. 91–115.

Cadenas, E. (1989) Biochemistry of oxygen toxicity. *Annu. Rev. Biochem.*, **58**, 79–110.

Callahan, F.E., Wergin, W.P., Nelson, N., Edelman, M., and Mattoo, A.K. (1989) Distribution of thylakoid proteins between stromal and granal lamellae in *Spirodela*; Dual location of photosystem II components. *Plant Physiol.*, **91**, 629–635.

Canvin, D.T., Berry, J.A., Badger, M.R., Foch, H., and Osmond, C.B. (1980) Oxygen exchange in leaves in the light. *Plant Physiol.*, **66**, 302–307.

Casano, L.M., Gómez, L.D., Lascano, H.R., González, C.A., and Trippi, V.S. (1997) Inactivation and degradation of CuZn-SOD by active oxygen species in wheat chloroplasts exposed to photooxidative stress. *Plant Cell Physiol.*, **38**, 433–440.

Chamnongpol, S., Willekens, H., Langebartels, C., Van Montagu, M., Inzé, D., and Van Camp, W. (1996) Transgenic tobacco with a reduced catalase activity develops necrotic lesions and induces pathogenesis-related expression under high light. *Plant J.*, **10**, 491–503.

Charles, S.A. and Halliwell, B. (1980) Effect of hydrogen peroxide on spinach (*Spinacea oleracea*) chloroplast fructose bisphosphatase. *Biochem. J.*, **189**, 373–376.

Chen, G.X. and Asada, K. (1989) Ascorbate peroxidases in tea leaves: Occurrence of two isozymes and the differences in their enzymatic and molecular properties. *Plant Cell Physiol.*, **30**, 987–998.

Chen, W., Chao, G., and Singh, K.B. (1996) The promoter of a H_2O_2-inducible *Arabidopsis* glutathione S-transferase gene contains closely linked OBF- and OBP1-binding sites. *Plant J.*, **10**, 955–966.

Chen, Z., Silva, H., and Klessig, D. (1993) Active oxygen species in the induction of plant systemic acquired resistance by salicylic acid. *Science*, **262**, 1883–1886.

Conklin, P.L., Pallanca, J.E., Last, R.L., and Smirnoff, N. (1997) L-ascorbic acid metabolism in the ascorbate-deficient *Arabidopsis* mutant *vtc1*. *Plant Physiol.*, **115**, 1277–1285.

Conklin, P.L., Williams, E.H., and Last, R.L. (1996) Environmental stress sensitivity of an ascorbic acid-deficient *Arabidopsis* mutant. *Proc. Natl. Acad. Sci. USA*, **93**, 9970–9974.

Cornic, G. and Louason, G. (1980) The effects of O_2 on net photosynthesis at low temperature (5°C). *Plant Cell Environ.*, **3**, 149–157.

Cseke, C. and Buchanan, B.B. (1986) Regulation of the formation and utilisation of photosynthate in leaves. *Biochim. Biophys. Acta*, **853**, 43–63.

Danon, A. and Mayfield, S.P. (1994) Light-regulated translation of chloroplast messenger RNAs through redox potential. *Science*, **266**, 1717–1719.

Demmig-Adams, B. (1990) Carotenoids and photoprotection in plants: A role for the xanthophyll zeaxanthin. *Biochim. Biophys. Acta*, **1020**, 1–24.

Demmig-Adams, B. and Adams, W.W. (1992) Photoprotection and other responses of plants to high light stress. *Annu. Rev. Pl. Physiol. Plant Mol. Biol.*, **43**, 599–626.

Demmig-Adams, B. and Adams, W.W. III. (1996) The role of xanthophyll cycle carotenoids in the protection of photosynthesis. *Trends Plant Sci.*, **1**, 21–26.

Dempsey, D.A. and Klessig, D.F. (1994) Salicylic acid, active oxygen species and systemic acquired resistance in plants. *Trends Cell Biol.*, **4**, 334–338.

Doke, N., Miura, Y., Leandro, M.S., and Kawakita, K. (1994) Involvement of superoxide in signal-transduction: Responses to attack by pathogens, physical and chemical shock and UV radiation. In C.H. Foyer, and P. Mullineaux, (eds.), *Causes of Photooxidative Stress and Amelioration of Defence Systems in Plants*, CRC Press, Boca Raton, pp. 177–197.

Doulis, A.G., Debian, N., Kingston-Smith, A.H., and Foyer, C.H. (1997) Differential localisation of antioxidants in maize leaves. *Plant Physiol.*, **114**, 1031–1037.

Droillard, M.J. and Paulin, A. (1990) Isozymes of superoxide dismutase in mitochondria and peroxisomes isolated from petals of carnation (*Dianthus caryophyllus*) during senescence. *Plant Physiol.*, **94**, 1187–1192.

Eckert, H.-J., Geiken, B., Bernarding, J., Napiwotzki, A., Eichler, H.-J., and Renger, G. (1991) Two sites of photoinhibition of the electron transfer in oxygen evolving and tris-treated PSII membrane fragments from spinach. *Photosynthesis Res.*, **27**, 97–108.

Egneus, H., Heber, U., Matthiesen, U., and Kirk, M. (1975) Reduction of oxygen by the electron transport chain of chloroplasts during assimilation of carbon dioxide. *Biochim. Biophys. Acta*, **408**, 252–268.

Eisenstark, A., Yallaly, P., Ivanova, A., and Miller, C. (1995) Genetic mechanisms involved in cellular recovery from oxidative stress. *Arch. Insect Biochem. Physiol.*, **29**, 159–173.

Elstner, E.F., Saran, M., Bors, W., and Lengfelder, E. (1978) Oxygen activation in isolated chloroplasts. Mechanism of ferredoxin-dependent ethylene formation from methionine. *Eur. J. Biochem.*, **89**, 61–66.

Escoubas, J-M., Lomas, M., LaRoche, J., and Falkowski, P.G. (1995) Light intensity regulates *cab* gene expression via the redox state of the plastoquinone pool in the green alga *Dunaliella tertiolecta. Proc. Natl. Acad. Sci. USA*, **92**, 10237–10241.

Eshdat, Y., Holland, D., Faltin, Z., and Ben-Hayyim, G. (1997) Plant glutathione peroxidases. *Physiol. Plant.*, **100**, 234–240.

Feierabend, J. and Dehne, S. (1996) Fate of the porphyrin cofactors during the light-dependent turnover of catalase and of the photosystem II reaction-center protein D1 in mature rye leaves. *Planta*, **198**, 413–422.

Fine, P.L. and Frisch, W.D. (1990) The mechanism of H$_2$O$_2$ production by the S$_2$ state of the oxygen-evolving complex. In M. Baltscheffsky, (ed.), *Current Research in Photosynthesis I*, Kluwer Academic Publishers, Dordrecht, pp. 905–908.

Firl, J., Frommeyer, D., and Elstner, E.F. (1981) Isolation and identification of an oxygen reducing cofactor (ORF) from isolated chloroplast lamellae. *Z. Naturforsch*, **36C**, 284–294.

Forti, G. and Elli, G. (1995) The function of ascorbic acid in photosynthetic phosphorylation. *Plant Physiol.*, **109**, 1207–1211.

Foyer, C.H. (1984) *Photosynthesis*, (Cell Biology: A Series of Monographs, Vol. 1), John Wiley & Sons, New York.

Foyer, C.H. (1993a) Ascorbic acid. In R.G. Alscher and J.L. Hess, (eds.), *Antioxidants in Higher Plants*, CRC Press, Boca Raton, pp. 31–58.

Foyer, C.H. (1993b) Interactions between electron transport and carbon assimilation in leaves: Co-ordination of activities and control. In Y. Abrol, P. Mohanty, and Govindjee, (eds.), *Photosynthesis: Photoreactions to Plant Productivity*, Oxford and IBH Publishing Co., New Delhi, pp. 199–224.

Foyer, C.H. and Halliwell, B. (1976) The presence of glutathione and glutathione reductase in chloroplasts: a proposed role in ascorbic acid metabolism. *Planta*, **133**, 21–25.

Foyer, C.H., and Harbinson, J. (1994) Oxygen metabolism and the regulation of photosynthetic electron transport. In C.H. Foyer, P.M. Mullineaux, (eds.), *Causes of Photooxidative Stress and Amelioration of Defence Systems in Plants.* CRC Press, Boca Raton, pp. 1–42.

Foyer, C.H., and Harbinson, J. (1997) The photosynthetic electron transport system: efficiency and control. In C.H. Foyer, and W.P. Quick, (eds.), *A molecular approach to primary metabolism in higher plants*, Taylor and Francis, London, UK, pp. 3–39.

Foyer, C.H., and Lelandais, M. (1995) Ascorbate transport into protoplasts, chloroplasts and thylakoid membranes of pea leaves. In P. Mathis, (ed.), *Photosynthesis: From Light to Biosphere*, Vol. V, Kluwer Academic Publishers, Dordrecht, pp. 511–514.

Foyer, C.H., and Lelandais, M. (1996) A comparison of the relative rates of transport of ascorbate and glucose across the thylakoid, chloroplast and plasmalemma membranes of pea leaf meosphyll cells. *J. Plant Physiol.*, **148**, 391–398.

Foyer, C.H., and Noctor, G. (2000) Oxygen processing in photosynthesis: regulation and signalling. *New Phytol.*, **146**, 359–388.

Foyer, C.H., Rowell, J., and Walker, D. (1983) Measurement of the ascorbate content of spinach leaf protoplasts and chloroplasts during illumination. *Planta*, **157**, 239–244.

Foyer, C.H., Furbank, R., Harbinson, J., and Horton, P. (1990) The mechanisms contributing to photosynthetic control of electron transport by carbon assimilation in leaves. *Photosynthesis Res.*, **25**, 83–100.

Foyer, C.H., Lelandais, M., Galap, C., and Kunert, K.J. (1991) Effects of elevated glutathione reductase activity on the cellular glutathione pool and photosynthesis in leaves under normal and stress conditions. *Plant Physiol.*, **97**, 863–872.

Foyer, C.H., Lelandais, M., and Harbinson, J. (1992) Control of the quantum efficiencies of photosystems I and II, electron flow and enzyme activation following dark-to-light transitions in pea leaves. *Plant Physiol.*, **99**, 979–986.

Foyer, C.H., Descourviéres, P., and Kunert, K.-J. (1994) Protection against oxygen radicals: an important defence mechanism studied in transgenic plants. *Plant Cell Environ.*, **17**, 507–523.

Foyer, C.H., Souriau, N., Perret, S., Lelandais, M., Kunert, K.J., Pruvost, C., and Jouanin, L. (1995) Over-expression of glutathione reductase but not glutathione synthetase leads to increases in antioxidant capacity and improved photosynthesis in poplar (*Populus tremula* x *P. alba*) trees. *Plant Physiol.*, **109**, 1047–1057.

Foyer, C.H., Lopez-Delgado, H., Dat, J.F., and Scott, I.M. (1997) Hydrogen peroxide- and glutathione-associated mechanisms of acclimatory stress tolerance and signalling. *Physiol. Plant.*, **100**, 241–254.

Frank, H.A., Cua, A., Chynwat, V., Young, A., Gosztola, D., and Wasielewski, M.R. (1994) Photophysics of the carotenoids associated with the xanthophyll cycle in photosynthesis. *Photosynthesis Res.*, **41**, 389–395.

Fridovich, I. (1995) Superoxide radical and superoxide dismutases. *Annu. Rev. Biochem.*, **64**, 97–112.

Furbank, R.T. and Foyer, C.H. (1988) C$_4$ plants as model experimental systems for the study of photosynthesis. *New Phytol.*, **109**, 265–277.

Gerbaud, A. and Andre, M. (1980) Effect of CO$_2$, O$_2$ and light on photosynthesis and respiration in wheat. *Plant Physiol.*, **66**, 1032–1036.

Gerst, U., Schönknecht, G., and Heber, U. (1994) ATP and NADPH as the driving force of carbon reduction in leaves in relation to thylakoid energetization. *Planta*, **193**, 421–429.

Giardi, M.T., Masojídek, J., and Godde, D. (1997) Effects of abiotic stresses on the turnover of the D$_1$ reaction centre II protein. *Physiol. Plant.*, **101**, 635–642.

Gillham, D.J. and Dodge, A.D. (1986) Hydrogen-peroxide-scavenging systems within pea chloroplasts. A quantitative study. *Planta*, **67**, 246–251.

Gillham, D.J. and Dodge, A.D. (1987) Chloroplast superoxide and hydrogen peroxide scavenging systems from pea leaves: seasonal variations. *Plant Sci.*, **50**, 105–109.

Gilmore, A.M. (1997) Mechanistic aspects of xanthophyll cycle-dependent photoprotection in higher plant chloroplasts and leaves. *Physiol. Plant.*, **99**, 197–209.

Gilmore, A.M., Mohanty, J., and Yamamoto, H.Y. (1994) Epoxidation of zeaxanthin and antheraxanthin reverses nonphotochemical quenching of photosystem II chlorophyll *a* fluorescence in the presence of a transthylakoid ΔpH. *FEBS Lett.*, **350**, 271–274.

Gilmore, A.M. and Yamamoto, H.Y. (1992) Dark induction of zeaxanthin-dependent nonphotochemical fluorescence quenching mediated by ATP. *Proc. Natl. Acad. Sci. USA*, **89**, 1899–1903.

Gilmore, A.M. and Yamamoto, H.Y. (1993) Linear models relating xanthophyll and lumen acidity to nonphotochemical fluorescence quenching. Evidence that antheraxanthin explains zeaxanthin-independent quenching. *Photosynthesis Res.*, **35**, 67–78.

Goetze, D.C. and Carpentier, R. (1994) Ferredoxin NADP$^+$ reductase is the site of oxygen reduction in pseudocyclic electron transport. *Can. J. Bot.*, **72**, 256–260.

Grace, S.C. and Logan, B.A. (1996) Acclimation of foliar antioxidant systems to growth irradiance in three broad-leaved evergreen species. *Plant Physiol.*, **112**, 1631–1640.

Grace, S., Pace, R., and Wydrzynski, T. (1995) Formation and decay of monodehydroascorbate radicals in illuminated thylakoids as determined by ERR spectroscopy. *Biochim. Biophys. Acta*, **1229**, 155–165.

Green, R. and Fluhr, R. (1995) UV-B-induced PR-1 accumulation is mediated by active oxygen species. *Plant Cell*, **7**, 203–212.

Groden, D. and Beck, E. (1979) H$_2$O$_2$ destruction by ascorbate-dependent systems from chloroplasts. *Biochim. Biophys. Acta*, **546**, 426–435.

Guan, L. and Scandalios, J.G. (1995) Developmentally related responses of maize catalase genes to salicylic acid. *Proc. Natl. Acad. Sci. USA*, **92**, 5930–5934.

Gueta-Dahan, Y., Yaniv, Z., Zilinskas, B.A., and Ben-Hayyim, G. (1997) Salt and oxidative stress; similar and specific responses and their relation to salt tolerance in Citrus. *Planta*, **203**, 460–469.

Guy, R.D., Fogel, M.L., and Berry, J.A. (1993) Photosynthetic fractionation of the stable isotopes of oxygen and carbon. *Plant Physiol.*, **107**, 37–47.

Haehnel, W. (1984) Photosynthetic electron transport in higher plants. *Annu. Rev. Plant. Physiol.*, **35**, 659–693.

Hager, A. (1969) Lichtbedingte pH-erniederigung in einem chloroplasten-kompartiment als Ursache der enzymatischen Violaxanthin $\rightarrow$ Zeaxanthin-Umwandlung, Besiehungen zur Photophosphorylierung. *Planta*, **89**, 224–243.

Hager, A. and Holocker, K. (1994) Localisation of the xanthophyll cycle enzyme violaxamthin deepoxidase within the thylakoid lumen and abolition of its mobility by a (light-dependent) pH decrease. *Planta*, **192**, 581–589.

Halliwell, B. (1987) Oxidative damage, lipid peroxidation and antioxidant protection in chloroplasts. *Chem. Phys. Lipids*, **44**, 327–340.

Harbinson, J. and Foyer, C.H. (1991) Relationships between the efficiencies of photosystems I and II and stromal redox state in CO$_2$-free air. Evidence for cyclic electron flow *in vivo*. *Plant Physiol.*, **97**, 41–49.

Harbinson, J. and Hedley, C.L. (1989) The kinetics of P$_{700}^+$ reduction in leaves: a novel *in situ* probe of thylakoid functioning. *Plant Cell Environ.*, **12**, 357–369.

Havaux, M. and Davaud, A. (1994) Photoinhibition of photosynthesis in chilled potato leaves is not correlated with a loss of Photosystem II activity. *Photosynthesis Res.*, **40**, 75–92.

Hayakawa, T., Kanematsu, S., and Asada, K. (1985) Purification and characterization of thylakoid-bound Mn-superoxide dismutase in spinach chloroplasts. *Planta*, **166**, 111–116.

Heber, U., Bligny, R., Streb, P., and Douce, R. (1996) Photorespiration is essential for the protection of the photosynthetic apparatus of C_3 plants against photoinactivation under sunlight. *Bot. Acta*, **109**, 307–315.

Heber, U., Egneus, H., Hanch Jensen, M., and Köster, S. (1978) Regulation of photosynthetic electron transport and photophosphorylation in intact chloroplasts and leaves of *Spinacia oleracea* L. *Planta*, **143**, 41–49.

Heber, U., Neimanis, S., and Dietz, K-J. (1988) Fractional control of photosynthesis by the Q_B protein, the cytochrome f/b_6 complex and other components of the photosynthetic apparatus. *Planta*, **173**, 267–274.

Heber, U., Neimanis, S., Dietz, K.J., and Vill, J. (1986) Assimilatory power as a driving force in photosynthesis. *Biochim. Biophys. Acta*, **852**, 144–155.

Heber, U., Schreiber, U., Siebke, K., and Dietz, K.J. (1990) Relationships between light-driven electron transport, carbon reduction and carbon oxidation in photosynthesis. In I. Zelitch, (ed.), *Perspectives in Biochemical and Genetic Regulation of Photosynthesis*, Alan R. Liss, New York, pp. 17–37.

Henkow, L., Strid, Å., Berglund, T., Rydstöm, J., and Ohlsson, A.B. (1996) Alteration of gene expression in *Pisum sativum* tissue cultures caused by the free radical-generating agent 2,2_-azobis dihydrochloride. *Physiol. Plant.*, **96**, 6–12.

Herbert, S.K., Samson, G., Fork, D.C., and Laudenbach, D.E. (1992) Characterisation of damage to photosystems I and II in a cyanobacterium lacking detectable iron superoxide dismutase activity. *Proc. Natl. Acad. Sci. USA*, **89**, 8716–8720.

Hertwig, B., Streb, P., and Feierabend, J. (1992) Light dependence of catalase synthesis and degradation in leaves and the influence of interfering stress conditions. *Plant Physiol.*, **100**,1547–1553.

Hideg, E., Spetea, C., and Vass, I. (1994) Singlet oxygen and free radical production during acceptor- and donor-side-induced photoinhibition. Studies with spin trapping ERR spectroscopy. *Biochim. Biophys. Acta*, **1186**, 143–152.

Hodgson, E.K. and Fridovich, I. (1975) The interaction of bovine erythrocyte superoxide dismutase with hydrogen peroxide:inactivation of enzyme. *Biochemistry*, **14**, 5294–5299.

Hope, A.B., Valente, P., and Matthews, D.B. (1994) Effects of pH on the kinetics of redox reactions in and around the cytochrome bf complex in an isolated system. *Photosynthesis Res.*, **42**, 111–120.

Horemans, N., Asard, H., and Caubergs, R.J. (1997) The ascorbate carrier of higher plant plasma membranes preferentially translocates the fully oxidised (dehydroascorbate) molecule. *Plant Physiol.*, **114**, 1247–1253.

Horemans, N., Foyer, C.H., and Asard, H. (2000) Transport and action of ascorbate at the plant plasma membrane. *Trends Plant Sci.*, **5**, 263–267.

Hormann, H., Neubauer, C., Asada, K., and Schreiber, U. (1993) Intact chloroplasts display pH 5 optimum of O_2-reduction in the absence of methyl viologen. Indirect evidence for a regulatory role of superoxide protonation. *Photosynthesis Res.*, **37**, 69–89.

Horton, P. (1989) Interactions between electron transport and carbon assimilation: regulation of light-harvesting and photochemistry. In W.R. Briggs, (ed.), *Photosynthesis*, (Plant Biology Series, Vol. 8), Alan R. Liss, New York, pp. 393–406.

Horton, P. and Lee, P. (1985) Phosphorylation of chloroplast membrane proteins partially protects against photoinhibition. *Planta*, **165**, 37–42.

Horton, P. and Ruban, A.V. (1992) Regulation of photosystem II. *Photosynthesis Res.*, **34**, 375–385.

Horton, P., Ruban, A.V., and Walters, R.G. (1994) Regulation of light-harvesting in green plants; indication by non-photochemical quenching of chlorophyll fluorescence. *Plant Physiol.*, **106**, 415–420.

Hosein, B. and Palmer, G. (1983) The kinetics and mechanism of reaction of reduced ferredoxin by molecular oxygen and its reduced products. *Biochim. Biophys. Acta*, **723**, 383–390.

Hossain, M.A. and Asada, K. (1985) Monodehydroascorbate reductase from cucumber is a flavin adenine dinucleotide enzyme. *J. Biol. Chem.*, **260**, 12920–12926.

Hossain, M.A., Nakano, Y., and Asada, K. (1984) Monodehydroascorbate reductase in spinach chloroplasts and its participation in regeneration of ascorbate for scavenging of hydrogen peroxide. *Plant Cell. Physiol.*, **25**, 385–395.

Hundal, T., Forsmark-Andrée, P., Ernster, L., and Andersson, B. (1995) Antioxidant activity of reduced plastoquinone in chloroplast thylakoid membranes. *Arch. Biochem. Biophys.*, **324**, 117–122.

Huner, N.P.A., Maxwell, D.P., Gray, G.R., Savitch, L.V., Krol, M., Ivanov, A.G., and Falk, S. (1996) Sensing environmental temperature change through imbalances between energy sypply and energy consumption: Redox state of photosystem II. *Physiol. Plant.*, **98**, 358–364.

Jakob, B., and Heber, U. (1996) Photoproduction and detoxification of hydroxyl radicals in chloroplasts and leaves in relation to photoinactivation of photosystems I and II. *Plant Cell Physiol.*, **37**, 629–635.

Jegerschöld, C., Virgin, I., and Styring, S. (1990) Light-dependent degradation of the D_1-protein in photosystem II is accelerated after inhibition of the water-splitting reaction. *Biochemistry*, **29**, 6179–6186.

Joliot, P., Lavergne, J., and Béal, D. (1992) Plastoquinone compartmentation in chloroplasts. I. Evidence for domains with different rates of photoreduction. *Biochim. Biophys. Acta*, **1101**, 1–12.

Jolliffe, P.A., and Tregunna, E.B. (1972) Environmental regulation of the oxygen effect on apparent photosynthesis in wheat. *Can. J. Bot.*, **51**, 841–853.

Jones, L.W., and Kok, B. (1966) Photoinhibition of chloroplast reactions. Kinetics and action spectra. *Plant Physiol.*, **41**, 1037–1043.

Kaiser, W.M. (1976) The effect of hydrogen peroxide on CO_2 fixation of isolated intact chloroplasts. *Biochim. Biophys. Acta*, **440**, 476–482.

Kaiser, W.M. (1979) Reversible inhibition of the Calvin cycle and activation of the oxidative pentose phosphate cycle in isolated intact chloroplasts by hydrogen peroxide. *Planta*, **145**, 377–382.

Kangasjärvi, J., Talvinen, J., Utrianinen, M., and Karajalainen, R. (1994) Plant defence systems induced by ozone. *Plant Cell Environ.*, **17**, 783–794.

Karpinski, S., Escobar, C., Karpinska, B., Crerssen, G., and Mullineaux, P.M. (1997) Photosynthetic electron transport regulates the expression of cytosolic ascorbate peroxidases genes in *Arabidopsis* during excess light stress. *Plant Cell*, **9**, 627–640.

Karpinski, S., Reynolds, H., Karpinska, B., Wingsle, G., Creissen, G., and Mullineaux, P. (1999) Systemic signaling and acclimation in response to excess excitation energy in *Arabidopsis*. *Science*, **284**, 654–657.

Kato, Y., Urano, J., Maki, Y., and Ushimaru T. (1997) Purification and characterisation of dehydroascorbate reductase from rice. *Plant Cell Physiol.*, **38**, 173–178.

Kelly, G.J. and Latzko, E. (1997) Carbon Metabolism: The carbon metabolisms of unstressed and stressed plants. *Progr. Bot.*, **58**, 187–220.

Kendall, A.C., Keys, A.J., Turner, J.C., Lea, P.J., and Miflin, B.J. (1983) The isolation and characterisation of a catalase-deficient mutant in barley (*Hordeum vulgare* L.). *Planta*, **159**, 505–511.

Keys, A.J. (1980) Synthesis and interconversion of glycine and serine. In B.J. Miflin, (ed.), *Amino Acids and Derivatives* (The Biochemistry of Plants, a comprehensive treatise, Vol. 5), Academic Press, New York, pp. 359–374.

Khan, A.U., and Kasha, M. (1994) Singlet molecular oxygen in the Haber-Weiss reaction. *Proc. Natl. Acad. Sci. USA*, **91**, 12365–12367.

Kingston-Smith, A.H. and Foyer, C.H. (2000) Bundle sheath proteins are more sensitive to oxidative damage than those of the mesophyll in maize leaves exposed to paraquat or low temperatures. *J. Exp. Bot.*, **51**, 123–130.

Koyama, Y. (1991) Structures and functions of carotenoids in photosynthetic systems. *J. Photochem. Photobiol. B. Biol.*, **9**, 265–280.

Kozaki, A., and Takeba, G. (1996) Photorespiration protects C_3 plants from photooxidation. *Nature*, **384**, 557–560.

Krause, G.H. and Weis, E. (1991) Chlorophyll fluorescence and photosynthesis: the basics. *Annu. Rev. Plant Physiol. Plant Mol. Biol.*, **42**, 313–349.

Krieger, A., Moya, I., and Weis, E. (1992) Energy-dependent quenching of chlorophyll *a* fluoresence: Effect of pH on stationary fluorescence and picosecond-relaxation kinetics in thylakoid membranes and photosystem II preparations. *Biochim. Biophys. Acta*, **1102**, 167–176.

Kurepa, J., Hérouart, D., Van Montagu, M., and Inzé, D. (1997) Differential expression of CuZn and Fe-superoxide dismutase genes of tobacco during development, oxidative stress, and hormonal treatments. *Plant Cell Physiol.*, **38**, 463–470.

Kyle, D.J. (1987) The biochemical basis for photoinhibition of photosystem II. In D.J. Kyle, C.B. Osmond, and C.J. Arntzen, (eds.), *Photoinhibition* (Topics in Photosynthesis, Vol. 9), Elsevier, Scientific Press, Amsterdam, pp. 197–226.

Kyle, D.J., Ohad, I., and Arntzen, C.J. (1985) Molecular mechanisms of compensation to light stress in chloroplast membranes. In J.L. Key, and T. Kosuge T., (eds.), *Cellular and Molecular Biology of Plant Stress*, Alan R. Liss, New York, pp. 51–69.

Lavergne, J., Bouchaud, J-P., and Joliot, P. (1992) Plastoquinone compartmentation in chloroplasts. II. Theoretical aspects. *Biochim. Biophys. Acta*, **1101**, 13–22.

Leegood, R.C. and Walker, D.A. (1982) Regulation of fructose 1,6-bisphosphatase activity in leaves. *Planta*, **156**, 449–456.

Leegood, R.C., Walker, D.A., and Foyer, C.H. (1985) Regulation of the Benson-Calvin cycle. In J. Barber, and N.R. Baker, (eds.), *Photosynthetic Mechanisms and the Environment*, Elsevier Science Publishers, Amsterdam, pp. 189–258.

Levine, A., Tenhaken, R., Dixon, R., and Lamb, C. (1994) H_2O_2 from the oxidative burst orchestrates the plant hypersensitive disease resistance response. *Cell*, **79**,583–593.

Levine, R.L. (1983) Oxidative modification of glutamine synthetase. *J. Biol. Chem.*, **258**, 11828–11833.

Levings, C.S. III and Siedow, J.N. (1995) Regulation of redox poise in chloroplasts. *Science*, **268**, 695–696.

Link, G., Tiller, K., and Baginsky, S. (1997) Glutathione is a regulator of chloroplast transcription. In K.K. Hatzios, (ed.), *Regulation of Enzymatic Systems Detoxifying Xenobiotics in Plants*, (NATO Sciences Partnership Sub-Series, Vol. 37), Kluwer Academic Publishers, Dordrecht, pp. 125–138.

Long, S.P., Humphries, S., and Falkowski, P.G. (1994) Photoinhibition of photosynthesis in nature. *Annu. Rev. Plant Physiol. Plant Mol. Biol.*, **45**, 633–662.

Lovelock, C.E., and Winter, K. (1996) Oxygen-dependent electron transport and protection from photoinhibition in leaves of tropical plant species. *Planta*, **198**, 580–587.

Macpherson, A.N., Telfer, A., Barber, J., and Truscott, T.G. (1993) Direct detection of singlet oxygen from isolated photosystem II reaction centres. *Biochim. Biophys. Acta*, **1143**, 301–309.

Mäenpää, P., Anderson, B., and Sundby, C. (1987) Difference in sensitivity to photoinhibition between photosystem II in the appressed and non-appressed thylakoid regions. *FEBS Lett.*, **215**, 31–36.

Marsho, T.V., Behrens, P.W., and Radmer, R.J. (1979) Photosynthetic oxygen reduction in isolated intact chloroplasts and cells from spinach. *Plant Physiol.*, **64**, 656–659.

Matsuda, Y., Okuda, T., and Sagisaka, S. (1994) Regulation of protein synthesis by hydrogen peroxide in the crowns of winter wheat. *Biosci. Biotech. Biochem.*, **58**, 906–909.

Mattoo, A.K., Marder, J.B., and Edelman, M. (1989) Dynamics of the photosystem II reaction center. *Cell*, **56**, 241–246.

Maxwell, D.P., Laudenbach, D.E., and Huner, N.P.A. (1995) Redox regulation of light-harvesting complex II and cab mRNA abundance in *Dunaliella salina*. *Plant Physiol.*, **109**, 787–795.

Mayfield, S.P. and Taylor, W.C. (1987) Chloroplast photooxidation inhibits the expression of a set of nuclear genes. *Mol. Gen. Genet.*, **208**, 309–314.

McKersie, B.D., Bowley, S.R., Harjanto, E., and Leprince, O. (1996) Water-deficit tolerance and field performance of transgenic alfalfa overexpressing superoxide dismutase. *Plant Physiol.*, **111**, 1177–1181.

McKersie, B.D., Chen, Y., de Beus, M., Bowley, S.R., Bowler, C, Inzé, D., D'Halluin, K., and Botterman, J. (1993) Superoxide dismutase enhances tolerance of freezing stress in transgenic alfalfa (*Medicago sativa* L.). *Plant Physiol.*, **103**, 1155–1163.

McVetty, P.B.E. and Canvin, D.T. (1981) Inhibition of photosynthesis by low oxygen concentrations. *Can. J. Bot.*, **59**, 721–725.

Mehler, A.H. (1951) Studies on reactions of illuminated chloroplasts. I. Mechanisms of the reduction of oxygen and other Hill reagents. *Arch. Biochem. Biophys.*, **33**, 65–77.

Mehler, A.H. and Brown, A.H. (1952) Studies on reactions of illuminated chloroplasts. III. Simultaneous photoproduction and consumption of oxygen studied with oxygen isotopes. *Arch. Biochem. Biophys.*, **38**, 365–370.

Miller, N.J., Sampson, J., Candeias, L.P., Bramley, P.M., and Rice-Evans, C.A. (1996) Antioxidant activities of carotenes and xanthophylls. *FEBS Lett.*, **384**, 240–242.

Mishra, N.P., Mishra, R.K., and Singhal, G.S. (1993) Involvement of active oxygen species in photoinhibition of photosystem II: protection of photosynthetic efficiency and inhibition of lipid peroxidation by superoxide dismutase and catalase. *J. Photochem. Photobiol. B. Biol.*, **19**, 19–24.

Misra, H.P. and Fridovich, I. (1971) The generation of superoxide radical during the autooxidation of ferredoxin. *J. Biol. Chem.*, **246**, 6886–6890.

Miyake, C. and Asada, K. (1992a) Thylakoid-bound ascorbate peroxidase in spinach chloroplasts and photoreduction of its primary oxidation product monodehydroascorbate radicals in thylakoids. *Plant Cell Physiol.*, **33**, 541–553.

Miyake, C. and Asada, K. (1992b) Thylakoid-bound ascorbate peroxidase scavenges hydrogen peroxide photoproduced-Photoreduction of monodehydroascorbate radical. In N. Murata, (ed.), *Research in Photosynthesis, Vol. II*, Kluwer Academic Publishers, Dordrecht, pp. 563–566.

Miyake, C. and Asada, K. (1994) Ferredoxin-dependent photoreduction of the monodehydroascorbate radical in spinach thylakoids. *Plant Cell Physiol.*, **35**, 539–549.

Miyake, C., Cao, W-H., and Asada, K. (1993) Purification and molecular properties of the thylakoid-bound ascorbate peroxidase in spinach chloroplasts. *Plant Cell Physiol.*, **34**, 881–889.

Miyake, C., Michihata, F., and Asada, K. (1991) Scavenging of hydrogen peroxide in ptokaryotic and eukaryotic algae: Acquisition of ascorbate peroxidase during the evolution of cyanobacteria. *Plant Cell Physiol.*, **32**, 33–43.

Miyake, C., Schreiber, U., Hormann, H., Sano, S., and Asada, K. (1996) Monodehydroascorbate radical reductase-dependent photoreduction of oxygen in chloroplast thylakoids. *Plant Cell Physiol.*, **37**, s284.

Miyao, M., Ikeuchi, M., Yamamoto, N., and Ono, T. (1995) Specific degradation of the D1 protein of photosystem II by treatment with hydrogen peroxide in darkness; implication for the mechanism of degradation of the D1 protein under illumination. *Biochemistry*, **34**, 10019–10026.

Morrell, S., Follmann, H., De Tullio, M., and Haberlein, I. (1997) Dehydroascorbate and dehydroascorbate reductase are phentom indicators of oxidative stress in plants. *FEBS Lett.*, **414**, 567–570.

Mullineaux, P.M., Karpinski, S., Jimenéz, A., Cleary, S.P., Robinson, C., and Creissen, G.P. (1997) Plastid-targeted glutathione peroxidase. *Plant J.*, in press.

Nakano, Y. and Asada, K. (1980) Spinach chloroplasts scavenge hydrogen peroxide on illumination. *Plant Cell Physiol.*, **21**, 1295–1307.

Neubauer, C. and Schreiber, U. (1989) Photochemical and non-photochemical quenching of chlorophyll fluorescence induced by hydrogen peroxide. *Z. Naturforsch*, **44C**, 262–270.

Neubauer, C. and Yamamoto, H.Y. (1992) Mehler-peroxidase reduction mediates zeaxanthin formation and zeaxanthin-related fluorescence quenching in intact chloroplasts. *Plant Physiol.*, **99**, 1354–1361.

Noctor, G.D., Rees, D., Young, A., and Horton, P. (1991) The relationship between zeaxanthin, energy-dependent quenching of chlorophyll fluorescence and trans-thylakoid pH gradient in isolated chloroplasts. *Biochim. Biophys. Acta*, **1057**, 320–330.

Noctor, G.D., Strohm, M., Jouanin, L., Kunert, K.-J., Foyer, C.H., and Rennenberg, H. (1996) Synthesis of glutathione in leaves of transgenic poplar (*Populus tremula* x *P. alba*) overexpressing γ-glutamylcysteine synthetase. *Plant Physiol.*, **112**, 1071–1078.

Noctor, G., Jouanin, L., Arisi, A.-C.M., Valadier, M.-H., Roux, Y., and Foyer, C.H. (1997a) Light-dependent modulation of foliar glutathione synthesis and associated amino acid metabolism in transformed poplar. *Planta*, **202**, 357–369.

Noctor, G., Arisi, A.-C.M., Jouanin, L., Valadier, M.-H., Roux, Y., and Foyer, C.H. (1997b) The role of glycine in determining the rate of glutathione synthesis in poplars. Possible implications for glutathione production during stress. *Physiol. Plant.*, **100**, 255–263.

Noctor, G., Jouanin, L., and Foyer, C.H. (1997c) The biosynthesis of glutathione explored in transformed plants. In K.K. Hatzios, (ed.), *Regulation of Enzymatic Systems Detoxifying Xenobiotics in Plants*, (NATO Sciences Partnership Sub-Series, Vol. 37), Kluwer Academic Publishers, Dordrecht, pp. 109–124.

Noctor, G., Arisi, A-C.M., Jouanin, L., and Foyer, C.H., (1998a) Manipulation of glutathione and amino acid biosynthesis in the chloroplast. *Plant Physiol.*, **118**, 471–482.

Noctor, G., Arisi, A-C.M., Jouanin, L., Kunert, K-J., Rennenberg, H., and Foyer C.H. (1998b) Glutathione: Biosynthesis and metabolism explored in transformed poplar. *J. Exp. Bot.*, **49**, 623–647.

Noctor, G., Veljovic-Jovanovic, S., and Foyer, C.H. (2000) Peroxide processing in photosynthesis: antioxidant coupling and redox signalling. *R. Soc. Phil. Trans. In Press.*

Ogawa, K., Kanematsu, S., Takabe, K., and Asada, K. (1995) Attachment of CuZn-superoxide dismutase to thylakoid membranes at the site of superoxide generation (PSI) in spinach chloroplasts: Detection by immuno-gold labelling after rapid freezing and substitution method. *Plant Cell Physiol.*, **36**, 565–573.

Ogren, W.L. (1984) Photorespiration: Pathways, regulation and modification. *Annu. Rev. Plant Physiol.*, **35**, 415–442.

Okuda, S., Kanematsu, S., and Asada, K. (1979) Intracellular distribution of manganese and ferric superoxide dismutases in blue-green algae. *FEBS Lett.*, **103**, 106–110.

Okuda, T., Matsuda, Y., Yamanaka, A., and Sagisaka, S. (1991) Abrupt incrrease in the level of hydrogen peroxide in leaves of winter wheat is caused by cold treatment. *Plant Physiol.*, **97**, 1265–1267.

Ornran, R.J. (1980) Peroxide levels and activities of catalase peroxidase and indoleacetic acid oxidase during and after chilling cucumber seedlings. *Plant Physiol.*, **65**, 407–408.

Osmond, C.B. (1994) What is photoinhibition? Some insights from the comparison of shade and sun plants. In N.R. Baker, and J.R. Boyer, (eds.), *Photoinhibition: Molecular Mechanisms to the Field*, BIOS Scientific Publications, Oxford, pp. 1–24.

Osmond, C.B., and Björkman, O. (1972) Simultaneous measurement of O_2 effects on net photosynthesis and glycolate metabolism in C_3 and C_4 species of *Atriplex. Carnegie Inst. Washington Year Book*, **71**, 141–148.

Osmond, C.B. and Grace, S.C. (1995) Perspectives on photoinhibition and photorespiration in the field: quintessential inefficiencies of the light and dark reactions of photosynthesis? *J. Exp. Bot.*, **46**, 1351–1362.

Pastori, G., Mullineaux, P., and Foyer, C.H. (1999) Post-transcriptional regulation prevents accumulation of glutathione reductase protein and activity in the bundle sheath cells of maize. Implications on the sensitivity of maize to low temperatures. *Plant J.*, submitted.

Payton, P., Allen, R.D., Trolinder, N., and Holaday, A.S. (1997) Overexpression of chloroplast-targeted Mn Superoxide dismutase in cotton (*Gossypium hirsutum* L. cv. Coker 312) does not alter the reduction of photosynthesis after short exposures to low temperature and high light intensity. *Photosynthesis Res.*, **52**, 233–244.

Pearson, C.K., Wilson, S.B., Schaffer, R., and Ross, A.W. (1993) NAD turnover and utilisation of metabolites for RNA synthesis in a reaction sensing the redox state of the cytochrome b_6/f complex in isolated chloroplasts. *Eur. J. Biochem.*, **218**, 397–404.

Perl, A., Perl-Treves, R., Galili, S., Aviv, D., Shalgi, E., Malkin, S., and Galun, E. (1993) Enhanced oxidative stress defense in transgenic potato expressing tomato CuZn superoxide dismutases. *Theor. Appl. Genet.*, **85**, 568–576.

Peterson, R.B. (1991) Effects of O_2 and CO_2 concentrations on quantum yields of photosystems I and II in tobacco leaf tissue. *Plant Physiol.*, **97**, 1383–1395.

Pfannschmidt, T., Nilsson, A., and Allen, J.F. (1999) Photosynthesis control of chloroplast gene expression. *Nature*, **397**, 625–628.

Polle, A. (1996) Mehler Reaction: Friend or Foe in Photosynthesis. *Bot. Acta*, **109**, 84–89.

Powles, S.B. (1984) Photoinhibition of photosynthesis is induced by visible light. *Annu. Rev. Plant Physiol.*, **35**, 15–44.

Prasad, T.K., Anderson, M.D., Martin, B.A., and Stewart, C.R. (1994) Evidence for chilling-induced oxidative stress in maize seedlings and a regulatory role for hydrogen peroxide. *Plant Cell*, **6**, 65–74.

Prasil, O., Adir, N., and Ohad, I. (1992) Dynamics of photosystem II. Mechanism of photoinhibition and recovery processes. In J. Barber, (ed.), *The Photosystems: Structure, Function, and Molecular Biology*, (Current Topics in Photosynthesis, Vol. 11), Elsevier, Amsterdam, pp. 220–250.

Price, A.H., Taylor, A., Ripley, S.J., Griffiths, A., Trewavas, A.J., and Knight M.R. (1994) Oxidative signals in tobacco increase cytosolic calcium. *Plant Cell*, **6**, 1301–1310.

Robinson, M.J. (1997) The influence of elevated foliar carbohydrate levels on the ascorbate: dehydroascorbate redox ratios in nitrogen-limited spinach and soybean plants. *Int. J. Plant Sci.*, **158**, 442–450.

Robinson, J.M. (1988) Does O_2 photoreduction occur within chloroplasts *in vivo. Physiol. Plant.*, **72**, 666–680.

Rockholm, D.C. and Yamamoto, H.Y. (1996) Violaxanthin de-poxidase. Purification of a 43-kilodalton lumenal protein from lettuce by lipid-affinity precipitation with monogalactosyl-diacylglyceride. *Plant Physiol.*, **110**, 697–703.

Rose, R.C. and Bode, A.M. (1993) Biology of free radical scavengers: an evaluation of ascorbate. *FASEB J.*, **7**, 1135–1142.

Roxas, V.P., Smith, R.K., Allen, E.R., and Allen, R.D. (1997) Overexpression of glutathione S-transferase/ glutathione peroxidase enhances the growth of transgenic tobacco seedlings during stress. *Nature Biotechnol.*, **15**, 988–991.

Ruban, A.V., Phillip, D., Young, A.J., and Horton, P. (1997) Carotenoid-dependent oligomerisation of the major chlorophyll *a/b* light-harvesting complex of Photosystem II of plants. *Biochemistry*, **36**, 7855–7859.

Ruban, A.V., Walters, R.G., and Horton, P. (1992) The molecular mechanism of the control of excitation energy dissipation in chloroplasts membranes. Inhibition of ΔpH-dependent quenching of chlorophyll fluorescence by dicyclohexylcarbodiimide. *FEBS Lett.*, **309**, 175–179.

Salin, M.L. (1988) Toxic oxygen species and protective systems of the chloroplast. *Physiol. Plant.*, **72**, 681–689.

Sandermann, H. (1996) Ozone and plant health. *Annu. Rev. Phytopathol.*, **34**, 347–366.

Sano, S., Miyake, C., Mikami, B., and Asada, K. (1995) Molecular characterization of monodehydroascorbate radical reductase from cucumber highly expressed in *Escherichia coli. J. Biol. Chem.*, **270**, 21354–21361.

Sapper, H., Kang, S-O., Paul, H-H., and Lohmann, W. (1982) The reversibility of the vitamin C redox system: electrochemical reasons and biological aspects. *Z. Naturforsch.*, **37**, 942–946.

Scandalios, J.G. (1983) Molecular varieties of isozymes and their role in studies of gene regulation and expression during eukaryote development. In M.C. Rattazzi, J.G. Scandalios, and S.S. Whitt, (eds.), *Isozymes* (Current Topics in Biological and Medical Research, Vol. 9), Alan R. Liss, New York, pp. 1–31.

Scandalios, J.G. (1994) Regulation and properties of plant catalases. In C.H. Foyer and P.M. Mullineaux, (eds.), *Causes of Photooxidative Stress and Amelioration of Defence Systems in Plants*, CRC Press, Boca Raton, pp. 275–315.

Scandalios, J.G. (1997) Molecular genetics of superoxide dismutases in plants. In J.D. Scandalios, (ed.), *Oxidative Stress and the Molecular Biology of Antioxidant Defenses*, (Monograph Series, Vol. 34), Cold Spring Harbor Laboratory Press, Cold Spring Harbor, pp. 527–568.

Schreiber, U., Reising, H., and Neubauer, C. (1992) Contrasting pH optima of light-driven O_2 and H_2O_2 reduction in spinach chloroplasts as measured via chlorophyll fluorescence quenching. *Z. Naturforsch*, **46C**, 173–181.

Schreiber, U., Bilger, W., and Neubauer, C. (1994) Chlorophyll fluorescence as a non-intrusive indicator for rapid assessment *in vivo* photosynthesis. In E.D. Schulze, M.M. Caldwell, (eds.), *Ecophysiology of Photosynthesis*, Springer-Verlag, Berlin, pp. 49–70.

Schreiber, U., Hormann, H., Asada, K., and Neubauer, C. (1995) O_2-dependent electron flow in intact spinach chloroplasts: properties and possible regulation of the Mehler-ascorbate peroxidase cycle. In P. Mathis, (ed.), *Photosynthesis: from Light to Biosphere*, Vol. II, Kluwer Academic Publishers, Dordrecht, pp. 813–818.

Sen Gupta, A., Heinen, J.L., Holaday, A.S., Burke, J.J., and Allen R.D. (1993a) Increased resistance to oxidative stress in transgenic plants that overexpress chloroplastic Cu/Zn superoxide dismutase. *Proc. Natl. Acad. Sci. USA*, **90**, 1629–1633.

Sen Gupta, A., Webb, R.P., Holaday, A.S., and Allen, R.D. (1993b) Oiverexpression of superoxide dismutase protects plants from oxidative stress. *Plant Physiol.*, **103**, 1067–1073.

Setlik, I., Allakverdieu, S.I., Nedodal, L., Setlikova, E., and Klimov, V.V. (1990) Three types of photosystem II photoinactivation 1. Damaging processes on the acceptor side. *Photosynthesis Res.*, **23**, 39–48.

Shen, B., Jensen, R.G., and Bohuent, H.J. (1997) Mannitol protects against oxidation byhydroxyl radicals. *Plant Physiol.*, **115**, 527–532.

Shipman, L.L. (1980) A theoretical study of excitons in chlorophyll *a* photosystems on a picosecond timescale. *Photochem. Photobiol.*, **31**, 157–167.

Siebke, K., Laisk, A., Neimanis, S., and Heber, U. (1991) Regulation chloroplast metabolism in leaves: evidence that NADP dependent glyceraldehyde phosphate dehydrogenase but not ferredoxin NADP reductase controls electron flow to phosphoglycerate in the dark to light transition. *Planta*, **185**, 337–343.

Siefermann-Harms, D. (1987) The light-harvesting and protective functions of carotenoids in photosynthetic membranes. *Physiol. Plant.*, **69**, 561–568.

Slooten, L., Capiau, K., Van Camp, W., Van Montagu, M., Sybesma, C., and Inzé, D. (1995) Factors affecting the enhancement of oxidative stress tolerance in transgenic tobacco over-expressing manganese superoxide dismutase in the chloroplasts. *Plant Physiol.*, **107**, 737–750.

Smirnoff, N. (1996) The function and metabolism of ascorbic acid in plants. *Ann. Bot.*, **78**, 661–669.

Staehelin, L.A. and Arntzen, C.J. (1983) Regulation of chloroplast membrane function: protein phosphorylation changes the spatial organisation of membrane components. *J. Cell Biol.*, **97**, 1327–1337.

Steiger, H-M., and Beck, E. (1981) Formation of hydrogen peroxide and oxygen dependence of photosynthetic CO_2 assimilation by intact chloroplasts. *Plant Cell Physiol.*, **22**,561–576.

Streb, P., Michael-Knauf, A., and Feierabend, J. (1993) Preferential photoinactivation of catalase and photoinhibition of photosystem II are common early symptoms under various osmotic and chemical stress conditions. *Physiol. Plant.*, **88**, 590–598.

Takahama, U., Shimizu-Takahama, M., and Heber, U. (1981) The redox state of the NADP system in illuminated chloroplasts. *Biochim. Biophys. Acta*, **637**, 530–539.

Takahashi, M. and Asada, K. (1982) Dependence of oxygen affinity for Mehler reaction on photochemical activity of chloroplast thylakoids. *Plant Cell Physiol.*, **23**, 1457–1461.

Takahashi, M. and Asada, K. (1988) Superoxide production in the aprotic interior of chloroplast thylakoids. *Arch. Biochem. Biophys.*, **267**,714–722.

Telfer, A., Cammack, R., and Evans, M.C.W. (1970) Hydrogen peroxide as the product of autooxidation of ferredoxin:reduced either chemically or by illuminated chloroplasts. *FEBS Lett.*, **10**, 21–24.

Tel-Or, E., Huflejt, M.E., and Packer, L. (1985) Hydroperoxide metabolism in cyanobacteria. *Arch Biochem. Biophys.*, **246**, 396–402.

Tepperman, J.M. and Dunsmuir, P. (1990) Transformed plants with elevated levels of chloroplastic SOD are not more resistant to superoxide toxicity. *Plant Mol. Biol.*, **14**, 501–511.

Terashima, I., Funayama, S., and Sonoike, K. (1994) The site of photoinhibition in leaves of *Cucumis sativus* L. at low temperatures is photosystem I, not photosystem II. *Planta*, **193**, 300–306.

Thalmair, M., Bauw, G., Thiel, S., Döhring, T., Langebartels, C., and Sandermann, H. (1996) Ozone and ultraviolet-B effects on the defense-related proteins B-1, 3-glucanase and chitinase in tobacco. *J. Plant Physiol.*, **148**, 222–228.

Theg, S.M., Filar, L.J., and Dilley, R.A. (1986) Photoinactivation of chloroplasts already inhibited on the oxidising side of photosystem II. *Biochim. Biophys. Acta*, **849**, 104–111.

Tourneaux, C. and Peltier, G. (1995) Effect of water deficit on photosynthetic oxygen exchange measured using $^{18}O_2$ and mass spectrometry in *Solanum tuberosum* L. leaf discs. *Planta*, **195**, 570–577.

Trumper, S., Follmann, H., and Häberlein, I. (1994) A novel dehydroascorbate reductase from spinach chloroplasts homologous to plant trysin inhibiter. *FEBS Lett.*, **352**, 159–162.

Tschiersch, H. and Ohmann, E. (1993) Photoinhibition in *Euglena grocilis*. *Planta*, **191**, 316–323.

Tyystjärvi, E., Kettunen, R., and Aro, E-M. (1994) The rate constant of photoinhibition *in vitro* is independent of the antenna size of photosystem II but depends on temperature. *Biochim. Biophys. Acta*, **1186**, 177–185.

Van Camp, W., Capiau, K., Van Montagu, M., Inzé, D., and Slooten, L. (1996) Enhancement of oxidative stress tolerance in transgenic tobacco plants overexpressing Fe-superoxide dismutase in chloroplasts. *Plant Physiol.*, **112**, 1703–1714.

Van Camp, W., Willekens, H., Bowler, C., Van Montagu, M., Inzé, D., Langebartels, C., and Sandermann, H. (1994) Elevated levels of superoxide dismutase protect transgenic plants against ozone damage. *Bio/Technology*, **12**, 165–168.

Van Ginkel, G. and Brown, J.S. (1978) Endogenous catalase and superoxide dismutase activities in photosynthetic membranes. *FEBS Lett.*, **94**, 284–286.

Vas, I., Stryring, S., Hundal, T., Koivuniemi, A., Aro, E.-M., and Andersson, B. (1992) Reversible and irreversible intermediates during photoinhibition of photosystem II. Stable reduced QA species promote chlorophyll triplet formation. *Proc. Natl. Acad. Sci. USA*, **89**, 1408–1412.

Weis, E., Ball, J.T., and Berry, J. (1987) Photosynthetic control of electron transport in leaves of *Phaseolus vulgaris*: evidence for regulation of photosystem II by the proton gradient. In J. Biggins, (ed.), *Progress in Photosynthesis Research*, Vol. II, Martinus Nijhoff, Dordrecht, pp. 553–556.

Wells, W.W., Xu, D.P., Yang, Y., and Rocque P.A. (1990) Mammalian thioltransferase (glutaredoxin) and protein disulfide isomerase have dehydroascorbate reductase activity. *J. Biol. Chem.*, **265**, 15361–15364.

West, K.R. and Wiskich, J.T. (1968) Photosynthetic control by isolated pea chloroplasts. *Biochem. J.*, **109**, 527–532.

Wheeler, G.L., Jones, M.A., and Smirnoff, N. (1998) The biosynthetic pathway of vitamin C in higher plants. *Nature*, **393**, 365–368.

Willekens, H., Van Camp, W., Van Montagu, M., Inzé, D., Langebartels, C., and Sandermann, H. (1994) Ozone, sulfur dioxide, and ultraviolet-B have similar effects on mRNA accumulation of antioxidant genes in *Nicotiana plumbaginifolia*. *Plant Physiol.*, **106**, 1007–1014.

Willekens, H., Chamnongpol, S., Davey, M., Schraudner, M., Langebartels, C., Van Montagu, M., Inzé, D., and Van Camp, W. (1997) Catalase is a sink for H_2O_2 and is indispensible for stress defence in C_3 plants. *EMBO J.*, **16**, 4806–4816.

Witt, H.T. (1979) Energy conversion in the functional membrane of photosynthesis. Analysis by light pulse and electric pulse methods. The central role of the electric field. *Biochim. Biophys. Acta*, **505**, 355–427.

Wu, G., Shortt, B.J., Lawrence, E.B., Léon, J., Fitzsimmons, K.C., Levine, E.B., Raskin, I., and Shah, D.M. (1997) Activation of host defense mechanisms by elevated production of H_2O_2 in transgenic plants. *Plant Physiol.*, **115**, 427–435.

Wu, J., Neimanis, S., and Heber, U. (1991) Photorespiration is more effective than the Mehler reaction in protecting the photosynthetic apparatus against photoinhibition. *Bot. Acta*, **104**, 283–291.

Wydrzynski, T., Ångström, J., and Vänngård, T. (1989) H_2O_2 formation by photosystem II. *Biochim. Biophys. Acta*, **973**, 23–28.

Yamasaki, H., Heshiki, R., Yamasu, T., Sakihama, Y., and Ikehara, N. (1995) Physiological significance of the ascorbate regenerating system for the high-light tolerance of chloroplasts. In P. Mathis, (ed.), *Photosynthesis: From Light to Biosphere*, Vol. IV, Dordrecht, The Netherlands, pp. 291–294.

Youngman, R.J. and Elstner, E.F. (1981) Oxygen species in paraquat toxicity: the crypto-OH radical. *FEBS Lett.*, **129**, 265–268.

Zelitch, I. (1973) Plant productivity and the control of photorespiration. *Proc. Natl. Acad. Sci. USA*, **10**, 579–584.

Zelitch, I. (1990) Oxygen resistant photosynthesis in tobacco plants selected for oxygen resistant growth. In I. Zelitch, (ed.), *Perspectives in Biochemical and Genetic Regulation of Photosynthesis*, Alan R. Liss, New York, pp. 239–252.

Zelitch, I. and Ochoa, S. (1953) Oxidation and reduction of glycolic and glyoxylic acids in plants. I. Glycolic acid oxidase. *J. Biol. Chem.*, **201**, 707–718.

Ziem-Hanck, K. and Heber, U. (1980) Oxygen requirement of photosynthetic CO_2 assimilation. *Biochim. Biophys. Acta*, **951**, 266–274.

3 Low-temperature Stress and Antioxidant Defense Mechanisms in Higher Plants

Stanislaw Karpinski, Gunnar Wingsle, Barbara Karpinska and Jan-Erik Hällgren

INTRODUCTION

Temperature is a major abiotic factor that limits agricultural and forestry production and the natural distribution of plants. In many habitats, plants are subjected to large seasonal and diurnal variations in temperature. The general effects are fairly well understood. However, the primary effects of low temperature appear at the molecular level and are not well known. The consequences of low temperature for the whole plant, for example on growth, are more complex and the literature has presented a number of interpretations dealing with both structural and metabolic functions (Levitt, 1980; Steponkus, 1984; Sakai and Larcher, 1987; Kacperska, 1989; Guy, 1990; Hällgren and Öquist, 1990; Hällgren *et al.*, 1991). More recently, comprehensive review articles also give mechanistic explanations for growth impairment or developmental processes, phenotypic modifications, or changes based on alterations of physiological, biochemical, and molecular processes (Thomashow, 1990, 1994, 1998; Alberdi and Corcuera, 1991; Bowler *et al.*, 1992; Palva, 1994; Wise, 1995; Bohnert *et al.*, 1995; Hughes and Dunn, 1996; Nishida and Murata, 1996; Gilmour *et al.*, 1998; Huner *et al.*, 1998).

Growth is certainly the best indicator of the metabolic activities of a plant, and one of the first symptoms of stress is evident in altered growth. Light intensity, photoperiod, and temperature are major environmental factors that quantitatively and qualitatively affect growth and growth rhythms in plants. However, growth is not the only temperature-sensitive event: morphogenetic changes, flowering, seed formation, seed germination, bud break, colour etc. are all affected by temperature. The sensitivity of the plants' organs differs according to growth stages, age, and, hence, season. Temperature has direct and indirect effects on plants. One good example of the direct effects is changes in the production rate of reactive oxygen intermediates (ROIs).

In this chapter, first a brief introduction will be given on general physiological and molecular responses as well as on the role of cryoprotectants, osmolytes, and membranes in low-temperature stress. Then, emphasis will be put on the evidence for the involvement of ROI metabolism, the roles of the enzymatic and non-enzymatic ROI-scavenging and antioxidant systems, and the avoidance mechanism of ROI production in chloroplasts during low-temperature-induced photooxidative stress. By considering the different antioxidant defense mechanisms in different plants, the signalling and regulation of genes encoding enzymatic ROI-scavenging systems and the consequences of engineering their expression with transgenic plants, we will be able to evaluate the role of ROI metabolism on the amelioration of low-temperature stress tolerance in higher plants.

LOW-TEMPERATURE STRESS

Cold and Freezing Acclimation

Plants can adjust to diurnal and seasonal changes of the temperature. Adaptation is a stable genotypic long-term response to sesonal changes in the environment, whereas acclimation is an environmentally induced short-term response that causes the phenotypic alterations with the underlying physiological, biochemical, and molecular changes. Not only higher plants but also mosses, lichens, algae, and cyanobacteria have the ability to acclimate to low temperatures.

Many of the important crops such as wheat, rye, rice, and maize, which are usually species of tropical or subtropical origin, exhibit a poor tolerance to low temperature and fail to grow or are damaged by exposure to temperatures in the range of approximately 0°C to +15°C. Growth and germination or reproduction can be restricted in this temperature range. The term "chilling injury" has been used to distinguish this type of injury from damage caused by freezing temperatures (Lyons, 1973). Chilling injury is a complex phenomenon and appears in different forms, such as loss of vigour, wilting, chlorosis, sterility, and numerous cellular and metabolic dysfunctions. Plants under chilling stress show impaired photosynthesis, altered respiration, cessation of protoplasmic streaming, and changes in membrane integrity.

It has long been known that photoperiod and low temperature are the main environmental stimuli responsible for initiating cold hardening, the process that leads to seasonal adaptation of plants to low temperature (Sakai and Larcher, 1987). Photoperiodic regulation of growth involves perception of light by the phytochrome pigment family and the transduction of the light signal. However, cold hardening is a two-step process in which the first stage, shortening of day length, is followed by a second stage, exposure to low temperature, which is needed to obtain full cold hardening. Photomorphogenetic processes controlled by phytochrome are, as a rule, also affected by plant hormones. Studies on *Salix* strongly suggest involvement of gibberellins in the regulation of growth cessation (Junttila, 1990).

The ability to acclimate to low temperature is under genetic control (Thomashow, 1990). The two major processes, cold and freezing acclimation, are different. Cold acclimation differentiates chilling-sensitive from chilling-tolerant species, whereas freezing acclimation discriminates freezing-sensitive from freezing-tolerant species within chilling-tolerant species. At present it is not known how plants sense changes in temperature.

Recently, it has been demonstrated that rapid increases in light intensity and/or chilling can also create an imbalance, so that the energy absorbed through the light-harvesting complex exceeds what can be dissipated or transduced by photosystem II (PSII). Such imbalance or excess excitation energy (EEE) can be strongly enhanced by a combination with other factors, such as limitations in nutritional and water status. Redox changes in the proximity of PSII, because of such an imbalance, might also be an environmental sensor that controls acclimation to low temperature and excess light (Karpinski *et al.*, 1997, 1999; Huner *et al.*, 1998; Pfannschmidt *et al.*, 1999). Moreover, such imbalance induces the genes encoding ROI-scavenging enzymes (Karpinski *et al.*, 1997, 1999).

There are striking similarities among the processes induced by different stress factors and most stresses also share their effect on water status of plants. During acclimation to cold, modifications of water relations have been demonstrated to occur and are a prerequisite for the adjustment of growth (Weiser, 1970; Kacperska, 1993). Plant water relations affect growth processes either directly or indirectly. The plant hormone abscisic acid (ABA) plays a key role in the plant's responses to drought. Evidence is accumulating that implies also a central role for ABA in cold acclimation: ABA exhibits a transient increase during cold acclimation (Daie and Campbell, 1981; Chen *et al.*, 1983; Lalk and Dorffling, 1985) and ABA treatment can alternate the low-temperature stimulus (Chen and Gusta, 1983). ABA mutants and ABA-insensitive mutants are impaired in cold acclimation (Heino *et al.*, 1990). ABA generally upregulates low-temperature-induced genes. However, in winter wheat, ABA cannot induce freezing tolerance, and appears, at least in this plant, not to be essential in regulating cold acclimation (Dallaire *et al.*, 1994; for a comprehensive review, see Palva, 1994). The involvement of ABA in the regulation of genes encoding ROI-scavengers will be discussed below.

The capacity to survive freeze-induced cellular dehydration varies among different plant species, and a marked seasonal-acclimation can be seen in woody, evergreen plants, and in crops such as winter rape, cereals, and alfalfa. Tolerance to low temperature and freezing seems to be the primary mechanism for plants to survive low temperatures, although avoidance of freezing stress can also be achieved by super-cooling of tissue liquids in some plant species (Ashworth, 1993).

Cryoprotectants

Osmolytes

Production of cryoprotectants and/or osmolytes is a general way to stabilize membranes and maintain protein conformation at low water potentials. It is therefore not surprising that one of the responses to low temperature is synthesis of metabolites, such as osmolytes. These compounds are simple solutes that usually accumulate in high concentrations without disturbing cellular enzyme-based processes. Osmolytes include nitrogen-containing compounds, such as proline and other amino acids, polyamines, and quaternary ammonium compounds (glycine-betaine). Other osmolytes are sucrose, polyols, sugar alcohols (pinitol) and oligosaccharides. The production of osmolytes varies among plant species, but accumulation of a single metabolite is not restricted to taxonomic groupings, indicating that these are evolutionarily old traits (Bohnert *et al.*, 1995; Ingram and Bartels, 1996). Osmolytes play a role in osmotic adjustment, however the amounts are not always sufficient to alter water potential gradients. Osmolytes may also have other functions, such as the protection of cellular structures, possibly by scavenging of ROIs or preventing ROI production (Bohnert and Jensen, 1996). Polyols, such as mannitol, sorbitol, or myo-inositol and its methylated derivatives, together with their metabolism have been comprehensively described by Bohnert *et al.* (1995).

Trehalose, in relatively small amounts, has been shown to protect tobacco seedlings against drought stress (Holmstrom *et al.*, 1996). However, the roles of osmolytes, such as trehalose, raffinose, mannitol, fructans, and other substances, such as glycerol and proline, in low-temperature stress are more or less unknown, even if some strong arguments for an involvement in protection against freezing stress have been put forward (Bohnert *et al.*, 1995; Ingram and Bartels, 1996). Recently, it has been demonstrated that proline is involved in reducing photodamage in the thylakoid membranes by scavenging and/or reducing the production of singlet oxygen (Alia and Mohanty, 1996).

Santarius (1982) showed that carbohydrates protect membranes against the effects of high concentrations of electrolytes. The protection of membranes by carbohydrates is now thought to be more specific (Crowe *et al.*, 1992), and involves the so-called "compatible solutes" (Rhodes and Hanson, 1993; Bartels and Nelson, 1994). Cold hardening affects the soluble carbohydrate content as a result of excess photosynthates in relation to respiratory activities (Smirnoff and Pallanca, 1996); altered activities of several enzymes of the photosynthetic carbon-partitioning reactions between starch and sucrose have been observed. In non-photosynthetic tissues, such as roots, cereal seeds, tubers, or tree stems, the exported carbon is deposited as starch in amyloplasts for long-term storage (Kleczkowski, 1996). The capacity to develop large carbohydrate reserves during cold hardening and, hence, the capacity to maintain active photosynthesis at low temperature, appears to be critical to the development of freezing tolerance and to the winter survival of cereals (Levitt, 1980; Guy, 1990; Hurry *et al.*, 1995). The accumulation of sugars during exposure to low temperature may reflect an adaptive response, enabling cold-tolerant plants to maximize the synthesis of compounds that are essential for the maintenance of basal metabolism during over-wintering and that have important cryoprotective functions (Savitch *et al.*, 1997). However, the actual roles of these compounds in protection against ROIs or preventing the production of ROIs remains to be proven.

Proteins

Weiser (1970) originally proposed that cold acclimation requires transcriptional activity of the specific genes. Plants increase their capacity for protein synthesis during cold acclimation (Cloutier, 1983), and more specific changes occur in apoplastic proteins (Griffith *et al.*, 1992; Marentes *et al.*, 1993). Changes in ribosome structure and polysome composition have been observed (Guy, 1990; Hughes and Dunn, 1996). Although the role of protein alteration is unclear, it has been demonstrated that protein synthesis is required for increasing cold tolerance (Chen *et al.*, 1983; Lalk and Dorffling, 1985; Thomashow, 1994; Hughes and Dunn, 1996).

Low-temperature-induced proteins are referred to as cold-regulated (COR) proteins, cold acclimation proteins (CAPs), and anti-freeze proteins (AFPs), and proteins involved in membrane phospholipids (Hughes *et al.*, 1992). Little information is available about the identity and function of low-temperature-induced proteins (Thomashow, 1994; Palva, 1994; Hughes and Dunn, 1996). Several different roles have been suggested for COR and CAPs, including chaperone-like activity, proteolytic activity, and as protein import factors. Recently, a cold binding factor (*CBF1*), a transcriptional activator that binds to the C-repeat/drought-responsive element (CRT/DRE) DNA sequence, has been demonstrated to induce *COR* gene expression and to increase the freezing tolerance of non-acclimated *Arabidopsis* plants (Jaglo-Ottosen *et al.*, 1998). *CBF1* was concluded to be a probable regulator of the cold acclimation response, controlling the level of *COR* gene expression, which in turn promotes tolerance to freezing (Jaglo-Ottosen *et al.*, 1998).

Volger and Heber (1975) and Hincha *et al.* (1990) have isolated small, soluble polypeptides that seem to have cryoprotectant properties, and are approximately 100-fold more effective than sucrose in protecting the chloroplast membranes against freeze-thaw damage. An example of protective proteins are dehydrins that have been suggested to protect cytoplasmic proteins against denaturation (Close, 1996). Proteins of the dehydrin family are also of interest because they exhibit a high affinity for metals (Mantyla, 1997), which are well known that metals play a major role in the production of ROIs (Halliwell and Gutteridge, 1989, 1993). Many other proteins related to desiccation tolerance have been described (for reviews, see Palva, 1994; Thomashow,1994). Evidence that certain cold-acclimated plants synthesize proteins that show similarities to AFPs also exists (Kurkela and Frank, 1990).

The subcellular localisation of proteins involved in cold acclimation is of great interest, but very little is known so far (Guy, 1990). Interesting studies have been presented on proteins in the intercellular spaces where ice formation takes place (Griffith *et al.*, 1992; Marentes *et al.*, 1993). These proteins have been shown to be AFPs. Recently, six AFPs have been isolated and these show similarities to pathogen response proteins (Hon *et al.*, 1995). These glucanase-like and chitinase-like proteins exhibit anti-freeze activity, suggesting a relation between disease resistance and freezing tolerance.

Membranes

It is widely accepted that the plasma membrane plays a key role in freeze-thaw stress (Steponkus, 1984; Lynch and Steponkus, 1987; Palta and Weiss, 1993). Cold acclimation increases the stability of the plasma membrane to withstand stresses associated with freezing. Several factors may be responsible for membrane damage, such as the physical effect of temperature itself, mechanical effects, and reduction in the surface area of cells and changes in pH and ionic strength. However, very little information exists on the interference between lipids and plasma membrane proteins and the changes these may undergo during cold acclimation. Palta *et al.* (1977) hypothesized that the plasma membrane ATPase is the primary target. Some plasma membrane proteins disappear and new ones appear during cold acclimation (Uemura and Yoshida, 1984; Zhou *et al.*, 1994).

The extent of the injury to the plasma membrane may depend both on the lipid composition of the membranes (Nishida and Murata, 1996) and the presence of specific cryoprotectants. It was stated earlier that one of the most critical elements in chilling-sensitive plants was related to the thermal influences on the physiological properties of membranes and membrane integrity (Lyons, 1973; Steponkus, 1984). The ratio of the bulk thylakoid lipids monogalactosyldiaglycerol (MGDG) to digalactosyl-diaglycerol (DGDG) during low-temperature treatment decreases in many plants, but the level of fatty acid unsaturation generally increases. The importance of such changes has been demonstrated in the fatty acid desaturases (*fad6*) mutant of *Arabidopsis thaliana* and in other transgenic plants (Browse and Somerville, 1994; Moon *et al.*, 1995; Somerville, 1995; Wu and Browse, 1995) These experiments demonstrated that the biological significance of changes in thylakoid lipid unsaturation in response to low temperature is not clear. A comprehensive review of how membrane lipid composition affects the sensitivity of plants to chilling stress has recently been presented (Nishida and Murata, 1996).

LOW-TEMPERATURE-INDUCED OXIDATIVE STRESS AND PROTECTIVE SYSTEMS

Reactive Oxygen Intermediates

The injuries manifested by low-temperature stress resemble the injuries caused by the strong oxidative stress inducer methyl viologen (Bowler *et al.*, 1992). Benson *et al.* (1992) also showed that early post-freeze recovery of thawed rice cells is limited by oxidative stress. Therefore, it has been suggested that these common injuries are caused by ROIs (McKersie, 1991; Bridger *et al.*, 1994). ROIs have been shown to increase proteolysis by activating peptidases and by facilitating attacks by proteases (Casano *et al.*, 1994). Oxidation by ROIs has also been considered to play a role in protection from oxidative stress (Del Corso *et al.*, 1994) via an oxidation of thiol groups that regulate major metabolic pathways, and provide increased levels of NADPH. For example, the inactivation of the thioredoxin system influences ribulose-1,5-bisphosphate carboxylase oxygenase (Byrd *et al.*, 1995) as well as several other enzyme systems and, hence, cellular metabolism (Wise, 1995).

The evidence for a higher production rate of ROIs during low-temperature stress in plants is mostly indirect and is based on observations of changes in the levels of different ROI-scavengers and antioxidants. A few experiments show direct evidence for ROI formation during low-temperature stress. For example, electron spin resonance studies suggest that microsomes from lethally frozen winter wheat contain significantly higher superoxide ($O_2^{\bullet-}$) levels than unstressed or sub-lethally stressed controls (Kendall and McKersie, 1989). The amplitude of an electron spin resonance signal increased with a step-wise decrease in temperature to $-40°C$ (Tao *et al.*, 1992, 1998). This result suggests higher organic free radicals levels in Scots pine needles because of the decrease in temperature. Thylakoids from cold-sensitive plants produce $O_2^{\bullet-}$ radicals at a higher rate than thylakoids from cold-acclimated plants (Hodgson and Raison, 1991). A well characterized site for $O_2^{\bullet-}$ production in a plant is the thylakoid-membrane-bound primary electron acceptor of PSI and of the peripherally reduced ferredoxin (Asada *et al.*, 1974; Furbank and Badger, 1983; Asada, 1994). One molecule of $O_2^{\bullet-}$ is assumed to be produced per electron by a single-turnover flash of light. Most of the $O_2^{\bullet-}$ produced in the thylakoid membrane is converted into an $O_2^{\bullet-}$-mediated cyclic electron flow to O_2 and by non-catalytic dismutation to H_2O_2, before it reaches the stroma or lumenal space (Asada, 1994; Ogawa *et al.*, 1995).

The electron transfer chains of chloroplasts, mitochondria and peroxisomes are well-documented sources of H_2O_2 (Cadenas, 1989; Del Rio *et al.*, 1992; Asada, 1994). Okuda *et al.* (1991) observed higher H_2O_2 levels in winter wheat subjected to cold treatment as well as in chilled cucumber plants (Omran, 1980). Increased levels of H_2O_2 have been shown to be a general response to low-temperature stress in chilling-sensitive plants. Chloroplasts are thought to be the major H_2O_2 producers and have been calculated to generate approximately 15 to 20 mmol (mg chlorophyll)$^{-1}$ hr^{-1} of H_2O_2 during photosynthesis (Asada and Takahashi, 1987; Asada, 1994). H_2O_2 induces membrane energization that probably leads to the more efficient electron transport through PSII and, in consequence, can provide protection against photodamage (Karpinski *et al.*, 1999). H_2O_2 is also a strong nucleophilic-oxidizing agent and has been reported to react with SH groups. For instance, on isolated chloroplasts, H_2O_2 inhibits SH enzymes in the Calvin cycle at relatively low (10 mM) concentrations (Kaiser, 1979) and induces oxidative stress. On the contrary, treatment of intact leaves with 1 to 100 mM H_2O_2, a few hours before the photooxidative stress, protects the chloroplast and the cell against EEE induced photooxidative damage (Karpinski *et al.*, 1999; S. Karpinski unpublished data; Figure 1).

It is a well-known fact that H_2O_2 and $O_2^{\bullet -}$ can react together in biochemical systems to form the hydroxyl radical ($OH^{\bullet}$). Pure H_2O_2 and $O_2^{\bullet -}$ do not react at significant rates *in vitro* unless traces of iron or copper salts are present (Halliwell and Gutteridge, 1989, 1993). $O_2^{\bullet -}$ acts as a reductant of Fe(III) or Cu(II) and is a source of H_2O_2 (Halliwell and Gutteridge, 1989; 1993). However, other reductants, such as semiquinone radicals, thiols, and ascorbate (AsA) are also able to reduce these metals. In addition, there are other metal-catalyzed reactions involving H_2O_2 that produce OH. Thus, the cellular location of metals and reductants, such as thiols and AsA, and the site of production of both $O_2^{\bullet -}$ and H_2O_2 will determine the significance of the $OH^{\bullet}$ toxicity. No specific direct scavengers of $OH^{\bullet}$ radicals have been found in any organism (Asada and Takahashi, 1987; Asada, 1994).

Other forms of ROIs are the singlet species. Singlet chlorophyll ($^1Chl^*$) is generated by light excitation. The carotenoid pigments appear to play a dual protective role in

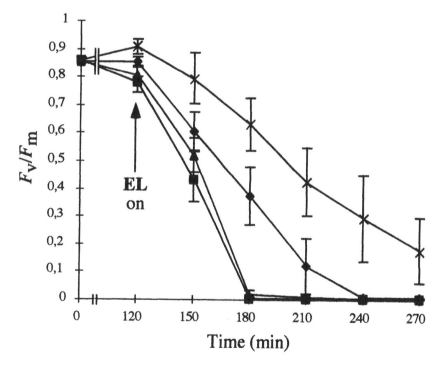

Figure 1. Protective role of excess of H_2O_2, and destructive role of excess of oxidized and reduced glutathione (GSSG and GSH) in *Arabidopsis* leaf tissue. F_v/F_m in detached leaves treated with water (black lozange, control), 30 mM H_2O_2 (black triangle), 10 mM GSSG (multiplier), and 10 mM GSH (black square) for 120 minutes in low light (LL; 150 mmol m^{-2} s^{-1}) in 20°C and then exposed to EEE, induced by excess light (1700 mmol m^{-2} s^{-1}) and low temperature (2 °C) for up to 150 minutes (parameters were measured in five different leaves obtained from three independent experiments, $n = 15 \pm$ SD).

quenching both ^{1}Chl* and singlet oxygen ($^1O_2$). The chloroplast membranes are particularly susceptible to $^1O_2$-induced lipid peroxidation because approximately 90% of the fatty acids of the thylakoid glycolipids, phospholipids, and sulpholipids are the unsaturated fatty acids a-linolenate (Knox and Dodge, 1985). According to our current knowledge there is no direct proof that the singlet species increase during low-temperature stress.

Low temperature causes an increase in ROI levels and induces oxidative stress in plants. However, the precise mechanisms remain to be established. There are many other free radicals, such as the monodehydroascorbate radical, the glutathione radical, thiyl radicals, other organic radicals, and peroxides, whose roles in relation to low-temperature stress are still not understood.

Lipid peroxidation

Experimental evidence suggests that lipid peroxidation reactions of the cellular membranes may play an important role in radical-mediated cell injury. For example, lipid peroxidation of biological membranes has been shown to lead to structural alterations (Bhaumik *et al.*, 1995). The mechanism of free radical-mediated lipid peroxidation involves at least three different phases. The initiation step occurs when a free radical, for example OH$^\bullet$ or $O_2^{\bullet-}$, interacts with polyunsaturated fatty acids and extracts a proton, to form a fatty acid radical. This step is followed by a propagation phase, in which the fatty acid radical reacts with O_2 and forms a fatty acid peroxy radical. This radical can react with other lipids and/or proteins, perpetuating the transfer of protons with subsequent oxidation of substrates (Bhaumik *et al.*, 1995). There are only a few examples of lipid peroxidation during chilling and freezing. Lipid peroxidation was observed in chilled suspension cultures of *Arabidopsis* based on measurements of thiobarbituric acid-reactive substances and in spruce subjected to frost events during the spring (Polle *et al.*, 1996). Lipid peroxidation products were also detected when *Arabidopsis* plants were subjected to low temperatures for 8 days (O'Kane *et al.*, 1996) and in developing maize seedlings (Prasad *et al.*, 1994, 1994b; Prasad, 1996). In young cucumber seedlings chilled at 4°C in the dark for 6 days, fluorescent lipid pigments accumulated in both phospholipid and glycolipid fractions of thylakoids (Hariyadi and Parkin, 1993).

On the one hand polyunsaturation of the membrane lipids because of low temperature would increase the potential for oxidative stress damage, but on the other hand, it provides new mechanistic features for membranes (De Vrije *et al.*, 1988; Kusters *et al.*, 1991). These new features have a strong influence on the recovery of photosynthesis, particularly the regeneration of the D1-protein after low-temperature-photooxidative stress (Nishida and Murata, 1996).

Antioxidants

It is thought that under optimal conditions, about 1% of the total O_2 consumed by the plant cell is converted to ROIs (Asada and Takahashi, 1987; Asada, 1994). During evolution, plants have developed different types of protection mechanisms against oxidative stress. Protective mechanisms can be divided into two separate categories: those involved in removing ROIs and those in reducing ROI production. The relative roles of these two processes are not known and may differ among plant species and cultivars. For example, plants can use O_2 "safely", without releasing any ROIs (for instance as a substrate in respiration and oxygenase reaction). Plant cells can also prevent oxidative damage by keeping substances that can react separately; e.g., keeping metals or O_2 bound to different proteins prevents the Haber-Weiss reaction or oxidation, respectively. Generally, the defense system against ROIs in plant cells is a net result of suppression mechanisms, scavenging, and repair systems. These systems interact to protect plant cell molecules and compartments against oxidation and they are active in different cellular physiological states and under both aerobic and anaerobic conditions. However, if oxidative stresses reach the maximum capacity of the defense system, ROIs can initiate a cascade of cellular damage leading to cell death.

Higher plants contain numerous enzymatic and non-enzymatic ROI-scavengers and antioxidants, both water- and lipid-soluble, localized in different cellular compartments (Larsson, 1988; Dalton, 1995; Asada, 1994; Wise, 1995). Non-enzymatic antioxidants include pigments, reduced glutathione (GSH), AsA, vitamin E, and many others, which we will not discuss in this chapter. The membrane-localized antioxidants mainly scavenge $^1O_2$, whereas water-soluble antioxidants usually scavenge $O_2^{\bullet -}$ and H_2O_2. For reviews on the interactions between different ROI-scavengers and antioxidants, see Dalton (1995) and Winkler *et al.* (1994).

Non-enzymatic antioxidants

One of the most acknowledged antioxidants in biological systems is α-tocopherol or vitamin E (Larsson, 1988; Polle and Rennenberg, 1992; Hess ,1993). This lipid-soluble vitamin functions as a ROI scavenger and plays an important role in protecting and maintaining the integrity of cell membranes, especially in the chloroplasts (Tappel, 1972). This vitamin also protects membrane proteins from oxidation. During scavenging of ROIs, α-tocopherol is oxidized to α-chromanoxyl radicals that can be regenerated by ascorbate or glutathione. α-Tocopherol is the most abundant tocopherol of the four forms found in plants (α-, β-, γ-, and δ-tocopherols) and its main location is within the chloroplast. In Scots pine, the content in α-tocopherol of needles from different age were investigated and older needles were found to contain higher levels (Wingsle and Hällgren, 1993). Only a small increase in α-tocopherol content was detected in autumn, whereas in spring (the end of April) a lower content was observed in Scots pine needles (Karpinski *et al.*, 1994). Chilling of tomatoes resulted in a decrease in α-tocopherol content (Wise and Naylor, 1987; Walker and McKersie, 1993). To our knowledge, no comprehensive data on seasonal changes in α-tocopherol levels are available for evergreens and perennials.

Glutathione is the most abundant thiol in higher plants (Foyer and Halliwell, 1976; Foyer, 1997; Mullineaux and Creissen, 1997). The general picture is that the levels of glutathione in its reduced form (GSH) increase several-fold during the wintertime in evergreens (Esterbauer and Grill, 1978; Anderson *et al.*, 1992). The increase in GSH levels due to cold acclimation, seasonal changes, and diurnal variation is well documented for several plant species (Esterbauer and Grill, 1978; Smith *et al.*, 1990; Anderson *et al.*, 1992; Walker and McKersie, 1993; Wingsle and Hällgren, 1993; Polle and Rennenberg, 1994; Wildi and Lutz, 1996). These data indicate that the nature of the changes in the glutathione pool is similar to those observed in the ascorbate pool.

Plants normally have a low oxidized glutathione (GSSG) level, for example in Scots pine it is approximately 20-fold lower than the GSH content (Wingsle *et al.*, 1989). Many factors, including low-temperature and other environmental stresses, have been shown to change the ratio or redox status of glutathione [GSH/(GSSG + GSH)] in different organisms (Huerta and Murphy, 1989; Gilbert, 1990; O'Kane *et al.*, 1996; Karpinski *et al.*, 1997) and an accumulation of GSSG can be an indicator of higher oxidative stress (Smith *et al.*, 1990). The roles of GSH have also been discussed as a potential cryoprotectant during frost injury, as a possible compound that prevents the formation of irreversible sulphur bonds, and as keeper of proteins in an active state, thereby increasing freezing tolerance (Levitt, 1980). Several authors who performed experiments in which the GSH concentration was increased without the plants becoming more chilling or frost tolerant (Schupp and Rennenberg, 1992; Stuvier *et al.*, 1992) have questioned this hypothesis. Moreover, artificially increased GSH and GSSG levels in *Arabidopsis* paradoxically caused higher susceptibility of leaves to photooxidative stress brought on by EEE, whereas H_2O_2 protected against this stress (Figure 1). In this context, the precise roles of glutathione and H_2O_2 in the oxidative stress response still remain to be established and recent data indicate that GSH and H_2O_2 levels play a fundamental role in the regulation of the photosynthetic electron transport (Karpinski *et al.*, 1997, 1999). The regulatory impact of glutathione and/or the redox status of the glutathione pool on oxidative stress response of plants is discussed below (see also Chapter 10, this volume).

The pivotal roles of AsA and dehydroascorbic acid (DHA) in several physiological processes in plants and mammalian cells have been thoroughly reviewed (Lewin, 1987; Loewus and Loewus, 1987; Foyer, 1993; Arrigoni, 1994; Asada, 1994; Dalton, 1995; Polle, 1997). The roles of AsA as an ROI scavenger in the aqueous phase of cells and as a cofactor in structural protein organization are well known and the redox status of AsA in different tissues has been shown to be an indicator of oxidative stress in plants (Sgherri *et al.*, 1994; Luwe and Heber, 1995). Ascorbate, and enzymes that metabolize AsA-related compounds, are involved in the control of several plant growth processes, such as the biosynthesis of hydroxyproline-rich proteins required for the progression into G1 and G2 phases of the cell cycle, the cross-linking of cell wall glycoproteins and other polymers, and redox reactions at the plasma membrane (Cordoba and Gonzalez-Reyes, 1994).

Seasonal changes in AsA have been documented in several investigations in other frost-tolerant species (Polle and Rennenberg, 1994). For example, Anderson *et al.* (1992) found a 4-fold increase in AsA in spruce needles from approximately 6 mmol g^{-1} FW in the summer to 22 mmol g^{-1} FW in the middle of the winter. Anderson *et al.* (1992) also measured DHA, and also found a lower redox state of AsA in the winter. Chloroplasts isolated from alpine plant species showed a much higher concentrations of AsA than those from lowland plants (Streb *et al.*, 1997). The AsA and DHA levels were analyzed in actively growing Scots pine needles, from greenhouse-cultivated plants, and dormant needles collected in the late winter (Wingsle and Moritz, 1997). The dormant needles showed a significantly higher content of both AsA and DHA, although the ratio of AsA/DHA was significantly lower in dormant needles. Such results indicate that DHA accumulates to a higher level in dormant than in actively growing needles and changes the redox status of the ascorbate pool to a more reduced state in dormant needles. Furthermore, AsA levels increased in illuminated wheat at low temperature (Mishra *et al.*, 1993). When chilling-sensitive plants were subjected to low temperature, a general decline in AsA levels was found (Hariyadi and Parkin, 1993; Walker and McKersie, 1993). A depletion of AsA was also found in chilled maize seedlings without changes in the DHA levels (Anderson *et al.*, 1995). Increased AsA levels may result from an increase in carbon supply during cold acclimation (Smirnoff and Pallanca, 1996). These data indicate that AsA metabolism play an important role in low-temperature-induced oxidative stress.

Other antioxidants, such as chlorogenic acid, have also been suggested to play an important role in the scavenging of ROIs during high light and low-temperature stresses (Grace *et al.*, 1998).

The enzymatic ROI-scavenging system

The enzymatic ROI-scavenging system in different compartments of plant cells consists of enzymes such as superoxide dismutase (SOD), catalases (CAT), ascorbate peroxidase (APX), monodehydroascorbate reductase (MDAR), dehydroascorbate reductase (DHAR), glutathione peroxidase (GPX), and glutathione reductase (GR) (Foyer and Halliwell, 1976; Bowler *et al.*, 1992; 1994; Foyer, 1993, 1997; Asada, 1994; Inzé and Van Montagu, 1995; Mullineaux and Creissen, 1997). Generally, chilling-sensitive and chilling-tolerant plants increase and/or try to compensate for losses in the enzymatic ROI-scavenging capacity during low-temperature-induced oxidative stress (Karpinski *et al.*, 1993). However, when plants are rapidly subjected to low temperature without cold acclimation, damages to the enzymatic ROI-scavengers might be too high and excess ROIs can initiate cell death. This process can be rapid when plants are subjected to additional stress factors, such as high light or air pollutants. Below, we present some examples of how the expression of genes that encode different members of the enzymatic ROI-scavenging system, is regulated during low-temperature-induced oxidative stress.

SOD has evolved to be one of the fastest enzymes known ($V_{max} = 2 \times 10^9$ M^{-1} s^{-1}) with an optimum close to the diffusion rate of $O_2^{\bullet-}$. SOD converts $O_2^{\bullet-}$ to H_2O_2 and constitutes the first link in the enzymatic scavenging system of ROIs (McCord and Fridovich, 1969; Foyer and Halliwell, 1976; Asada, 1994). Thirty years of research on this enzyme have shown that SOD plays an important role in numerous physiological, biochemical, and molecular processes in aerobic and anaerobic organisms. In humans and other animals, the involvement of SOD in the programming of neurone cell death has been considered (Deng *et al.*, 1993; Kane *et al.*, 1993; Raff *et al.*, 1993).

In prokaryotes and eukaryotes, three different types of SOD have been found containing either Mn, Fe, or Cu and Zn as prosthetic metals (Asada, 1994). The above types of SOD can be distinguished by their differential sensitivities to KCN and H_2O_2. Different SOD isoforms in plants are differentially expressed and also localized in different compartments within and outside the cell (Perl-Treves and Galun, 1991; Tsang *et al.*, 1991; Wingsle *et al.*, 1991; Bowler *et al.*, 1992; Karpinski *et al.*, 1992a, 1992b, 1993; Streller and Wingsle, 1994; Wingsle and Karpinski, 1996; Schinkel *et al.*, 1998; B. Karpinska, unpublished data).

Increases in SOD mRNA levels have been observed during recovery from naturally established winter stress, a combination of high light and low-temperature stress (Karpinski *et al.*, 1993; 1994; Figure 2). In that experiment, higher mRNA levels of chloroplastic and cytosolic isoforms of Cu/ZnSOD were observed in needles protruding above snow than in those covered by snow. Changes in transcript levels were not reflected in a corresponding increase in protein activities. Moreover, Cu/ZnSOD activity levels were

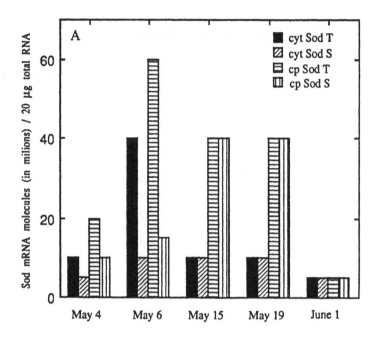

Figure 2. The estimated relative transcripts levels of cytosolic Cu/ZnSOD (cyt SOD) and chloroplastic Cu/ZnSOD (cp SOD) in top (T) and side (S) shoot needles of *Pinus sylvestris* (L.) seedlings during recovery from winter stress (May 4, 6, 15, 19, and June 1). The Cu/ZnSOD mRNA relative levels were estimated at 20 mg of total RNA and normalized to poly (A+) RNA content (after Karpinski *et al.*, 1993).

similar in covered and protruding needles. These results may suggest higher turnover rates of Cu/ZnSOD in needles protruding above the snow. The lack of correlation between mRNA levels and protein activity for Cu/ZnSODs in response to oxidative stress, has been observed before and was also suggested to be a result of higher turnover rates of Cu/ZnSODs during oxidative stress (Karpinski *et al.*, 1992b). Recently, evidence has been presented that higher oxidative stress leads to an increase in proteolytic degradation, initiated by the OH• radical (Casano *et al.*, 1994). Theoretically, during high oxidative stress, Cu/ZnSOD and FeSOD have the potential to produce the OH• radical, which can lead to a self-inactivation of the enzyme.

Different SOD isoforms are differentially expressed during recovery from winter stress. A comparison of chloroplastic and cytosolic Cu/ZnSOD mRNA levels showed a 4-fold higher transcript level for the chloroplastic form until mid-May (Figure 2; Karpinski *et al.*, 1993). This higher transcript level was also associated with a higher chloroplastic Cu/ZnSOD activity. Transcript levels were reduced for both chloroplastic and cytosolic Cu/ZnSODs and reached similar low levels after the repair process of the photosynthetic apparatus was completed and photosynthetic capacity had fully recovered from winter stress (Karpinski *et al.*, 1993; 1994). These data indicate that chloroplasts in evergreens play a major role in the generation of ROIs during low-temperature-induced oxidative stress.

That SOD plays an important role in low-temperature-induced oxidative stress response, has been observed earlier in chilling-temperature-sensitive plants. Cold- and dark-stored and then illuminated tomato leaves were shown to have a significantly lower activity of Cu/ZnSOD

(Michalski and Kaniuga 1981). In spinach and maize, SOD activity increased because of low-temperature stress (Schöner and Krause, 1990; Jahnke *et al.*, 1991). Evidence for a role of SOD in low-temperature-induced oxidative stress was obtained from transgenic plants (McKersie *et al.*, 1993; Gupta *et al.*, 1993a, 1993b).

Responses and regulatory mechanisms that differ from those of *SOD* genes have been observed for chloroplastic glutathione reductase (*GR*). In Scots pine needles exposed to naturaly established EEE-induced photooxidative stress (Karpinski *et al.*, 1993), GR activity was induced but the transcript levels of chloroplastic *GR* gene was not changed. Similar regulation of *GR* gene expression was observed in pea (Edwards *et al.*, 1994). Later it was demonstrated that GR activity in Scots pine needles can be upregulated by redox intraconversion of the enzyme without change in its mRNA and protein levels (Wingsle and Karpinski, 1996).

The annual total GR activity was also measured in SO_2- and NO_2-fumigated Scots pine needles (Wingsle and Hällgren, 1993), and a well-known pattern was observed with higher and lower activities in winter and summer, respectively (Esterbauer and Grill, 1978; Anderson *et al.*, 1992; Polle and Rennenberg, 1994). However, fumigated trees had a lower increase in GR activity than non-fumigated (Figure 3). This difference occurred during winter when trees were not exposed to air pollutants. The chlorophyll *a* fluorescence and other photosynthetic parameters showed also lower values in fumigated trees (Strand, 1993). A decreased frost tolerance in trees fumigated with NO_2 and SO_2 has been correlated with a higher impairment of the photosynthetic apparatus and lower activities of GR during winter. Changes in GR isozyme profiles because of exposure to low temperature have been also observed in deciduous plants (Guy and Carter, 1984), in other conifers (Hausladen and Alscher, 1994a, 1994b; Harris and Dalton, 1997), chilling-tolerant annual crops, such as spinach and pea (Guy and Carter, 1984; Edwards *et al.*, 1994), and in chilling-sensitive maize (Anderson *et al.*, 1995). In some plants, these changes were also associated with an increased affinity of GR for its substrates (Edwards *et al.*, 1994; Mullineaux and Creissen, 1997).

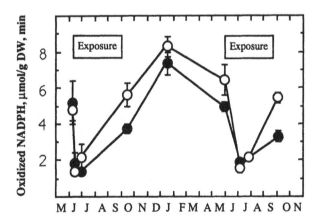

Figure 3.   Seasonal changes (from May (M) to November (N) next year) in glutathione reductase activity in one-year-old Scots pine needles from trees exposed to low levels of SO_2 and NO_2. Control (open symbol) and exposed (closed symbol). Each value represent mean ±SE ($n = 6$, after Wingsle and Hällgren 1993).

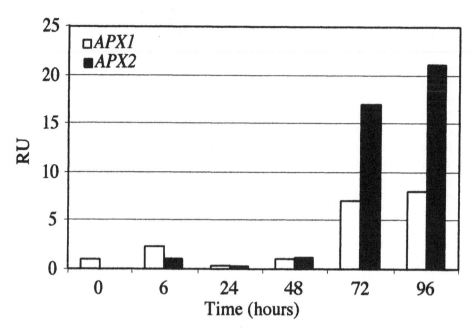

Figure 4. The estimated relative transcripts levels of the cytosolic *APX1* and *APX2* in *Arabidopsis* leaf tissue exposed to chilling (2°C) and low light (LL). The *APX1* and *APX2* relative transcripts levels were estimated at 20 mg of total RNA (S. Karpinski unpublished results).

Are the levels of other members of the enzymatic ROI-scavenging system critical for the defense against low-temperature-induced oxidative stress? The general answer is yes. In the light, the key enzyme involved in H_2O_2 scavenging is APX, which catalyzes the reaction: $2 AsA + H_2O_2 = 2$ monodehydroascorbate (MDHA) $+ 2H_2O$. The *APX1* and *APX2* genes are induced during chilling (Figure 4). Catalases can also scavenge H_2O_2, but they are peroxisomal enzymes and do not require a reducing substrate. Recently, the protective role of CAT in the photooxidative stress has been demonstrated (Willekens *et al.*, 1998). Chloroplasts photoregenerate AsA from MDHA or DHA. MDHA is converted into AsA either by reduced ferredoxin or NAD(P)H with MDHAR. DHAR is thought to regenerate AsA by using GSH as an electron donor. In plants and in animal systems, GPX has generated much attention as an important enzyme in the scavenging of H_2O_2 or the products of lipid peroxidation. Recently, a plastidial GPX has been identified (Mullineaux *et al.*, 1998), indicating that enzymatic oxidation of GSH to GSSG in the chloroplast can be made without involvement of DHAR. It will be very interesting to find the role and function of the chloroplastic GPX during cold hardening and low-temperature-induced oxidative stress in plants.

The correlation between collapsed antioxidant systems and damage to plant tissues by low temperature has been observed in several studies (Sagisaka, 1985; Wise and Naylor, 1987; Kuroda *et al.*, 1992; Kuroda and Sagisaka, 1992; Walker and McKersie, 1993). Late frost events in the spring and the emergence of the new flush is a particularly sensitive period for evergreens and deciduous trees. This was demonstrated in an experiment in which spruce seedlings were exposed to an artificial frost event of –5°C for one night in the spring (Polle *et al.*, 1996). APX activity declined before any visual damage to the needles could be detected and was accompanied by an increase in lipid peroxidation. In needles that were later found to be severely damaged, the enzymatic antioxidant system had collapsed. Surviving needles showed a transient increase in all antioxidant components, suggesting a general response of the enzymatic ROI-scavenging system. However, even when the ROI-scavenging system was transiently induced in such needles, they still had a lower level of antioxidants compared with controls, indicating a "memory effect" of the spring frost. This experiment clearly implies that the level of other members of the enzymatic ROI-scavenging system are critical for the integrity of the plant cells.

Is expression of genes encoding different enzymes of the ROI-scavenging system coregulated during low-temperature-induced oxidative stress? There is no clear answer to this question. Different plant species and different plant organs and tissues may differ in the regulation of expression of genes encoding these enzymes. Moreover, expression of genes that code for different isoforms of the same ROI-scavenging enzymse are regulated differently in response to low-temperature-induced oxidative stress (Bowler *et al.*, 1992; Karpinski *et al.*, 1993). Total activities of SOD, APX, MDAR, DHAR, and GR have been shown to be higher in chilling-tolerant than in chilling-sensitive ecotypes (Jahnke *et al.*, 1991; Walker and McKersie, 1993). In *Arabidopsis* leaves, *APX1* and *APX2* genes are induced after a shift to 4°C (Figure 3). Conflicting results have been presented for seasonal changes in total APX activities in spruce (Anderson *et al.*, 1992; Polle and Rennenberg 1994; Polle *et al.*, 1996). MDAR showed elevated levels in the needles during autumn and winter. During bud break, both APX and MDAR had higher activity levels (Polle *et al.*, 1996). In Scots pine, activities of enzymes such as SOD, MDR, APX, and DHAR increased during cold acclimation (Tao *et al.*, 1998). However, in many other experiments, total SOD activities did not vary seasonally (Hausladen *et al.*, 1990; Madamanchi *et al.*, 1991; Kröninger *et al.*, 1993; Wingsle and Hällgren, 1993).

Catalases have also received much attention in respect of the responses of plants to low-temperature oxidative stress. Inactivation of catalases (CAT) is observed below 10°C and 15°C in chilling-tolerant and chilling-sensitive species, respectively (Taylor *et al.*, 1974; Omran 1980; Feierabend *et al.*, 1992; Mishra *et al.*, 1993). Generally, chilling stress causes at least a 50% reduction in CAT activities in different plants, but in cold-acclimated spinach leaves, only a 25% reduction was observed. The transcript levels for *CAT* genes are also affected during chilling. In the chilling-sensitive *Nicotiana plumbaginifolia*, *CAT1* transcript levels were strongly reduced within 2 hours after stress, *CAT2* mRNA levels remained constant, while *CAT3* mRNA levels showed a slight induction 6 hours after the onset of the stress. After 6 hours of recovery, the *CAT1* transcript level was restored to control levels (Willekens *et al.*, 1994). In maize, Prasad *et al.* (1994a, 1994b) found evidence for a dual role of H_2O_2 in low-temperature stress response. When seedlings, grown at 27°C, were exposed to a temperature of 4°C they did not survive unless they were transiently exposed to a temperature of 14°C. It was demonstrated that H_2O_2 could induce chilling acclimation at a higher temperature (27°C) yet fail to induce this at 4°C. Details of this experiment are discussed below.

Another important question in determining the significance of the enzymatic ROI-scavenging system in plants is how does low-temperature-induced oxidative stress influences different ROI-scavenging enzymes localized in different tissues, outside and inside the cell. An experiment performed with pre-emergent maize seedlings acclimated (14°C) and chilled (4°C) in darkness showed that the primary source of active oxygen in the dark is probably localized in mitochondria (Prasad *et al.*, 1994b; Anderson *et al.*, 1995; Prasad, 1996). H_2O_2 accumulated in different tissues. In the mesocotyl, which is the most chilling-sensitive tissue, *CAT3* mRNA levels were elevated in acclimated seedlings and were suggested to be the first line of response towards H_2O_2 generated by mitochondria. Nine isoforms of the predominant peroxidases (POXs) were induced by cold acclimation. Two of these POXs, located in the cell wall, correlated with an increase in lignin formation, probably improving the mechanical strength of the mesocotyl.

The enzymatic ROI-scavenging system plays an important role in both chilling-tolerant and chilling-sensitive plants during low-temperature-induced oxidative stress. However, chilling-sensitive plants seem to differ from chilling-tolerant plants in their ability to increase and/or compensate for the losses in their ROI-scavenging capacity at chilling temperatures (Karpinski *et al.*, 1993; Baker 1994; Wise 1995). This observation also led Stassart *et al.*, (1995) to conclude that in chilling-sensitive plants, high levels of the enzymatic ROI-scavengers are indicative of oxidative stress rather than of stress tolerance.

Signalling and regulation of genes encoding enzymatic ROI scavengers

Many compounds have been nominated as agents involved in signalling both in biotic and abiotic stress responses, including salicylic acid (SA), H_2O_2 (Chen *et al.*, 1993; Levine *et al.*, 1994; Prasad *et al.*, 1994a, 1994b; Bi *et al.*, 1995; Neuenschwander *et al.*, 1995; Alvarez *et al.*, 1998; Karpinski *et al.*, 1999), NO (Delledonne *et al.*, 1998), $O_2^{\bullet-}$ (Tsang *et al.*, 1991, Jabs *et al.*, 1996); GSH and GSSG (Wingate *et al.*, 1988; Hérouart *et al.*, 1993; Wingsle and Karpinski, 1996; Karpinski *et al.*, 1997); calcium (Ca^{2+}) (Price *et al.*, 1994; Monroy and Dhindsa, 1995; Knight *et al.*, 1996); photoreceptors with Ca^{2+} (Neuhaus *et al.*, 1993; Millar *et al.*, 1995); ABA (Giraudat *et al.*, 1994; Giraudat, 1995), and recently the redox status of plastoquinone pool (Karpinski *et al.*, 1997; 1999). However, very little is known about the signalling cascades initiated by these molecules. ROIs are known to be involved in the regulation of processes as diverse as the hypersensitive response and systemic acquired resistance (SAR) (Lamb and Dixon, 1997); chilling responses (Prasad *et al.*, 1994a, 1994b), cross tolerance to different abiotic stresses (Bowler *et al.*, 1992), regulation of the photosynthetic electron transport, and systemic acquired acclimation (SAA) (Karpinski *et al.*, 1997, 1999). Accumulating evidence suggests that there are specific transcription factors, which regulate gene expression in response to oxidative stress (Lu *et al.*, 1996; Cao *et al.*, 1997; Dietrich *et al.*, 1997).

Generally, Ca^{2+} is considered to act as a secondary messenger in the oxidative stress response of plants (Price *et al.*, 1994; Monroy and Dhindsa, 1995; Neuhaus *et al.*, 1993; Millar *et al.*, 1995; Knight *et al.*, 1996), whereas Ca^{2+} channels might function as temperature-sensors during cold acclimation (Ding and Pickard, 1993). Cytosolic Ca^{2+} levels increased transiently as a result of oxidative stress induced by cold shock (Knight *et al.*, 1991). It has been suggested before that cold damage in plants may be due to lack of Ca^{2+} homeostasis and subsequent Ca^{2+} toxicity (Minorsky, 1985). However, more recently, changes in Ca^{2+} levels are considered a necessary step in sensing and signalling low temperatures in plants. Two cDNAs for Ca^{2+}-dependent protein kinases (CDPK) have been isolated from *Arabidopsis* (Urao *et al.*, 1994; 1998) and the gene coding for CDPK from pea is induced by cold stress. In animals, phospholipase generates two secondary messengers, inositol 3-phosphate and diacylglycerol. Inositol 3-phosphate induces the release of Ca^{2+} in the cytoplasm, and it has been suggested that a similar mechanism could exist in plants and inositol 3-phosphate and Ca^{2+} could function as secondary messengers during drought and cold conditions. Monroy and Dhindsa (1995) demonstrated that elevated levels of Ca^{2+} are needed to initiate the induction of genes involved in the cold acclimation process in alfalfa but were not enough to sustain this induction for a longer time. These results suggest that other signals are necessary to complete cold acclimation. Later, changes in the levels of Ca^{2+} could be detected within seconds of cold shock treatment (Knight *et al.*, 1996). This immediate increase in the cytosolic free Ca^{2+} concentration was observed in both chilling-sensitive tobacco and chilling-tolerant *Arabidopsis*. That Ca^{2+} can regulate enzymatic ROI scavengers and the oxidative stress response was demonstrated by Price *et al.*, (1994).

Protein phosphorylation and dephosphorylation, catalyzed by protein kinases and phosphatases respectively, play key roles in regulating many aspects of plant growth, such as development, metabolism, and stress responses. Protein phosphorylation has been suggested to be involved in a signal transduction pathway that regulate the response to low-temperature stress. It has been shown that changes in the pattern of protein phosphorylation occur during cold acclimation of alfalfa cell suspension cultures (Monroy *et al.*, 1993), whereas changes induced by low-temperature stress phosphorylated existing proteins and were inhibited by Ca^{2+} channel blockers and by an antagonist of calmodulin and CDPK (Monroy *et al.*, 1993). Several genes involved in the mitogen-activated protein kinase kinase kinase (MAPKKK) cascade are induced by low-temperature stress (Mizoguchi *et al.*, 1996; 1998). The MAPKKK and adenosine-triphosphate kinase 19 (ATPK19) are induced at low temperature, suggesting that they might be involved in signal transduction pathways under low-temperature stress (Mizoguchi *et al.*, 1996; 1998). Other genes encoding factors involved in signal transduction, such as transcription factors (Kusano *et al.*, 1995; 1998; Jaglo-Ottosen *et al.*, 1998), are also induced by cold and drought and have been discussed by Shinozaki and Yamaguchi-Shinozaki (1996). Kusano *et al.*, (1995) showed that the gene encoding a bZIP DNA-binding factor is induced by low temperature. This protein binds to the histone motif ACGTCA and the promoter region of a cold-inducible gene, which suggests that it may control some cold-inducible genes in chilling-sensitive plants. Our data suggest that induction of the *APX1* and *APX2* genes in *Arabidopsis* depends on calcium and, therefore, could involve a kinase-mediated cascade (S. Karpinski, unpublished results).

ABA plays an important role in signalling of drought and low-temperature stress and its role in low-temperature-stress response has been discussed above. Zhu and Scandalios (1994) demonstrated that different members of the Mn*SOD* gene family in maize respond differently to ABA and high osmoticum. Mn*SOD3.1* transcript levels do not respond to ABA and high osmoticum, whereas others members of the Mn*SOD* gene family increase transcript levels after ABA treatment. Of the maize Cu/Zn*SOD* gene family, only Cu/Zn*SOD4* transcript levels were higher after ABA treatment. ABA has been shown to increase both GR and APX activities in *Arabidopsis* (O'Kane *et al.*, 1996).

A regulatory role for H_2O_2 as a signalling molecule in different secondary messenger systems in humans and in animals is well documented (Ramasasrma, 1982; Meyer *et al.*, 1993; Ginnpease and Whisler, 1996). In plants, the ability to control H_2O_2, $O_2^{\bullet-}$, and GSH levels is an important factor in biotic and abiotic stress responses. Exogenous application of H_2O_2 or menadione, an $O_2^{\bullet-}$-generating compound, to maize seedlings, suggested that mild oxidative stress at 27°C might induce chilling acclimation (Prasad *et al.*, 1994a, 1994b). Both these compounds caused an increase in *CAT3* and *POX* transcript levels and protein activities, but SOD activity levels were constant in these experiments. However, the exogenously applied and endogenously accumulated H_2O_2 failed to increase the activity of these enzymes at 4°C. Hence, higher levels of H_2O_2 that could cause cell death at a lower temperature induced cold-acclimation at a higher temperature. The mechanism by which H_2O_2 or menadione induce *CAT3* and *POX* gene expression in maize is sensitive to chilling. Moreover, in non-acclimated seedlings, chilling injury was partly due to the excess of ROIs, which promoted protein and lipid oxidation. In chilling-acclimated seedlings, the enhanced enzymatic ROI-scavenging system prevented the accumulation of ROIs and, therefore, prevented damage to lipids and proteins at 4°C (Prasad *et al.*, 1994a, 1994b; Prasad 1996).

In tobacco, increased mRNA levels for Fe*SOD* was triggered by $O_2^{\bullet-}$ generated in the proximity of PSI due to paraquat treatment (Tsang *et al.*, 1991). Recently, Cu/Zn*SOD4* and Cu/Zn*SOD4A* transcript levels in maize were shown to increase in response to H_2O_2 treatment (Kernodle and Scandalios, 1996). A similar response of cytosolic and chloroplastic Cu/Zn*SOD* transcript levels was observed in Scots Pine (S. Karpinski, unpublished results).

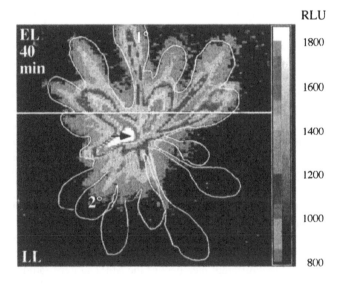

Figure 5. Systemic induction of *APX2-LUC* expression in transgenic *Arabidopsis* leaf tissue. Image of luciferase activity in relative light units (RLU). A part of the whole rosette (as shown) exposed to excess light (EL; 2700 mmol m^{-2} s^{-1} for 40 minutes); the arrow indicates the apical region of the rosette. A typical primary (1°) EL-exposed leaf and secondary (2°) LL-exposed leaf are shown (after Karpinski *et al.*, 1999 with permition of Science). Such plants are also acclimated to low temperature stress {combination of EL (1700 mmol m^{-2} s^{-1}) and LT (2°C) S. Karpinski unpublished results}.

Recently, we have demonstrated that information regarding PSII function in stressed chloroplasts can be transduced systemically by H_2O_2 to change the PSII function in non-stressed chloroplasts of remote cells that did not experience photooxidative stress conditions provoked by EEE (Karpinski *et al.*, 1999). These systemic redox changes in the proximity of PSII are prerequisite for systemic activation of antioxidant defenses and triggering of the SAA (Figure 5).

The most relevant functions of GSH in the context of oxidative stress are those in which GSH participates in redox reactions. Therefore, oxidized glutathione (GSSG) is generated (Foyer and Halliwell, 1976). Recently, other important roles for glutathione in oxidative stress responses have been considered; for example, GSH and H_2O_2 have been proven to inactivate and activate, respectively, the stress-responsive nuclear factor NF-κB in mammalian cells (Meyer *et al.*, 1993; Ginnpease and Whisler, 1996). In plants, high concentrations of GSH, but not GSSG, enhanced the expression of genes encoding enzymes involved in phytoalexin and lignin biosynthesis, suggesting a general role for GSH in signalling systems in biological stress (Wingate *et al.*, 1988). Different thiols, such as GSH, cysteine, and dithiothreitol, increased the transcript level of a reporter gene under the control of the cytosolic Cu/Zn*SOD* promoter in transgenic tobacco protoplasts (Hérouart *et al.*, 1993). Recently, changes in the glutathione levels and/or redox status of glutathione pool have been found to have a regulatory impact on the expression of genes

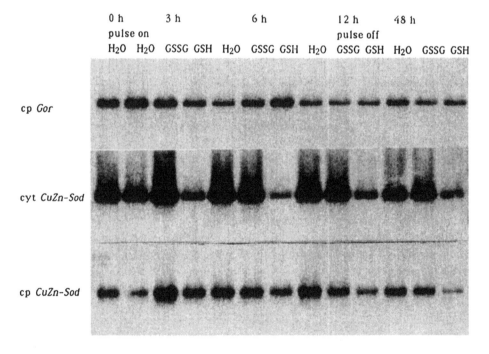

Figure 6. RNA gel blot hybridization analysis of chloroplastic (cp) *GOR*, cytosolic (cyt), and cp Cu/ZnSOD mRNA levels in poly (A$^+$) RNA. Poly (A$^+$) RNA (10 mg per lane) was isolated on five different occasions (0, 3, 6, 12, and 48 hours) from needles of Scots pine shoots treated with water, 5 mM GSSG, or 5 mM GSH. The RNA was separated by gel electrophoresis, transferred to a filter, and hybridized with heterologous (pea) cp *GOR*, homologous cyt and cp Cu/ZnSOD cDNA probes. Pulses of glutathione were applied at time 0 hour and terminated after 12 hours (after Wingsle and Karpinski 1996).

encoding cytosolic and chloroplastic isoforms of Cu/ZnSOD in Scots pine (Wingsle and Karpinski, 1996) and cytosolic *APX* in *Arabidopsis* (Karpinski *et al.*, 1997).

The regulatory impact of glutathione on the transcript levels of Cu/Zn*SOD* and *GOR*, genes are presented in Figure 6. Our results that GSH reduced the cytosolic Cu/Zn*SOD* transcript level, are in agreement with findings for human Cu/Zn*SOD* and Mn*SOD* genes, which were found to be downregulated by thiols (Suzuki *et al.*, 1993), but are in contrast to those reported by Hérouart *et al.* (1993). The difference may be due to the fact that different systems were used. The levels of GSH and GSSG, or the redox state of the glutathione pool, might play an important role in the *in vivo* regulation of the expression of genes encoding the enzymatic ROI-scavenging enzymes in plants. We concluded that the mechanisms regulating the expression of *SOD* and *GOR* genes respond differently to altered levels of GSH and GSSG in Scots pine needles (Wingsle and Karpinski, 1996). The activity of GR increased *per se* (but not the *GOR* transcript level) in response to higher levels of GSSG, suggesting that the enzyme itself undergoes redox intraconversion *in vivo*. However, the transcript levels of cytosolic and chloroplastic Cu/ZnSOD were lowered by GSH. Recently, we have demonstrated that exogenous (10 mM) GSH and GSSG can also inhibit *APX1* and *APX2* gene expression in *Arabidopsis* leaves.

The network of signalling pathways that regulate expression of genes encoding the enzymatic ROI-scavenging system in plant cells is complex. One gene can be regulated by more than one signalling pathway. Interactions between different signalling pathways are not understood. However, the redox mechanisms of the oxidative stress response plays a primary role in regulation of the expression of these genes.

Photosynthesis in Low Temperature and ROIs

A physiological process, such as photosynthesis is well known for generating ROIs, but it can also be involved in the removal of and protection against ROIs. Several excellent reviews exist on photosynthesis, covering the role of oxygen in photoinhibition (Krause, 1994), oxygen metabolism (Foyer and Harbinson, 1994), and chilling stress (Baker, 1994; Wise, 1995). Asada (1994) described the production of ROIs in chloroplasts, whereas Horton *et al.* (1996) described comprehensively a biophysical mechanism for the regulation of light harvesting and energy dissipation in PSII, in which the D-pH across thylakoid membranes controls the energy dissipation.

Chilling in combination with light causes EEE and, thus, can induce photooxidative stress and consequent O_2-dependent bleaching of photosynthetic pigments (Figure 1). Chlorophyll bleaching in conifers is much greater in sun-exposed than shaded habitats (Karpinski *et al.*, 1994). In Scots pine, during the winter time, the chlorophyll concentration is lower and the carotenoid levels remain equal, or slightly increase. At the end of the winter, when the quantum flux density is relatively high, the pigment levels are lowest (Linder, 1972; Karpinski *et al.*, 1994). This phenomenon coincides with a very low PSII efficiency (Strand and Öquist, 1985; Lundmark *et al.*, 1988; Karpinski *et al.*, 1994; Ottander *et al.*, 1995). It has been suggested that the fall and winter reorganization of the photosynthetic apparatus allow Scots pine to maintain a large fraction of chlorophyll in a quenched, photo-protected state (Öquist *et al.*, 1992), by which photosynthesis is rapidly recovered in the spring (Lundmark *et al.*, 1988; Karpinski *et al.*, 1993, 1994; Ottander *et al.*, 1995).

The reaction centre in PSII, including the D1 protein, is generally described as the most sensitive part of the photosynthetic apparatus when plants are subjected to high light and low-temperature stress. The so-called "acceptor side photoinhibition", which is associated with proteolytic degradation of the D1 protein and induction of photooxidative stress can ultimately lead to photodamage of leaf tissue. The other type of photoinhibition, the so-called "donor side photoinhibition", is not associated with degradation of the D1 protein, does not induce photooxidative stress, and, thus, prevents photodamage (Aro *et al.*, 1993; Barber, 1995; Russell *et al.*, 1995; Van Wijk *et al.*, 1997). Photoinactivation of PSI has been reported to be more severe than that of PSII during chilling stress (Sonoike, 1996). Chilling and low temperature overproduce ROIs through leakage of absorbed energy from the electron transport chains. In the chloroplasts, light reactions will continue while the energy-consuming biochemical reactions are more limited in low temperature. Chloroplasts subjected to chilling may reduce the generation of ROIs by dissipating energy through a number of mechanisms. Leaf and chloroplast movements, increased energy dissipation, and increased use of reducing equivalents in other chloroplast processes are examples of such mechanisms. Increased energy dissipation can be achieved by decreasing the photochemical PSII activity, by increasing the photorespiratory activity, by the Mehler-peroxidase reaction (Schreiber and Neubauer, 1990), and by an increased conversion of absorbed light into heat (Horton *et al.*, 1996). The thermal dissipation process occurs within the antenna and the [(violaxanthin (V) + antheraxanthin (A) + zeaxanthin (Z); VAZ] cycle is suggested to play a major role (Demmig-Adams and Adams III, 1994; Horton *et al.*, 1996). Between October and January, the VAZ cycle pigments in Scots pine changed their epoxidation state from 0.9 to 0.1 and the D1 proteins content decreased. An increased use of anabolic pathways in the stroma also limits ROI production (Wise, 1995).

In relation to adaptation of photosynthesis to low temperature, two different strategies seem to have evolved in over-wintering plants. One is to maintain photosynthetic capacity throughout the winter by different adjustments in the photosynthetic apparatus, and the other is to photosynthesize during warm periods and downregulate photosynthesis during winter. A correlation between photosynthetic capacity at low temperature and freezing tolerance in winter cereals results from photosynthesis that provides energy for cellular metabolism (Öquist *et al.*, 1993). Cold acclimation, which induces significant freezing resistance of winter cereals and spinach, does increase resistance to photoinhibition. In contrast, cold acclimation of spring cereals which lack the ability to frost harden, does not induce increased resistance to photoinhibition (Somersalo and Krause, 1989; Öquist *et al.*, 1992; Hurry and Huner, 1992). The temperature-induced resistance to photoinhibition in winter rye is not a pure temperature effect, but reflects adjustments to high PSII pressure and modulation by light (Gray *et al.*, 1996).

Although cold acclimation does not affect the susceptibility of photosynthesis to photoinhibition in Scots pine, there is a distinct increase in resistance to photoinhibition at the level of PSII reaction centres, limiting photoinhibition despite suppression of the capacity for photosynthesis (Krivosheeva *et al.*, 1996). Figure 7 shows the photochemical quenching (q_p) as a function of photoinhibition of photosynthesis in non-acclimated and cold-acclimated Scots pine needles. The experiment was designed to generate photoinhibition in Scots pine needles of cold-acclimated and non-acclimated seedlings under the same PSII excitation pressures, for example at similar values of q_p. After cold acclimation, roughly 50% of light was needed to reach a certain value of q_p. Clearly, under

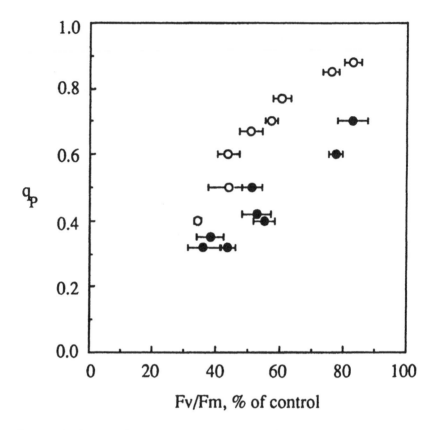

Figure 7. Photochemical quenching (q_p), as the function of photoinhibition of photosynthesis in non-acclimated (open symbols) and cold-acclimated (closed symbols) Scots pine needles. Measurements were obtained after 8 hr exposure to various photon flux densities at 20°C. Each value represent mean ±SE ($n = 6$; after Krivosheeva *et al.*, 1996).

the similar excitation pressures of PSII as defined by q_p, needles of cold-acclimated Scots pine were much more resistant to photoinhibition than needles of non-hardened pine. Unlike the winter varieties of rye and wheat, which respond to cold acclimation by increased capacities for photosynthesis, seedlings of Scots pine inhibited photosynthesis by 25% as response to cold acclimation over the studied range of absorbed photon flux density, together with increased activities and levels of several enzymes and metabolites of the enzymatic ROI-scavenging system (Krivosheeva *et al.*, 1996).

Two major oxygen-consuming reactions are associated with photosynthesis, the oxygenase reaction leading to photorespiration and the donation of electrons to oxygen to form superoxide in a pseudocyclic electron flow. Several studies have emphasized the significance of electron transport to the O_2 (Osmond and Grace, 1995). Biehler and Fock (1996) concluded that the Mehler-peroxidase reaction increased in wheat during drought stress when the availability of CO_2 was limited. The reaction of AsA with H_2O_2 is efficient (Asada, 1994) and there is accumulating evidence that the Mehler-peroxidase reaction serves as an important sink for excess electrons (Krivosheeva *et al.*, 1996; Willekens *et al.*, 1998), although the role is not well understood. Krivosheeva *et al.* (1996) hypothesized that the H_2O_2-scavenging system plays two roles in the protection of cold-acclimated needles from photoinhibition: defense against ROIs formed upon EEE in general and permission for O_2 to function as an electron acceptor, thus opening a fraction of PSII reaction centres and consuming electrons in excess of the requirements of CO_2 fixation.

Protection *via* the photorespiratory pathway is theoretically possible, however, to our knowledge, no data can be found in the literature to support the hypothesis that increased photorespiration during chilling or low-temperature stress would protect the photosynthetic machinery in plant leaves. The argument is that low and chilling temperatures generally slow down the enzymatic reactions more than the photosynthetic electron transport rate, and hence photorespiration would not serve as an effective protective mechanism. Krause (1994) has comprehensively described the role of oxygen in photoinhibition of photosynthesis and the possible protective role of photorespiration during stress. This protective role cannot be explained simply in quantitative terms by energy dissipation. The limited rate of coupled electron flow that is facilitated by photorespiration protects the photosynthetic machinery in two ways (Wu *et al.*, 1991): by maintaining the primary electron acceptor, plastoquinone, in a partly oxidized state and by building up a high proton gradient over the thylakoid membrane. This highly-energized state of the thylakoid membrane may be dependent on cyclic electron flow in the proximity of PSI (Wu *et al.*, 1991). Acidification of the thylakoid interior is supposed to be a control mechanism in PSII and will lead to an increased dissipation of excitation energy via chlorophyll-fluorescence. This energy-dependent quenching mechanism is known to protect against photoinhibition (Krause and Weis, 1991; Horton *et al.*, 1996).

The relative roles of avoidance mechanisms and antioxidant systems remain to be established for plants subjected to unfavourable temperatures and light. The above data indicate that plants can adjust the defense systems against low-temperature-induced oxidative stress, which depends on a number of factors that should be considered in further studies on improvement of the oxidative stress tolerance of plants.

GENETIC MANIPULATION OF LOW-TEMPERATURE STRESS TOLERANCE

Low-temperature resistance of cyanobacteria and chilling-sensitive higher plants can be significantly enhanced by elevated levels of polyunsaturated fatty acids (Wada *et al.*, 1990, 1994) or by reducing the saturated fatty acid content in most membrane lipids (Nishida and Murata, 1996; Ishizaki-Nishizawa *et al.*, 1996). In essence, unsaturation of membrane lipids is correlated with chilling sensitivity of plants and cyanobacteria. Recently, the genes coding for key enzymes in the biosynthesis (Slabas and Fawcett, 1992) and desaturation (Murata and Wada, 1995) of fatty acids were determined. Also, mutants with defective *FAD* have been isolated (Miguel *et al.*, 1993) and the results indicate that *fad2* mutants of *Arabidopsis* require polyunsaturated lipids for normal growth and survival at low temperatures.

A genetic alteration of chilling sensitivity by a change in saturation of the phospholipid, phospatidylglycerol (PG) a minor constituent of membrane lipids that are critical to chilling tolerance, was first described by Murata *et al.* (1992). The level of PG fatty acid unsaturation was altered by transforming tobacco plants with the glycerol-3-phosphate acyltransferase (GPAT) from different plants. Overexpression of a cDNA encoding GPAT in tobacco plants demonstrated that the chilling sensitivity can be engineered through this enzyme. Furthermore, experiments performed with the fatty acid desaturase A (*desA*) gene of *Synechocystis* sp PCC6803 and transformed to defective cells of *Synechococcus* sp PCC7942 showed enhanced low-temperature tolerance (Wada *et al.*, 1990).

Transgenic plants that overproduce AFPs from flounder have not been convincing with respect to improving plant cold hardiness, and Antikainen (1996) suggested that transformation of genes encoding plant AFPs may prove more promising.

Generally, all reports, except that by Tepperman and Dunsmuir (1990), show that transgenic plants with moderately enhanced activities of MnSOD or Cu/ZnSOD localized in the organelles have a higher tolerance to oxidative stress. Transgenic tobacco plants with increased expression of MnSOD, localized in chloroplasts and mitochondria, have reduced cellular damage after paraquat treatment (Bowler *et al.*, 1991). The same plants were more resistant against ozone damage (Van Camp *et al.*, 1994), whereas transgenic potato overproducing Cu/ZnSOD were also more tolerant to paraquat (Perl *et al.*, 1993). Gupta *et al.*, (1993a; 1993b), reported a rapid recovery of photosynthesis after photooxidative stress in combination with low temperatures, in transgenic plants overproducing Cu/ZnSOD in chloroplasts compared to wild types. McKersie *et al.* (1993) reported that overproduction of MnSOD in transgenic *Medicago sativa* enhanced tolerance to freezing stress. The above results suggest that SOD (and/or the SOD product, H_2O_2) might play a role in the mechanism of tolerance to a variety of environmental stresses, including low-temperature stress. However, in our opinion, these results are not fully convincing and needs more confirmation. Our criticisms of all of these studies are the following. Firstly, the genetic significance and the segregation of enhanced tolerance to oxidative stress in progenies that overproduce SOD were not analyzed. Secondly, many SOD transgenic lines were generated but only a few lines demonstrated higher tolerance to oxidative stress. Thirdly, to our knowledge, no report described that SOD transgenic plants demonstrated higher tolerance to the naturally established EEE and consequent photooxidative stress in the field.

In general, it is difficult to understand how a single gene that encodes SOD can induce tolerance in transgenic plants to a wide diversity of stresses, such as freezing, chilling, O_3, photoinhibition, or paraquat, because many other genes and mechanisms are involved in different stress responses and tolerances. One possible explanation could be that all the above stresses are associated with higher levels of $O_2^{\bullet-}$. Since SOD scavenges $O_2^{\bullet-}$ and produces H_2O_2 (both signalling molecules), its activity, besides having a protective function, is of great importance in regulatory mechanisms that control the responses to a wide diversity of biotic and abiotic stresses, including low-temperature stress. This interpretation is in agreement with the signalling role of H_2O_2 in low-temperature-induced oxidative stress (Prasad *et al.*, 1994a, 1994b), in SAR (Alvarez *et al.*, 1998), and in SAA (Karpinski *et al.*, 1999).

It was thought that transgenic plants overproducing different SOD isoforms gave a true evaluation of the effects of increasing the SOD activity alone (Bowler *et al.*, 1992). However, in transgenic plants overproducing Cu/ZnSOD in the chloroplast the cytosolic APX was more active (Gupta *et al.*, 1993b) and the SOD/APX activity ratio in SOD transformants was similar to that in control plants. The levels of other enzymes, such as GR and DHAR, were also similar in transgenic and control plants. The mechanisms by which cytosolic APX mRNA and APX activity had increased in SOD transformants are not known, although it was speculated that increased cytosolic APX activity could be a result of higher H_2O_2 levels in SOD transformants (Gupta *et al.*, 1993b). This speculation has been recently confirmed by discovery of the SAA mechanism (Karpinski *et al.*, 1999) and mechanism of induction of *APX1* and *APX2* genes that are regulated by H_2O_2 and redox status of the plastoquinone pool (Figure 5; Karpinski *et al.*, 1997, 1999). Higher H_2O_2 levels in the chloroplast could cause an increase in transthylakoid ΔpH, therefore could modulate the redox status of the plastoquinone pool (Horton *et al.*, 1996; Karpinski *et al.*, 1999) and trigger the induction of *APX1* and *APX2* genes.

Transgenic tobacco plants with higher levels of GR activity in chloroplasts and mitochondria did not show improved tolerance to low-temperature stress (Broadbent *et al.*, 1995). Foyer *et al.* (1995) demonstrated that transgenic poplar trees (primary transformants) that overproduced a bacterial GR in the chloroplasts slightly improved their tolerance to chilling and photooxidative stresses; they suggested that the improvement of chilling and photooxidative stress tolerance in such transgenic trees was due to moderately higher GSH and AsA levels. However, the genetic significance of this transformation was also not analyzed.

Improvement of ROI metabolism may play an important role in ameliorating the effects of chilling and freezing stress, but, in our opinion, engineering of one component of the enzymatic ROI-scavenging system alone cannot give satisfactory results. Theoretically, better results can be obtained for the amelioration of chilling and freezing stress tolerance in the future by engineering the saturation of phospholipids, the light-harvesting complex antenna size for PSII, together with an enhanced scavenging capacity of ROIs, including $^1O_2$ and fatty acid radicals. Moreover, we are still far from understanding the nature of the redox signalling and the regulatory mechanisms of oxidative stress response as well as the roles of the ROIs and ROI-scavenging systems in the low-temperature stress. Oxidative stress avoidance mechanism in higher plants also play a major role in acclimation to EEE (B. Karpinska, V. Klimyuk and S. Karpinski, unpublished results).

CONCLUDING REMARKS

In different plant species, different strategies have evolved to acclimate to low-temperature stress. An understanding of these strategies may pave the way to create improved stress tolerance in some plants. The signal transduction pathways that link day length or temperature perception to increased or decreased contents of phytohormones, such as ABA, and subsequent gene activation or deactivation mechanisms, remain to be elucidated.

At present, it is still difficult to understand how H_2O_2 can induce such diverse processes as the hypersensitive response, SAR, SAA, and cold acclimation in different plant species, and how glutathione can generate an opposite regulatory effect on expression of genes encoding the enzymatic ROI-scavenging system. To elucidate how the oxidative-stress responses are induced, for example, by low temperatures in plants, we have to pinpoint the subcellular compartments and processes, which initiate the specific signalling cascades. Interactions between signals mediated by Ca^{2+}, ROIs, NO, ethylene, glutathione, plastoquinone, phytohormones, and phosphorylation of proteins during cold acclimation remain also to be elucidated. In addition, we have to identify and isolate regulatory genes and the redox-activated transcription factors involved in these signalling cascades. Recently, isolation of such regulatory genes controlling biotic stress responses (Cao *et al.*, 1997; Dietrich *et al.*, 1997) provided a great potential to engineer SAR in plants. Similarly, regulatory genes for SAA should be detected and isolated (Karpinski *et al.*, 1999).

REFERENCES

Alberdi, M. and Corcuera, L.J. (1991) Cold acclimation in plants. *Phytochemistry*, **30**, 3177–3184.

Alia, P.P.S. and Mohanty, P. (1996) Involvement of proline in protecting thylakoid membranes against free radical-induced photodamage. *J. Phytochem. Phytobiol.*, **38**, 253–257.

Alvarez, M.E., Pennell, R.I., Meijer, P-J., Ishikawa, A., Dixon, R.A., and Lamb, C. (1998) Reactive oxygen intermediates mediate a systemic signal network in the establishment of plant immunity. *Cell*, **92**, 773–784.

Anderson, J.V., Chevone, B.I., and Hess, J.L. (1992) Seasonal variation in the antioxidant system of Eastern white pine needles. Evidence for thermal dependence. *Plant Physiol.*, **98**, 501–508.

Anderson, M.D., Prasad, T.K., and Stewart, C.R. (1995) Changes in isozyme profiles of catalase, peroxidase, and glutathione reductase during acclimation to chilling in mesocotyls of maize seedlings. *Plant Physiol.*, **109**, 1247–1257.

Antikainen, M. (1996) Cold acclimation in winter rye (*Secale cereale* L.): Identification and characterization of proteins involved in freezing tolerance. Ph.D. Thesis, *Annales Universitatis Turkuensis*, pp. 1–46.

Aro, E-M., Virgin, I., and Andersson, B. (1993) Photoinhibition of photosystem II. Inactivation, protein damage and turnover. *Biochim. Biophys. Acta*, **1143**, 113–134.

Arrigoni, O. (1994) Ascorbate system in plant development. *J. Bioenerg. Biomembr.*, **26**, 407–419.

Asada, K. (1994) Production and action of active oxygen species in photosynthetic tissues. In C.H. Foyer and P.M. Mullineaux, (eds.), *Causes of Photooxidative Stress and Amelioration of Defense Systems in Plants*, CRC Press, Boca Raton, pp. 77–103.

Asada, K. and Takahashi, M. (1987) Production and scavenging of active oxygen in photosynthesis. In D.J. Kyle C.B. Osmond, and C.J. Arntzen, (eds.), *Photoinhibition*, Elsevier, Amsterdam, pp. 227–287.

Asada, K., Kiso, K., and Yoshikawa, K. (1974) Univalent reduction of molecular oxygen by spinach chloroplasts on illumination. *J. Biol. Chem.*, **249**, 2175–2179.

Ashworth, E.N. (1993) Deep supercooling in woody plant tissues. In P.H. Li and L. Christersson, (eds.), *Advances in Plant Cold Hardiness*, CRC Press, Boca Raton, pp. 203–213.

Baker, N. (1994) Chilling stress and photosynthesis. In C.H. Foyer and P.M. Mullineaux, (eds.), *Causes of Photooxidative Stress and Amelioration of Defence Systems in Plants*, CRC Press, Boca Raton, pp. 127–154.

Barber, J. (1995) Molecular basis of the vulnerability of photosystem II to damage by light. *Aust. J. Plant Physiol.*, **22**, 201–208.

Bartels, D. and Nelson, D. (1994) Approaches to improve stress tolerance using molecular genetics. *Plant Cell Environ.*, **17**, 659–667.

Benson, E.E., Lynch, P.T., and Jones J. (1992) Detection of lipid peroxidation products in cryo-protected and frozen rice cells: consequences for post-thaw survival. *Plant Sci.*, **85**, 107–114.

Bhaumik, G., Srivastava, K.K., and Selvamurthy, W. (1995) The role of free-radicals in cold injuries. *Int. J. Biomet.*, **38**, 171–175.

Bi, Y-M., Kenton, P., Mur, L., Darby, R., and Draper, J. (1995) Hydrogen peroxide does not function downstream of salicylic acid in the induction of PR protein expression. *Plant J.*, **8**, 235–245.

Biehler, K. and Fock, H. (1996) Evidence for the contribution of the Mehler-peroxidase reaction in dissipating excess electrons in drought-stressed wheat. *Plant Physiol.*, **112**, 265–272.

Bohnert, H.J., Nelson, D.E., and Jensen, R.G. (1995) Adaptations to environmental stresses. *Plant Cell*, **7**, 1099–1111.

Bohnert, H.J. and Jensen, R.G. (1996) Strategies for engineering water stress tolerance in plants. *Trends Biotechnol.*, **14**, 89–97.

Bowler, C., Slooten, L., Vandenbranden, S., Derycke, R., Botterman, J., Sybesma, C., Van Montagu, M., and Inzé, D. (1991) Manganese superoxide dismutase can reduce cellular damage mediated by oxygen radicals in transgenic plants. *EMBO J.*, **10**, 1723–1732.

Bowler, C., Van Montagu, M., and Inzé, D. (1992) Superoxide dismutase and stress tolerance. *Annu. Rev. Plant Physiol. Plant Mol. Biol.*, **43**, 83–116.

Bowler, C., Van Camp, W., Van Montagu, M., and Inzé, D. (1994) Superoxide dismutase in plants. *Crit. Rev. Plant Sci.*, **13**, 199–218.

Bridger, G.M., Yang, W., Falk, D.E., and Mckersie, B.D. (1994) Cold acclimation in-creases tolerance of activated oxygen in winter cereals. *J. Plant Physiol.*, **144**, 235–240.

Broadbent, P., Creissen, G.P., Kular, B., Wellburn, A.R., and Mullineaux, P.M. (1995) Oxidative stress responses in transgenic tobacco containing altered levels of glutathione reductase activity. *Plant J.*, **8**, 247–255.

Browse, J. and Somerville, C.R. (1994) Glycerolipids. In E.M Meyerowitz and C.R. Somerville, (eds.), *Arabidopsis* (Monograph Series, Vol. 27), Cold Spring Harbor Laboratory Press, Cold Spring Harbor, pp. 881–912.

Byrd, G.T., Ort, D.R., and Ogren, W.L. (1995) The effects of chilling in the light on ribulose-1,5-bisphosphate carboxylase/oxygenase activation in tomato (*Lycopersicon esculentum* Mill.). *Plant Physiol.*, **107**, 585–591.

Cadenas, E. (1989) Biochemistry of oxygen toxicity. *Annu. Rev. Biochem.*, **58**, 79–110.

Cao, H., Glazebrook, J., Clarke, J.D., Volko, S., and Dong, X.N. (1997) The *Arabidopsis NPR1* gene that controls systemic acquired resistance encodes a novel protein containing ankyrin repeats. *Cell*, **88**, 57–63.

Casano, L.M., Lascano, H.E., and Trippi, V.S. (1994) Hydroxyl radicals and a thylakoid-bound endopeptidase are involved in light- and oxygen-induced proteolysis in oat chloroplasts. *Plant Cell Physiol.* 35, 145–152.

Chen, H.H. and Gusta L.V. (1983) Abscisic acid-induced freezing resistance in cultured plant cells. *Plant Physiol.*, **73**, 71–75.

Chen, H.H., Li, P.H., and Brenner, M.L. (1983) Involvement of abscisic acid in potato cold acclimation. *Plant Physiol.*, **71**, 362–365.

Chen, Z., Silva, H., and Klessig, D.F. (1993) Active oxygen species in the induction of plant systematic acquired resistance by salicylic acid. *Science*, **262**, 1883–1885.

Close, T.J. (1996) Dehydrins: Emergence of a biolochemical role of a family of plant dehydration proteins. *Physiol. Plant.*, **92**, 969–717.

Cloutier, Y. (1983) Changes in the electrophoretic patterns of the soluble proteins of winter wheat and rye following cold acclimation and desiccation stress. *Plant Physiol.*, **69**, 256–258.

Cordoba, F. and Gonzalez-Reyes, J.A. (1994) Ascorbate and plant-cell growth. *J. Bioenerg. Biomembr.*, **26**, 399–405.

Crowe, J.H., Hoekstra, F.A., and Crowe, L.M. (1992) Anhydrobiosis. *Annu. Rev. Physiol.*, **54**, 579–599.

Daie, J. and Campbell, W.F. (1981) Response of tomato plants to stressful temperatures. *Plant Physiol.*, **135**, 351–354.

Dallaire, S., Houde, M., Gagne, Y., Saini, H.S., Boileau, S., Chevrier, N. and Sarhan, F. (1994) ABA and low temperature induce freezing tolerance *via* distinct regulatory pathways in wheat. *Plant Cell Physiol.*, **35**, 1–9.

Dalton, D.A. (1995) Antioxidant defences of plants and fungi. In S. Ahman, (ed.), *Oxidant-induced Stress and Antioxidant Defences in Biology*, Chapman and Hall, New York, pp. 298–355.

Del Corso, A., Cappiello, M., and Mura, U. (1994) Thiol-dependent oxidation of enzymes: The last chance against oxidative stress. *Int. J. Biochem.*, **26**, 745–750.

Delledonne, M., Xia, Y., Dixon, R.A., and Lamb, C. (1998) Nitric oxide functions as a signal in plant disease resistance. *Nature*, **394**, 585–587.

Del Río, L.A., Sandalio, L.M., Palma, J.M., Bueno, P., and Corpas, F.J. (1992) Metabolism of oxygen radicals in peroxisomes and cellular implications. *Free Rad. Biol. Med.*, **13**, 557–580.

Demmig-Adams, B. and Adams III, W.W. (1994) Light stress and photoprotection related to the xanthopyll cycle. In C.H. Foyer and P.M. Mullineaux, (eds.), *Causes of Photooxidative Stress and Amelioration of Defense Systems in Plants*, CRC Press, Boca Raton, pp. 105–126.

Deng, H.-X., Hentati, A., Tainer, J.A., Iqbal, Z., Cayabyab, A., Hung, W.-Y., Getzoff, E.D., Hu, P., Herzfeldt, B., Roos, R.P., Warner, C., Deng, G., Soriano, E., Smyth, C., Parge, H.E., Ahmed, A., Roses, A.D., Hallewell, R.A., Pericak-Vance, M.A., and Siddique, T. (1993) Amyotrophic lateral sclerosis and structural defects in Cu,Zn superoxide dismutase. *Science*, **261**, 1047–1051.

De Vrije, T., De Swart, R.L., Dowhan, W. Tommassen, J. and Dekruijff, B. (1988) Phosphatidylglycerol is involved in protein translocation across *Escherichia coli* inner membranes. *Nature*, **334**, 173–75.

Dietrich, R.A., Richberg, M.H. Schmidt, R. Dean, C., and Dangl, J.L. (1997) A novel zinc finger protein is encoded by the *Arabidopsis* LSD1 gene and functions as a negative regulator of plant cell death. *Cell*, **88**, 685–694.

Ding, J.P. and Pickard, B.G. (1993) Modulation of mechanosensitive calcium-selective channels by temperature. *Plant J.*, **3**, 713–720.

Edwards, E.A., Enard, C., Creissen, G.P., and Mullineaux, P.M. (1994) Synthesis and properties of glutathione reductase in stressed peas. *Planta*, **192**, 137–143.

Esterbauer, H. and Grill, D. (1978) Seasonal variation of glutathione and glutathione reductase in needles of *Picea abies*. *Plant Physiol.*, **61**, 119–121.

Feierabend, J., Schaan, C., and Hertwig, B. (1992) Photoinactivation of catalase occurs under both high- and low-temperature stress conditions and accompanies photoinhibition of photosystem II. *Plant Physiol.*, **100**, 1554–1561.

Foyer, C.H. (1993) Ascorbic acid. In R.G. Alscher and J.L. Hess, (eds.), *Antioxidants in Higher Plants*, CRC press, Boca Raton, pp. 31–58.

Foyer, C.H. (1997) Oxygen metabolism and electron transport in photosynthesis. In J.G. Scandalios, (ed.), *Oxidative Stress and the Molecular Biology of Antioxidant Defenses*, (Monograph Series, Vol. 34), Cold Spring Harbor Laboratory Press, Cold Spring Harbor, pp. 587–622.

Foyer, C.H. and Halliwell, B. (1976) The presence of glutathione and glutathione reductase in chloroplasts: A proposed role in ascorbic acid metabolism. *Planta*, **133**, 21–25.

Foyer, C.H. and Harbinson, J. (1994) Oxygen metabolism and the regulation of photosynthetic electron transport. In C.H. Foyer and P.M. Mullineaux, (eds.), *Causes of Photooxidative Stress and Amelioration of Defense Systems in Plants*, CRC Press, Boca Raton, pp. 1–42.

Foyer, C.H., Souriau, N., Perret, S., Lelandais, M., Kunert, K.J., Pruvost, C., and Jouanin, L. (1995) Overexpression of glutathione reductase but not glutathione synthetase leads to increases in antioxidants capacity and resistance to photoinhibition in poplar trees. *Plant Physiol.*, **109**, 1047–1057.

Furbank, R.T. and Badger, M.R. (1983) Oxygen exchange associated with electron transport and photophosphorylation in spinach chloroplasts. *Biochim. Biophys. Acta*, **723**, 400–405.

Gilbert, H.F. (1990) Molecular and cellular aspects of thiol-disulphide exchange. *Adv. Enzymol.*, **63**, 69–172.

Gilmour, S.J., Zarka, D.G., Stockinger, E.J., Salazar, M.P., Houghton, J.M., and Thomashow, M.F. (1998) Low temperature regulation of the Arabidopsis CBF family of AP2 transcriptional activators as an early step in cold-induced COR gene expression. *Plant J.*, **16**, 433–442.

Ginnpease, M.E. and Whisler, R.L. (1996) Optimal NF-kB mediated transcriptional responses in Jurkat T-cells exposed to oxidative stress are dependent on intracellular glutathione and co-stimulatory signals. *Biochem. Biophys. Res. Commun.*, **226**, 695–702.

Giraudat, J. (1995) Abscisic acid signalling. *Curr. Opin. Cell Biol.*, **7**, 232–238.

Giraudat, J., Parcy, F., Bertauche, N., Gosti, F., Leung, J., Morris, P.C., Bouvierdurand, M., and Vartanian, N. (1994) Current advances in abscisic acid action and signalling. *Plant Mol. Biol.*, **26**, 1557–1577.

Grace, S.C., Logan, B.A., and Adams, W.W. III. (1998) Seasonal differences in foliar content of chlorogenic acid, a phenylpropanoid antioxidant, in *Mahonia repens. Plant Cell Environ.*, **21**, 513–521.

Gray, G.R., Savitch, L.V., Ivanov, A.G., and Huner, N.P.A. (1996) Photosystem II excitation pressure and development of resistance to photoinhibition. II. Adjustment to photosynthetic capacity in winter wheat and winter rye. *Plant Physiol.*, **110**, 61–71.

Griffith, M., Ala, P., Yang, D.S.C., Hon, W.C., and Moffatt, B.A. (1992) Antifreeze protein produced endogenously in winter rye leaves. *Plant Physiol.*, **100**, 593–596.

Gupta, A.S., Heinen, J.L., Holaday, S.A., Burke, J.J., and Allen, R.D. (1993a) Increased resistance to oxidative stress in transgenic plants that overexpress chloroplastic Cu/Zn superoxide dismutase. *Proc. Natl. Acad. Sci. USA*, **90**, 1629–1633.

Gupta, A.S., Webb, R.P., Holaday, A.S., and Allen R.D. (1993b) Overexpression of superoxide dismutase protects plants from oxidative stress. Induction of ascorbate peroxidase in superoxide dismutase-overexpressing plants. *Plant Physiol.*, **103**, 1067–1073.

Guy, C.L. (1990) Cold acclimation and freezing tolerance: role of protein metabolism. *Annu. Rev. Plant Physiol. Plant Mol. Biol.*, **41**, 187–223.

Guy, C.L. and Carter, J.V. (1984) Characterization of partially purifiedglutathione reductase from cold-hardened spinach leaf tissue. *Cryobiology*, **21**, 454–464.

Halliwell, B. and Gutteridge, J.M.C. (1989) *Free Radicals in Biology and Medicine*. Oxford University Press, Oxford.

Halliwell, B. and Gutteridge, J.M.C. (1993) Biologically relevant metal ion-dependent hydroxyl radical generation. *FEBS Lett.* **307**, 108–112.

Hällgren, J-E. and Öquist, G. (1990) Adaptations to low temperatures. In R.G. Alscher and G.R. Cummings, (eds.), *Stress Responses in Plants: Adaptation and Acclimation Mechanisms*, Wiley-Liss, New York, pp. 265–293.

Hällgren, J-E., Strand, M. and Lundmark, T. (1991) Temperature stress. In A.S. Raghavendra, (ed.), *Physiology of Trees*, John Wiley and Sons, New York, pp. 301–334.

Hariyadi, P. and Parkin, K.L. (1993) Chilling-induced oxidative stress in cucumber (*Cucumis sativus* L. cv. Calypso) seedlings. *J. Plant Physiol.*, **141**, 733–738.

Harris, W.N. and Dalton, D.A. (1997) Seasonal variation in glutathione reductase activity in coastal and montane populations of lodgepole pine (*Pinus contorta*). *Northwest Science*, **71**, 205–213.

Hausladen, A. and Alscher, R.G. (1994a) Purification and characterization of glutathione reductase isozymes specific for the state of cold hardiness of red spruce. *Plant Physiol.*, **105**, 205–214.

Hausladen, A. and Alscher, R.G. (1994b) Cold-hardiness-specific glutathione reductase isozymes in red spruce. *Plant Physiol.*, **105**, 215–223

Hausladen, A., Madamanchi, N.R., Fellows, S., Alscher, R.G. and Amundson, R.G. (1990) Seasonal changes in antioxidants in red spruce as affected by ozone. *New Phytol.*, **115**, 447–458.

Heino, P., Sandman, G., Lång, V., Nordin, K. and Palva, E.T. (1990) Abscisic acid deficiency prevents development of freezing tolerance in *Arabidopsis thaliana* (L.) Heynh. *Theor. Appl. Genet.*, **79**, 801–806.

Hérouart, D., Van Montagu, M., and Inzé, D. (1993) Redox-activated expression of the cytosolic copper/zinc superoxide dismutase gene in *Nicotiana*. *Proc. Natl. Acad. Sci. USA*, **90**, 3108–3112.

Hess, J.L. (1993) Vitamin E, α-tocopherol. In R.G. Alscher and J.L. Hess, (eds.), *Antioxidants in Higher Plants*, CRC Press, Boca Raton, pp. 11–134.

Hincha, D.K., Heber, U., and Schmitt, J.M. (1990) Proteins from frost-hardy leaves protect thylakoids against mechanical freeze-thaw damage *in vitro*. *Planta*, **180**, 416–419.

Hodgson, R.A. and Raison, J.K. (1991) Superoxide production by thylakoids during chilling and its implication in the susceptibility of plants to chilling-induced photoinhibition. *Planta*, **183**, 222–228.

Holmstrom, K-O., Mäntylä, E., Welin, B., Mandal, A., and Palva, E.T. (1996) Drought tolerance in tobacco. *Nature*, **379**, 683–684.

Hon, W.C., Griffith, M., Mlynarz, A., Kwok, Y.C., and Yang, D.S.C. (1995) Antifreeze proteins in winter rye are similar to pathogenesis-related proteins. *Plant Physiol.*, **104**, 971–980.

Horton, P., Ruban, A.V., and Walters, R.G. (1996) Regulation of light harvesting in green plants. *Annu. Rev. Plant Physiol. Plant Mol. Biol.*, **47**, 656–684.

Huerta, A.J. and Murphy T.M. (1989) Control of intracellular glutathione and its effect on ultraviolet radiation-induced K^+ efflux in cultured rose cells. *Plant Cell Environ.*, **12**, 825–830.

Hughes, M.A. and Dunn, A.M. (1996) The molecular biology of plant acclimation to low temperature. *J. Exp. Bot.*, **47**, 291–305.

Hughes, M.A., Dunn, M.A., Pearce, R.S., White, A.J. and Zhang, L. (1992) An abscisic-acid-responsive, low-temperature barley gene has homology with a maize phospholip transfer protein. *Plant Cell Environ.*, **15**, 861–865.

Huner, N.P.A., Öquist, G., and Sarhan, F. (1998) Energy balance and acclimation to light and cold. *Trends Plant Sci.*, **3**, 224–230.

Hurry, V.M. and Huner, N.P.A. (1992) Effects on cold hardening on sensitivity of winter and spring wheat leaves to short term photoinhibition and recovery of photosynthesis. *Plant Physiol.*, **100**, 1283–1290.

Hurry, V.M., Strand, A., Tobiaeson, M., Gardestrom, P., and Öquist, G. (1995) Cold hardening of spring and winter-wheat and rape results in differential-effects on growth, carbon metabolism, and carbohydrate content. *Plant Physiol.*, **109**, 697–706.

Ingram, J. and Bartels, D. (1996) The molecular basis of dehydration tolerance in plants. *Annu. Rev. Plant Physiol. Plant Mol. Biol.*, **47**, 377–403.

Inzé, D. and Van Montagu, M. (1995) Oxidative stress in plants. *Curr. Opin. Biotechnol.*, **6**, 153–158.

Ishizaki-Nishizawa, O., Fujii, T., Azuma, M., Sekiguchi, K., Murata, N., Ohtani, T., and Toguri, T. (1996) Low-temperature resistance of higher plants is significantly enhanced by a non-specific cyanobacterial desaturase. *Nature Biotechnol.*, **14**, 1003–1006.

Jabs, T., Dietrich, R.A., and Dangl, J.L. (1996) Initiation of runaway cell death in an Arabidopsis mutant by extracellular superoxide. *Science*, **273**, 1853–1856.

Jaglo-Ottosen, K.R., Gilmour, S.J., Zarka, D.G., Schabenberger, O., and Thomashow, M.F. (1998) Low temperature regulation of the Arabidopsis CBF family of AP2 transcriptional activators as an early step in cold-induced COR gene expression. *Science*, **280**, 104–106.

Jahnke, L.S., Hull, M.R., and Long, S.P. (1991) Chilling stress and oxygen metabolising enzymes in *Zea mays* and *Zea diploperennis*. *Plant Cell Environ.*, **14**, 97–104.

Junttila, O. (1990) Gibberellins and regulation of shoot elongation in woody plants. In N. Takahashi, B.O. Phinney, and J. MacMillan, (eds.), *Gibberellins*, Springer-Verlag, Berlin, pp. 199–210.

Kacperska, A. (1989) Metabolic consequences of low temperature stress in chilling-insensitive plants. In P.H. Li, (ed.), *Low Temperature Stress Physiology in Crops*, CRC Press, Boca Raton, pp. 27–40.

Kacperska, A. (1993) Water potential alterations-a prerequisite or a triggering stimulus for the development of freezing tolerance in overwintering herbaceous plants. In P.L. Steponkus, (ed.), *Advances in Low Temperature Biology, Vol. 2*, JAI Press, London, pp. 73–92.

Kaiser, W.M. (1979) Reversible inhibition of the Calvin cycle and activation of oxidative pentose phosphate cycle in isolated intact chloroplasts by hydrogen peroxide. *Planta,* **145**, 377–382.

Kane, D.J. Srafian, T.A., Hahn, R.-A.H., Gralla, E.B., Valentine, J.S., Örd, T., and Bredesen, D.E. (1993) Bcl-2 inhibition of neural death: Decreased generation of reactive oxygen species. *Science,* **262**, 1274–1277.

Karpinski, S., Wingsle, G., Olsson, O., and Hällgren, J.-E. (1992a) Characterization of cDNAs encoding CuZn-superoxide dismutases in Scots pine. *Plant Mol. Biol.,* **18**, 545–555.

Karpinski, S., Wingsle, G., Karpinska, B., and Hällgren, J-E. (1992b) Differential expression of CuZn-superoxide dismutases in *Pinus sylvestris* (L.) needles exposed to SO_2 and NO_2. *Physiol. Plant.,* **85**, 689–696.

Karpinski, S., Wingsle, G., Karpinska, B., and Hällgren, J-E. (1993) Molecular responses to photooxidative stress in *Pinus sylvestris* (L.). II. Differential expression of CuZn-superoxide dismutases and glutathione reductase. *Plant Physiol.,* **103**, 1385–1391.

Karpinski, S., Karpinska, B., Wingsle, G., and Hällgren, J-E. (1994) Molecular responses to photooxidative stress in *Pinus sylvestris*. I. Differential expression of nuclear and plastid genes in relation to recovery from winter stress. *Physiol. Plant.,* **90**, 358–366.

Karpinski, S., Escobar, C., Karpinska, B., Creissen, G., and Mullineaux, P. (1997) Photosynthetic electron transport regulates the expression of cytosolic ascorbate peroxidase genes in *Arabidopsis* during excess light stress. *Plant Cell,* **9**, 627–642.

Karpinski, S., Reynolds, H., Karpinska, B., Wingsle, G., Creissen, G., and Mullineaux, P. (1999) Systemic signalling and acclimation in response to excess excitation energy in *Arabidopsis*. *Science,* **284**, 654–657.

Kendall, E.J. and McKersie, B.D. (1989) Free radical and freezing injury to cell membranes in winter wheat (superoxide, phospholipide-esterification). *Physiol. Plant.,* **76**, 86–94.

Kernodle, S.P. and Scandalios, J.G. (1996) A comparison of the structure and function of the highly homologous maize antioxidant Cu/Zn superoxide dismutase genes, *Sod4* and *Sod4A*. *Genetics,* **143**, 317–328.

Kleczkowski, L.A. (1996) Back to the drawing board—redefining starch synthesis in cereals. *Trends Plant Sci.,* **11**, 363–364.

Knight, M.R., Campbell, A.K., Smith, S.M., and Trewavas, A.J. (1991) Transgenic plant aquorin reports the effects of touch and cold-shock and elicitors on cytoplasmatic calcium. *Nature,* **352**, 524–526.

Knight, H., Trewavas, A.J., and Knight, M. (1996) Cold calcium signalling in *Arabidopsis* involves two cellular pools and change in calcium signature after acclimation. *Plant Cell,* **8**, 489–503.

Knox, P.J, and Dodge, A.D. (1985) Singlet oxygen and plants. *Phytochemistry,* **24**, 889–896.

Krause, G.H. and Weis, E. (1991) Chlorophyll fluorescence and photosynthesis: The basics. *Annu. Rev. Plant. Physiol. Plant Mol. Biol.,* **42**, 313–349.

Krause, H. (1994) The role of oxygen in photoinhibition of photosynthesis. In C.H. Foyer and P.M. Mullineaux, (eds.), *Causes of Photooxidative Stress and Amelioration of Defense Systems in Plants*, CRC Press, Boca Raton, pp. 43–76.

Krivosheeva, A., Tao, D.-L., Ottander, C., Öquist, G., and Wingsle, G. (1996) Cold acclimation and photoinhibition of photosynthesis in Scots pine. *Planta,* **200**, 296–305.

Kröninger, W., Rennenberg, H., and Polle, A. (1993) Developmental changes of CuZn- and Mn-superoxide dismutase isozymes in seedlings and needles of Norway spruce. *Plant Cell Physiol.,* **34**, 1145–1149.

Kurkela, S. and Frank, M. (1990) Cloning and characterization of a cold- and ABA-inducible *Arabidopsis* gene. *Plant Mol. Biol.,* **15**, 137–144.

Kuroda. H. and Sagisaka, S. (1992) Malfunction of enzyme systems involved in the regeneration of glutathione in perennials at low temperature. *Biosci. Biotech. Biochem.,* **56**, 712–715.

Kuroda, H., Sagisaka, S., and Chiba, K. (1992) Collapse of peroxide-scavenging systems in apple flower-buds associated with freezing injury. *Plant Cell Physiol.,* **33**, 743–750.

Kusano, T., Berberich, T., Harada, N., Suzuki, N., and Sugawara, K. (1995) A maize DNA-binding factor with a bZIP motif is induced by low temperature. *Mol. Gen. Genet.,* **248**, 507–517.

Kusano, T., Sugawara, K., Harada, M., and Berberich, T. (1998) Molecular cloning and partial characterization of a tobacco cDNA encoding a small bZIP protein. *Biochim. Biophys. Acta,* **1395**, 171–175.

Kusters, R., Dowhan, W., and de Kruijff, B. (1991) Negatively charged phospholipids restore prePhoE translocation across phosphatidylglycerol-depleted *Escherichia coli* inner membranes. *J. Biol. Chem.,* **266**, 8659–8662.

Lalk, I. and Dorffling, K. (1985) Hardening, abscisic acid, proline, and freezing resistance in two winter wheat varieties. *Physiol. Plant.,* **63**, 287–292.

Lamb, C. and Dixon, R.A. (1997) The oxidative burst in plant disease resistance. *Annu. Rev. Plant Phys. Plant. Mol. Biol.,* **48**, 251–275.

Larsson, R.A. (1988) The antioxidants of higher plants. *Phytochemistry*, **27**, 969–978.

Levine, A., Tenhaken, R., Dixon, R., and Lamb, C. (1994) H_2O_2 from the oxidative burst orchestrates the plant hypersensitive disease resistance response. *Cell*, **79**, 583–593.

Levitt, J. (1980) *Responses of Plants to Environmental Stress*, Academic Press, New York.

Lewin, S. (1987) Vitamin C: its molecular biology and medical potential. In P.H. Li and R. Liss-Alan, (eds.), *Plant Cold Hardiness*, Academic Press, New York, pp. 5–39.

Linder, S. (1972) Seasonal variation of pigments in needles: A study of Scots pine and Norway spruce seedlings grown under different nursery conditions. *Stud. For. Suec.*, **100**, 1–27.

Loewus, F.A. and Loewus, M.W. (1987) Biosynthesis and metabolism of ascorbic acid in plants. *Crit. Rev. Plant Sci.*, **5**, 101–119.

Lu, G., Paul, A.-L., McCarty, D.R., and Ferl, R.J. (1996) Transcription factor veracity: Is GBF3 responsible for ABA-regulated expression of *Arabidopsis* Adh? *Plant Cell*, **8**, 847–857.

Lundmark, T., Hällgren, J.-E., and Hedén, J. (1988) Recovery from winter depression of photosynthesis in pine and spruce. *Trees*, **2**, 110–114.

Luwe, M. and Heber, U. (1995) Ozone detoxification in the apoplasm and symplasm of spinach, broad bean and beech leaves at ambient and elevated concentrations of ozone in air. *Planta*, **197**, 448–455.

Lynch, D.V. and Steponkus, P.L. (1987) Plasma membrane lipid alterations associated with cold acclimation of winter rye seedlings (*Secale cereale* L. cv Puma). *Plant Physiol.*, **83**, 761–767.

Lyons, J.M. (1973) Chilling injury in plants. *Annu. Rev. Plant Physiol.*, **24**, 445–466.

Madamanchi, N.R., Hausladen, A., Alscher, R.G., Amundson, R.G., and Fellows, S. (1991) Seasonal changes in antioxidants in red spruce (*Picea rubens Sarg.*) from three field sites in the north eastern United States. *New Phytol.*, **118**, 331–338.

Mantyla, E. (1997) Molecular mechanisms of cold acclimation and drought tolerance in plants. Ph.D. Thesis, Swedish University of Agricultural Sciences, *Agraria*, **23**, pp. 1–52.

Marentes, E., Griffith, M., Mlynarz, A., and Brush, R.A. (1993) Proteins accumulation in the apoplast of winter rye leaves during cold acclimation. *Physiol. Plant.*, **87**, 499–507.

McCord, J.M. and Fridovich, I. (1969) Superoxide dismutase. An enzymatic function for erythrocuprein (*Hemocuprein*). *J. Biol. Chem.*, **244**, 6049–6055.

McKersie, B.D. (1991) The role of oxygen free radicals in mediating freezing and desiccation stress in plants. In E.J. Pell and K.L. Steffen, (eds.), *Active Oxygen, Oxidative Stress and Plant Metabolism*, American Society of Plant Physiologists, Rockville, pp. 107–118.

McKersie, B.D., Cen, Y., de Beus, M., Bowley, S.R., Bowler, C., Inzé, D., D'Halluin, K., and Botterman, J. (1993) Superoxide dismutase enhances tolerance of freezing stress in transgenic alfalfa (*Medicago sativa* L.). *Plant Physiol.*, **103**, 1155–1163.

Meyer, M., Schreck, R., and Baeuerle, P.A. (1993) H_2O_2 and antioxidants have opposite effects on activation of NF-kB and AP-1 in intact cells: AP-1 as secondary antioxidant-responsive factor. *EMBO J.*, **12**, 2005–2015.

Michalski, W.P. and Kaniuga, Z. (1981) Photosynthetic apparatus of chilling sensitive plants. X. Relationship between superoxide dismutase activity and photoperoxidation of chloroplasts lipids. *Biochem. Biophys. Acta*, **637**, 159–167.

Miguel, M., James, D., Dooner, H., and Browse, J. (1993) *Arabidopsis* requires polyunsaturated lipids for low-temperature survival. *Proc. Natl. Acad. Sci. USA*, **90**, 6208–6212.

Millar, A.J., Straume, M., Chory, J., Chua, N.H., and Kay, S.A. (1995) The regulation of circadian period by phototransduction pathways in *Arabidopsis*. *Science*, **267**, 1163–1166.

Minorsky, P.V. (1985) An heuristic hypothesis of chilling injury in plants: A role for calcium as the primary physiological transducer of injury. *Plant Cell Environ.*, **8**, 75–94.

Mishra, N.P., Mishra, R.K., and Singhal, G.S. (1993) Changes in the activities of anti-oxidant enzymes during exposure of intact wheat leaves to strong visible light at different temperatures in the presence of protein synthesis inhibitors. *Plant Physiol.*, **102**, 903–910.

Mizoguchi, T., Irie, K., Hirayama, T., Hayashida, N., Yamaguchi-Shinozaki, K., Matsumoto, K., and Shinozaki, K. (1996) A gene encoding a MAP kinase kinase kinase is induced simultaneously with genes for a MAP kinase and an S6 kinase by touch, cold and water stress in *Arabidopsis thaliana*. *Proc. Natl. Acad. Sci. USA*, **93**, 765–769.

Mizoguchi, T., Ichimura, K., Irie, K., Morris, P., Giraudat, J., Matsumoto, K. and Shinozaki, K. (1998) Identification of a possible MAP kinase cascade in *Arabidopsis thaliana* based on pairwise yeast two-hybrid analysis and functional complementation tests of yeast mutants *FEBS Lett.*, **437**, 56–60.

Monroy, A.F. and Dhindsa, R.S. (1995) Low-temperature signal transduction: Induction of cold acclimation-specific genes of alfalfa by calcium at 25°C. *Plant Cell*, **7**, 321–331.

Monroy, A.F., Sarhan, F., and Dhindsa, R.S. (1993) Cold-induced changes in freezing tolerance, protein phosphorylation, and gene expression: Evidence for role of calcium. *Plant Physiol.*, **102**,1227–1235.

Moon, B.Y., Higashi, S.I., Gombos, Z., and Murata, N. (1995) Unsaturation of membrane lipids of chloroplasts stabilises the photosynthetic machinery against low-temperature photoinhibition in transgenic tobacco plants. *Proc. Nat. Acad. Sci. USA*, **92**, 6219–6223.

Mullineaux, P.M. and Creissen, G.P. (1997) Glutathione reductase: regulation and role in oxidative stress. In J.G. Scandalios, (ed.), *Oxidative Stress and the Molecular Biology of Antioxidant Defenses*, (Monograph Series, Vol. 34), Cold Spring Harbor Laboratory Press, Cold Spring Harbor, pp. 667–714.

Mullineaux, P.M., Karpinski, S., Jimenéz, A., Cleary, S.P., Robinson, C., and Creissen, G. (1998) Identification of cDNAs encoding plastid-targeted glutathione peroxidase. *Plant J.*, **13**, 375–379.

Murata, N., Ishizaki-Nishizawa, O., Higashi, S., Hayashi, H., Tasaka, Y., and Nishida, I. (1992) Genetically engineered alteration in the chilling sensitivity of plants. *Nature*, **356**, 710–713.

Murata, N. and Wada, H. (1995) Acyl-lipid desaturases and their importance in the tolerance and acclimatisation to cold of cyanobacteria. *Biochem. J.*, **308**, 1–8.

Neuenschwander, U., Vernooij, B., Friedrich, L., Uknes, S., Kessmann, H., and Ryals, J. (1995) Is hydrogen peroxide a second messenger of salicylic acid in systemic acquired resistance? *Plant J.*, **8**, 227–233.

Neuhaus, G., Bowler, C., Kern, R., and Chua, N.H. (1993) Calcium/calmodulin-dependent and calcium/calmodulin-independent phytochrome signal-transduction pathways. *Cell*, **73**, 937–952.

Nishida, I. and Murata, N. (1996) Chilling sensitivity in plants and cyanobacteria: The crucial contribution of membrane lipids. *Annu. Rev. Plant Physiol. Plant Mol. Biol.*, **47**, 541–568.

Ogawa, K., Kanematsu, S., Takabe K., and Asada, K. (1995) Attachment of CuZn-superoxide dismutase to thylakoid membranes the side of superoxide generation (PSI) in spinach chloroplasts: Detection of immunno-gold labelling after rapid freezing and substitution method. *Plant Cell Physiol.*, **36**, 565–573.

O'Kane, D., Gill, V., Boyd, P., and Burdon, B. (1996) Chilling, oxidative stress and antioxidant responses in *Arabidopsis thaliana* callus. *Planta*, **198**, 371–377.

Okuda, T., Matsuda, Y., and Yamanaka, A. (1991) Abrupt increase in the level of hydrogen peroxide in leaves of winter wheat is caused by cold treatment. *Plant Physiol.*, **97**, 1265–1267.

Omran, R.G. (1980) Peroxide levels and activities of catalase, peroxidase and indoleacetic acid oxidase during and after chilling cucumber seedlings. *Plant Physiol.*, **65**, 407–408.

Öquist, G., Chow, W.S., and Andersson, J.M. (1992) Photoinhibition of photosynthesis represents a mechanism for the long term regulation of photosystem II. *Planta*, **186**, 450–460.

Öquist, G., Hurry, V.M., and Huner, N.P.A. (1993) Low temperature effects on photosynthesis and correlation with freezing tolerance in spring and winter cultivars of wheat and rye. *Plant Physiol.*, **101**, 245–250.

Osmond, C.B. and Grace, S.C. (1995) Perspective of photoinhibition and photorespiration in the field: Quintessential inefficiencies of the light and dark reactions in the terrestrial oxygenic photosynthesis? *J. Exp. Bot.*, **46**, 1351–1362.

Ottander, C., Campbell, D., and Öquist, G. (1995) Seasonal changes in photosystem II: organisation and pigment composition in *Pinus sylvestris*. *Planta*, **197**, 176–183.

Palta, J.P. and Weiss, L.S. (1993) Ice formation and freezing injury: An overview on the survival mechanisms and molecular aspects of injury and cold acclimation in herbaceous plants. In P.H. Li and L. Christersson, (eds.), *Advances in Plant Cold Hardiness*, CRC Press, Boca Raton, pp. 143–176.

Palta, J.P., Levitt, J., and Stadelmann, E.J. (1977) Freezing injury in onion bulb cell. I. Evaluation of the conductivity method for an analysis of ion and sugar efflux from injured cells. *Plant Physiol.*, **60**, 393–397.

Palva, T.E. (1994) Gene expression under low temperature stress. In A.S. Basra, (ed.), *Stress-Induced Gene Expression in Plants*, Hardwood Academic Publishers, Chur, pp. 103–130.

Perl-Treves, R. and Galun E. (1991) The tomato Cu,Zn superoxide dismutase genes are developmentally regulated and respond to light and stress. *Plant Mol. Biol.*, **17**, 745–760.

Perl, A., Perl-Treves, R., Galili, S., Aviv, D., Shalgi, E., Malkin, S., and Galun, E. (1993) Enhanced oxidative stress defence in transgenic potato expressing tomato Cu,Zn superoxide dismutases. *Theor. Appl. Genet.*, **85**, 568–576.

Pfannschmidt, T., Nilsson, A., and Allen, J.F. (1999) Photosynthetic control of chloroplast gene expression. *Nature*, **397**, 625–628.

Polle, A. (1997) Defense against photooxidative damage in plants. In J.G. Scandalios, (ed.), *Oxidative Stress and the Molecular Biology of Antioxidant Defenses*, (Monographs Series, Vol. 34), Cold Spring Harbor Laboratory Press, Cold Spring Harbor, pp. 623–666.

Polle, A. and Rennenberg, H. (1994) Field studies on Norway spruce trees at high altitudes. II Defence systems against oxidative stress in needles. *New Phytol.*, **121**, 635–642.

Polle, A., Kroniger, W., and Rennenberg, H. (1996) Seasonal fluctuations of ascorbate-related enzymes: acute and delayed effects of late frost in spring on antioxidative systems in needles of Norway spruce (*Picea abies* L.). *Plant Cell Physiol.*, **37**, 717–725.

Prasad, T.K. (1996) Mechanisms of chilling-induced oxidative stress injury and tolerance in developing maize seedlings: changes in antioxidant system, oxidation of proteins and lipids, and protease activities. *Plant J.*, **10**, 1017–1026.

Prasad, T.K., Anderson, M.D., Martin, B.A., and Stewart, C.R. (1994a) Evidence for chilling-induced oxidative stress in maize seedlings and regulatory role of hydrogen peroxide. *Plant Cell*, **6**, 65–74.

Prasad, T.K., Anderson, M.D., and Stewart, C.R. (1994b) Acclimation, hydrogen peroxide, and abscisic acid protect mitochondria against irreversible chilling injury maize seedlings. *Plant Physiol.*, **105**, 619–627.

Price, A.H., Taylor, A., Ripley, S.J., Cuin, T. Tomos, D., and Ashenden, T. (1994) Oxidative signals in tobacco increase cytosolic calcium. *Plant Cell*, **6**, 1301–1310.

Raff, M.C., Barres, B.A., Burne, J.F., Coles, H.S., Ishizaki, Y., and Jacobson, M. D. (1993) Programmed cell death and the control of cell survival: lessons from the nervous system. *Science*, **262**, 695–700.

Ramasarma, T. (1982) Generation of H2O2 in biomembranes. *Biochem. Biophys. Acta*, **694**, 69–93.

Rhodes, D. and Hanson, A.D. (1993) Quaternary ammonium and tertiary sulfonium compounds in higher plants. *Annu. Rev. Plant Physiol. Plant Mol. Biol.*, **44**, 357–384.

Russell, A.W., Critchley, C., Robinson, S.A., Franklin, L.A., Seaton, G.G.R., Chow, W-S., Anderson, J., and Osmond, C.B. (1995) Photosystem II regulation and dynamics of the chloroplast D1 protein in *Arabidopsis* leaves during photosynthesis and photoinhibition. *Plant Physiol.*, **107**, 943–952.

Sagisaka, S. (1985) Injuries of cold acclimatised polar twigs resulting from enzyme inactivation and substrate depression during frozen storage at ambient temperatures for a long period. *Plant Cell Physiol.*, **26**, 1135–1145.

Sakai, A. and Larcher, W. (1987) *Frost Survival of Plants*, Springer-Verlag, Berlin.

Santarius, K.A. (1982) The mechanism of cryoprotection of biomembrane systems by carbohydrates. In P.H. Li and A. Sakai, (eds.), *Plant Cold Hardiness and Freezing Stress: Mechanisms and Crop Implications*, Vol. 2, Academic Press, New York, pp. 457–486.

Savitch, L.V., Gray, G.R., and Huner, N.P.A. (1997) Feedback-limited photosynthesis and regulation of sucrose-starch accumulation during cold acclimation and low-temperature stress in a spring and winter wheat. *Planta*, **201**, 18–26.

Schinkel, H., Streller, S., and Wingsle, G. (1998) Multiple forms of extracellular superoxide dismutase in needles, stem tissues and seeds of Scots pine. *J. Exp. Bot.*, **49**, 931–936.

Schöner, S. and Krause, H. (1990) Protective systems against active oxygen species in spinach: Response to cold acclimation in excess light. *Planta*, **180**, 383–389.

Schreiber, U. and Neubauer, C. (1990) O_2-dependent electron flow, membrane energization and the mechanism of non-photochemical quenching of chlorophyll fluorescence. *Photosynthesis Res.*, **25**, 279–293.

Schupp, R. and Rennenberg, H. (1992) Changes in sulphur metabolism during needles development of Norway spruce. *Bot. Acta*, **105**, 180–189.

Sgherri, C.L.M., Loggini, B., Puliga, S., and Navariizzo, F. (1994) Antioxidant system in *Sporobolus stapfianus*: Changes in response to desiccation and rehydration. *Phytochemistry*, **35**, 561–565.

Shinozaki, K. and Yamaguchi-Shinozaki, K. (1996) Molecular responses to drought and cold stress. *Curr. Opin. Biotechnol.*, **7**, 161–167.

Skriver, K. and Mundy, J. (1990) Gene expression in response to abscisic acid and osmotic stress. *Plant Cell*, **2**, 503–512.

Slabas, A.R. and Fawcett, T. (1992) The biochemistry and molecular-biology of plant lipid biosynthesis. *Plant Mol. Biol.*, **19**, 169–191.

Smirnoff, N. and Pallanca, J.E. (1996) Ascorbate metabolism in relation to oxidative stress. *Bioch. Soc. Trans.*, **24**, 472–478.

Smith, I., Polle, A., and Rennenberg, H. (1990) Glutathione. In R.G. Alscher and J. Cumming, (eds.), *Stress Responses in Plants: Adaptation and Acclimation Mechanisms*, Wiley-Liss, New York, pp. 201–217.

Somersalo, S. and Krause, G.H. (1989) Photoinhibition at chilling temperature. Fluorescence characteristics of unhardened and cold acclimated spinach leaves. *Planta*, **177**, 409–416.

Somerville, C.R. (1995) Direct test of the role of membrane lipid composition in low-temperature-induced photoinhibition and chilling sensitivity in plants and cyanobacteria. *Proc. Natl. Acad. Sci. USA*, **92**, 6215–6218.

Sonoike, K. (1996) Photoinhibition of photosystem I; Its physiological significance in the chilling sensitivity of plants. *Plant Cell Physiol.*, **37**, 239–247.

Steponkus, P.L. (1984) Role of the plasma membrane in freezing injury and cold acclimation. *Annu. Rev. Plant Physiol.*, **35**, 543–584.

Strand, M. (1993) Photosynthetic activity of Scots pine (*Pinus sylvestris* L.) needles during winter is affected by exposure to SO_2 and NO_2 during summer. *New Phytol.*, **123**, 133–141.

Strand, M. and Öquist, G. (1985) Inhibition of photosynthesis by freezing temperatures and high light levels in cold-acclimated seedlings of Scots pine (*Pinus sylvestris*). I. Effect on the light limited and light saturated rates of CO_2 assimilation. *Physiol. Plant.*, **64**, 425–430.

Stassart, J.-M., Slooten, L., Botterman, J., Van Breusegem, F., Meray, N., and Inzé, D. (1995) Cold-tolerance, oxidative stress tolerance and anti-oxidant enzyme levels in non-transgenic and Mn-Superoxide dismutase overexpressing maize. In P. Mathis, (ed.), *Photosynthesis: from Light to Biosphere*, Vol. IV, Kluwer Academic Publishers, Dordrecht, pp. 873–876.

Streb, P., Feierabend, J., and Bligny, R. (1997) Resistance to photoinhibition of photosystem II and catalase and antioxidant protection in high mountain plants. *Plant Cell Environ.*, **20**, 1030–1040.

Streller, S. and Wingsle, G. (1994) *Pinus sylvestris* (L.) needles contain extracellular CuZn superoxide dismutase. *Planta*, **192**, 195–201.

Stuvier, E., De Kok, L., and Kuiper P.C.J. (1992) Freezing tolerance and biochemical changes in wheat shoots as affected by H_2S fumigation. *Plant Physiol. Biochem.*, **30**, 47–55.

Suzuki, H., Matsumori, A., Matoba, Y., Kyu, B., Tanaka, A., Fujita, J., and Sasayama, S. (1993) Enhanced expression of superoxide dismutase messenger RNA in viral myocarditis: an SH-dependent reduction of its expression and myocardial injury. *J. Clin. Invest.*, **6**, 2727–2733.

Tao, D.L., Jin, Y.H., and Du, Y.J. (1992) Response of organic-free-radicals production in overwintering conifer needles to low temperature and light. *Chinese J. App. Ecol.*, **3**, 120–124.

Tao, D.L., Öquist, G., and Wingsle, G. (1998) Active oxygen scavengers during cold acclimation of Scots pine seedlings in relation to freezing tolerance. *Cryobiology*, **37**, 38–45.

Tappel, A.L. (1972) Vitamin E and free radical peroxidation of lipids. *Ann. New York Acad. Sci.*, **203**, 12–28.

Taylor, A.O., Slack, C.R., and McPherson, H.G. (1974) Plants under climatic stress. VI Chilling and light effect on photosynthesis enzymes of sorghum and maize. *Plant Physiol.*, **54**, 696–701.

Tepperman, J.M. and Dunsmuir, P. (1990) Transformed plants with elevated levels of chloroplast superoxide dismutase are not more resistant to superoxide toxicity. *Plant Mol. Biol.*, **14**, 501–511.

Thomashow, M.F. (1990) Molecular genetics of cold acclimation in higher plants. *Adv. Genet.*, **28**, 99–131.

Thomashow, M.F. (1994) *Arabidopsis thaliana* as a model for studying mechanisms of plant cold tolerance. In E.M. Meyerowitz and C.R. Somerville, (eds.), *Arabidopsis*, (Monograph Series, Vol. 27), Cold Spring Harbor Laboratory Press, Cold Spring Harbor, pp. 807–834.

Thomashow, M.F. (1998) Role of cold-responsive genes in plant freezing tolerance. *Plant Physiol.*, **118**, 1–7.

Tsang, E.W.T., Bowler, C., Hérouart, D., Van Camp, W., Villarroel, R., Genetello, C., Van Montagu, M., and Inzé, D. (1991) Differential regulation of superoxide dismutases in plants exposed to environmental stress. *Plant Cell*, **3**, 783–792.

Uemura, M. and Yoshida, S. (1984) Involvement of plasma membrane alterations in cold acclimation of winter rye seedlings (*Secale cerale* L cv Puma). *Plant Physiol.*, **75**, 818–826.

Urao, T., Katagiri, T., Mizoguchi, T., Yamaguchi-Shinozaki, K., Hayashida, N. and Shinozaki, K. (1994) Two genes that encode Ca^{2+}-dependent protein kinases are induced by drought and high salt stresses in *Arabidopsis thaliana*. *Mol. Gen. Genet.*, **224**, 331–340.

Urao, T., Yakubov, B., Yamaguchi-Shinozaki, K., and Shinozaki, K. (1998) Stress-responsive expression of genes for two-component response regulator-like proteins in *Arabidopsis thaliana*. *FEBS Lett.*, **427**, 175–178.

Van Camp, W., Willekens, C., Bowler, C., Van Montagu, H., Inzé, D., Reupold-Popp, P., Sandermann, H. Jr, and Langebartels, C. (1994) Elevated levels of superoxide dismutase protect transgenic plants against ozone damage. *Bio/technology*, **12**, 165–168.

Van Wijk, K.J., Roobol-Boza, M., Kettunen, R., Andersson, B., and Aro, E.M. (1997) Synthesis and assembly of the D1 protein into photosystem II: Processing of the C-terminus and identification of the initial assembly partners and complexes during photosystem II repair. *Biochemistry*, **36**, 6178–6186.

Volger, H.G. and Heber, U. (1975) Cryoprotective leaf proteins. *Biochem. Biophys. Acta*, **412**, 335–349.

Wada, H., Gombos, Z., and Murata, N. (1990) Enhancement of chilling tolerance of a cyanobacterium by genetic manipulation of fatty acids desaturation. *Nature*, **347**, 200–203.

Wada, H., Gombos, Z., and Murata, N. (1994) Contribution of membrane lipids to the ability of the photosynthetic machinery to tolerate temperature stress. *Proc. Natl. Acad. Sci. USA*, **91**, 4273–4277.

Walker, M. and McKersie, B. (1993) Role of the ascorbate-glutathione antioxidant system in chilling resistance of tomato. *J. Plant Physiol.*, **141**, 234–239.

Weiser, C.J. (1970) Cold resistance and injury in woody plants. *Science*, **169**, 1269–1278.

Wildi, B. and Lutz, C. (1996) Antioxidant composition of selected high alpine plant species from different altitudes. *Plant Cell Environ.*, **19**, 138–146.

Willekens, H., Langebartels, C., Tiré, C., Van Montagu, M., and Inzé, D., and Van Camp, W. (1994) Differential expression of catalase genes in Nicotiana plumbaginifolia. *Proc. Natl. Acad. Sci. USA*, **91**, 10450–10454.

Willekens, H., Chamnongpol, S., Davey, M., Schraudner, M., Langebartels, C., Van Montagu, M., Inzé, D., and Van Camp, W. (1998) Catalase is a sink for H_2O_2 and is indispensable for stress defence in C-3 plants. *EMBO J.*, **16**, 4806–4816.

Wingate, V.P.M., Lawton, M.A., and Lamb, C.J. (1988) Glutathione causes a massive and selective induction of plant defence genes. *Plant Physiol.*, **87**, 206–210.

Wingsle, G., Sandberg, G., and Hällgren, J.-E. (1989) Determination of glutathione in Scots pine needles by high-performance liquid chromatography as its monobromobimane derivative. *J. Chromatogr. A*, **479**, 335–344.

Wingsle, G., Gardeström, P., Hällgren, J.-E., and Karpinski, S. (1991) Isolation, purification, and subcellular localization of isozymes of superoxide dismutase from Scots pine (*Pinus sylvestris* L.) needles. *Plant Physiol.*, **95**, 21–28.

Wingsle, G. and Hällgren, J.-E. (1993) Influence of SO_2 and NO_2 exposure on glutathione, superoxide dismutase and glutathione reductase activities in Scots pine needles. *J. Exp. Bot.*, **44**, 463–470.

Wingsle, G. and Karpinski, S. (1996) Differential redox regulation by glutathione of glutathione reductase and CuZn superoxide dismutase genes expression in *Pinus sylvestris* (L.) needles. *Planta*, **198**, 151–157.

Wingsle, G. and Moritz, T. (1997) Analysis of ascorbate and dehydroascorbate in plant extracts by high-resolution selected ion monitoring gas chromatography-mass spectrometry. *J. Chromatogr. B*, **782**, 95–103.

Winkler, B.S., Orselli, S., and Rex, T.S. (1994) The redox couple between glutathione and ascorbic acid: a chemical and physiological perspective. *Free Rad. Biol. Med.*, **17**, 333–349.

Wise, R.R. (1995) Chilling-enhanced photooxidation: The production, action and study of reactive oxygen species produced during chilling in the light. *Photosynthesis Res.*, **45**, 79–97.

Wise, R. and Naylor, A. (1987) Chilling enhanced photooxidation. Evidence for the role of singlet oxygen and superoxide in the breakdown of pigments and endogenous antioxidants. *Plant Physiol.*, **83**, 278–282.

Wu, J., Neimanis, S., and Heber, U. (1991) Photorespiration is more effective than the Mehler reaction in protecting the photosynthetic apparatus against photoinhibition. *Bot. Acta*, **104**, 283–291.

Wu, J. and Browse, J. (1995) Elevated levels of high-melting-point phosphatidylglycerols do not induce chilling sensitivity in *Arabidopsis* mutants. *Plant Cell*, **7**, 17–27.

Zhou, B.L., Arakawa, K., Fujikawa, S., and Yoshida, S. (1994) Cold induced alterations in plasma membrane proteins that are specifically related to the development of freezing tolerance in cold-hardy winter wheat. *Plant Cell Physiol.*, **35**, 175–182.

Zhu, D. and Scandalios, J.G. (1994) Differential accumulation of manganese-superoxide dismutases in maize in response to abscisic acid and high osmoticum. *Plant Physiol.*, **106**, 173–178.

4 Oxidative Stress and Defense Reactions in Plants Exposed to Air Pollutants and UV-B Radiation

Christian Langebartels, Martina Schraudner, Werner Heller, Dieter Ernst and Heinrich Sandermann Jr.

INTRODUCTION: AIR POLLUTANTS AND ELEVATED UV-B RADIATION AS FACTORS OF GLOBAL CLIMATE CHANGE

Apparent changes in the levels of air pollutants and of ultraviolet (UV) radiation in the UV-B range (280–315 nm) at the earth's surface are components of the widely discussed "global changes". Thinning of the stratospheric ozone layer is projected to increase UV radiation reaching the terrestrial surface (Kerr and McElroy, 1993; Madronich *et al.*, 1998). On the other hand, tropospheric ozone increased over the past years because of the release of precursor substances such as NO_x and hydrocarbons by human activities (Stockwell *et al.*, 1997; Kley *et al.*, 1999). Recent models predict that ozone and UV-B radiation will simultaneously increase in the troposphere in the future (for a review, see Runeckles and Krupa 1994; Kley *et al.*, 1999). At present, both stresses occur in combination at elevated sites, e.g. mountains of Central Europe and North America.

Ozone is found as lead substance of photooxidants together with lower amounts of peroxyacetyl nitrate, H_2O_2, and organic peroxides, such as methyl and hydroxymethyl hydroperoxide (Cape, 1997). Adverse effects of elevated ozone levels on plant productivity and competitiveness have been described (Reich, 1987; Sandermann *et al.*, 1997). New critical levels for ozone injury in plants have been established using accumulated hourly doses over a threshold of 40 nl/l (AOT40) as a basis (Kärenlampi and Skärby, 1996). The proposed AOT40 values for crops and trees are often exceeded in Central Europe and North America (Stockwell *et al.*, 1997; Kley *et al.*, 1999), suggesting that ozone is a potential threat for plants. Nitrogen- or sulfur-containing air pollutants may affect plant growth at acute toxic levels (Hippeli and Elstner 1996). Lower concentrations, on the other hand, may add to the plant's productivity when the respective elements are limiting. Adverse effects of artificially enhanced UV-B radiation have been demonstrated for sensitive cultivars of soybean and rice (Tevini, 1993; Rozema *et al.*, 1997) and include reductions in photosynthetic activity, biomass production, and yield, as well as effects upon morphology, flowering, and reproduction (Runeckles and Krupa, 1994; Jenkins, 1998).

The present interest in ozone and UV-B radiation has not only arisen from their ecological significance but also from their potential as defined model stresses to elicit plant responses at the molecular level. Increased UV-B radiation and air pollutants are assumed to influence plants via elevated levels of reactive oxygen species (ROS). Ozone is taken up through stomata and is rapidly decomposed to secondary ROS, such as superoxide anion radicals ($O_2^{\bullet-}$), H_2O_2, hydroxyl radicals ($OH^{\bullet}$), and other species in the leaf apoplast (Heath and Taylor, 1997; Mudd, 1997). It is therefore assumed that the internal ozone dose in the gas space of leaves is close to zero (Laisk *et al.*, 1989). Evidence for ozone-derived

ROS was obtained in an electron paramagnetic resonance spectrometry study by Runeckles and Vaartnou (1997) where a signal with similarity to $O_2^{\bullet-}$ and with putative chloroplastic localization occurred during and up to 15 min after the end of ozone exposure. SO_2 is known to produce free radicals and to initiate lipid peroxidation in cells whereas NO and NO_2 are free radicals on their own right (Hippeli and Elstner 1996). UV-B radiation stimulates the generation of ROS in plants, possibly via NAD(P)H oxidase activation (Green and Fluhr 1995; Rao et al., 1996; Dai et al., 1997).

A number of excellent reviews on the effects of air pollutants and UV-B on plants are available, including overviews on ozone and plant defense systems (Kangasjärvi et al., 1994; Sandermann 1996; Pell et al., 1997; Sharma and Davis 1997; Sandermann et al., 1998), physiological responses (Heath and Taylor 1997; Kley et al., 1999), critical ozone levels (Kärenlampi and Skärby 1996), and ozone in relation to forest decline (Sandermann et al., 1997). UV-B effects on plants have been summarized at the levels of gene expression (Strid et al., 1994; Jenkins 1997; Jansen et al., 1998) and plant physiology and biochemistry (Tevini 1993). Impacts of ozone, UV-B radiation, and elevated CO_2 levels on plant biochemistry as well as interactions of these factors were reviewed by Runeckles and Krupa (1994), whereas Manning and von Tiedemann (1995) gave an overview on interactions of air pollutants and UV-B radiation with pathogens. Methods used to measure ozone effects on plant defense have recently been summarized (Langebartels et al., 2000). This chapter will describe certain aspects of oxidative stress caused by ozone, SO_2, and UV-B radiation as well as molecular plant reactions to these environmental stresses. Emphasis is also placed on potential signalling mechanisms and on the use of transgenic plants overexpressing antioxidant genes in research on air pollutant and general stress tolerance.

EFFECTS OF OZONE AND OTHER AIR POLLUTANTS ON ANTIOXIDANT AND PATHOGEN DEFENSE SYSTEMS

Various biochemical pathways were found to respond to ozone exposure at the levels of genes, proteins and stress metabolites (Pell et al., 1997; Sharma and Davis 1997; Sandermann et al., 1998). The reactions occur between hours and days of sub-acute ozone exposure and include induction of ethylene and polyamine metabolism, polyphenol and lignin biosynthesis, defense-related proteins as well as antioxidant compounds and enzymes. Some of these reactions increase the antioxidative potential of plants while others promote senescence and drive plants into a prooxidative state.

Ethylene and Polyamines

Emission of the gaseous growth regulator, ethylene, as well as the level of its precursor, 1-aminocyclopropane-1-carboxylic acid (ACC; Figure 1) rapidly increase upon ozone exposure (Tingey et al., 1976) as well as under other types of stress (Abeles et al., 1992). The extent of stress ethylene formation is correlated with visible ozone injury. Inhibition of ethylene biosynthesis or perception leads to elevated ozone tolerance of certain species (Mehlhorn and Wellburn 1987; Langebartels et al., 1991; Bae et al., 1996; Tuomainen et al., 1997). The reaction of ethylene with ozone is thought to produce a number of highly

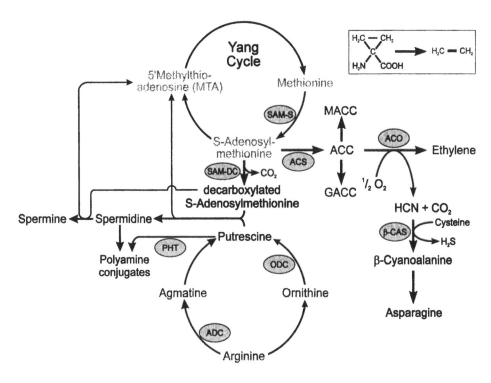

Figure 1. Scheme of ethylene and polyamine metabolism.
Abbreviations: ACO, ACC oxidase; ACS, ACC synthase; ADC, arginine decarboxylase; β-CAS, β-cyanoalanine synthetase; ODC, ornithine decarboxylase; PHT, putrescine hydroxycinnamoyl transferase; SAM-S, S-adenosyl methionine synthetase; SAM-DC, SAM decarboxylase.

reactive, water-soluble free oxyradicals and aldehydes, which may contribute to plant damage (Elstner *et al.*, 1985). In a survey of ozone-tolerant and ozone-sensitive plant pairs, Wellburn and Wellburn (1996) consistently found increased ethylene formation in all sensitive plants, together with even reduced levels in the tolerant ones compared to controls. In the latter plants, mechanisms that reduce the generation of free radicals are hypothesized to be involved during the initial phase of acclimation (Wellburn and Wellburn, 1996).

Ethylene biosynthesis in ozone-treated potato and tomato plants proceeds via the normal ACC pathway (Figure 1), namely transcriptional activation of ACC synthase (ACS) and ACC oxidase (ACO) together with *S*-adenosyl-L-methionine synthetase (SAM-S; Schlagnhaufer *et al.*, 1995, 1997; Bae *et al.*, 1996; Tuomainen *et al.*, 1997). All three proteins are encoded by divergent multigene families; their genes show differential expression during growth and respond differentially to environmental cues. Only one or two out of up to seven ACC synthase genes (*ACS2* and *ACS1b* in tomato, *ACS4* and *ACS5* in potato, *ACS6* in *Arabidopsis thaliana*) and one out of three SAM-S genes (*SAM3* in tomato) were affected by ozone (Schlagnhaufer *et al.*, 1997; Tuomainen *et al.*, 1997; Vahala *et al.*, 1998; C. Betz, J. Kangasjärvi, and C. Langebartels, unpublished results) and the respective isoforms in tomato resembled pathogen- rather than wound-responsive isoforms. The main leaf isoform of ACC oxidase, *ACO1*, was highly induced by ozone whereas the senescence-related *ACO3* as well as *ACO2* and *ACO4* were moderately affected at the transcript level (J. Kangasjärvi, D. Grierson, and C. Langebartels, unpublished results), suggesting that, in addition to the established role of ACS as rate-limiting step, ACO has regulatory functions in ethylene biosynthesis. ACS was additionally activated at the protein level as demonstrated by the use of inhibitors of protein kinases and phosphatases (Tuomainen *et al.*, 1997). The changes in ethylene biosynthesis of tobacco occurred without other phytohormones (cytokinins, indole-6-acetic acid, abscisic acid) being significantly affected (Yin *et al.*, 1994).

Ozone tolerance of plants has been postulated to depend on the relative amounts of antioxidant compounds, e.g. ascorbate and polyamines, present in the leaf tissue (Wellburn and Wellburn, 1996; Sharma and Davis, 1997). The key enzyme in polyamine biosynthesis, arginine decarboxylase, as well as putrescine and spermidine markedly respond to ozone and SO_2 exposure (Priebe *et al.*, 1978; Bors *et al.*, 1989; Rowland-Bamford *et al.*, 1989; Langebartels *et al.*, 1991; Reddy *et al.*, 1993) as well as to other types of stress (for a review, see Bouchereau *et al.*, 1999). Free and conjugated polyamines were proposed to contribute to ozone tolerance as they may exert several functions that counteract the effects of oxidative stress (Bors *et al.*, 1989; Heath and Taylor 1997). Polyamines are present in the apoplastic fluid of leaves and have been implicated in the inhibition of lipid peroxidation of membranes, the activation of membrane-bound ATPases and K^+ channels, and in the reduction of ethylene formation (Bouchereau *et al.*, 1999). The hydroxycinnamoyl derivatives of polyamines, but not the free polyamines, are efficient scavengers of oxyradicals (Bors *et al.*, 1989). In this line of observations, external application of polyamines conferred protection to tobacco and tomato plants against ozone injury (Ormrod and Beckerson, 1986; Bors *et al.*, 1989).

Ethylene and polyamines share the same biosynthetic precursor, SAM, and mutually inhibit their own biosynthesis (Figure 1; Zarembinski and Theologis, 1994). A metabolic switch to ethylene or polyamine biosynthesis has therefore been postulated as a major factor in ozone tolerance of tobacco (Langebartels *et al.*, 1991). In the future, the use of transgenic plants altered in ethylene and polyamine contents should allow us to clarify the role of these compounds in ozone and general oxidative stress tolerance of plants.

Polyphenolic Metabolites

Polyphenols are widespread plant metabolites largely derived from the amino acid phenylalanine (Figure 2; Heller, 1994; Forkmann and Heller, 1999). They were described as a second line of defense in pathogen-infected plants, and are usually formed in cells close to the infection site. Soluble compounds may be stored in the vacuole or secreted into the apoplastic space where they occur transiently in the apoplastic fluid, and may subsequently be incorporated into the cell wall matrix. Polyphenols are either constitutively present (e.g. the complex lignins) or form part of the inducible plant reactions during abiotic and biotic stress.

The key enzyme of the phenylpropanoid pathway, phenylalanine ammonia lyase (PAL; Figure 2), is transiently triggered by ozone in herbaceous and tree species (Sandermann *et al.*, 1998). In parsley, furanocoumarins and flavone *O*-malonyl glucosides are induced by ozone (Eckey-Kaltenbach *et al.*, 1994). Furanocoumarin levels have previously been shown to be elevated by elicitor treatments while the flavone derivatives were

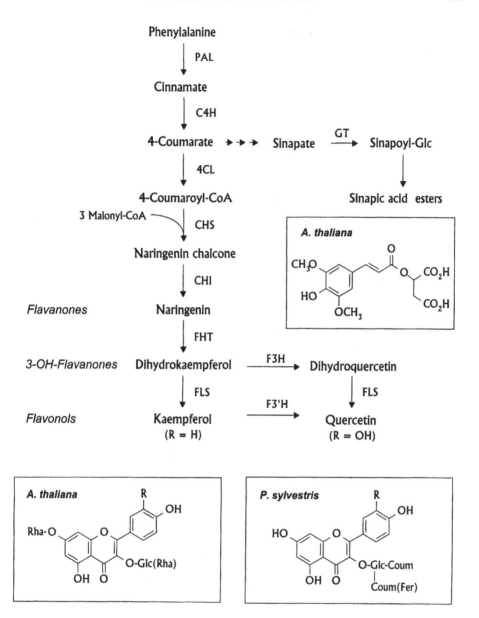

Figure 2. Scheme of phenylpropanoid metabolism in plants leading to sinapate esters and flavonoids. Abbreviations: C4H, cinnamate 4-hydroxylase; 4CL, 4-coumarate:CoA ligase; CHS, chalcone synthase; CHI, chalcone flavanone isomerase; F3H, flavanone 3-hydroxylase; FS, flavonol synthase; F3′H, flavanoid 3′-hydroxylase; 3GT, Flavonoid 3-*O*-glucosyl transferase; PAL, phenylalanine ammonia lyase

UV-B responsive. Therefore, ozone seems to act as a ''cross-inducer'' for both pathways in this system. Ozone-induced polyphenols of tree species have been summarized by Langebartels *et al.* (1997). Among these, several known antimicrobial compounds were detected to respond to ozone, e.g. stilbenes in Scots pine (Rosemann *et al.*, 1991), and a biphenyl derivative in European beech (Zielke and Sonnenbichler, 1990). Several

Table 1. Ozone-induced gene expression

Gene	Plant species
Ethylene biosynthesis	
SAM synthetase***	*Lycopersicon esculentum* (Tuomainen *et al.*, 1997)
ACC synthase***	*Solanum tuberosum* (Schlagnhaufer *et al.*, 1995, 1997), *L. esculentum* (Tuomainen *et al.*, 1997); *Arabidopsis thaliana* (Vahala *et al.*, 1998)
ACC oxidase***	*L. esculentum* (Tuomainen *et al.*, 1997)
Phenylpropanoid metabolism	
Phenylalanine ammonia-lyase*,**,***	*A. thaliana* (Sharma and Davis 1994), *Petroselinum crispum* (Eckey-Kaltenbach *et al.*, 1994), *Nicotiana tabacum* (Bahl *et al.*, 1995), *Betula pendula* (Tuomainen *et al.*, 1996; Pääkönen *et al.*, 1998), hybrid poplar (Koch *et al.*, 1998)
4-Coumarate: CoA ligase**, Chalcone synthase**	*P. crispum* (Eckey-Kaltenbach *et al.*, 1994, 1997; Bahl *et al.*, 1995)
Stilbene synthase*	*Pinus sylvestris, Vitis vinifera* (Schubert *et al.*, 1997; Zinser *et al.*, 1998)
Cinnamyl alcohol dehydrogenase*	*Picea abies* (Galliano *et al.*, 1993), *P. sylvestris* (Zinser *et al.*, 1998)
O-Methyltransferase*	hybrid poplar (Koch *et al.*, 1998)
Pathogenesis-related proteins	
Basic *β*-1,3-glucanase***	*N. tabacum* (Schraudner *et al.*, 1992; Ernst *et al.*, 1992; Bahl *et al.*, 1995)
Basic** and acidic* chitinase, PR1b*	*N. tabacum* (Ernst *et al.*, 1992; Bahl *et al.*, 1995; Thalmair *et al.*, 1996)
PR 1–1***, PR1–4***, PR 2***, Eli16***	*P. crispum* (Eckey-Kaltenbach *et al.*, 1994, 1997)
PR1**	*A. thaliana* (Sharma *et al.*, 1996), hybrid poplar (Koch *et al.*, 1998)
PR10*	*B. pendula* (Pääkönen *et al.*, 1998)
8.6-kDa basic PR protein	*A. thaliana* (Sharma and Davis 1995)
Antioxidant enzymes	
Catalase 2***, 3**	*N. plumbaginifolia* (Willekens *et al.*, 1994b), *N. tabacum* (Örvar *et al.*, 1997)
Glutathione peroxidase**	*N. plumbaginifolia* (Willekens *et al.*, 1994b), *N. tabacum* (Schraudner *et al.*, 1998)
Glutathione *S*-transferase***	*A. thaliana* (Sharma and Davis 1994; Conklin and Last 1995; Sharma *et al.*, 1996)
Glutathione reductase	*Brassica campestris* (Lee *et al.*, 1998)
Ascorbate peroxidase*	*N. tabacum* (Willekens *et al.*, 1994b; Örvar *et al.*, 1997), *A. thaliana* (Conklin and Last 1995; Kubo *et al.*, 1995)
Peroxidase*,**	*P. crispum* (Eckey-Kaltenbach *et al.*, 1994); *A. thaliana* (Sharma and Davis 1994); *Ipomoea batatas* (Kim *et al.*, 1999)
Cu/Zn superoxide dismutase*, Mn superoxide dismutase*	*A. thaliana* (Sharma and Davis 1994; Conklin and Last 1995; Kliebenstein *et al.*, 1998)

Table 1. (*continued*)

Gene	Plant species
Others	
Extensin*,**	*Fagus sylvatica, P. abies, P. sylvestris* (Schneiderbauer *et al.*, 1995), *P. crispum* (Eckey-Kaltenbach *et al.*, 1994)
Glycine-rich protein*	*Atriplex canescens* (No *et al.*, 1997)
Small heat shock protein*	*P. crispum* (Eckey-Kaltenbach *et al.*, 1997)
Lipoxygenase**	*Glycine max* (Maccarrone *et al.*, 1992), *Lens culinaris* (Maccarrone *et al.*, 1997)
Short-chain alkohol dehydrogenase**,***	*P. abies* (Bauer *et al.*, 1993, Miller *et al.*, 1999)
3-Hydroxymethylglutaryl-CoA synthase**	*P. sylvestris* (Wegener *et al.*, 1997a)
Polyubiquitin**	*P. sylvestris* (Wegener *et al.*, 1997b)
Thiol protease*, proteinase inhibitor*	*A. canescens* (No *et al.*, 1997)
Mitochondrial phosphate translocator**	*B. pendula* (Kiiskinen *et al.*, 1997)
Wound-induced gene *WIN3.7*	hybrid poplar (Koch *et al.*, 1998)
Phosphoribosylanthranilate transferase**	*A. thaliana* (Conklin and Last 1995)
Reticuline: oxygen oxidoreductase*	*A. thaliana* (Richards *et al.*, 1998)
Blue copper binding protein*	*A. thaliana* (Richards *et al.*, 1998; Miller *et al.*, 1999)
Copper chaperone*	*A. thaliana* (Himelblau *et al.*, 1998; Miller *et al.*, 1999)
Metallothionin*, Porin*	*P. abies* (Buschmann *et al.*, 1998; Miller *et al.*, 1999)
Ozone-related proteins, unknown function	*P. sylvestris* (Wegener *et al.*, 1997a), *P. crispum* (Eckey-Kaltenbach *et al.*, 1997)
Senescence-associated genes, unknown function	*A. thaliana* (Miller *et al.*, 1999)

***, rapid response (0.5–3 h); **, intermediate response (4–8 h); *, late response, usually occurring together with lesion development (9–48 h).

polyphenols, such as flavonoids, are effective radical scavengers (Bors *et al.*, 1995, 1998) and may therefore play a role in the elimination of ozone-derived oxyradicals. The levels of salicylic acid, a signal molecule in the resistance of plants against pathogens (Durner *et al.*, 1997), were induced in ozone-exposed tobacco (Yalpani *et al.*, 1994; Heiden *et al.*, 1999) and *A. thaliana* (Sharma *et al.*, 1996). It is now generally accepted that salicylic acid is not only one of several stress-induced phytoalexins (see Rüffer *et al.*, 1995), but mediates the expression of many defense-related genes, including those for defined pathogenesis-related (PR) proteins in ozone-exposed plants (Sharma and Davis, 1997; see below).

"Ozone-Related" Genes and Proteins

Ozone exposure of plants results in the accumulation of transcripts for various "ozone-related" genes (Table 1). The "ozone-related" genes were identified primarily in tobacco and *A. thaliana*, but are now also found in a growing number of other species, including deciduous and coniferous trees. It is obvious that a number of defense-related genes, particularly those encoding PR proteins, is induced, which had been identified before in pathogen-infected plants (Sharma and Davis, 1997; Sandermann *et al.*, 1998). PR proteins can be activated by abiotic factors, such as UV radiation (see below) or heavy metals

(Kombrink and Somssich, 1995). Ozone was found to be a major elicitor of PR proteins (β-1,3-glucanase, chitinase, and PR1) in tobacco (Ernst *et al.*, 1992; Schraudner *et al.*, 1992; Yalpani *et al.*, 1994; Thalmair *et al.*, 1996), *A. thaliana* (Sharma *et al.*, 1996), parsley (Eckey-Kaltenbach *et al.*, 1997), hybrid poplar (Koch *et al.*, 1998), and Norway spruce (Kärenlampi *et al.*, 1994; Table 1). This ozone induction of genes coding for PR proteins is in contrast to those associated with photosynthesis (e.g., ribulose-1,5-bisphosphate carboxylase/oxygenase, chlorophyll *a/b*-binding protein) which are generally reduced (Pell *et al.*, 1997). Only ozone, but not exposure to SO_2 or NO_2, leads to β-1,3-glucanase induction, and NO_2 is a strong synergist when applied in combination with ozone (Schraudner *et al.*, 1994). Recent findings that the primary air pollutant nitric oxide (NO) also plays a major role in signalling and cell damage in plants (Delledonne *et al.*, 1998; Durner *et al.*, 1998; Durner and Klessig, 1999) require that effects of NO as primary compound in combustion processes have to be reassessed.

Predominantly, basic isoforms of the glycosidases β-1,3-glucanase and chitinase which are also known to be induced by ethylene, are ozone responsive. This observation supports the view that ozone may act in part via stress ethylene (Schraudner *et al.*, 1992; Sandermann *et al.*, 1998). A second phase of ozone responses in tobacco and *A. thaliana* comprised the induction of acidic isoforms of PR1 (Ernst *et al.*, 1992; Sharma *et al.*, 1996), which are characteristic of systemic acquired resistance (SAR; Lamb and Dixon, 1997). In the field, β-1,3-glucanase and chitinase activity in tobacco correlated with the ozone levels occurring during the respective exposure period (Thalmair, 1996). This result further strenghthens the hypothesis that tropospheric ozone is a major abiotic elicitor of plant responses.

Ozone-induced genes from trees include the cell wall protein extensin (Schneiderbauer *et al.*, 1995), 3-hydroxy-3-methyl-glutaryl-CoA synthase, an enzyme involved in isoprenoid biosynthesis (Wegener *et al.*, 1997a), a new mitochondrial phosphate translocator (Kiiskinen *et al.*, 1997), and polyubiquitin, which represents the longest (with 10 repeats of 228 nucleotides) polyubiquitin cDNA found so far in plants (Wegener *et al.*, 1997b). As degradation of proteins is a major ozone effect in plants (Pell *et al.*, 1997), the involvement of ubiquitin may be a general feature in pollutant-treated plants. Ubiquitin may also contribute to the development of ozone damage, which in tobacco and other species resembles hypersensitive response (HR)-like lesions. Sharma and Davis (1995) reported a novel stress-induced gene that encodes a basic 8.6-kDa protein likely to represent a new class of pathogenesis-related proteins. Differential cDNA screening and *in vitro* translation revealed 13 independent cDNA clones for induced and 11 for repressed genes in parsley (Eckey-Kaltenbach *et al.*, 1997). Among these, a new member of PR1 (PR1-4) and a small heat shock protein were found (Table 1). Analysis of these and other "ozone-related" genes and proteins not identified so far (Kärenlampi *et al.*, 1994; Wegener *et al.*, 1997a) will most probably contribute to our understanding of general plant responses to oxidative stress.

Antioxidant Systems

Plants respond to elevated ROS levels by activating a number of antioxidative defense mechanisms including enzymes protecting from ROS and constituents of the ascorbate-glutathione pathway (see Chapters by Foyer and Mano, this volume). Two cellular compartments, the apoplast and the chloroplast, are of primary interest in context with air pollutants (Heath and Taylor, 1997). Both are well protected against elevated ROS levels by a variety of antioxidant systems, probably as they naturally experience internal ROS production. An "oxidative burst" in the apoplast is frequently observed upon pathogen infection (Baker and Orlandi, 1995; Lamb and Dixon, 1997; Morel and Dangl 1997). In addition, $O_2^{\bullet-}$ and H_2O_2 are involved in cross-linking of cell wall components and in lignin formation (Showalter, 1993; Whetten and Sederoff, 1995). Highly elevated ROS levels in chloroplasts are inherent as well as stress-induced consequences of photosynthesis, for instance when carbon assimilation and/or transport is impaired (Hippeli and Elstner, 1996; Foyer, 1997).

Ascorbate has been proposed to play a pivotal role in ozone tolerance because it reacts readily with ozone and with pollutant-derived ROS (Mudd, 1997). Other antioxidant compounds include glutathione, α-tocopherol, polyamines, and various polyphenols (Foyer *et al.*, 1997; Heath and Taylor, 1997). A semi-dominant ozone-sensitive *Arabidopsis* mutant (*soz1*; now renamed *vtc1*) exhibited decreased levels of ascorbate (approximately 30% of the wild type; Conklin *et al.*, 1996, 1997) demonstrating that at least one of the genes involved in ozone tolerance is connected with ascorbate biosynthesis. *vtc1* was also sensitive to other types of oxidative stress, namely exposure to SO_2 and UV-B radiation. Ascorbate levels as well as the reduction state of ascorbate and glutathione were not altered by the stress treatments in this line. The localization of ascorbate in *vtc1* has not yet been identified, but previously the pool of reduced ascorbate in the apoplastic fluid was postulated as a first-line factor in ozone tolerance of plants (Luwe *et al.*, 1993). Recent studies, however, have questioned the effectiveness of apoplastic ascorbate in removing ozone during the exposure period and propose a heterogeneous distribution of ozone action and plant reactions in leaves (Jakob and Heber, 1998; Schraudner *et al.*, 1998; Ranieri *et al.*, 1999).

Glutathione is present in high amounts in chloroplasts and the cytosol whereas its apoplastic localization is uncertain (Noctor and Foyer, 1998). Recent studies have shown that leaf necrosis in catalase-deficient plants correlates with an accumulation of oxidized glutathione as well as increased ascorbate peroxidase and glutathione peroxidase levels (Chamnongpol *et al.*, 1998), and it was postulated that glutathione acts as a signal molecule in HR (May *et al.*, 1998). It has to be shown in the future whether an altered redox balance during ozone stress participates in the initiation of defense responses.

Key antioxidant enzymes include superoxide dismutases (SOD), which convert superoxide anion radicals to hydrogen peroxide and free oxygen, H_2O_2-detoxifying enzymes, such as ascorbate peroxidase (APX), other peroxidases (POD) and catalases (CAT), as well as enzymes of the Asada-Halliwell cycle for the turnover of reduced ascorbate and glutathione (Inzé and Van Montagu, 1995; Foyer *et al.*, 1997; Noctor and Foyer, 1998). A number of studies have focused on correlations between antioxidant enzymes and ozone as well as SO_2 tolerance, and they strongly suggest that plants regulate against excessive ROS levels (Kangasjärvi *et al.*, 1994; Noctor and Foyer, 1998). It is noteworthy that the majority of the antioxidant enzymes only respond with small amplitudes and that only certain isoforms are highly stress responsive.

In a thorough study of the expression of nine antioxidant genes (Willekens *et al.*, 1994b), two main conclusions were drawn: (1) the responses to ozone, SO_2, and UV-B exposure are remarkably similar and (2) antioxidant enzymes can be classified into two groups, one reacting in an early phase of oxidative stress (CATs, GPX), the other responding rather late, shortly before visible damage starts to develop (APX, SODs). Cu/ZnSOD mRNA was 4- to 5-fold increased only concomitant with visible injury in *Nicotiana tabacum* cv. PBD6 while no changes were observed in the tolerant *N. plumbaginifolia*. Marked ozone- and vehicle exhaust-induced increases in mRNA for MnSOD and cytosolic Cu/ZnSOD were also found in *N. tabacum* cv. SR1, whereas FeSOD was only slightly responsive (Bahl *et al.*, 1995).

A similar reaction pattern was also found for antioxidant enzymes in *A. thaliana* after short-term exposure to ozone. Transcript levels for cytosolic Cu/ZnSOD and a neutral peroxidase increased after 10 to 15 h (Sharma and Davis, 1994). These responses occurred without visible injury, but correlated in time with significant changes in fresh to dry weight ratios. Kliebenstein *et al.* (1998) found one early (*CSD1*) and one late (*CSD3*) responding isoform of Cu/ZnSOD, whereas two FeSODs (*FSD1*, *FSD2*) and MnSOD (*MSD1*) did not react to the exposure conditions. Conklin and Last (1995) described a preferential accumulation of transcripts for cytosolic antioxidant enzymes (glutathione S-transferase (GST1), APX, and Cu/ZnSOD) whereas chloroplastic FeSOD and glutathione reductase (GR) declined. Sharma and Davis (1994) as well as Conklin and Last (1995) found the most prominent reaction to ozone and UV-B radiation with GST1 mRNA that had been previously identified as ethylene- and pathogen-induced isoform. GST1 thus may play an important role in the recognition and/or detoxification of ROS resulting from these environmental stresses. Exposure to ozone or UV-B radiation for several days finally led to an induction of all SOD isoforms together with guaiacol peroxidase (Rao and Ormrod, 1995; Rao *et al.*, 1996). No significant changes were observed in the activities of enzymes of the Asada-Halliwell pathway under non-acute ozone concentrations (Kubo *et al.*, 1995), confirming the highly regulated state of this pathway.

Three differentially regulated catalases are present in tobacco and other species (Willekens *et al.*, 1995, 1997). Their transcript levels reacted with rapid decreases (CAT1) or increases (CAT2) to ozone, SO_2, and UV-B in *N. plumbaginifolia* (Willekens *et al.*, 1994b) whereas in *A. thaliana* the levels of catalase mRNA were not different from controls (Sharma and Davis, 1994). It has been shown that CAT1 is the major isoform in leaves of *N. plumbaginifolia* comprising 80% of total activity. *CAT1* is predominantly expressed in the palisade parenchyma cells and exhibits a diurnal course with lowest activity during light periods (Willekens *et al.*, 1994a), thereby confirming CAT sensitivity to light-induced inactivation. In contrast, the ozone-induced CAT2, with 20% of total leaf activity, was preferentially found in and around vascular bundles and was postulated to play a (localized) role in defense rather than lignification processes. The subcellular localization of CAT2 is not yet clear, and both peroxisomes and mitochondria are being discussed (Willekens *et al.*, 1995).

Together with the induction of CAT2, transcript levels for a glutathione peroxidase (GPX) rapidly responded to the stress treatments (Willekens *et al.*, 1994b; Schraudner *et al.*, 1998). This isoform had approximately 50% sequence identity to a mammalian phospholipid hydroperoxide glutathione peroxidase (PHGPX) that has a broad substrate specificity and can reduce membrane lipid peroxides (Eshdat *et al.*, 1997). Until now, GPXs in plants have only rarely been demonstrated, and do not contain the selenocysteine codon typical for mammalian PHGPX (Inzé and Van Montagu, 1995; Eshdat *et al.*, 1997). Detoxification of lipid peroxidation products, such as lipid hydroperoxides, aldehydes, and alcohols, is mainly performed by glutathione peroxidases in animals and by GSTs in animals and plants. As shown above, GST1 transcripts specifically respond to ozone treatment in *A. thaliana* (Sharma and Davis, 1994; Conklin and Last, 1995; Rao and Davis, 1999). Tolerant plants seem to react to the initially elevated ROS and peroxide levels with an increase in peroxide-degrading enzymes, such as CAT, GPX, and GSTs, thus avoiding propagation of the damaging species. Lipid hydroperoxide formation has been observed in tobacco plants treated with ozone and pathogens (Schraudner *et al.*, 1996); furthermore, a variety of volatile lipid peroxidation product, mainly C_6 aldehydes and alcohols with *cis*-3-hexenol as dominating compound, were emitted from ozone-sensitive tobacco before visible damage was observed (Heiden *et al.*, 1999). Presently, the contribution of these compounds to the activation of defence reactions and to cell death in ozone-sensitive plants is being investigated.

OZONE TOLERANCE OF TRANSGENIC PLANTS ALTERED IN ANTIOXIDANT GENES

Plants respond to pollutant gases with elevated levels of transcripts and activity of antioxidant enzymes. The activity of distinct isoforms may thus be rate limiting for the removal of ROS generated by ozone or SO_2 stress. Consequently, the possibility to improve stress tolerance in transgenic plants overproducing antioxidant enzymes has aroused considerable interest (Table 2; see earlier reviews by Foyer *et al.*, 1994, 1997; Rennenberg and Polle, 1994; Allen, 1995; Lea *et al.*, 1998).

Transgenic plants overproducing 15-fold higher levels of chloroplastic Cu/ZnSOD did not exhibit increased ozone tolerance (Pitcher *et al.*, 1991), probably because of an excess production of H_2O_2 that could react with superoxide anions and Fe^{2+} to produce highly reactive hydroxyl radicals. Alternatively, Cu/ZnSOD, which is sensitive to H_2O_2, may have been inactivated by the accumulated H_2O_2. In contrast, 2- to 4-fold overproduction of cytosolic Cu/ZnSOD (Pitcher and Zilinskas 1996), as well as of MnSOD (Van Camp *et al.*, 1994) or FeSOD in chloroplasts (W. Van Camp, H. Willekens, C. Langebartels, and D. Inzé, unpublished results) protects plants from ozone injury. The latter studies stress the importance of chloroplasts for ozone tolerance because overproduction of MnSOD in mitochondria was less effective (Van Camp *et al.*, 1994). The initial site of attack of ozone and ozone-derived ROS is thought to be the apoplastic space and the plasma membrane (Heath and Taylor, 1997). Constitutive overexpression of cytosolic Cu/ZnSOD therefore may help to detoxify superoxide anion radicals as byproducts from ozone reactions at a location accessible for SOD. It is, however, unclear whether superoxide anion radicals can enter the cytosol because of their high reactivity with cell wall and membrane components.

At a first glance, the reported beneficial effect of MnSOD and FeSOD, but not Cu/ZnSOD, in chloroplasts is even more surprising. Ozone or ozone-derived ROS are hardly capable of entering this organelle and, in any case, may not significantly alter the preexisting levels of ROS produced by the Mehler reaction or other photosynthetic processes (Foyer 1997). Ozone has been suggested to be degraded during its entry into the plant cell, but to elicit an accumulation of ROS in the apoplast and possibly also inside the cell, which overrides the existing antioxidant systems (Schraudner *et al.*, 1998; see also below). Several findings also point to an involvement of the chloroplast in ozone-induced injury: chloroplast DNA from ozone-exposed bean and pea contains elevated levels of 8-hydroxyguanine, a product resulting from hydroxyl radical attack of guanine (Floyd *et al.*, 1989), and distinct changes in chloroplast lipids were found in pea plants (Hellgren

Table 2. Overexpression of enzymes contributing to ozone and paraquat tolerance

Enzyme overexpressed	Subcellular localization	Increase in activity (fold)	Ozone treatment (nl l^{-1}), duration	Tolerance to Ozone	Paraquat	Reference
Nicotiana tabacum						
GR*	Cytosol	2–4	500, 4 h	no		Aono et al. (1991)
	Chloroplast	3	500, 18 h	no[+]	yes	Aono et al. (1993)
	Cytosol, chloroplast	5 –	200, 2 d	no	yes	Broadbent et al. (1995)
	Chloroplast, mitochondria	5 –	200, 2 d	yes	no	Broadbent et al. (1995)
GR + APX			250, 2 d	yes		Aono et al. (1997)
Cu/Zn SOD	Chloroplast	15	300, 6 h	no		Pitcher et al. (1991)
	Cytosol	2–5	200–300, 4.5–6 h	yes		Pitcher and Zilinskas (1996)
APX	Cytosol		250, 3 d	no		Örvar and Ellis (1997)
	Chloroplast	10	250, 4.5 h	no		Torsethaugen et al. (1997)
			80, 7 d	no		Torsethaugen et al. (1997)
Nicotiana plumbaginifolia						
Mn SOD	Mitochondria	8	90–120, 7 d	moderate		Van Camp et al. (1994)
	Chloroplast	2–4	90–120, 7 d	yes	yes	Van Camp et al. (1994)
	Chloroplast	2–4	600, 6 h	no		C. Bowler et al. (unpublished data)
Arabidopsis thaliana						
NahG	Cytosol	–	300, 6 h	yes		Sharma et al. (1996)
Populus tremula x P. alba						
GS	Cytosol	200–300	100/200/300, 3 d	no		Strohm et al. (1999)
GR	Cytosol, chloroplast	5, 200	100/200/300,3 d	no		Strohm et al. (1999)

* Abbreviations: APX, ascorbate peroxidase; GR, glutathione reductase; GS, glutathione synthetase; SOD, superoxide dismutase; NahG, bacterial salicylate hydroxylase.
+ Increased tolerance to SO_2 (1000 nl l^{-1}, 2 d).

et al., 1995). Therefore, ozone, similar to other environmental stresses, may produce transport or metabolic "blocks" and thereby enhance the transfer of electrons to oxygen, rather than NADP$^+$, with resulting high superoxide anion radical levels in the chloroplast (Hippeli and Elstner, 1996).

Overexpression of antioxidant genes conferred partial tolerance only to ozone concentrations that are between ambient and twice-ambient summer levels in Central Europe and North America. There was no or only moderate protection to higher levels as exemplified by exposing tobacco plants overproducing chloroplastic MnSOD to 300 or 600 nl l^{-1} ozone (C. Bowler, C. Langebartels, and W. Van Camp, unpublished results; Table 2). In addition, plant damage was only significantly reduced when occurring after one to several days after the onset of exposure in the wild type. Therefore, plant defense responses are thought to be activated as a prerequisite for ozone tolerance in this initial period (Willekens *et al.*, 1994b). This situation is similar to chilling stress where an acclimation phase with increased antioxidant enzyme activities is needed to protect maize plants (Prasad *et al.*, 1994). Synergistic effects of antioxidant enzymes were found for paraquat tolerance in plants that overproduce APX as well as GR (Aono *et al.*, 1997). Parallel overproduction of GST and GPX led to increased cold and salt tolerance in tobacco (Roxas *et al.*, 1997) whereas ozone tolerance has not been investigated until now.

Several enzymes of the glutathione biosynthetic pathway have been overproduced in tobacco, *A. thaliana*, and poplar (Table 2). However, the glutathione content was found to be highly regulated, possibly via feed-back regulation (Strohm *et al.*, 1995). Overexpression of genes that encode glutathione synthetase (*GSHII*) in the cytosol or GR either in the cytosol or in the chloroplast did not lead to increased tolerance to near-ambient ozone concentrations (Strohm *et al.*, 1999). Overexpression of the gene coding for γ-glutamylcysteine synthetase (*GSHI*) resulted in an accumulation of γ-glutamylcysteine and necrotic lesion development possibly because of a perturbed oxidative stress response of the plants (Creissen *et al.*, 1996). Tobacco expressing high levels of bacterial GR genes showed no increased tolerance to ozone, but to paraquat within a certain concentration range (Aono *et al.*, 1991; Foyer *et al.*, 1994). Accordingly, tobacco plants expressing the *GOR* gene for GR in cytosol, mitochondria or chloroplasts contained between 4- and 10-fold elevated GR activity and were generally more tolerant to paraquat, but only two lines were more tolerant to ozone (Creissen *et al.*, 1996). GR overexpression in tobacco also led to elevated tolerance to SO$_2$ and paraquat, but not to ozone (Aono *et al.*, 1993).

Genetically modified lines for H$_2$O$_2$-scavenging enzymes (POX, CAT) have been analyzed in recent years. Antisense lines for cytosolic APX exhibited increased ozone sensitivity to ambient and elevated ozone levels in the ozone-sensitive tobacco cv. Bel W3, whereas overexpression of the enzyme did not alter ozone tolerance (Örvar and Ellis, 1997). In line with this result, tobacco lines with 10-fold higher chloroplastic APX activity were not significantly more tolerant to ozone than the wild type (Torsethaugen *et al.*, 1997). Overexpression of *CAT1* or *CAT2* did not lead to higher ozone tolerance in tobacco (H. Willekens, W. Van Camp, and C. Langebartels, unpublished results). On the other hand, lowering catalase activity in tobacco to 10–20% of the wild type (*CAT1* antisense and co-suppression lines) resulted in severe sensitivity to high light, probably because of impaired detoxification of photorespiratory H$_2$O$_2$, as well as increased sensitivity to ozone, paraquat, and salt stress (Chamnongpol *et al.*, 1996, 1998; Willekens *et al.*, 1997). In conclusion, H$_2$O$_2$-scavenging enzymes are present in wild-type tobacco in sufficient amounts, while reductions result in significantly decreased stress tolerance. Cross-tolerance to several forms of oxidative stress has been reported to occur once the antioxidant defenses have been increased (Sandermann 1996; Sharma and Davis, 1997). Therefore, transgenic lines for antioxidant genes will remain highly interesting for studies on general stress tolerance of plants and as a strategy for crop improvement.

EFFECTS OF UV-B RADIATION ON ANTIOXIDANT AND UV-B SCREENING MECHANISMS

The possible increase in UV-B radiation has gained interest in recent years as a major environmental factor affecting plant performance. Tevini (1993), Runeckles and Krupa (1994) as well as Jordan (1996) have summarized the impacts of UV-B on plant growth and photosynthesis and have pointed out that adverse effects on sensitive species and cultivars are qualitatively well established, but that quantitative relations are far from being clear, particularly in view of exposure regimes which are relevant for the field situation. Three major features have to be considered: (i) the solar spectrum exhibits a spectral edge of approximately 300 nm because of the filtering effect of the stratospheric ozone layer, and plants do not experience in their natural environment UV-B radiation below 291 nm (Doehring *et al.*, 1996); (ii) the range of UV-B currently encountered during the growing period in Northern mid -latitudes is in the range of 1 kJ to 5 kJ m^{-2} d^{-1} UV-B$_{BE}$ (Madronich *et al.*, 1998); (iii) a balanced UV-B/UV-A (315–400 nm)/photosynthetically active radiation (PAR) ratio is a prerequisite for a realistic and ecologically sound study of plant responses (Caldwell and Flint, 1994). Fiscus and Booker (1995) reviewed the results from UV-B experiments and stated that studies under balanced UV-B/UV-A/PAR ratios (typically 1:20:200) did not provide evidence for impaired photosynthesis, stomatal conductance and carboxylation efficiency. Below, several examples for UV-B activation of metabolites, proteins and genes will be given. It has to be stressed, however, that in the light of high UV-B doses and imbalanced UV-B/PAR ratios used, effects may have been overestimated in some of the studies.

Higher plants can reduce UV-B radiation that reaches the mesophyll cells by producing a variety of UV-B-absorbing secondary metabolites, which typically accumulate in cuticle, cell walls, and cytosol of the upper epidermal layer (Tevini, 1993; Jansen *et al.*, 1998). These compounds include hydroxycinnamate esters, flavones, flavonols, and anthocyanins, often esterified with hydroxycinnamic acids. Plants grown under conditions that induce flavonoid biosynthesis are more tolerant to UV-B radiation as demonstrated for various crop plants as well as for species growing in mountainous habitats (reviewed by Tevini, 1993). Two different classes of phenolic compounds are currently discussed as screening pigments, namely soluble hydroxycinnamic acid derivatives and flavonoids (Figure 2). At present, more than 4000 flavonoids are known (Forkmann and Heller, 1999), and this structural diversity is reflected in a variety of biological functions, such as screening pigments, antioxidants, phytoalexins, and allelochemicals.

Major progress in UV-B responses of plants has been achieved by the analysis of gene loci for flavonoid biosynthesis in *A. thaliana*. More than 10 loci have been identified on the basis of altered seed color (recessive *transparent testa* or *tt* mutants; Shirley *et al.*, 1995; Shirley, 1996). *Arabidopsis tt* mutants are unable to synthesize the normal pattern of flavonoids because of defects in biosynthetic genes, such as chalcone synthase (*tt4*), chalcone isomerase (*tt5*), and dihydroflavonol 4-reductase (*tt3*; Shirley *et al.*, 1995; see also Figure 2). These lines have been shown to be more sensitive to UV-B radiation than wild-type plants (Li *et al.*, 1993). *A. thaliana* accumulates kaempferol and quercetin di- and triglycosides, with rhamnose and glucose as sugar moieties (Mittal *et al.*, 1995; Figure 2). The difference in UV-B sensitivity of *tt4* and *tt5* mutants correlates with the accumulation of sinapate esters such as *O*-sinapoyl-L-malate (Figure 2) and 1-*O*-sinapoyl-β-D-glucose. A sinapate ester mutant deficient in ferulate 5-hydroxylase *(fah1)* was described by Landry *et al.*, (1995). Its extreme UV-B sensitivity has led to the conclusion that hydroxycinnamate esters are more effective screening pigments in *A. thaliana* than flavonol derivatives. It will be interesting in the future to study possible feedback responses between hydroxycinnamic acid and flavonoid pathways, as well as to calculate the relative importance of both types of metabolites in protecting DNA against UV damage. DNA photoproducts, cyclobutane dimers, and pyrimidine[6,4]pyrimidone adducts, were found at elevated levels in plants lacking flavonoids (summarized by Strid *et al.*, 1994). Low shielding efficiency in *tt* mutants resulted in increased lipid peroxidation and protein oxidation (Landry *et al.*, 1995).

A detailed phytochemical analysis of UV-B-screening pigments in the coniferous species, Scots pine and Norway spruce, has recently been performed (Jungblut *et al.*, 1995; Schnitzler *et al.*, 1996; Fischbach *et al.*, 1999). Scots pine was shown to contain at least six structurally closely related diacylated flavonol monoglucosides (Figure 2). These molecules contained either two coumaroyl or one coumaroyl and one feruloyl residue attached to the glucose moiety, shifting the absorption maximum from 350 nm (UV-A range, typical for flavonols) to 315 nm in the UV-B range (Jungblut *et al.*, 1995). Norway spruce also accumulated diacylated flavonol glucosides, in particular kaempferol 3-*O*-(3″,6″-*O*-di-*p*-coumaroyl)-glucoside, during needle development (Fischbach *et al.*, 1999). These species as well as the major European deciduous tree, European beech (J. Rothenburger and W. Heller, unpublished results), therefore produce UV-B-screening compounds that combine flavonoids and hydroxycinnamic acids in one molecule. The soluble acylated flavonol glucosides were induced by UV-B radiation (Schnitzler *et al.*, 1997) and were exclusively found in the epidermal layer of pine needles as demonstrated with isolated epidermal strips (Schnitzler *et al.*, 1996). When the UV-screening efficiency of the soluble (acylated flavonol glucosides) and the wall-bound metabolites (p-coumaric and ferulic acids, kaempferol glucoside) was calculated at 300 nm, UV-B radiation was found shielded by the epidermal layer by more than 99.99% (Schnitzler *et al.*, 1996; Fischbach *et al.*, 1999). Conifers therefore seem to be well protected from UV-B radiation. Similar analyses and calculations of the shielding efficiency of individual compounds have so far been performed for a few herbaceous plants (Tevini, 1993).

The additional function of flavonoids as antioxidant molecules has also to be pointed out in the context of this chapter. Flavonoids may interfere with oxidative processes both by chelating metal ions or by scavenging oxyradicals, forming less reactive "antioxidant radicals" that disappear by dismutation, recombination or reduction (Bors *et al.*, 1992, 1998). Results from pulse radiolytic and photolytic studies with more than 30 flavonoids indicated that two structural elements are required for optimal antioxidative potential: (i) conjugation of the 2,3-double bond and the 4-oxo group and (ii) a 3′,4′-dihydroxy (catechol) structure in the B-ring (Bors *et al.*, 1992). With this respect, quercetin was considered as an optimal flavonoid antioxidant. Such compounds, however, have a higher redox potential than ascorbate and consequently may lead to the formation of ascorbyl radicals (Bors *et al.*, 1995). It will be highly interesting to examine this prooxidative behavior of flavonoids in those cellular localizations where flavonoids and ascorbate may occur together, i.e. the apoplastic space.

UV-B EFFECTS ON GENE EXPRESSION AND PROTEIN ACCUMULATION

UV radiation and blue light play key roles in plant morphogenesis, phototropism as well as flowering, and affect the expression of a variety of genes (Jenkins, 1998; Kim *et al.*, 1998). Apart from the phytochromes, several photoreceptors have been identified that mediate various responses to UV radiation and blue light, e.g. cryptochrome 1 and 2 (CRY1, CRY2), a putative UV-B photoreceptor, and a photoreceptor for stomatal opening (Jenkins 1998). A wealth of literature exists on UV-B impaired photosynthetic processes (for reviews, see Jordan, 1996; Allen *et al.*, 1998). Photosystem II, in particular the reaction center polypeptide D1, is a major target under artificially increased UV-B radiation. Nuclear (chlorophyll *a/b*-binding protein, ribulose-1,5-bisphosphate carboxylase small subunit, subunit γ of CF_1-ATPase) as well as chloroplast-encoded transcripts (ribulose-1,5-bisphosphate carboxylase large subunit, psbA, subunits β and ε of the CF_1-ATPase) are downregulated within a few hours in pea and *A. thaliana* (Strid *et al.*, 1994; Allen *et al.*, 1998). Reduced protein levels and enzyme activities as well as lowered pigment contents are subsequently observed within one and several days. These effects were less drastic when plants were exposed to both UV-B and high PAR. In addition, mRNA levels returned to control levels after some days, demonstrating acclimation (Allen *et al.*, 1998; Jansen *et al.*, 1998). Interestingly, cell cycle-related genes are also repressed by UV-B radiation in parsley suspension cells and leaf buds (Logemann *et al.*, 1995). Brosché *et al.*, (1999) have recently divided UV-B regulated genes of *Pisum sativum* into four groups, (i) genes for chloroplast-localized proteins were down-regulated, (ii) genes encoding proteins of the ubiquitin protein degradation pathway were up- or downregulated, (iii) expression of genes for proteins involved in intracellular signaling were specifically altered and (iv) genes for phenylpropanoid or flavonoid biosynthesis were up-regulated.

UV-B radiation is known to induce plant secondary metabolism at the transcript level (Hahlbrock and Scheel, 1989). A rapid and coordinated increase in PAL and chalcone synthase (CHS), the key enzyme in flavonoid biosynthesis, was observed in parsley and a range of other plants (Table 3; Forkmann and Heller, 1999). Transcripts for individual *CHS* isoforms were transiently increased in *A. thaliana*, pea and *Petunia hybrida*, exposed to UV-B radiation (Jenkins, 1998). UV-inducible elements have been described in promoters from various *CHS* isoforms, and both, UV-B and UV-A/blue light phototransduction pathways have been reported for CHS activation (Christie and Jenkins 1996; Jenkins, 1998). The expression of *CHS* transcripts in UV-A/blue light seems to be mediated by CRY1 whereas different photoreceptors are responsible for UV-B induction (Fuglevand *et al.*, 1996; Long and Jenkins, 1998). Chalcone synthase as well as PR1 protein, unlike the acylated flavononol glucosides in Scots pine, are not confined to the epidermal layer of cells but accumulate in the total exposed leaf area of tobacco and pine (Green and Fluhr 1995; Schnitzler *et al.*, 1997). It is currently unclear how this differential accumulation of proteins and metabolites is regulated.

The growth reduction observed in UV-B-treated plants points to alterations in cell division and/or elongation, produced by changes in growth regulating substances. Oxidative degradation of indole-3-acetic acid and inhibiting effects of the resulting

Table 3. UV-B-induced gene expression

Gene	Plant species
Phenylpropanoid metabolism	
Phenylalanine ammonia-lyase*,**	*Arabidopsis thaliana* (Shirley *et al.*, 1995; Long and Jenkins 1998)
Chalcone synthase**	*Petroselinum sativum* (Strid *et al.*, 1994); *A. thaliana* (Jordan 1996; Christie and Jenkins 1996; Fuglevand *et al.*, 1996), *Pinus sylvestris* (Schnitzler *et al.*, 1996)
Chalcone isomerase*	*A. thaliana* (Kubasek *et al.*, 1992; Brosché *et al.*, 1999)
(+)6a-hydroxymaackiain 3-*O*-methyltransferase	*A. thaliana* (Brosché *et al.*, 1999)
Pathogenesis-related proteins	
Acidic PR1**	*Nicotiana tabacum* (Green and Fluhr, 1995; Yalpani *et al.*, 1994)
Acidic PR-1*, PR-2*, PR-5**	*A. thaliana* (Surplus *et al.*, 1998)
PR-1**, PDF1–2***	*A. thaliana* (Mackerness *et al.*, 1999)
Antioxidant enzymes	
Catalase 2***, 3*	*N. plumbaginifolia* (Willekens *et al.*, 1994b)
Glutathione peroxidase***	*N. plumbaginifolia* (Willekens *et al.*, 1994b)
Glutathione reductase*	*Pisum sativum* (Strid *et al.*, 1994)
Others	
Ubiquitin-conjugating enzyme***,[1]	*A. thaliana* (Brosché *et al.*, 1999)
Polyubiquitin***,[1]	*A. thaliana* (Brosché *et al.*, 1999)

***, rapid response (0.5–3 h); **, intermediate response (4–8 h); *, late response (9–48 h).
[1] One out of two isoforms.

photoproducts have been implicated in reduced hypocotyl growth in sunflower seedlings (Tevini, 1993). In the same species, UV-B radiation caused an increase in ethylene biosynthesis in the leaves. Further analysis of growth regulators is clearly needed to understand growth responses and alterations in flowering in plants exposed to UV-B.

UV-B radiation and ozone share common features in affecting specific antioxidant gene and protein levels (see below). UV-B and ozone treatments induced the transcripts for distinct catalase isoforms, *CAT2* and *CAT3*, while *CAT1* was reduced (Table 3; Willekens *et al.*, 1994b). SOD isoforms were not affected in this study which used balanced UV-B/UV-A/PAR ratios that did not lead to growth reduction and visible damage in the experimental plant, *N. plumbaginifolia*. Subsequent studies on *A. thaliana* demonstrated that mainly Cu/ZnSOD was affected by UV-B while MnSOD was not responsive (Rao and Ormrod, 1995; Rao *et al.*, 1996). Abundance of GR transcripts was elevated by UV-B exposure in pea together with a decrease of cytosolic Cu/ZnSOD (Strid *et al.*, 1994). Wild-type plants and *tt5* mutants showed a comparable pattern of protein activation for APX, CAT, GR, POD and SOD, but maximum levels were reached much earlier in the flavonoid mutant because of the higher effective fluence rates (Rao *et al.*, 1996).

PR proteins react to UV-C and UV-B radiation. Mainly acidic isoforms of β-1,3-glucanase and chitinase as well as PR1 protein seem to be responsive. UV-C/UV-B treatment caused an induction of various PR proteins together with lesion development in tobacco leaves (Brederode *et al.*, 1991; Yalpani *et al.*, 1994). This response may be due to wounding effects of the radiation rather than receptor-mediated responses. UV-B radiation was described as a potent inducer of three isoforms of PR1 at the transcript and protein level (Green and Fluhr, 1995). The responses were fluence- and wavelength-dependent. Antioxidant compounds reduced PR1 accumulation (Green and Fluhr, 1995), suggesting that oxidative stress was responsible for the induction. Supplementary UV-B radiation led to decreased transcript levels of photosynthesis-associated genes and a concomitant increase in acidic PR1, PR2 and PR5 in *A. thaliana* (Surplus *et al.*, 1998). On the other hand, enzyme activities and protein levels of basic β-1,3-glucanase (PR2) and chitinase (PR3) were not affected by UV-B treatment (cut-off at 291 nm; Thalmair *et al.*, 1996). Responses were only detected at cut-off wavelengths below 290 nm which are not experienced by field-grown plants, and which caused the typical glazing symptoms on tobacco. It will be necessary for future experiments to properly characterize the light regime for PR induction in order to separate primary effects by radiation from those resulting from cellular damage.

INTERFERENCE OF AIR POLLUTANTS AND UV-B RADIATION WITH SIGNALLING PATHWAYS IN PLANTS

A number of studies have now given evidence that ozone induces the expression of defense-related genes that are also induced in response to pathogen infection (for a review, see Sandermann *et al.*, 1998). This phenomenon has been termed "cross-induction", and different stresses have been proposed to activate the same or overlapping signalling pathways. Ozone-induced necrotic lesion development has much in common with the HR response that results from infection with incompatible pathogens (Goodman and Novacky, 1994; Morel and Dangl, 1997). The HR occurs as a localized cell death at the site of infection and is thought to prevent invading pathogens from spreading throughout the leaf (see Chapter 5, this volume). High levels of H_2O_2, together with other unknown factors, being produced at the site of infection, may cause host cell death (Tenhaken *et al.*, 1995; Alvarez *et al.*, 1998). Following exposure to acute ozone levels, HR-like responses starting with the localized death of a small number of mesophyll cells are found in tobacco, tomato, poplar, and birch, whereas other species, such as potato, cereals, and conifers typically respond with symptoms of chlorosis and accelerated senescence (Heath and Taylor 1997; Pell *et al.*, 1997; Sandermann *et al.*, 1998). In the following paragraphs, we will concentrate on ozone symptoms of the HR-type and discuss possible signalling molecules activated before the appearance of visible damage.

Ethylene emission is one of the earliest responses to ozone exposure and other stress factors and has been implicated in lesion formation of *Arabidopsis* mutants (Abeles *et al.*, 1992; Greenberg and Ausubel, 1993; Rate *et al.*, 1999). Part of the mutant lines exhibiting uncontrolled cell death required a stimulus, such as ethylene, for the phenotype to appear, suggesting that ethylene could modulate HR and control cell death in plants (Greenberg, 1996). There is also evidence for an involvement of ethylene in cell death processes during pea carpel senescence (Orzáez and Granell, 1997), hypoxia-induced aerenchyma formation in maize root cortex (He *et al.*, 1996) and maize endosperm development (Young *et al.*, 1997). The oxidative burst of ROS is an integral and essential part in the initiation and propagation of programmed cell death in plant cells (Lamb and Dixon, 1997; Bolwell, 1999). The spatial location of ethylene synthesis (M. Utriainen, W. Moeder, J. Kangasjärvi, and C. Langebartels, unpublished results) and H_2O_2 production (Schraudner *et al.*, 1998; see below) correlate closely with cell death in tobacco and tomato. Inhibition of ethylene synthesis (Tuomainen *et al.*, 1997) and perception (Bae *et al.*, 1996) reduced lesion formation in these species. These results suggest that both ethylene and ROS could be factors required for cell death in ozone-treated plants, but more experiments are needed to clarify this point. It may turn out that visible injury does not result directly from ozone or ozone-derived ROS, but indirectly from the induction of a cell death programme mediated by ROS, salicylic acid and ethylene. Neither ethylene nor H_2O_2 alone were sufficient to produce the typical necrotic lesions. Therefore it seems that at least two components are needed to trigger programmed cell death, both of which being induced by ozone in sensitive species and cultivars.

Several signalling molecules, such as ethylene, H_2O_2, products of a lipoxygenase pathway, jasmonate, and salicylate have been implicated in the defense of plants towards pathogens. As will be shown below, exposure of plants to ozone or UV-B radiation results in the activation of multiple signalling pathways. The signals include ethylene and ACC for ozone, while salicylic acid, elevated ROS levels, and jasmonate have been implicated in the response to both stressors. Figure 3 summarizes possible signalling pathways involved in ozone and UV-B responses of plants. Ethylene seems to be a primary response in ozone, whereas its UV-B responsiveness has only been demonstrated in sunflowers (Tevini, 1993). Ethylene is thought to act through a kinase cascade (for a review, see Solano and Ecker, 1998; Chang and Shockey, 1999), and possibly mediates the activation of basic PR proteins (Schraudner *et al.*, 1992) and stilbene synthase (Schubert *et al.*, 1997). Ethylene-responsive promoter elements have been analyzed for PR1b, basic PR2, and stilbene synthase, and these studies have recently provided compelling evidence for the role of ethylene in ozone responses of tobacco and other plants (Ernst *et al.*, 1999).

Hydrogen peroxide may act as a diffusible signal molecule in the induction of distinct antioxidant genes, such as GST and GPX (Lamb and Dixon, 1997; Morel and Dangl, 1997). By analogy, ozone-derived ROS, such as H_2O_2 and superoxide, may initiate

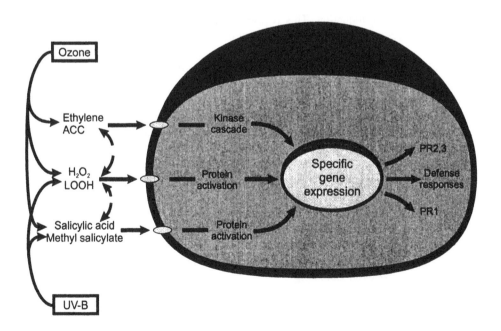

Figure 3. Interference of ozone and UV-B radiation with signalling pathways in plants.

defense responses. Elevated levels of ROS were found after exposure to UV-B (Green and Fluhr, 1995) and ozone (Schraudner *et al.*, 1998). The levels of apoplastic H_2O_2 were similar in ozone-sensitive and ozone-tolerant tobacco cultivars during the ozone exposure period, suggesting that this initial response is not the basis of differential ozone tolerance in this species. Transcript levels for antioxidant genes (*CAT2*, *GPX*) were elevated shortly thereafter (Willekens *et al.*, 1994b; Schraudner *et al.*, 1998), possibly as a response to increased H_2O_2 levels. A second peak of highly localized H_2O_2 production was demonstrated in ozone-sensitive, but not ozone-tolerant tobacco (Schraudner *et al.*, 1998). This peak occurred during post-cultivation in pollutant-free air, suggesting a cellular origin of the accumulated ROS via activation of NAD(P)H oxidase, oxalate oxidase, or diamine oxidase activity. The areas of H_2O_2 accumulation corresponded in location, pattern and size with those of localized cell death and visible injury observed later (Schraudner *et al.*, 1998). Hydrogen peroxide or superoxide accumulation (an ozone-induced "oxidative burst") was subsequently also found in ozone-treated *A. thaliana* (Overmyer *et al.*, 1998; Rao and Davis, 1999) and birch (Pellinen *et al.*, 1999). More work is needed to unravel the putative ROS-producing enzymes and to demonstrate the occurrence of this ozone-induced "burst" in other sensitive plants.

Green and Fluhr (1995) as well as Surplus *et al.*, (1998) demonstrated the involvement of ROS in UV-B induction of PR1 proteins. This induction was abolished by antioxidant compounds, and could be achieved in the absence of UV-B by the use of ROS generating systems. Thus, ozone and UV-B that induce PR1, the marker protein for SAR, seem to lead to an "oxidative burst" as a primary signal for defense responses, possibly via activation of NAD(P)H oxidase (Rao *et al.*, 1996). Recently, ozone effects on the stilbene biosynthetic enzyme, stilbene synthase, were studied in transgenic tobacco under the regulation of the Vst1 promoter from grapevine (Grimmig *et al.*, 1997; Schubert *et al.*, 1997). Deletion mutants showed that a promotor region between −430 and −280 bp was responsible for ozone induction. This region differed from the pathogen- and ethylene-responsive regions and possibly contained new oxidant-responsive elements (Ernst *et al.*, 1999).

It has also been suggested that defense responses are mediated by a lipid-based signal transduction cascade under both ozone and UV-B exposure. Key steps seem to be the liberation of linolenic (18:3) and linoleic (18:2) acids from cell membranes (Conconi *et al.*, 1996a, 1996b) and the subsequent formation of lipid hydroperoxides (LOOH; Schraudner *et al.*, 1996). Lipid hydroperoxides are then transformed into a variety of products, such as cyclopentanones, aldehydes, alcohols, and jasmonate (Farmer *et al.*, 1998). Jasmonic acid, a 12-carbon cyclopentanone metabolite and a known transcriptional activator, accumulates after wounding, ozone or UV-B exposure (Conconi *et al.*, 1996b; Örvar *et al.*, 1997). Jasmonate together with ethylene was implicated in the UV-B induction of the plant pathogen-related gene *PDF1-2* in *A. thaliana* (Mackerness *et al.*, 1999). It was further shown that UV-B activation of proteinase inhibitors was blocked by an inhibitor of the octadecanoid pathway and did not occur in a tomato mutant defective in this pathway. These findings led the authors to conclude that the response to UV-B exposure was mediated by octadecanoid signalling.

Salicylic acid is a major component of the signalling pathways leading to HR and SAR (Draper, 1997; Durner *et al.*, 1997). The levels of salicylic acid respond to ozone (Yalpani *et al.*, 1994; Sharma *et al.*, 1996; Moeder *et al.*, 1999) and UV-B radiation (Yalpani *et al.*, 1994) in tobacco and *A. thaliana*. UV-B radiation induced salicylic acid levels as well as transcript levels of PR1, PR2, and PR5 in *A. thaliana*. It was postulated that UV-B effects on acidic PR genes are mediated by ROS and salicylic acid (Surplus *et al.*, 1998). The use of transgenic plants that express a bacterial salicylate hydroxylase gene (*NahG*) provided evidence that mRNA levels of PR1 and, partly, of GST, but not of PAL, were induced via the salicylate pathway (Sharma and Davis, 1997). The ozone-sensitive tobacco Bel W3 accumulated salicylic acid and emitted methyl salicylate to high levels (Heiden *et al.*, 1999; W. Moeder and C. Langebartels, unpublished results) suggesting that salicylic acid potentiates cell death in ozone-treated tobacco. Salicylic acid, ROS, cell death, and ethylene have recently been proposed by Van Camp *et al.*, (1998) to act in a self-amplifying "oxidative cell death cycle": similar to the pathogen response, this cycle appears to act in ozone-sensitive tobacco, as well.

CONCLUSIONS AND PERSPECTIVES

Ozone as well as UV-B radiation act as abiotic elicitors of responses that are characteristic of HR and SAR in pathogen-infected plants. These responses include antioxidant as well as antimicrobial metabolites and proteins, and therefore seem to be multifunctional, protecting tissues from both, oxidative and pathogen stress. Part of the ozone responses is probably mediated by interference with at least three different signalling pathways, depending on ethylene, ROS/LOOH, and salicylate (Figure 3). ROS, octadecanoid, and salicylate pathways are thought to be involved in UV-B responses. These pathways may be activated in parallel or in sequence (Van Camp *et al.*, 1998; Van Loon *et al.*, 1998), and antagonistic interactions of salicylic acid, jasmonic acid and ethylene have been described (Li *et al.*, 1992). Therefore, "cross-talk" may be assumed to exist between the above pathways, and combinations of signals might be necessary to trigger the expression of antioxidant and pathogen defense responses. This assumption is also reflected in the multitude of putative regulatory sequences in the promoter regions of ozone-induced genes including Myb-like recognition elements as well as elements responsive to ethylene, salicylic acid, jasmonic acid, and elicitors (Ernst *et al.*, 1999). It will be shown in the future which of the responses are primary reactions to the environmental stress, and which are found in relation with the development of lesions and resulting wounded cells. The most promising way to look at these processes is by carefully analyzing the kinetics of both phases in mutant lines and wild-type plants, under adequately simulated outdoor conditions.

In ozone-treated plants, little is known on the mechanisms of ethylene, ROS and salicylic acid perception, and control of gene expression. Major progress has been achieved in the identification of ethylene receptors in *A. thaliana* and tomato (Hua and Meyerowitz, 1998; McGrath and Ecker, 1998; Chang and Shockey, 1999), which were shown to be similar in sequence to "histidine-aspartate phosphorelais" (formally known as "communication modules") in bacterial signaling (Parkinson and Kofoid, 1992). These receptors fulfill a variety of functions in bacteria by sensing osmolarity (EnvZ), levels of nutrients and heavy metals (NtrB, CutR) as well as the redox state (ArcA, ArcB) of the surrounding medium. They are also involved in host detection and invasion (NodV, VirA, VirG; Pierson *et al.*, 1998). Recent studies demonstrate that additional components of "phosphorelais" exist in cells of higher plants apart from the ethylene and cytokinin receptors (Imamura *et al.*, 1998; Kakimoto, 1998). These modules or other sensor proteins could monitor the apoplastic fluid surrounding mesophyll cells. It may, for example, be speculated that a redox-sensitive module is also responsible for ozone perception in sensitive plants.

The intracellular and tissue localization of ozone- and UV-B-responsive genes, proteins, and metabolites is also interesting to note. Typical UV-B responses occur in epidermal layers of needles and leaves, increasing their antioxidative and antimicrobial capacity, by flavonoid accumulation, for instance. Ozone responses, such as the accumulation of ascorbate, polyamine and tyramine conjugates, acidic PR proteins, and extensin are typically directed to the cell wall and apoplastic fluid (Sandermann *et al.*, 1998; Moeder *et al.*, 1999). In the apoplast, they may constitute a first-line defense towards air pollutants once entered into the intercellular space. The responses, however, are not homogeneously distributed in the leaf, but occur preferentially in the periveinal regions. This observation has been exemplified for the ozone-responsive catalase, CAT2, tyramine integration into cell wall material, PR proteins, the ethylene-forming enzyme, ACC oxidase, as well as the "burst initiation" sites exhibiting H_2O_2 accumulation (Willekens *et al.*, 1994b; Schraudner *et al.*, 1996, 1998; W. Moeder, J. Kangasjärvi, and C. Langebartels, unpublished results). In addition, the occurrence of visible ozone symptoms in the vicinity of leaf veins has already been described in early studies by Hill *et al.*, (1961) and Lucas (1975). Preferential accumulation of stress proteins and genes near veins has also been found for β-1,3-glucanase and chitinase proteins after ethylene treatment (Mauch *et al.*, 1992) and for transcripts for ACC oxidase following wounding (Bowles, 1992). In addition, the recently discovered "micro-bursts" and "micro-HRs" in systemic leaves of *A. thaliana* infected with a necrotizing pathogen also occur in periveinal regions (Alvarez *et al.*, 1998). Both locations, apoplast and periveinal regions, are sites for the attack and the spreading of pathogens growing intercellularly within the mesophyll tissue or using the vascular system to spread within the plant. Ozone responses may thus occur at "preformed" sites for pathogen defense, and the plant seems to misinterpret ozone as invading pathogens. Whether the above responses are individually regulated at these sites or are an indication of a general mechanism of local defense is currently unknown.

Exposure of plants to ozone and UV-B radiation is known to affect subsequent pathogen infections (Manning and von Tiedemann, 1995). Ozone exposure may lead to increased tolerance or susceptibility, depending on the plant and pathogen studied, exposure regime, and timing of exposure and infection. The identification of ozone as a major elicitor of plant defense reactions may help to explain that infection by biotrophic pathogens is initially reduced, whereas infections by necrotrophic pathogens are supported after long-term exposure (Sandermann, 1999). At the moment, it is not clear how far the analogy between ozone and pathogen responses can be carried. Striking similarities, though, exist not only in herbaceous plants but also in deciduous and coniferous trees (Langebartels *et al.*, 1997). Studies on ozone and UV-B radiation have also led to the identification of new responses possibly involved in general plant defense. As plants with enhanced antioxidative capacity often show "cross-tolerance" to other types of stress, these studies may help resolve basic problems in plant defense, because of the relative ease of experimentation. Finally, the studies will also be valuable in setting new standards for ozone and UV-B radiation in plants as the above characterized plant responses have not been taken into account in the presently discussed critical levels.

Acknowledgements

The authors wish to thank all colleagues for their suggestions and helpful discussions, and for providing manuscripts before publication. We are also grateful to our co-workers for their important contributions during the past years. This work was supported by grants from DFG (Sonderforschungsbereich 607), BMBF, BStMLU (PBWU, BayFORKLIM), Fondation Limagrain (Chappes, France) and Fonds der Chemischen Industrie.

REFERENCES

Abeles, F.B., Morgan, P.W., and Saltveit, M.E. (1992) *Ethylene in Plant Biology*, Academic Press, New York.

Allen, D.J., Nogués, S., and Baker, N.R. (1998) Ozone depletion and increased UV-B radiation: Is there a real threat to photosynthesis? *J. Exp. Bot.*, **49**, 1775–1788.

Allen, R.D. (1995) Dissection of oxidative stress tolerance using transgenic plants. *Plant Physiol.*, **107**, 1049–1054.

Alvarez, M., Pennell, R., Meijer, P.-J., Ishikawa, A., Dixon, R., and Lamb, C. (1998) Reactive oxygen intermediates mediate a systemic signal network in the establishment of plant immunity. *Cell*, **92**, 773–784.

Aono, M., Kubo, A., Saji, H., Natori, T., Tanaka, K., and Kondo, N. (1991) Resistance to active oxygen toxicity of transgenic *Nicotiana tabacum* that express the gene for glutathione reductase from *Escherichia coli*. *Plant Cell Physiol.*, **32**, 691–697.

Aono, M., Kubo, A., Saji, H., Tanaka, K., and Kondo, N. (1993) Enhanced tolerance to photooxidative stress of transgenic *Nicotiana tabacum* with high chloroplastic glutathione reductase activity. *Plant Cell Physiol.*, **34**, 129–135.

Aono, M., Ando, M., Nakajima, N., Kubo, A., Kondo, N., Tanaka, K., and Saji, H. (1997) Response to photooxidative stress of transgenic tobacco plants with altered activities of antioxidant enzymes. *Plant Physiol.* (Suppl.), **114**, 101.

Bae, G.Y., Nakajima, N., Ishizuka, K., and Kondo, N. (1996) The role in ozone phytotoxicity of the evolution of ethylene upon induction of 1-aminocyclopropane-1-carboxylic acid synthase by ozone fumigation in tomato plants. *Plant Cell Physiol.*, **37**, 129–134.

Bahl, A., Loitsch, S., and Kahl, G. (1995) Transcriptional activation of plant defence genes by short-term air pollution stress. *Environ. Pollut.*, **89**, 221–227.

Baker, C., and Orlandi, E. (1995) Active oxygen in plant pathogenesis. *Annu. Rev. Phytopathol.*, **33**, 299–321.

Bauer, S., Galliano, H., Pfeiffer, F., Messner, B., Sandermann, H., and Ernst, D. (1993) Isolation and characterization of a cDNA clone encoding a novel short-chain alcohol dehydrogenase from Norway spruce (Picea abies L. Karst). *Plant Physiol.*, **103**, 1479–1480.

Bolwell, G.P. (1999) Role of active oxygen species and NO in plant defence responses. *Curr. Opin. Plant Biol.*, **2**, 287–294.

Bors, W., Langebartels, C., Michel, C., and Sandermann, H. (1989) Polyamines as radical scavengers and protectants against ozone damage. *Phytochemistry*, **28**, 1589–1595.

Bors, W., Heller, W., Michel, C., and Saran, M. (1992) Structural principles of flavonoid antioxidants. In G. Csomó, and J. Fehér, (eds.), *Free Radicals and the Liver*, Springer, Berlin, pp. 77–95.

Bors, W., Michel, C., and Schikora, S. (1995) Interaction of flavonoids with ascorbate and determination of their univalent redox potentials: A pulse radiolysis study. *Free Rad. Biol. Med.*, **19**, 45–52.

Bors, W., Heller, W., and Michel, C. (1998) The chemistry of flavonoids. In C.A. Rice-Evans and L. Packer, (eds.), *Flavonoids in Health and Disease*, Marcel Dekker, New York, pp. 111–136.

Bouchereau, A., Aziz, A., Larher, F., and Martin-Tanguy, J. (1999) Polyamines and environmental challenges: Recent development. *Plant Sci.*, **140**, 103–125.

Bowles, D. (1992) Signals in the wounded plant. In P.G. Ayres, (ed.), *Pests and Pathogens-Plant Responses to Foliar Attack*, BIOS Scientific Publishers, Oxford, pp. 33–38.

Brederode, F., Linthorst, H., and Bol, J. (1991) Differential induction of acquired resistance and PR gene expression in tobacco by virus infection, ethylene treatment, UV light and wounding. *Plant Mol. Biol.*, **17**, 1117–1125.

Broadbent, P., Creissen, G., Kular, B., Wellburn, A., and Mullineaux, P. (1995) Oxidative stress responses in transgenic tobacco containing altered levels of glutathione reductase activity. *Plant J.*, **8**, 247–255.

Brosché, M., Fant, C., Bergkvist, S.W., Strid, H., Svensk, A., Olsson, O., and Strid, Å. (1999) Molecular markers for UV-B stress in plants: Alteration of the expression of four classes of genes in *Pisum sativum* and the formation of high molecular mass RNA adducts. *Biochim. Biophys. Acta*, **1447**, 185–198.

Buschmann, K., Etscheid, M., Riesner, D., and Scholz, F. (1998) Accumulation of a porin-like mRNA and a metallothionein-like mRNA in various clones of Norway spruce upon long-term treatment with ozone. *Eur. J. For. Pathol.* **28**, 307–322.

Caldwell, M.M. and Flint, S.D. (1994) Solar ultraviolet radiation and ozone layer change: Implications for crop plants. In K.J. Boote, J.M. Bennett, T.R. Sinclair, and G.M. Paulson, (eds.), *Physiology and Determination of Crop Yield*, American Society of Agronomy, Madison, pp. 487–507.

Cape, J.N. (1997) Photochemical oxidants-What else is in the atmosphere besides ozone? *Phyton*, **37**, 45–58.

Chamnongpol, S., Willekens, H., Langebartels, C., Van Montagu, M., Inzé, D., and Van Camp, W. (1996) Transgenic tobacco with a reduced catalase activity develops necrotic lesions and induces pathogenesis-related expression under high light. *Plant J.*, **10**, 491–503.

Chamnongpol, S., Willekens, H., Moeder, W., Langebartels, C., Sandermann, H., Van Montagu, M., Inzé, D., and Van Camp, W. (1998) Defense activation and enhanced pathogen tolerance induced by H_2O_2 in transgenic tobacco. *Proc. Natl. Acad. Sci. USA*, **95**, 5818–5823.

Chang, C. and Shockey, J.A. (1999) The ethylene-response pathway: Signal perception to gene regulation. *Curr. Opin. Plant Biol.*, **2**, 352–358.

Christie, J. and Jenkins, G. (1996) Distinct UV-B and UV-A blue light signal transduction pathways induce chalcone synthase gene expression in *Arabidopsis* cells. *Plant Cell*, **8**, 1555–1567.

Conconi, A., Miquel, M., Browse, J.A., and Ryan, C.A. (1996a) Intracellular levels of free linolenic and linoleic acids increase in tomato leaves in response to wounding. *Plant Physiol.*, **111**, 797–803.

Conconi, A., Smerdon, M.J., Howe, G.A., and Ryan, C.A. (1996b) The octadecanoid signalling pathway in plants mediates a response to ultraviolet radiation. *Nature*, **383**, 826–829.

Conklin, P.L. and Last, R.L. (1995) Differential accumulation of antioxidant mRNAs in *Arabidopsis thaliana* exposed to ozone. *Plant Physiol.*, **109**, 203–212.

Conklin, P.L., Williams, E.H., and Last, R.L. (1996) Environmental stress sensitivity of an ascorbic acid-deficient *Arabidopsis* mutant. *Proc. Natl. Acad. Sci. USA*, **93**, 9970–9974.

Conklin, P.L., Pallanca, J., Last, R., and Smirnoff, N. (1997) L-Ascorbic acid metabolism in the ascorbate-deficient *Arabidopsis* mutant vtc1. *Plant Physiol.*, **115**, 1277–1285.

Creissen, G., Broadbent, P., Stevens, R., Wellburn, A.R., and Mullineaux, P. (1996) Manipulation of glutathione metabolism in transgenic plants. *Biochem. Soc. Trans.*, **24**, 465–469.

Dai, Q., Yan, B., Huang, S., Liu, X., Peng, S., Miranda, M.L.M., Chavez, A.Q., Vegara, B.S., and Olszyk, D. (1997) Response of oxidative stress defence systems in rice (*Oryza sativa*) leaves with supplemental UV-B radiation. *Physiol. Plant.*, **101**, 301–308.

Delledonne, M., Xia, Y., Dixon, R.A., and Lamb, C. (1998) Nitric oxide signal functions in plant disease resistance. *Nature*, **394**, 585–588.

Doehring, T., Köfferlein, M., Thiel, S., and Seidlitz, H.K. (1996) Spectral shaping of artificial UV-B irradiation for vegetation stress research. *J. Plant Physiol.*, **148**, 115–119.

Draper, J. (1997) Salicylate, superoxide synthesis and cell suicide in plant defence. *Trends Plant Sci.*, **2**, 162–166.

Durner, J. and Klessig, D.F. (1999) Nitric oxide as a signal in plants. *Curr. Opin. Plant Biol.*, **2**, 369–374.

Durner, J., Shah, J., and Klessig, D.F. (1997) Salicylic acid and disease resistance in plants. *Trends Plant Sci.*, **2**, 266–274.

Durner, J., Wendehenne, D., and Klessig, D. (1998) Defense gene induction in tobacco by nitric oxide, cyclic GMP, and cyclic ADP-ribose. *Proc. Natl. Acad. Sci. USA*, **95**, 10328–10333.

Eckey-Kaltenbach, H., Ernst, D., Heller, W., and Sandermann, H. (1994) Biochemical plant responses to ozone: IV. Cross-induction of defensive pathways in parsley (*Petroselinum crispum* L.) plants. *Plant Physiol.*, **104**, 67–74.

Eckey-Kaltenbach, H., Kiefer, E., Grosskopf, E., Ernst, D., and Sandermann, H. (1997) Differential transcript induction of parsley pathogenesis-related proteins and of a small heat shock protein by ozone and heat shock. *Plant Mol. Biol.*, **33**, 343–350.

Elstner, E.F., Osswald, W., and Youngman, R.J. (1985) Basic mechanisms of pigment bleaching and loss of structural resistance in spruce (*Picea abies*) needles: advances in phytomedical diagnostics. *Experientia*, **41**, 591–597.

Ernst, D., Schraudner, M., Langebartels, C., and Sandermann, H. (1992) Ozone-induced changes of mRNA levels of β-1,3-glucanase, chitinase and 'pathogenesis-related' protein 1b in tobacco plants. *Plant Mol. Biol.*, **20**, 673–682.

Ernst, D., Grimmig, B., Heidenreich, B., Schubert, R., and Sandermann, H. (1999) Ozone-induced genes: Mechanisms and biotechnological applications. In M.F. Smallwood, C.M. Calvert, and D.J Bowles, (eds.), *Plant Responses to Environmental Stress*, BIOS, Oxford, pp. 33–41.

Eshdat, Y., Holland, D., Faltin, Z., and Ben-Hayyim, G. (1997) Plant glutathione peroxidases. *Physiol. Plant.*, **100**, 234–240.

Farmer, E.E., Weber, H., and Vollenweider, S. (1998) Fatty acid signaling in *Arabidopsis*. *Planta*, **206**, 167–174.

Fischbach, R.J., Kossmann, B., Panten, H., Steinbrecher, R., Heller, W., Seidlitz, H.K., Sandermann, H., Hertkorn, N., and Schnitzler, J.P. (1999) Seasonal accumulation of ultraviolet-B screening pigments in needles of Norway spruce (*Picea abies* (L.) Karst.). *Plant Cell Environ.*, **22**, 27–37.

Fiscus, E.L. and Booker, F.L. (1995) Is increased UV-B a threat to crop photosynthesis and productivity? *Photosynthesis Res.*, **43**, 81–92.

Floyd, R.A., West, M.S., Hogsett, W.E., and Tingey, D.T. (1989) Increased 8-hydroxyguanine content of chloroplast DNA from ozone-treated plants. *Plant Physiol.*, **91**, 644–647.

Forkmann, G. and Heller, W. (1999) Biosynthesis of flavonoids. In D. Barton, K. Nakanishi, and O. Meth-Cohn, (eds.), *Comprehensive Natural Products Chemistry*, Elsevier, Amsterdam, pp. 713–748.

Foyer, C.H. (1997) Oxygen metabolism and electron transport in photosynthesis. In J.G. Scandalios, (ed.), *Oxidative Stress and the Molecular Biology of Antioxidant Defenses*, (Monograph Series, Vol. 34), Cold Spring Harbor Laboratory Press, Cold Spring Harbor, pp. 587–621.

Foyer, C.H., Descourvières, P., and Kunert, K.J. (1994) Protection against oxygen radicals: An important defence mechanism studied in transgenic plants. *Plant Cell Environ.*, **17**, 507–523.

Foyer, C.H., Lopez-Delgado, H., Dat, J.F., and Scott, I.M. (1997) Hydrogen peroxide- and glutathione-associated mechanisms of acclimatory stress tolerance and signalling. *Physiol. Plant.*, **100**, 241–254.

Fuglevand, G., Jackson, J.A., and Jenkins, G.I. (1996) UV-B, UV-A, and blue light signal transduction pathways interact synergistically to regulate chalcone synthase gene expression in *Arabidopsis*. *Plant Cell*, **8**, 2347–2357.

Galliano, H., Cabané, M., Eckerskorn, C., Lottspeich, F., Sandermann, H., and Ernst, D. (1993) Molecular cloning, sequence analysis and elicitor-/ozone-induced accumulation of cinnamyl alcohol dehydrogenase from Norway spruce (*Picea abies* L.). *Plant Mol. Biol.*, **23**, 145–156.

Goodman, R.N. and Novacky, A.J. (1994) *The Hypersensitive Reaction in Plants to Pathogens: A Resistance Phenomenon*. American Phytopathological Society Press, St. Paul.

Green, R. and Fluhr, R. (1995) UV-B-induced PR-1 accumulation is mediated by active oxygen species. *Plant Cell*, **7**, 203–212.

Greenberg, J.T. (1996) Programmed cell death: A way of life for plants. *Proc. Natl. Acad. Sci. USA*, **93**, 12094–12097.

Greenberg, J.T. and Ausubel, F.M. (1993) *Arabidopsis* mutants compromised for the control of cellular damage during pathogenesis and aging. *Plant J.*, **4**, 327–341.

Grimmig, B., Schubert, R., Fischer, R., Hain, R., Schreier, P.H., Betz, C., Langebartels, C., Ernst, D., and Sandermann, H. (1997) Ozone- and ethylene-induced regulation of a grapevine resveratrol synthase promoter in transgenic tobacco. *Acta Physiol. Plant.*, **19**, 467–474.

Hahlbrock, K. and Scheel, D. (1989) Physiology and molecular biology of phenylpropanoid metabolism. *Annu. Rev. Plant Physiol. Plant Mol. Biol.*, **40**, 347–369.

He, C.-J., Morgan, P.W., and Drew, M.C. (1996) Transduction of an ethylene signal is required for cell death and lysis in the root cortex of maize during aerenchyma formation induced by hypoxia. *Plant Physiol.*, **112**, 463–472.

Heath, R.L. and Taylor, G.E. (1997) Physiological processes and plant responses to ozone exposure. In H. Sandermann, A.R. Wellburn, and R.L. Heath, (eds.), *Ozone and Forest Decline: A Comparison of Controlled Chamber and Field Experiments*, (Ecological Studies, Vol. 127), Springer, Berlin, pp. 317–368.

Heiden, A.C., Hoffmann, T., Kahl, J., Kley, D., Klockow, D., Langebartels, C., Mehlhorn, H., Sandermann, H., Schraudner, M., Schuh, G., and Wildt, J. (1999) Emission of volatile organic compounds from ozone-exposed plants. *Ecol. Appl.*, **9**, 1160–1167.

Heller, W. (1994) Topics in the biosynthesis of plant phenols. *Acta Horticulturae*, **381**, 46–73.

Hellgren, L.I., Carlsson, A.S., Selldén, G., and Sandelius, A.S. (1995) *In situ* leaf lipid metabolism in garden pea (*Pisum sativum* L.) exposed to moderately enhanced levels of ozone. *J. Exp. Bot.*, **46**, 221–230.

Hill, A.C., Pack, M.R., Treshow, M., Downs, R.J., and Transtrum, L.G. (1961) Plant injury induced by ozone. *Phytopathology*, **51**, 356–363.

Himelblau, E., Mira, H., Lin, S.J., Culotta, V.C., Penarrubia, L., and Amasino, R.M. (1998) Identification of a functional homolog of the yeast copper homeostasis gene *ATX1* from *Arabidopsis*. *Plant Physiol.*, **117**, 1227–1234.

Hippeli, S. and Elstner, E.F. (1996) Mechanisms of oxygen activation during plant stress: Biochemical effects of air pollutants. *J. Plant Physiol.*, **148**, 249–257.

Hua, J. and Meyerowitz, E.M. (1998) Ethylene responses are negatively regulated by a receptor gene family in *Arabidopsis thaliana*. *Cell*, **94**, 261–271.

Imamura, A., Hanaki, N., Umeda, H., Nakamura, A., Suzuki, T., Ueguchi, C., and Mizuno, T. (1998) Response regulators implicated in His-to-Asp phosphotransfer signaling in *Arabidopsis*. *Proc. Natl. Acad. Sci. USA*, **95**, 2691–2696.

Inzé, D. and Van Montagu, M. (1995) Oxidative stress in plants. *Curr. Opin. Biotechnol.*, **6**, 153–158.

Jakob, B. and Heber, U. (1998) Apoplastic ascorbate does not prevent the oxidation of fluorescent amphiphilic dyes by ambient and elevated concentrations of ozone in leaves. *Plant Cell Physiol.*, **39**, 313–322.

Jansen, M.A.K., Gaba, V., and Greenberg, B.M. (1998) Higher plants and UV-B radiation: Balancing damage, repair and acclimation. *Trends Plant Sci.*, **3**, 131–135.

Jenkins, G.I. (1997) UV and blue light signal transduction in *Arabidopsis*. *Plant Cell Environ.*, **20**, 773–778.

Jenkins, G.I. (1998) UV and blue light signal transduction in the regulation of flavonoid biosynthesis gene expression in *Arabidopsis*. In F. Lo Schiavo, R.L. Last, G. Morelli, and N.V. Raikhel, (eds.), *Cellular Integration of Signalling Pathways in Plant Development*, Springer, Berlin, pp. 71–82.

Jordan, B.R. (1996) The effects of ultra violet-B radiation on plants: A molecular perspective. In J.A. Callow, (ed.), *Advances in Botanical Research*, Vol. 22, Academic Press, San Diego, pp. 97–162.

Jungblut, T.P., Schnitzler, J.P., Hertkorn, N., Heller, W., and Sandermann, H. (1995) Structure of UV-B induced sunscreen pigments of Scots pine (Pinus sylvestris L.). *Angew. Chem. Int. Ed. Engl.*, **34**, 312–314.

Kakimoto, T. (1998) Cytokinin signaling. *Curr. Opin. Plant Biol.*, **1**, 399–403.

Kangasjärvi, J., Talvinen, J., Utriainen, M., and Karjalainen, R. (1994) Plant defence systems induced by ozone. *Plant Cell Environ.*, **17**, 783–794.

Kärenlampi, L. and Skärby, L. (1996) Critical levels for ozone in Europe: Testing and finalizing the concepts. UN-ECE workshop report, University of Kuopio, Kuopio, Finland.

Kärenlampi, S.O., Airaksinen, K., Miettinen, A.T.E., Kokko, H.I., Holopainen, J.K., Kärenlampi, L.V., and Karjalainen, R.O. (1994) Pathogenesis-related proteins in ozone-exposed Norway spruce (*Picea abies* [Karst] L.). *New Phytol.*, **126**, 81–89.

Kerr, J.B. and McElroy, C.T. (1993) Evidence for large upward trends of ultraviolet-B radiation linked to ozone depletion. *Science*, **262**, 1032–1034.

Kiiskinen, M., Korhonen, M., and Kangasjärvi, J. (1997) Isolation and characterization of cDNA for a plant mitochondrial phosphate translocator (Mpt1). Ozone stress induces Mpt1 mRNA accumulation in birch (*Betula pendula* Roth). *Plant Mol. Biol.*, **35**, 271–279.

Kim, B.C., Tennessen, D.J., and Last, R.L. (1998) UV-B-induced photomorphogenesis in *Arabidopsis thaliana*. *Plant J.*, **15**, *667–674.*

Kim, K.Y., Huh, G.H., Lee, H.S., Kwon, S.Y., and Kwak, S.S. (1999) Molecular characterization of cDNAs for two anionic peroxidases from suspension cultures of sweet potato. *Mol. Gen. Genet.*, **261**, 941–947.

Kley, D., Kleinman, M., Sandermann, H., and Krupa, S. (1999) Photochemical oxidants: State of the science. *Environ. Pollut.*, **100**, 19–42.

Kliebenstein, D.J., Monde, R.A., and Last, R.L. (1998) Superoxide dismutase in *Arabidopsis*: An eclectic enzyme family with disparate regulation and protein localization. *Plant Physiol.*, **118**, 637–650.

Koch, J.R., Scherzer, A.J., Eshita, S.M., and Davis, K.R. (1998) Ozone sensitivity in hybrid poplar is correlated with a lack of defense-gene activation. *Plant Physiol.*, **118**, 1243–1252.

Kombrink, E. and Somssich, I.E. (1995) Defense responses of plants to pathogens. *Adv. Bot. Res.*, **21**, 1–34.

Kubasek, W., Shirley, B. McKillop, A., Goodman, H., Briggs, W., and Ausubel, F. (1992) Regulation of flavonoid biosynthetic genes in germinating *Arabiopsis* seedlings. *Plant Cell*, **4**, 1229–1236.

Kubo, A., Saji, H., Tanaka, K., and Kondo, N. (1995) Expression of *Arabidopsis* cytosolic ascorbate peroxidase gene in response to ozone or sulfur dioxide. *Plant Mol. Biol.*, **29**, 479–489.

Laisk, A., Kull, O., and Moldau, H. (1989) Ozone concentration in leaf intercellular air spaces is close to zero. *Plant Physiol.*, **90**, 1163–1167.

Lamb, C. and Dixon, R.A. (1997) The oxidative burst in plant disease resistance. *Annu. Rev. Plant Physiol. Plant Mol. Biol.*, **48**, 251–275.

Landry, L.G., Chapple, C.C.S., and Last, R.L. (1995) *Arabidopsis* mutants lacking phenolic sunscreens exhibit enhanced ultraviolet-B injury and oxidative damage. *Plant Physiol.*, **109**, 1159–1166.

Langebartels, C., Kerner, K., Leonardi, S., Schraudner, M., Trost, M., Heller, W., and Sandermann, H. (1991) Biochemical plant responses to ozone. I. Differential induction of polyamine and ethylene biosynthesis in tobacco. *Plant Physiol.*, **95**, 882–889.

Langebartels, C., Ernst, D., Heller, W., Lütz, C., Payer, H.D., and Sandermann, H. (1997) Ozone responses of trees. Results from controlled chamber exposures at the GSF phytotron. In H. Sandermann, A.R. Wellburn, and R.L. Heath, (eds.), *Forest Decline and Ozone: A comparison of Controlled Chamber and Field Experiments*, (Ecological Studies, Vol. 127), Springer, Berlin, pp. 163–200.

Langebartels, C., Ernst, D., Kangasjärvi, J., and Sandermann, H. (2000) Ozone effects on plant defense. In L. Packer and H. Sies (eds.), *Singlet Oxygen, UV-A and Ozone* (Methods in Enzymology, Vol. 319), Academic Press, San Diego, pp. 520–535.

Lea, P.J., Wellburn, F.A.M., Wellburn, A.R., Creissen, G.P., and Mullineaux, P.M. (1998) Use of transgenic plants in the assessment of responses to atmospheric pollutants. In L.J. De Kok and I. Stulen (eds.), *Responses of Plant Metabolism to Air Pollution and Global Change*, Backhuys Publishers, Leiden, pp. 241–250.

Lee, H., Jo, J., and Son, D. (1998) Molecular cloning and characterization of the gene encoding glutathione reductase in *Brassica campestris. Biochim. Biophys. Acta*, **1395**, 309–314.

Li, J., Ou Lee, T.M., Raba, R., Amundson, R.G., and Last, R.L. (1993) *Arabidopsis* flavonoid mutants are hypersensitive to UV-B irradiation. *Plant Cell*, **5**, 171–179.

Li, N., Parsons, B.L., Liu, D., and Mattoo, A.K. (1992) Accumulation of wound-inducible ACC synthase transcript in tomato fruit is inhibited by salicylic acid and polyamines. *Plant Mol. Biol.*, **18**, 477–487.

Logemann, E., Wu, S.C., Schröder, J., Schmelzer, E., Somssich, I.E., and Hahlbrock, K. (1995) Gene activation by UV light, fungal elicitor or fungal infection in *Petroselinum crispum* is correlated with repression of cell cycle-related genes. *Plant J.*, **8**, 865–876.

Long, J.C. and Jenkins, G.I. (1998) Involvement of plasma membrane redox activity and calcium homeostasis in the UV-B and UV-A/blue light induction of gene expression in *Arabidopsis. Plant Cell*, **10**, 2077–2086.

Lucas, G.B. (1975) *Diseases of Tobacco*. Biological Consulting Associates, Raleigh.

Luwe, M.W.F., Takahama, U., and Heber, U. (1993) Role of ascorbate in detoxifying ozone in the apoplast of spinach (*Spinacea oleracea* L.) leaves. *Plant Physiol.*, **101**, 969–976.

Maccarrone, M., Veldink, G.A., and Vliegenthart, J.F.G. (1992) Thermal injury and ozone stress affect soybean lipoxygenases expression. *FEBS Lett.*, **309**, 225–230.

Maccarrone, M., Veldink, G.A., Vliegenthart, J.F.G., and Agrò, A.F. (1997) Ozone stress modulates amine oxidase and lipoxygenase expression in lentil (*Lens culinaris*) seedlings. *FEBS Lett.*, **408**, 241–244.

Mackerness, S.A.H., Surplus, S.L., Blake, P., John, C.F., Buchanan-Wollaston, V., Jordan, B.R., and Thomas, B. (1999) Ultraviolet-B-induced stress and changes in gene expression in *Arabidopsis thaliana*: Role of signalling pathways controlled by jasmonic acid, ethylene and reactive oxygen species. *Plant Cell Environ.*, **22**, 1413–1423.

Madronich, S., McKenzie, R.L., Bjorn, L.O., and Caldwell, M.M. (1998) Changes in biologically active ultraviolet radiation reaching the earth's surface. *J. Photochem. Photobiol. B*, **46**, 5–19.

Manning, W.J. and von Tiedemann, A. (1995) Climate change: Potential effects of increased atmospheric carbon dioxide (CO_2), ozone (O_3), and ultraviolet-B (UV-B) radiation on plant diseases. *Environ. Pollut.*, **88**, 219–245.

Mauch, F., Meehl, J.B., and Staehelin, L.A. (1992) Ethylene-induced chitinase and β-1,3-glucanase accumulate specifically in the lower epidermis and along vascular strands of bean leaves. *Planta*, **186**, 367–375.

May, M.J., Vernoux, T., Leaver, C., Van Montagu, M., and Inzé, D. (1998) Glutathione homeostasis in plants: Implications for environmental sensing and plant development. *J. Exp. Bot.*, **49**, 649–667.

McGrath, R.B. and Ecker, J.R. (1998) Ethylene signaling in *Arabidopsis*: Events from the membrane to the nucleus. *Plant Physiol. Biochem.*, **36**, 103–113.

Mehlhorn, H. and Wellburn, A.R. (1987) Stress ethylene formation determines plant sensitivity to ozone. *Nature*, **327**, 417–418.

Miller, J.D., Arteca, R.N., and Pell, E.J. (1999) Senescence-associated gene expression during ozone-induced leaf senescence in *Arabidopsis. Plant Physiol.*, **120**, 1015–1023.

Mittal, S., Graham, T.L., and Davis, K.R. (1995) Genetic dissection of the phenylpropanoid pathway in *Arabidopsis thaliana*. In D.L. Gustine and H.E. Flores, (eds.), *Phytochemicals and Health*, American Society of Plant Physiologists, Rockville, pp. 260–262.

Moeder, W., Anegg, S., Thomas, G., Langebartels, C., and Sandermann, H. (1999) Signal molecules in ozone activation of stress proteins in plants, In M.F. Smallwood, C.M. Calvert, and D.J Bowles, (eds.), *Plant Responses to Environmental Stress*, BIOS, Oxford, pp. 43–49.

Morel, J.-B. and Dangl, J.L. (1997) The hypersensitive response and the induction of cell death in plants. *Cell Death Differ.*, **4**, 671–683.

Mudd, J.B. (1997) Biochemical basis for the toxicity of ozone. In M. Yunus and M. Iqbal, (eds.), *Plant Responses to Air Pollution*, Wiley, New York, pp. 267–284.

No, E.G., Flagler, R.B., Swize, M.A., Cairney, J., and Newton, R.J. (1997) cDNAs induced by ozone from *Atriplex canescens* (saltbush) and their response to sulfur dioxide and water-deficit. *Physiol. Plant.*, **100**, 137–146.

Noctor, G. and Foyer, C.H. (1998) Ascorbate and glutathione: keeping active oxygen under control. *Annu. Rev. Plant Physiol. Plant Mol. Biol.*, **49**, 249–279.

Ormrod, D.P. and Beckerson, D.W. (1986) Polyamines as antiozonants for tomato. *Hortic. Sci.*, **21**, 1070–1071.

Örvar, B.L. and Ellis, B.E. (1997) Transgenic tobacco plants expressing antisense RNA for cytosolic ascorbate peroxidase show increased susceptibility to ozone injury. *Plant J.*, **11**, 1297–1305.

Örvar, B.L., McPherson, J., and Ellis, B.E. (1997) Pre-activating wounding response in tobacco prior to high-level ozone exposure prevents necrotic injury. *Plant J.*, **11**, 203–212.

Orzáez, D. and Granell, A. (1997) DNA fragmentation is regulated by ethylene during carpel senescence in *Pisum sativum. Plant J.*, **11**, 137–144.

Overmyer, K., Kangasjärvi, J., Kuittinen, T., and Saarma, M. (1998) Gene expression and cell death in ozone-exposed plants: Is programmed cell death involved in ozone damage in ozone-sensitive *Arabidopsis* mutants? In L.J. De Kok and I. Stulen, (eds.), *Responses of Plant Metabolism to Air Pollution and Global Change*, Backhuys Publishers, Leiden, pp. 403–406.

Pääkönen, E., Seppänen, S., Holopainen, T., Kokko, H., Kärenlampi, S., Kärenlampi, L., and Kangasjärvi, J. (1998) Induction of genes for the stress proteins PR-10 and PAL in relation to growth, visible injuries and stomatal conductance in birch (*Betula pendula*) clones exposed to ozone and/or drought. *New Phytol.*, **138**, 295–305.

Parkinson, J.S. and Kofoid, E.C. (1992) Communication modules in bacterial signaling proteins. *Annu. Rev. Genet*, **26**, 71–112.

Pell, E.J., Schlagnhaufer, C.D., and Arteca, R.N. (1997) Ozone-induced oxidative stress: Mechanisms of action and reaction. *Physiol. Plant.*, **100**, 264–273.

Pellinen, R., Palva, T., and Kangasjärvi, J. (1999) Subcellular localization of ozone-induced hydrogen peroxide production in birch (*Betula pendula*) leaf cells. *Plant J.*, **20**, 349–356.

Pierson, L.S.I., Wood, D.W., and Pierson, E.A. (1998) Homoserine lactone-mediated gene regulation in plant-associated bacteria. *Annu. Rev. Phytopathol.*, **36**, 207–225.

Pitcher, L.H. and Zilinskas, B.A. (1996) Overexpression of copper/zinc superoxide dismutase in the cytosol of transgenic tobacco confers partial resistance to ozone-induced foliar necrosis. *Plant Physiol.*, **110**, 583–588.

Pitcher, L.H., Brennan, E., Hurley, A., Dunsmuir, P., Tepperman, J.M., and Zilinskas, B.A. (1991) Overproduction of petunia chloroplastic copper/zinc superoxide dismutase does not confer ozone tolerance in transgenic tobacco. *Plant Physiol.*, **97**, 452–455.

Prasad, T.K., Anderson, M.D., Martin, B.A., and Stewart, C.R. (1994) Evidence for chilling-induced oxidative stress in maize seedlings and a regulatory role for hydrogen peroxide. *Plant Cell*, **6**, 65–74.

Priebe, A., Klein, H., and Jäger, H.J. (1978) Role of polyamines in SO_2-polluted pea plants. *J. Exp. Bot.*, **29**, 1045–1050.

Ranieri, A., Castagna, A., Padu, E., Moldau, H., Rahi, M., and Soldatini, G.F. (1999) The decay of O_3 through direct reaction with cell wall ascorbate is not sufficient to explain the different degrees of O_3-sensitivity in two poplar clones. *J. Plant Physiol.*, **154**, 250–255.

Rao, M.V. and Davis, K.R. (1999) Ozone-induced cell death occurs via two distinct mechanisms in *Arabidopsis*: The role of salicylic acid. *Plant J.*, **17**, 603–614.

Rao, M.V. and Ormrod, D.P. (1995) Impact of UV-B and O_3 on the oxygen free radical scavenging system in *Arabidopsis thaliana* genotypes differing in flavonoid biosynthesis. *Photochem. Photobiol.*, **62**, 719–726.

Rao, M.V., Paliyath, G., and Ormrod, D.P. (1996) Ultraviolet-B- and ozone-induced biochemical changes in antioxidant enzymes of *Arabidopsis thaliana*. *Plant Physiol.*, **110**, 125–136.

Rate, D.N., Cuenca, J.V., Bowman, G.R., Guttman, D.S., and Greenberg, J.T. (1999) The gain-of-function *Arabidopsis acd6* mutant reveals novel regulation and function of the salicylic acid signaling pathway in controlling cell death, defenses, and cell growth. *Plant Cell*, **11**, 1695–1708.

Reddy, G.N., Arteca, R.N., Dai, Y.R., Flores, H.E., Negm, F.B., and Pell, E.J. (1993) Changes in ethylene and polyamines in relation to mRNA levels of the large and small subunits of ribulose bisphosphate carboxylase/oxygenase in ozone-stressed potato foliage. *Plant Cell Environ.*, **16**, 819–826.

Reich, P.B. (1987) Quantifying plant response to ozone: A unifying theory. *Tree Physiol.*, **3**, 63–91.

Rennenberg, H. and Polle, A. (1994) Protection from oxidative stress in transgenic plants. *Biochem. Soc. Trans.*, **22**, 92–96.

Richards, K.D., Schott, E.J., Sharma, Y.K., Davis, K.R., and Gardner, R.C. (1998) Aluminum induces oxidative stress genes in *Arabidopsis thaliana*. *Plant Physiol.*, **116**, 409–418.

Rosemann, D., Heller, W., and Sandermann, H. (1991) Biochemical plant responses to ozone. II. Induction of stilbene biosynthesis in Scots pine (*Pinus sylvestris* L.) seedlings. *Plant Physiol.*, **97**, 1280–1286.

Rowland-Bamford, A.J., Borland, A.M., Lea, P.J., and Mansfield, T.A. (1989) The role of arginine decarboxylase in modulating the sensitivity of barley to ozone. *Environ. Pollut.*, **61**, 95–106.

Roxas, V.P., Smith, R.K., Allen, E.R., and Allen, R.D. (1997) Overexpression of glutathione S-transferase/glutathione peroxidase enhances the growth of transgenic tobacco seedlings during stress. *Nature Biotechnol.*, **15**, 988–991.

Rozema, J., Van de Staaij, J., Björn, L.O., and Caldwell, M. (1997) UV-B as an environmental factor in plant life: stress and regulation. *Trends Ecol. Evol.*, **12**, 22–28.

Rüffer, M., Steipe, B., and Zenk, M. (1995) Evidence against specific binding of salicylic acid to plant catalase. *FEBS Lett.*, **377**, 175–180.

Runeckles, V.C. and Krupa, S.V. (1994) The impact of UV-B radiation and ozone on terrestrial vegetation. *Environ. Pollut.*, **83**, 191–213.

Runeckles, V.C. and Vaartnou, M. (1997) EPR evidence for superoxide anion formation in leaves during exposure to low levels of ozone. *Plant Cell Environ.*, **20**, 306–314.

Sandermann, H. (1996) Ozone and plant health. *Annu. Rev. Phytopathol.*, **34**, 347–366.

Sandermann, H. (1999) Ozone/biotic disease interactions: Molecular biomarkers as a new experimental tool. *Environ. Pollut.*, **107**, 1–6.

Sandermann, H., Wellburn, A.R., and Heath, R.L. (1997) *Forest Decline and Ozone. A Comparison of Controlled Chamber and Field Experiments* (Ecological Studies, Vol. 127), Springer, Berlin.

Sandermann, H., Ernst, D., Heller, W., and Langebartels, C. (1998) Ozone: An abiotic elicitor of plant defense reactions. *Trends Plant Sci.*, **3**, 47–50.

Schlagnhaufer, C.D., Glick, R.E., Arteca, R.N., and Pell, E.J. (1995) Molecular cloning of an ozone-induced 1-aminocyclopropane-1-carboxylate synthase cDNA and its relationship with a loss of rbcS in potato (*Solanum tuberosum* L.) plants. *Plant Mol. Biol.*, **28**, 93–103.

Schlagnhaufer, C.D., Arteca, R.N., and Pell, E.J. (1997) Sequential expression of two 1-aminocyclopropane-1-carboxylate synthase genes in response to biotic and abiotic stresses in potato (*Solanum tuberosum* L.) leaves. *Plant Mol. Biol.*, **35**, 683–688.

Schneiderbauer, A., Back, E., Sandermann, H., and Ernst, D. (1995) Ozone induction of extensin mRNA in Scots pine, Norway spruce and European beech. *New Phytol.*, **130**, 225–230.

Schnitzler, J.-P., Jungblut, T. P., Heller, W., Köfferlein, M., Hutzler, P., Heinzmann, U., Schmelzer, E., Ernst, D., Langebartels, C., and Sandermann, H. (1996) Tissue localization of UV-B screening pigments and of chalcone synthase mRNA in needles of Scots pine (*Pinus sylvestris* L.) seedlings. *New Phytol.*, **132**, 247–258.

Schnitzler, J.-P., Jungblut, T. P., Feicht, C., Köfferlein, M., Langebartels, C., Heller, W., and Sandermann, H. (1997) UV-B induction of flavonoid biosynthesis in Scots pine (*Pinus sylvestris* L.) seedlings. *Trees*, **11**, 162–168.

Schraudner, M., Ernst, D., Langebartels, C., and Sandermann, H. (1992) Biochemical plant responses to ozone. III. Activation of the defense-related proteins β-1,3-glucanase and chitinase in tobacco leaves. *Plant Physiol.*, **99**, 1321–1328.

Schraudner, M., Graf, U., Langebartels, C., and Sandermann, H. (1994) Ambient ozone can induce plant defense reactions in tobacco. *Proc. R. Soc. Edinburgh*, **102B**, 55–61.

Schraudner, M., Langebartels, C., and Sandermann, H. (1996) Plant defence systems and ozone. *Biochem. Soc. Trans.*, **24**, 456–461.

Schraudner, M., Möder, W., Wiese, C., Van Camp, W., Inzé, D., Langebartels, C., and Sandermann, H. (1998) Ozone-induced oxidative burst in the ozone biomonitor plant, tobacco Bel W3. *Plant J.*, **16**, 235–245.

Schubert, R., Fischer, R., Hain, R., Schreier, P. H., Bahnweg, G., Ernst, D., and Sandermann, H. (1997) An ozone-responsive region of the grapevine resveratrol synthase promoter differs from the basal pathogen-responsive sequence. *Plant Mol. Biol.*, **34**, 417–426.

Sharma, Y.K. and Davis, K.R. (1994) Ozone-induced expression of stress-related genes in *Arabidopsis thaliana*. *Plant Physiol.*, **105**, 1089–1096.

Sharma, Y.K. and Davis, K.R. (1995) Isolation of a novel *Arabidopsis* ozone-induced cDNA by differential display. *Plant Mol. Biol.*, **29**, 91–98.

Sharma, Y.K. and Davis, K.R. (1997) The effects of ozone on antioxidant responses in plants. *Free Rad. Biol. Med.*, **23**, 480–488.

Sharma, Y.K., León, J., Raskin, I., and Davis, K.R. (1996) Ozone-induced responses in *Arabidopsis thaliana*: The role of salicylic acid in the accumulation of defense-related transcripts and induced resistance. *Proc. Natl. Acad. Sci. USA*, **93**, 5099–5104.

Shirley, B.W. (1996) Flavonoid biosynthesis: 'New' functions for an 'old' pathway. *Trends Plant Sci.*, **1**, 377–382.

Shirley, B.W., Kubasek, W.L., Storz, G., Bruggemann, E., Koornneef, M., Ausubel, F.M., and Goodman, H.M. (1995) Analysis of *Arabidopsis* mutants deficient in flavonoid biosynthesis. *Plant J.*, **8**, 659–671.

Showalter, A.M. (1993) Structure and function of cell wall proteins. *Plant Cell*, **5**, 9–23.

Solano, R. and Ecker, J.R. (1998) Ethylene gas: perception, signaling and response. *Curr. Opin. Plant Biol.*, **1**, 393–398.

Stockwell, W.R., Kramm, G., Scheel, H.-E., Mohnen, V.A., and Seiler, W. (1997) Ozone formation, destruction and exposure in Europe and the United States. In H. Sandermann, A.R. Wellburn, and R.L. Heath, (eds.), *Forest Decline and Ozone: A Comparison of Controlled Chamber and Field Experiments*, (Ecological Studies, Vol. 127), Springer, Berlin, pp. 1–38.

Strid, Å., Chow, W.S., and Anderson, J.M. (1994) UV-B damage and protection at the molecular level in plants. *Photosynthesis Res.*, **39**, 475–489.

Strohm, M., Jouanin, L., Kunert, K. J., Pruvost, C., Polle, A., Foyer, C.H., and Rennenberg, H. (1995) Regulation of glutathione synthesis in leaves of transgenic poplar (*Populus tremula* × *P. alba*) overexpressing glutathione synthetase. *Plant J.*, **7**, 141–145.

Strohm, M., Eiblmeier, M., Langebartels, C., Jouanin, L., Polle, A., Sandermann, H., and Rennenberg, H. (1999) Responses of transgenic poplar (*Populus tremula* × *P. alba*) overexpressing glutathione synthetase or glutathione reductase to acute ozone stress: Visible injury and leaf gas exchange. *J. Exp. Bot.*, **50**, 365–374.

Surplus, S.L., Jordan, B.R., Murphy, A.M., Carr, J.P., Thomas, B., and Mackerness, S.A.H. (1998) Ultraviolet-B-induced responses in Arabidopsis thaliana: Role of salicylic acid and reactive oxygen species in the regulation of transcripts encoding photosynthetic and acidic pathogenesis-related proteins. *Plant Cell Environ.*, **21**, 685–694.

Tenhaken, R., Levine, A., Brisson, L.F., Dixon, R.A., and Lamb, C. (1995) Function of the oxidative burst in hypersensitive disease resistance. *Proc. Natl. Acad. Sci. USA*, **92**, 4158–4163.

Tevini, M. (1993) Effects of enhanced UV-B radiation on terrestrial plants. In M. Tevini, (ed.), *UV-B Radiation and Ozone Depletion: Effects on Humans, Animals, Plants, Microorganisms, and Materials*, Lewis Publishers, Boca Raton, pp. 125–153.

Thalmair, M. (1996) Reaktionen von Streßproteinen auf Ozon und UV-B Strahlung bei Tabak (*Nicotiana tabacum* L. cv. Bel W3 and Bel B). PhD Thesis, University of Munich.

Thalmair, M., Bauw, G., Thiel, S., Döhring, T., Langebartels, C., and Sandermann, H. (1996) Ozone and ultraviolet B effects on the defense-related proteins β-1,3-glucanase and chitinase in tobacco. *J. Plant Physiol.*, **148**, 222–228.

Tingey, D.T., Standley, C., and Field, R.W. (1976) Stress ethylene evolution: A measure of ozone effects on plants. *Atmos. Environ.*, **10**, 969–974.

Torsethaugen, G., Pitcher, L.H., Zilinskas, B.A., and Pell, E.J. (1997) Overproduction of ascorbate peroxidase in the tobacco chloroplast does not provide protection against ozone. *Plant Physiol.*, **114**, 529–537.

Tuomainen, J., Pellinen, R., Roy, S., Kiiskinen, M., Eloranta, T., Karjalainen, R., and Kangasjärvi, J. (1996) Ozone affects birch (*Betula pendula* Roth) phenylpropanoid, polyamine and active oxygen detoxifying pathways at biochemical and gene expression level. *J. Plant Physiol.*, **148**, 179–188.

Tuomainen, J., Betz, C., Kangasjärvi, J., Ernst, D., Yin, Z.-H., Langebartels, C., and Sandermann, H. (1997) Ozone induction of ethylene emission in tomato plants: Regulation by differential accumulation of transcripts for the biosynthetic enzymes. *Plant J.*, **12**, 1151–1162.

Vahala, J., Schlagnhaufer, C.D., and Pell, E.J. (1998) Induction of an ACC synthase cDNA by ozone in light-grown *Arabidopsis thaliana* leaves. *Physiol. Plant.*, **103**, 45–50.

Van Camp, W., Willekens, H., Bowler, C., Van Montagu, M., Inzé, D., Reupold-Popp, P., Sandermann, H., and Langebartels, C. (1994) Elevated levels of superoxide dismutase protect transgenic plants against ozone damage. *Bio/Technology*, **12**, 165–168.

Van Camp, W., Van Montagu, M., and Inzé, D. (1998) H$_2$O$_2$ and NO: redox signals in disease resistance. *Trends Plant Sci.*, **3**, 330–334.

Van Loon, L.C., Bakker, P.A.H.M., and Pieterse, C.M.J. (1998) Systemic resistance induced by rhizosphere bacteria. *Annu. Rev. Phytopathol.*, **36**, 453–483.

Wegener, A., Gimbel, W., Werner, T., Hani, J., Ernst, D., and Sandermann, H. (1997a) Molecular cloning of ozone-inducible protein from *Pinus sylvestris* L. with high sequence similarity to vertebrate 3-hydroxy-3-methylglutaryl-CoA synthase. *Biochim. Biophys. Acta*, **1350**, 247–252.

Wegener, A., Gimbel, W., Werner, T., Hani, T., Ernst, D., and Sandermann, H. (1997b) Sequence analysis and ozone-induced accumulation of polyubiquitin mRNA in *Pinus sylvestris*. *Can. J. For. Res.*, **27**, 945–948.

Wellburn, F.A.M. and Wellburn, A.R. (1996) Variable patterns of antioxidant protection but similar ethene emission differences in several ozone-sensitive and ozone-tolerant plant selections. *Plant Cell Environ.*, **19**, 754–760.

Whetten, R. and Sederoff, R. (1995) Lignin biosynthesis. *Plant Cell*, **7**, 1001–1013.

Willekens, H., Langebartels, C., Tiré, C., Van Montagu, M., Inzé, D., and Van Camp, W. (1994a) Differential expression of catalase genes in *Nicotiana plumbaginifolia* (L.). *Proc. Natl. Acad. Sci. USA*, **91**, 10450–10454.

Willekens, H., Van Camp, W., Van Montagu, M., Inzé, D., Langebartels, C., and Sandermann, H. (1994b) Ozone, sulfur dioxide, and ultraviolet B have similar effects on mRNA accumulation of antioxidant genes in *Nicotiana plumbaginifolia* (L.) *Plant Physiol.*, **106**, 1007–1014.

Willekens, H., Inzé, D., Van Montagu, M., and Van Camp, W. (1995) Catalases in plants. *Mol. Breeding*, **1**, 207–228.

Willekens, H., Chamnongpol, S., Davey, M., Schraudner, M., Langebartels, C., Van Montagu, M., Inzé, D., and Van Camp, W. (1997) Catalase is a sink for H$_2$O$_2$ and is indispensable for stress defence in C$_3$ plants. *EMBO J.*, **16**, 4806–4816.

Yalpani, N., Enyedi, A.J., León, J., and Raskin, I. (1994) Ultraviolet light and ozone stimulate accumulation of salicylic acid, pathogenesis-related proteins and virus resistance in tobacco. *Planta*, **193**, 372–376.

Yin, Z.H., Langebartels, C., and Sandermann, H. (1994) Specific induction of ethylene biosynthesis in tobacco plants by the air pollutant, ozone. *Proc. R. Soc. Edinburgh*, **102B**, 127–130.

Young, T.E., Gallie, D.R., and DeMason, D.A. (1997) Ethylene-mediated programmed cell death during maize endosperm development of wild-type and *shrunken2* genotyes. *Plant Physiol.*, **115**, 737–751.

Zarembinski, T.I. and Theologis, A. (1994) Ethylene biosynthesis and action: A case of conservation. *Plant Mol. Biol.*, **26**, 1579–1597.

Zielke, H. and Sonnenbichler, J. (1990) Natural occurrence of 3,3′,4,4′-tetramethoxy-1,1′-biphenyl in leaves of stressed European beech. *Naturwissenschaften*, **77**, 384–385.

Zinser, C., Ernst, D., and Sandermann, H. (1998) Induction of stilbene synthase and cinnamyl alcohol dehydrogenase mRNAs in Scots pine (*Pinus sylvestris* L.) seedlings. *Planta*, **204**, 169–176.

5 Oxidative Burst and the Role of Reactive Oxygen Species in Plant-Pathogen Interactions

Dierk Scheel

INTRODUCTION

Plants are able to successfully defend themselves against most potential pathogens in their environment, because they are not an appropriate host plant for the majority of pathogens they encounter (species, non-host, or basic resistance). Only relatively few true host-pathogen pairs exist in which a particular pathogen has evolved the capability to resist or prevent a specific plant's defense response (species, host, or basic compatibility). Distinct cultivars of such host plant species have developed specific resistances against single pathogen races (cultivar resistance). In these plant-pathogen combinations resistance relies on the presence of corresponding plant resistance and pathogen avirulence genes according to the gene-for-gene hypothesis (Flor, 1955, 1971; Alfano and Collmer, 1996; Bent, 1996). In both, species and cultivar resistance, plants appear to recognize pathogens through receptors located intracellularly or on the plasma membrane (Alfano and Collmer, 1996; Knogge, 1996). Signal molecules either released from the plant surface during pathogen attack (endogenous elicitors) or originating directly from the pathogen (exogenous elicitors) appear to function as ligands for these receptors in non-host recognition and thereby initiate the plant's defense response (Figure 1) (Ebel and Scheel, 1992; Knogge, 1996). In host-incompatible interactions race-cultivar-specific elicitors are encoded by avirulence genes of the pathogen (De Wit, 1992; Alfano and Collmer, 1996; Bent, 1996; Knogge, 1996). The corresponding plant receptors are believed to be products of the matching plant resistance genes (Alfano and Collmer, 1996; Bent, 1996). Although several plant resistance genes have now been isolated, their receptor function remains to be demonstrated (Bent, 1996).

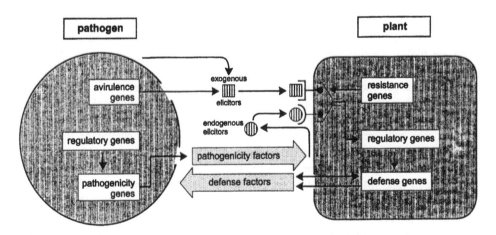

Figure 1. Hypothetical model for signal transduction processes in plant-pathogen interactions.

The plant's defense mechanisms appear to be similar in species and cultivar resistance. Upon pathogen recognition, a multicomponent defense response is initiated, whose individual elements are activated in a spatially and temporally complex pattern (Kombrink and Somssich, 1995). Typical components of such responses include the production of phytoalexins, the synthesis of pathogenesis-related (PR) proteins, the induction of systemic acquired resistance (SAR), reinforcement of the cell wall, local cell death, and the generation of reactive oxygen species (ROS) (Kombrink and Somssich, 1995; Mehdy et al., 1996). This oxidative burst is one of the earliest reactions detectable after pathogen attack. Since the pioneering work of Doke (Doke et al., 1987) it has been postulated to play a key role in plant defense. In spite of their potentially important role in plant defense, the chemical nature of the ROS involved, their exact subcellular location, the molecular mechanisms of their synthesis, and their biological function are still a matter of debate (Baker and Orlandi, 1995; Mehdy et al., 1996).

REACTIVE OXYGEN SPECIES OF THE OXIDATIVE BURST

The terms active oxygen species, reactive oxygen intermediates, or reactive oxygen species (ROS) are commonly used for oxygen derivatives formed as a result of the reduction of molecular oxygen, O_2, in biological systems (Sutherland, 1991; Baker and Orlandi, 1995; Mehdy et al., 1996). The first reduction product of O_2 is the superoxide anion radical, $O_2^{\bullet-}$, that exists in equilibrium with the hydroperoxyl radical, $HO_2^{\bullet}$ (Sutherland, 1991; Grisham, 1992; Baker and Orlandi, 1995; Able et al., 1998; Murphy et al., 1998). Whereas this reduction requires energy input, the next three one-electron reductions occur spontaneously in the appropriate environment and result in the stepwise formation of hydrogen peroxide, H_2O_2, hydroxyl radical, $OH^{\bullet}$, and water, H_2O.

Detection and identification of the ROS formed during the oxidative burst of plants are difficult and require specific techniques, because they exist only transiently at relatively low levels in small cellular and tissue areas (Tzeng and DeVay, 1993; Baker and Orlandi, 1995; Able et al., 1998; Murphy et al., 1998). Because of their high reactivity they rapidly interact with and modify most molecules present in life materials. Besides their chemical reactivity, however, lipophilicity and site of production are important for the biological activity of ROS (Baker and Orlandi, 1995). The superoxide anion and hydroperoxyl radicals differ in their lipophilicity and, therefore, in their capability of lipid peroxidation. Both are highly reactive and do not easily cross lipid bilayers. In contrast, hydrogen peroxide is perfectly able to penetrate biological membranes and, in addition, has the appropriate longer chemical half life to move intra- and intercellularly. The hydroxyl radical is extremely reactive; it is believed to be biologically active only in the direct vicinity of its site of generation and its appearance has been reported in suspension-cultured rice cells treated with an N-acetylchitooligosaccharide elicitor or the protein phosphatase inhibitor, calyculin A (Kuchitsu et al., 1995).

One site of ROS production during the oxidative burst is the extracellular surface of the plasma membrane (Mehdy et al., 1996). Results recently obtained with 2',7'-dichloro-fluorescin diacetate suggest that fungal infection of parsley cells may also stimulate intracellular production of ROS within the cell that is directly attacked by the pathogen before it undergoes programmed cell death (Naton et al., 1996). Local mechanical stimulation or elicitor treatment also resulted in intracellular ROS accumulation in parsley cells, but no cell death was observed (Gus-Mayer *et al.*, 1998). Intracellular ROS generation was also detected in tobacco epidermal cells treated with cryptogein, a protein elicitor derived from *Phytophthora cryptogea* (Allan and Fluhr, 1997). In powdery mildew-infected barley leaves H_2O_2 was detected with 3,3-diaminobenzidine first in epidermal cell walls and later inside those epidermal cells that subsequently underwent hypersensitive cell death (Thordal-Christensen *et al.*, 1997). Further experimental evidence on the nature, origin, and location of intracellular ROS is required, but technically difficult to obtain. So far, the possibility cannot be excluded that extracellularly generated H_2O_2 penetrates the plasma membrane and causes intracellular ROS accumulation.

OCCURRENCE OF THE OXIDATIVE BURST DURING PLANT-PATHOGEN INTERACTIONS

An oxidative burst of varying intensity and duration has been observed in plants or cultured plant cells in response to fungal and bacterial infection, as well as upon treatment with different elicitors (Sutherland, 1991; Baker and Orlandi, 1995; Low and Merida, 1996; Mehdy *et al.*, 1996). Most frequently, the oxidative burst is measured as the extracellular generation of H_2O_2 (Tzeng and DeVay, 1993; Baker and Orlandi, 1995). Interestingly, in non-host and host-incompatible interactions a long-lasting, often two-phasic production of ROS has been observed, whereas in compatible interactions the oxidative burst was either absent or only the early phase was stimulated (Figure 2) (Baker and Orlandi, 1995).

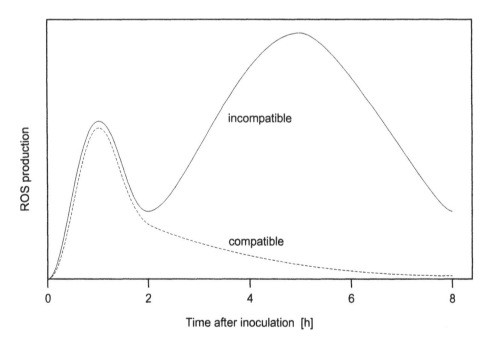

Figure 2. Typical time courses of ROS production in compatible and incompatible interactions of cultured plant cells and bacterial pathogens.

In their pioneering work, Doke *et al.* (1987) measured the production of superoxide anion radicals in potato tubers and leaves infected with virulent or avirulent races of the late blight pathogen, *Phytophthora infestans*. In aged and wounded tuber tissue, only avirulent pathogen races rapidly stimulated a significant oxidative burst. Intact potato leaves, however, which are the primary target of this devastating pathogen, responded with a rapid burst before penetration of the host tissue by avirulent as well as virulent races of the pathogen. Only in incompatible interactions, the early burst was followed by a massive second phase of $O_2^{\bullet-}$ production. This second phase of the oxidative burst was accompanied by local cell death, a typical component of the resistance response of potato against avirulent races of *Phytophthora infestans*. Based on these correlative data, Doke *et al.* (1987) proposed a causal relationship between the oxidative burst and hypersensitive cell death. Infection of rice leaves with avirulent races of the rice blast fungus, *Magnaporthe grisea*, also stimulated $O_2^{\bullet-}$ generation, whereas virulent races of the pathogen only caused a delayed and reduced response (Sekizawa *et al.*, 1987). In the early phase of the resistance response of barley leaves to powdery mildew, H_2O_2 was found to accumulate in developing papillae and around cell wall appositions, whereas later it was also detected intracellularly in epidermal cells undergoing hypersensitive cell death (Thordal-Christensen *et al.*, 1997).

During the interaction of *Cladosporium fulvum* with tomato, race-specific elicitors are synthesized by the fungus and secreted into the intercellular space (Scholtens-Toma and De Wit, 1988; Joosten *et al.*, 1994). These avirulence gene-encoded elicitors were found to be responsible for resistance of those tomato cultivars that carry the matching resistance gene against the corresponding fungal race (Van den Ackerveken *et al.*, 1992; Bent, 1996). Injection of race-specific elicitors from *Cladosporium fulvum* into tomato leaves that possessed the appropriate resistance gene stimulated a multicomponent defense response with the oxidative burst being the most rapid reaction (Hammond-Kosack and Jones, 1996; May *et al.*, 1996). When leaves from plants lacking the corresponding resistance gene were injected, a delayed formation of lower levels of $O_2^{\bullet-}$ was observed. Whereas the occurrence of the oxidative burst was found to be strictly dependent on the resistance gene, the rapidity of its initiation varies in plant cultivars with different resistance genes. Suspension-cultured tomato cells derived from plants that are near-isogenic for the *Cladosporium fulvum* resistance genes *Cf4* and *Cf5*, were described to retain their responsiveness to the appropriate race-specific elicitors including the oxidative burst (Vera-Estrella *et al.*, 1992). Most interestingly, transgenic suspension-cultured tobacco cells carrying the resistance gene *Cf9* from tomato respond with rapid oxidative burst to treatment with the corresponding avirulence gene product, Avr9 (Piedras *et al.*, 1998).

Inoculation of *Arabidopsis* leaves with an avirulent *Pseudomonas syringae* strain induced local accumulation of ROS, but in addition, ROS generation was also detected systemically at microscopically small localized areas in distant non-infected leaves (Alvarez *et al.*, 1998). Cells in both local and systemic burst areas subsequently underwent hypersensitive cell death. Bacterial plant pathogens frequently stimulated a complete two-phasic oxidative burst upon co-cultivation with cultured plant cells of non-host species or resistant host cultivars (Baker and Orlandi, 1995). Non-host and avirulent pathovars of *Pseudomonas syringae* elicited a two-phasic oxidative burst in cultured tobacco (Baker *et al.*, 1987, 1991, 1993a; Adam *et al.*, 1989; Keppler and Baker, 1989; Keppler *et al.*, 1989; Orlandi *et al.*, 1992), soybean (Baker *et al.*, 1993a; Baker and Mock, 1994; Levine *et al.*, 1994, 1996), potato (Baker and Orlandi, 1995), and tomato cells (Chandra *et al.*,

1996b), whereas only the first reaction occurred in compatible interactions. In tobacco cells, also *Erwinia amylovora* stimulated an oxidative burst (Baker *et al.*, 1993b). During non-host resistance of lettuce against *Pseudomonas syringae* pv *phaseolicola* prolonged H_2O_2 accumulation was detected cytochemically in plant cell walls adjacent to attached bacteria by its reaction with cerium chloride (Bestwick *et al.*, 1997). Infection of the same tissue with a *hrpD* mutant that did not cause a hypersensitive response (Bestwick *et al.*, 1995) did also not result in the accumulation of H_2O_2 (Bestwick *et al.*, 1997). The prolonged accumulation of H_2O_2 detected *in situ* in infected tissue probably resembles the second phase of the oxidative burst seen in the medium of cultured cells. Because of the close correlation between the induction of the second phase of the oxidative burst and the occurrence of plant cell death (Baker and Orlandi, 1995; Low and Merida, 1996; Mehdy *et al.*, 1996) and the ability of exogenously applied millimolar concentrations of H_2O_2 to cause plant cell death, a causal link was postulated to exist between biphasic and/ or prolonged ROS production during the oxidative burst and the hypersensitive disease resistance response (Levine *et al.*, 1994). This correlation was lost, however, when tobacco cells were co-cultivated with a *Pseudomonas syringae* strain carrying a mutation in the *hrmA* region (Glazener *et al.*, 1996). This strain induced a two-phasic oxidative burst in tobacco that was similar to that elicited by the wild-type strain. However, in contrast to the wild type the mutant strain did not cause cell death in cultured cells (Glazener *et al.*, 1996) and only a delayed and atypical hypersensitive response in leaves (Huang *et al.*, 1991), indicating that H_2O_2 from the oxidative burst may be necessary but is not sufficient for the induction of programmed plant cell death.

The induction of multicomponent defense responses in cultured plant cells or isolated plant tissues by non-race-specific elicitors of fungal, bacterial, or plant origin has frequently been used to investigate details of the plant's defense machinery (Alfano and Collmer, 1996; Hammond-Kosack and Jones, 1996; Knogge, 1996; Ebel and Scheel, 1997). In many of these experimental systems one of the earliest reactions was found to be an oxidative burst. However, in contrast to the biphasic and/or long-lasting ROS production observed upon co-cultivation of cultured plant cells with avirulent or non-host bacteria, in most of these experimental systems the duration of the elicitor-stimulated burst was significantly shorter (Schwacke and Hager, 1992; Vera-Estrella *et al.*, 1992; Levine *et al.*, 1994; Desikan *et al.*, 1996; Fauth *et al.*, 1996; Kauss and Jeblick, 1996; Otte and Barz, 1996; Mithöfer *et al.*, 1997) or it was only measured for a short period of time (Low and Heinstein, 1986; Legendre *et al.*, 1993a; Sanchez *et al.*, 1993; Bottin *et al.*, 1994; Rustérucci *et al.*, 1996). Only oligogalacturonides and oligoglucans in cucumber hypocotyl segments (Svalheim and Robertsen, 1993), an oligopeptide elicitor in cultured parsley cells (Jabs *et al.*, 1997), and the avirulence gene product, Avr9, from *Cladosporium fulvum* in transgenic tobacco cells carrying the *Cf9* gene from tomato have been reported so far to stimulate a long-lasting oxidative burst (Piedras *et al.*, 1998). ROS production in these experimental systems was found to be quantitatively and kinetically similar to that observed in the plant-bacterial cocultivation experiments. However, in none of the systems plant cell death was stimulated, suggesting again that ROS may be necessary but are not sufficient for its induction. Furthermore, these similarities between the oxidative bursts induced by avirulent or non-host bacteria and by non-race-specific or race-specific elicitors demonstrate that the presence of neither life pathogens (Low and Merida, 1996) nor avirulence gene products are required for stimulation of a long-lasting and/or biphasic burst (Mehdy *et al.*, 1996).

INITIATION OF THE OXIDATIVE BURST

In most cases, the oxidative burst is measured by determining of extracellularly formed H_2O_2 (Tzeng and DeVay, 1993; Baker and Orlandi, 1995). However, use of detection techniques that discriminate between H_2O_2 and $O_2^{\bullet-}$ as well as of superoxide dismutase (SOD) inhibitors demonstrated that different mechanisms of ROS production during the oxidative burst appear to exist in different plants. In cultured French bean cells treated with an elicitor from *Colletotrichum lindemuthianum*, a cell wall-located peroxidase was found to be primarily responsible for H_2O_2 generation (Bolwell *et al.*, 1998). Inhibitor experiments suggest that this is also the case for the oxidative burst in lettuce infected with *Pseudomonas syringae* pv *phaseolicola* (Bestwick *et al.*, 1997). In other plants, $O_2^{\bullet-}$ is the ROS species formed primarily during the oxidative burst, which is then dismutated by extracellular SOD into H_2O_2 and oxygen (Doke *et al.*, 1987; Jabs *et al.*, 1997; Lamb and Dixon, 1997; Bolwell *et al.*, 1998; Piedras *et al.*, 1998). This production of ROS is paralleled by transient increases in oxygen consumption (Bolwell *et al.*, 1995; Jacks and Davidonis, 1996; Piedras *et al.*, 1998). The plant oxidative burst therefore involves a massive activation of oxidative metabolism resembling the respiratory burst of mammalian neutrophils, eosinophils, and macrophages (Babior *et al.*, 1997; Wojtaszek, 1997).

In human macrophages, the respiratory burst is initiated by G protein-mediated assembly of the two membrane-bound (gp91[phox], p22[phox]) and three soluble subunits (Rac2, p47[phox], p67[phox]) of the NADPH oxidase, which also involves phosphorylation of at least p47[phox] (Babior *et al.*, 1997). The heterodimeric NADPH-binding flavocytochrome b558 consists of gp91[phox] and p22[phox], contains the electron transport chain from NADPH to oxygen (Babior *et al.*, 1997), and displays proton channel activity (Henderson *et al.*, 1995). Upon activation, the cytosolic subunits, p67[phox], p47[phox], and the small G protein, Rac2, translocate to the plasma membrane and form together with gp91[phox] and p22[phox] the active oxidase complex (Figure 3) (Babior *et al.*, 1997; Diekmann *et al.*, 1994).

In plants, the existence of a homologous enzyme complex is under debate (Low and Merida, 1996; Mehdy *et al.*, 1996; Lamb and Dixon, 1997; Wojtaszek, 1997). Antibodies raised against subunits or oligopeptides corresponding to subunit sequences of the human

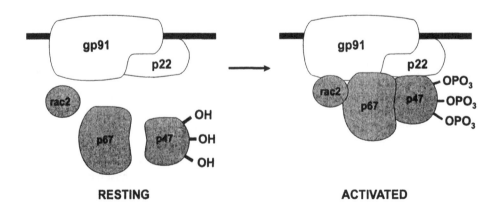

Figure 3.   Induction of subunit assembly of human NADPH oxidase to the active enzyme complex.

NADPH oxidase have been described to cross-react with plant proteins of similar size in membrane or cytosolic fractions (Levine *et al.*, 1994; Tenhaken *et al.*, 1995; Desikan *et al.*, 1996; Dwyer *et al.*, 1996; Kieffer *et al.*, 1997; Xing *et al.*, 1997). Xing *et al.* (1997) used such heterologous antibodies in protein gel blot analyses of cytosolic and plasma membrane fractions isolated from suspension-cultured tomato cell lines carrying different resistance genes against *Cladosporium fulvum*. Proteins that cross-reacted with antibodies against p47^{phox}, p67^{phox}, or Rac2 were detected in cytosolic fractions of untreated cells, but in plasma membrane fractions of cells treated with elicitor preparations containing the corresponding avirulence protein from *Cladosporium fulvum*. From these data, Xing *et al.* (1997) concluded that race-specific elicitors from *Cladosporium fulvum* stimulate assembly of an NADPH oxidase complex in tomato cells with the corresponding resistance gene and thereby remove the soluble subunits p47^{phox}, p67^{phox}, and Rac2 from the cytosol (Figure 3). In contrast, in a similar approach with suspension-cultured tobacco cells, a protein cross-reacting with Rac2 antibodies was detected at elevated levels in the cytosol of cryptogein-treated cells (Kieffer *et al.*, 1997). Antibodies raised against the different subunits of the human NADPH oxidase did not cross-react with similar proteins of parsley cells (M. Tschöpe and D. Scheel, unpublished results) that display a typical two-phasic oxidative burst in response to elicitor treatment (Jabs *et al.*, 1997). In soybean, however, antibodies against human p22^{phox}, p47^{phox}, and p67^{phox} nicely cross-reacted with proteins of the appropriate molecular mass in microsomal or cytosolic fractions (Tenhaken and Rübel, 1998). Extensive expression screening of a soybean cDNA library with these antibodies resulted in a single class of cDNA clones for each antibody. Sequence analyses clearly demonstrated that none of these cDNA clones encoded a protein homologous to the corresponding human NADPH oxidase subunit. In the light of these data results obtained with antibodies raised against the three subunits of the human NADPH oxidase must be interpreted more carefully.

Recently, plant homologs of the human gp91^{phox}-encoding gene have been isolated from rice (Groom *et al.*, 1996), *Arabidopsis* (Keller *et al.*, 1998; Torres *et al.*, 1998), and parsley (Heidi Zinecker and Dierk Scheel, unpublished results). Although the overall homology is rather low, important sequence motifs are indeed highly conserved. One *Arabidopsis* gp91^{phox} was shown to be localized in the plasma membrane (Keller *et al.* 1998). Since mammalian cell lines are available that are possibly suitable for complementation experiments (Henderson *et al.*, 1995; Yu *et al.*, 1997), functional analysis of these plant genes may indeed become feasible.

In contrast to mammalian systems, NAD(P)H oxidase activity is easily detectable in plant microsomes and plasma membrane preparations without induction of assembly (M. Tschöpe and D. Scheel, unpublished results) (Auh and Murphy, 1995; Mithöfer *et al.*, 1997; Van Gestelen *et al.*, 1997; Murphy *et al.*, 1998). Plasma membrane preparations from potato tuber slices either infected with an avirulent race of *Phytophthora infestans* or treated with a crude elicitor derived from hyphal cell walls were found to have increased activities of an NADPH-dependent $O_2^{\cdot-}$-generating enzyme that was absent in membrane fractions from untreated tissue or slices inoculated with virulent races (Doke and Miura, 1995). In the presence of cytosolic protein preparations, this system appeared to retain responsiveness to the elicitor. Heterologous reconstitution of a superoxide-generating activity has been reported upon combination of an *Arabidopsis* membrane preparation with human cytosol without, however, demonstrating homologous reconstitution (Desikan *et al.*, 1996). Parsley microsomal membranes harbor NADH and NADPH oxidases displaying similar sensitivities towards diphenylene iodonium (DPI) and diphenyl iodonium (IDP), respectively, as the mammalian NADPH oxidase (M. Tschöpe and D. Scheel, unpublished results). The fact that similar amounts of some NADPH oxidase inhibitors, such as DPI and IDP, are required to block bursts and enzyme activities in mammalian and plant systems (Hancock and Jones, 1987; Desikan *et al.*, 1996; Jabs *et al.*, 1997; Mithöfer *et al.*, 1997) may suggest, but does not prove the existence of a similar enzyme complex in plants, in particular, because this correlation does not hold true for other classes of inhibitors (Mithöfer *et al.*, 1997). In conclusion, the contradictory and incomplete data from plant systems available to date do not allow a more careful comparison of the plant and mammalian enzymes responsible for extracellular generation of $O_2^{\bullet-}$.

The picture of the signal transduction processes involved in elicitor-mediated activation of the oxidative burst oxidase of plants primarily originates from pharmacological studies with cultured cells and does not allow to develop a general model (Low and Merida, 1996; Wojtaszek, 1997). The existence of distinct signaling pathways has been suggested for different plant species, as well as for a single plant species in response to diverse elicitors (Low and Merida, 1996). Heterotrimeric GTP-binding proteins (Legendre *et al.*, 1992, 1993a; Vera-Estrella *et al.*, 1994), inositolphosphates (Legendre *et al.*, 1993b), ion channels (Schwacke and Hager, 1992; Baker *et al.*, 1993b; Nürnberger *et al.*, 1994a, 1994b; Doke and Miura, 1995; Miura *et al.*, 1995; Tavernier *et al.*, 1995; Salzer *et al.*, 1996; Jabs *et al.*, 1997; Piedras *et al.*, 1998), protein kinases and phosphatases (Schwacke and Hager, 1992; Baker *et al.*, 1993b; Levine *et al.*, 1994; Vera-Estrella *et al.*, 1994; Viard *et al.*, 1994; Chandra and Low, 1995; Kauss and Jeblick, 1995; Miura *et al.*, 1995; Desikan *et al.*, 1996; Salzer *et al.*, 1996; Ligterink *et al.*, 1997; Piedras *et al.*, 1998; Takahashi *et al.*, 1999), phospholipase C (Legendre *et al.*, 1993b), and phospholipase A (Chandra *et al.*, 1996a; Piedras *et al.*, 1998) have been suggested but hardly proven to represent signaling elements in elicitor-mediated ROS generation. In contrast to the mammalian system, phospholipase D does not appear to be involved in the signalling of the oxidation burst in plants (Taylor and Low, 1997). The receptor-mediated activation of plasma membrane located ion channels allowing, among others, the influx of Ca^{2+} was demonstrated to be necessary and sufficient for initiation of the oxidative burst and phytoalexin production in parsley cells (Jabs *et al.*, 1997; Zimmermann *et al.*, 1997). The use of aequorin-expressing tobacco cells furthermore allowed the detection of rapid and transient increases in cytoplasmic Ca^{2+} levels in response to an oligogalacturonic acid elicitor, the G protein-activating oligopeptide, Mastoparan, hypo-osmotic stress, and cold shock, whereas high concentrations of the *Erwinia amylovora*-derived elicitor, harpin, stimulated the oxidatve burst, but no Ca^{2+} transients (Chandra and Low, 1997). Oligogalacturonate-stimulated Ca^{2+} transients and oxidative burst were both inhibited by Ca^{2+} chelators and ion channel blockers and stimulated in the absence of elicitor by the Ca^{2+} ionophore, ionomycin, indicating again that Ca^{2+} influx and transient increases in cytosolic Ca^{2+} levels are seemingly involved in the initiation of the oxidative burst by elicitor. A distinct mitogen-activated type of protein kinase was shown to be an integral element of the elicitor signal transduction cascade in parsley that was located downstream of ion channel activation but upstream or independent of the oxidative burst (Ligterink *et al.*, 1997). Its involvement in the initiation of ROS production remains to be demonstrated.

The origin of ROS generated intracellularly in response to pathogen attack is yet unknown (Naton *et al.*, 1996; Allan and Fluhr, 1997). However, experiments with tobacco epidermal cells and the *Phytophthora cryptogea*-derived elicitor cryptogein suggested intracellular production of ROS by a flavin-containing oxidase (Allan and Fluhr, 1997). Apparently, after conversion to H_2O_2 these ROS rapidly enter the apoplastic space and neighboring cells.

ROLE OF ROS IN PLANT-PATHOGEN INTERACTIONS

The high reactivity of ROS results in initiation of a complex pattern of reactions. It is therefore difficult to identify individual effects of specific ROS during plant-pathogen interactions. The rapid generation of micromolar amounts of ROS in the apoplastic space of infected tissue probably affects directly the invading pathogen. Antimicrobial activity of ROS has been demonstrated against fungal (Peng and Kuc, 1992; Aver'yanov *et al.*, 1993; Baker and Orlandi, 1995) and bacterial plant pathogens (Minardi and Mazzucchi, 1988; Keppler and Baker, 1989; Baker and Orlandi, 1995). The toxicity of ROS to microbial pathogens *in vivo* depends on the pathogen's ability to rapidly detoxify these chemicals (Klotz and Hutcheson, 1992; Klotz, 1993; Klotz and Anderson, 1994; Baker and Orlandi, 1995) and, therefore, was found to be related to bacterial titer (Ma and Eaton, 1992; Baker and Orlandi, 1995). Because the oxidative burst is one of the most rapid defense rections of plant cells and the inoculum in nature will mostly be of low titer, direct ROS toxicity might indeed play an important role in disease resistance. This view is further supported by the finding that the peptide methionine sulfoxide reductase, an enzyme that repairs oxidized proteins, is required for full virulence of *Erwinia chrysanthemi* (Hassouni *et al.*, 1999). Bacteria lacking this enzyme were found to be more sensitive to oxidative stress, were less virulent, and showed no systemic invasion. The fungal tobacco pathogen, *Alternaria alternata*, produces mannitol in response to plant signals, which is a potent quencher of ROS (Jennigs *et al.*, 1998). On the other hand, fungal infection induces the synthesis of mannitol dehydrogenase in tobacco that does not produce mannitol. This enzyme may catabolize mannitol of fungal origin and thereby provide a mechanism by which the plant can counteract fungal suppression of ROS action.

By investigating early elicitor responses of cultured soybean and bean cells, rapid insolubilization was detected of two cell wall proteins, p33 and p100, which was accompanied by a dramatic decrease in protoplasting efficiency (Bradley *et al.*, 1992; Brisson *et al.*, 1994). This response, interpreted as cross-linking of these two proteins with other cell wall constituents, was also observed in non-host and incompatible interactions of bean and soybean cells with *Pseudomonas syringae* pv *tabaci* and *Pseudomonas syringae* pv *glycinea*, respectively, but never in compatible interactions (Brisson *et al.*, 1994). H_2O_2 was perfectly able to replace the elicitor in these reactions, whereas catalase and ascorbate were efficient inhibitors (Bradley *et al.*, 1992). A similar reaction was observed in cultured chickpea cells upon treatment with a yeast glucan elicitor (Otte and Barz, 1996). Two cell wall structural proteins, p80 and p190, were rapidly insolubilized upon treatment of the cells with elicitor or H_2O_2. The peroxidase inhibitor, salicylhydroxamic acid, blocked this reaction no matter whether it was induced by elicitor or H_2O_2, whereas DPI inhibited elicitor-stimulated cross-linking only, suggesting that the H_2O_2 originated from $O_2^{\bullet-}$, which was generated by the putative oxidative burst NAD(P)H oxidase, and that a peroxidase was involved downstream of H_2O_2 supply. ROS-catalyzed cross-linking of cell wall proteins may be an important component of early defense reactions, because it appears to increase the resistance of plant cells against pathogen-derived cell wall-degrading enzymes (Brisson *et al.*, 1994). Thereby, the infection process of pathogens that rely on degradation of the plant cell wall will be rapidly inhibited.

In plant cells, exogenously applied H_2O_2 transcriptionally activates genes that encode enzymes involved in protection against oxidative stress, such as glutathione *S*-transferase and glutathione peroxidase (Levine *et al.*, 1994). Stimulation of the protective enzymes was also observed in soybean cells separated from elicited cells by a dialysis membrane. This response was inhibited by both, DPI and catalase, indicating that H_2O_2 from the oxidative burst was the intercellular signal. Therefore, H_2O_2 may be perceived by healthy cells surrounding infected tissue and thereby activate protection mechanisms against ROS diffusing into healthy tissue from the infection site.

Extracellularly generated ROS from the oxidative burst have repeatedly been postulated to function as elements of the plant's defense signaling machinery (Low and Merida, 1996; Mehdy *et al.*, 1996). Pharmacological loss- and gain-of-function experiments have been the preferred approach to investigate this question. Based on the assumption that production of extracellularly accumulating ROS is catalyzed by a plasma membrane-located NAD(P)H oxidase, several inhibitors and effectors have been utilized to interfere with this and subsequent reactions (Figure 4). Elicitor- or pathogen-induced accumulation of phytoalexins, a defense response known to depend on the activation of genes that code for the corresponding biosynthetic enzymes (Kombrink and Somssich, 1995), was inhibited by antioxidant enzymes, ROS scavengers, or NAD(P)H oxidase inhibitors in some experimental systems (Doke, 1983; Apostol *et al.*, 1989; Jabs *et al.*, 1997), whereas in other cases no inhibitory effect was detectable (Levine *et al.*, 1994; Rustérucci *et al.*, 1996; Mithöfer *et al.*, 1997). In parsley, loss-of-function experiments were complemented by gain-of-function experiments demonstrating that $O_2^{\bullet-}$ radicals rather than H_2O_2 were the ROS involved in signaling phytoalexin formation (Jabs *et al.* 1997). Since neither $O_2^{\bullet-}$ nor $HO_2^{\bullet}$ can easily cross the plasma membrane, their interaction with membrane constituents is expected to generate unknown second messengers. Results obtained with the mutant lesions simulating disease resistance response 1 (*lsd1*) of *Arabidopsis thaliana*

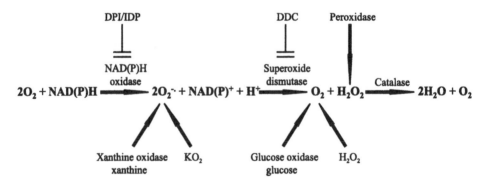

Figure 4. Chemical reactions believed to occur during the oxidative burst of plant cells. NAD(P)H oxidase is inhibited by diphenylene iodonium (DPI) and diphenyl iodonium (IDP). Sodium diethyldithiocarbamate (DDC) is an inhibitor of superoxide dismutase. For pharmacological studies, $O_2^{\bullet-}$ has been generated by xanthine oxidase and xanthine or by bringing KO_2 into aqueous solutions. H_2O_2 has been formed by glucose oxidase in the presence of glucose or directly added in aqueous solutions.

suggest similar signaling mechanisms in the initiation of programmed cell death and defense gene expression (Dietrich *et al.*, 1994; Jabs *et al.*, 1996). The *LSD1* gene, which appears to be required for confinement of programmed cell death (Dietrich *et al.*, 1994), encodes a zinc finger protein that monitors an $O_2^{\bullet-}$-dependent signal and negatively regulates a plant cell death pathway (Dietrich *et al.*, 1997). The *lsd1* mutant was unable to properly control cell death and defense gene expression in the absence of pathogens. In mutant, but not in wild-type plants extracellular $O_2^{\bullet-}$ rather than H_2O_2 was found to be necessary and sufficient to initiate lesion formation and defense gene expression (Jabs *et al.*, 1996).

Increase of intracellular H_2O_2 in transgenic tobacco with reduced catalase activity leads to elevated salicylate levels, expression of PR proteins, and reduced sensitivity towards pathogen attack (Takahashi *et al.*, 1997; Chamnongpol *et al.*, 1998). The fact that salicylate stimulates ROS production and *vice versa* that H_2O_2 induces increases in salicylate levels together with the finding that salicylate is necessary for local and one type of systemic resistance suggested that salicylate acts downstream of the oxidative burst with a potentiation loop stimulating at least the second phase of the burst (Draper, 1997; Van Camp *et al.*, 1998).

The role of ROS in SAR induction was recently studied in *Arabidopsis* that was locally inoculated with avirulent *Pseudomonas syringae* (Alvarez *et al.*, 1998). Local lesion development at the infection sites was accompanied by formation of small-sized systemic lesions and SAR. Systemic lesion and SAR establishment depend on H_2O_2 from both the local oxidative burst and secondary systemic microbursts at the sites of subsequent systemic lesion development. In addition, the local reaction, requires NO that is synthesized simultaneously with H_2O_2 upon infection of *Arabidopsis* leaves with avirulent bacteria (Delledonne *et al.*, 1998). In cultured soybean cells, H_2O_2 and NO in combination are necessary and sufficient for defense gene activation and induction of programmed cell death. A binary signal transduction system is postulated with separate pathways for ROS and NO generation that synergistically activate both responses (Delledonne *et al.*, 1998). The downstream signaling elements appear to be similar to those of mammalian systems and include cyclic GMP and cyclic ADP ribose (Durner *et al.*, 1998). In summary, ROS from the oxidative burst play a central role in local and systemic signaling of the multicomponent defense response of plants against pathogens.

PERSPECTIVES

ROS from the oxidative burst are key components in the plant's defense against pathogens. They act as direct toxicants, are involved in early reinforcement of physical barriers, and play a central role in signaling later reactions. Despite their importance for the plant's defense, little is known about site and mode of ROS formation, ROS compartmentation, and the immediate reaction products of their interaction with different biomolecules. Elucidation of these processes will not only increase our knowledge on basic principles of plant-pathogen interactions, but also allow the development of novel plant protection strategies. Constitutive apoplastic generation of H_2O_2 in transgenic potato, for example, activated several typical plant defense reactions and increased disease resistance of these plants (Wu *et al.*, 1995, 1997). A better understanding of the molecular mechanisms underlying the oxidative burst will open further ways to more precisely modulate this important regulator of plant defense.

ACKNOWLEDGEMENTS

The author thanks Heidi Zinecker and Markus Tschöpe for valuable discussions and critical reading of the manuscript and Christine Kaufmann for preparing the figures. Work on the oxidative burst was supported by the Deutsche Forschungsgemeinschaft (Sche 235/3–4 and Innovationskolleg, project number B6), the European Commission DGXII Biotechnology Program (contract number BIO4CT960101), and the Fonds der chemischen Industrie.

REFERENCES

Able, A.J., Guest, D.I., and Sutherland, M.W. (1998) Use of a new tetrazolium-based assay to study the production of superoxide radicals by tobacco cell cultures challenged with avirulent zoospores of *Phytophthora parasitica* var *nicotianae*. *Plant Physiol.*, **117**, 491–499.

Adam, A., Farkas, T., Somlya, G., Hevesi, M., and Kiraly, Z. (1989) Consequence of O_2^- generation during a bacterially induced hypersensitive reaction in tobacco: deterioration of membrane lipids. *Physiol. Mol. Plant Pathol.*, **34**, 13–26.

Alfano, J.R. and Collmer, A. (1996) Bacterial pathogens in plants: Life up against the wall. *Plant Cell*, **8**, 1683–1698.

Allan, A.C. and Fluhr, R. (1997) Two distinct sources of elicited reactive oxygen species in tobacco epidermal cells. *Plant Cell*, **9**, 1559–1572.

Alvarez, M.E., Pennell, R.I., Meijer, P.-J., Ishikawa, A., Dixon, R.A., and Lamb, C. (1998) Reactive oxygen intermediates mediate a systemic network in the establishment of plant immunity. *Cell*, **92**, 773–784.

Apostol, I., Heinstein, P.F., and Low, P.S. (1989) Rapid stimulation of an oxidative burst during elicitation of cultured plant cells: role in defense and signal transduction. *Plant Physiol.*, **90**, 109–116.

Auh, C.-K. and Murphy, T.M. (1995) Plasma membrane redox enzyme is involved in the synthesis of O^{2-} and H_2O_2 by *Phytophthora* elicitor stimulated rose cells. *Plant Physiol.*, **107**, 1241–1247.

Aver'yanov, A.A., Lapikova, V.P., and Djawakhia, V.G. (1993) Active oxygen mediates heat-induced resistance of rice plants blast disease. *Plant Sci.*, **92**, 27–34.

Babior, B.M., Benna, J.E., Chanock, S.J., and Smith, R.M. (1997) The NADPH oxidase of leukocytes: the respiratory burst oxidase. In J.G. Scandalios, (ed.), *Oxidative Stress and the Molecular Biology of Antioxidant Defenses*, (Monograph Series, Vol. 34), Cold Spring Harbor Laboratory Press, Cold Spring Harbor, pp. 737–783.

Baker, C.J., Atkinson, M.M., and Collmer, A. (1987) Concurrent loss in Tn5 mutants of *Pseudomonas syringae* pv *syringae* of the ability to induce the hypersensitive response and host plasma membrane K^+/H^+ exchange in tobacco. *Phytopathology*, **77**, 1268–1272.

Baker, C.J. and Mock, N.M. (1994) An improved method for monitoring cell death in cell suspension and leaf disc assays using Evans Blue. *Plant Cell Tissue Org. Cult.*, **39**, 7–12.

Baker, C.J., Mock, N.M., Glazener, J.A., and Orlandi, E.W. (1993a) Recognition responses in pathogen/nonhost and race/cultivar interactions involving soybean (*Glycine max*) and *Pseudomonas syringae* pathovars. *Physiol. Mol. Plant Pathol.*, **43**, 81–94.

Baker, C.J., O'Neill, N.R., Keppler, L.D., and Orlandi, E.W. (1991) Early responses during plant-bacteria interactions in tobacco cell suspensions. *Phytopathology*, **81**, 1504–1507.

Baker, C.J. and Orlandi, E.W. (1995) Active oxygen in plant pathogenesis. *Annu. Rev. Phytopathol.*, **33**, 299–321.

Baker, C.J., Orlandi, E.W., and Mock, N.M. (1993b) Harpin, an elicitor of the hypersensitive response in tobacco caused by *Erwinia amylovora* elicits active oxygen production in suspension cells. *Plant Physiol.*, **102**, 1341–1344.

Bent, A.F. (1996) Plant disease resistance genes: Function meets structure. *Plant Cell*, **8**, 1757–1771.

Bestwick, C.S., Bennett, M.H., and Mansfield, J.W. (1995) *Hrp* mutant of *Pseudomonas syringae* pv *phaseolicola* induces cell wall alterations but not membrane damage leading to the HR in lettuce (*Lactuca sativa*). *Plant Physiol.*, **108**, 503–516.

Bestwick, C.S., Brown, I.R., Bennett, M.H.R., and Mansfield, J.W. (1997). Localization of hydrogen peroxide accumulation during the hypersensitive reaction of lettuce cells to *Pseudomonas syringae* pv *phaseolicola*. *Plant Cell*, **9**, 209–221.

Bolwell, G.P., Butt, V.S., Davies, D.R., and Zimmerlin, A. (1995). The origin of the oxidative burst in plants. *Free Rad. Res.*, **23**, 517–532.

Bolwell, G.P., Davies, D.R., Gerrish, C., Auh, C.-K., and Murphy, T.M. (1998) Comparative biochemistry of the oxidative burst produced by rose and french bean cells reveals two distinct mechanisms. *Plant Physiol.*, **116**, 1379–1385.

Bottin, A., Véronési, C., Pontier, D., Esquerré-Tugayé, M.T., Blein, J.P., Rusterucci, C., and Ricci, P. (1994) Differential responses of tobacco cells to elicitors from two *Phytophthora* species. *Plant Physiol. Biochem.*, **32**, 373–378.

Bradley, D.J., Kjellbom, P., and Lamb, C.J. (1992) Elicitor- and wound-induced oxidative cross-linking of a proline-rich plant cell wall protein: a novel, rapid defense response. *Cell*, **70**, 21–30.

Brisson, L.F., Tenhaken, R., and Lamb, C. (1994) Function of oxidative cross-linking of cell wall structural proteins in plant disease resistance. *Plant Cell*, **6**, 1703–1712.

Chamnongpol, S., Willekens, H., Moeder, W., Langebartels, C., Sandermann, H., Van Montagu, M., Inzé, D., and Van Camp, W. (1998) Defense activation and enhanced pathogen tolerance induced by H_2O_2 in transgenic tobacco. *Proc. Natl. Acad. Sci. USA*, **95**, 5818–5823.

Chandra, S., Heinstein, P.F., and Low, P.S. (1996a) Activation of phospholipase A by plant defense elicitors. *Plant Physiol.*, **110**, 979–986.

Chandra, S. and Low, P.S. (1995) Role of phosphorylation in elicitation of the oxidative burst in cultured soybean cells. *Proc. Natl. Acad. Sci. USA*, **92**, 4120–4123.

Chandra, S. and Low, P.S. (1997) Measurement of Ca^{2+} fluxes during elicitation of the oxidative burst in aequorin-transformed tobacco cells. *J. Biol. Chem.*, **272**, 28274–28280.

Chandra, S., Martin, G.B., and Low, P.S. (1996b) The Pto kinase mediates a signaling pathway leading to the oxidative burst in tomato. *Proc. Natl. Acad. Sci. USA*, **93**, 13393–13397.

De Wit, P.J.G.M. (1992) Molecular characterization of gene-for-gene systems in plant-fungus interactions and the application of avirulence genes in control of plant pathogens. *Annu. Rev. Phytopathol.*, **30**, 391–418.

Delledonne, M., Xia, Y., Dixon, R.A., and Lamb, C. (1998) Nitric oxide signal functions as a signal in plant disease resistance. *Nature*, **394**, 585–588.

Desikan, R., Hancock, J.T., Coffey, M.J., and Neill, S.J. (1996) Generation of active oxygen in elicited cells of *Arabidopsis thaliana* is mediated by a NADPH oxidase-like enzyme. *FEBS Lett.*, **382**, 213–217.

Diekmann, E., Abo, A., Johnston, C., Segal, A.W., and Hall, A. (1994) Interaction of *Rac* with P67 *Phox* and regulation of the phagocytic NADPH oxidase. *Science*, **265**, 531–533.

Dietrich, R.A., Delaney, T.P., Uknes, S.J., Ward, E.R., Ryals, J.A., and Dangl, J.L. (1994) Arabidopsis mutants simulating disease resistance response. *Cell*, **77**, 565–577.

Dietrich, R.A., Richberg, M.H., Schmidt, R., Dean, C., and Dangl, J.L. (1997) A novel zinc finger protein is encoded by the Arabidopsis *LSD1* gene and functions as a negative regulator of plant cell death. *Cell*, **88**, 685–694.

Doke, N. (1983) Generation of superoxide anion by potato tuber protoplasts during the hypersensitive response to hyphal cell wall components of *Phytophthora infestans* and specific inhibition of the reaction by suppressors of hypersensitivity. *Physiol. Plant Pathol.*, **23**, 359–367.

Doke, N., Chai, H.B., and Kawaguchi, A. (1987) Biochemical basis of triggering and suppression of hypersensitive cell response. In S. Nishimura, G.P. Vance, and N. Doke, (eds.), *Molecular Determinants of Plant Diseases*, Springer-Verlag, Berlin, pp. 235–251.

Doke, N. and Miura, Y. (1995) *In vitro* activation of NADPH-dependent O_2-generating system in a plasma membrane-rich fraction of potato tuber tissues by treatment with an elicitor from *Phytophthora infestans* or with digitonin. *Physiol. Mol. Plant Pathol.*, **46**, 17–28.

Draper, J. (1997) Salicylate, superoxide synthesis and cell suicide in plant defence. *Trends Plant Sci.*, **2**, 162–165.

Durner, J., Wendehenne, D., and Klessig, D.F. (1998) Defense gene induction in tobacco by nitric oxide, cyclic GMP and cyclic ADP ribose. *Proc. Natl. Acad. Sci. USA*, **95**, 10328–10333.

Dwyer, S.C., Legendre, L., Low, P.S., and Leto, T.L. (1996) Plant and human neutrophil oxidative burst complexes contain immunologically related proteins. *Biochim. Biophys. Acta*, **1289**, 231–237.

Ebel, J. and Scheel, D. (1992) Elicitor recognition and signal transduction. In T. Boller and F. Meins, (eds.), *Genes Involved in Plant Defense*, Springer-Verlag, Wien, pp. 183–205.

Ebel, J. and Scheel, D. (1997) Signals in host-parasite interactions. In G.C. Carroll and P. Tudzynski, (eds.), *The Mycota, Vol. V, Plant Relationships, Part A*, Springer-Verlag, Berlin, pp. 85–105.

Fauth, M., Merten, A., Hahn, M.G., Jeblick, W., and Kauss, H. (1996) Competence for elicitation of H_2O_2 in hypocotyls of cucumber is induced by breaching the cuticle and is enhanced by salicylic acid. *Plant Physiol.*, **110**, 347–354.

Flor, H.H. (1955) Host-parasite interactions in flax rust—Its genetics and other implications. *Phytopathology*, **45**, 680–685.

Flor, H.H. (1971) Current status of the gene-for-gene concept. *Annu. Rev. Phytopathol.*, **9**, 275–296.

Glazener, J.A., Orlandi, E.W., and Baker, C.J. (1996) The active oxygen response of cell suspensions to incompatible bacteria is not sufficient to cause hypersensitive cell death. *Plant Physiol.*, **110**, 759–763.

Grisham, M.B. (1992) *Reactive Metabolites of Oxygen and Nitrogen in Biology and Medicine*, R.G. Landes Co., Austin.

Groom, Q.J., Torres, M.A., Fordham-Skelton, A.P., Hammond-Kosack, K.E., Robinson, N.J., and Jones, J.D.G. (1996) *rbohA* a rice homologue of the mammalian *gp91phox* respiratory burst oxidase. *Plant J.*, **10**, 515–522.

Gus-Mayer, S., Naton, B., Hahlbrock, K., and Schmelzer, E. (1998) Local mechanical stimulation induces components of the pathogen defense response in parsley. *Proc. Natl. Acad. Sci. USA*, **95**, 8398–8403.

Hammond-Kosack, K.E., and Jones, J.D.G. (1996) Resistance gene-dependent plant defense responses. *Plant Cell*, **8**, 1773–1791.

Hancock, J.T. and Jones, O.T.G. (1987) The inhibition by dipenylene iodonium and its analogues of superoxide generation by macrophages. *Biochem. J.*, **242**, 103–107.

Hassouni, M.E., Chambost, J.P., Expert, D., Van Gijsegem, F., and Barras, F. (1999) The minimal gene set member *msrA* encoding peptide methionine sulfoxide reductase, is a virulence determinant of the plant pathogen *Erwinia chrysanthemi*. *Proc. Natl. Acad. Sci. USA*, **96**, 887–892.

Henderson, L.M., Banting, G., and Chappell, J.B. (1995) The arachidonate-activable, NADPH oxidase-associated H^+ channel. Evidence that gp91-*phox* functions as an essential part of the channel. *J. Biol. Chem.*, **270**, 5909–5916.

Huang, H.C., Hutcheson, S.W., and Collmer, A. (1991) Characterization of the *hrp* cluster from *Pseudomonas syringae* pv. *syringae* 61 and Tn*phoA* tagging of genes encoding exported or membrane-spanning *hrp* proteins. *Mol. Plant-Microbe Interact.*, **4**, 469–476.

Jabs, T., Dietrich, R.A., and Dangl, J.L. (1996) Initiation of runaway cell death in an Arabidopsis mutant by extracellular superoxide. *Science*, **273**, 1853–1856.

Jabs, T., Tschöpe, M., Colling, C., Hahlbrock, K., and Scheel, D. (1997) Elicitor-stimulated ion fluxes and O^{2-} from the oxidative burst are essential components in triggering defense gene activation and phytoalexin synthesis in parsley. *Proc. Natl. Acad. Sci. USA*, **94**, 4800–4805.

Jacks, T.J. and Davidonis, G.H. (1996) Superoxide, hydrogen peroxide, and the respiratory burst of fungally infected plant cells. *Mol. Cell. Biochem.*, **158**, 77–79.

Jennigs, D.B., Ehrenshaft, M., Pharr, D.M., and Williamson, J.D. (1998) Roles for mannitol and mannitol dehydrogenase in active oxygen-mediated plant defense. *Proc. Natl. Acad. Sci. USA*, **95**, 15129–15133.

Joosten, M.H.A.J., Cozijnsen, A.J., and De Wit, P.J.G.M. (1994) Host resistance to a fungal tomato pathogen lost by a single base-pair change in an avirulence gene. *Nature*, **367**, 348–387.

Kauss, H. and Jeblick, W. (1995) Pretreatment of parsley suspension cultures with salicylic acid enhances spontaneous and elicited production of H_2O_2. *Plant Physiol.*, **108**, 1171–1178.

Kauss, H. and Jeblick, W. (1996) Influence of salicylic acid on the induction of competence for H_2O_2 elicitation. *Plant Physiol.*, **111**, 755–763.

Keller, T., Damude, H.G., Werner, D., Doerner, P., Dixon, R.A., and Lamb, C. (1998) A plant homolog of the neutrophil NADPH oxidase gp91^{phox} subunit gene encodes a plasma membrane protein with Ca^{2+} binding motifs. *Plant Cell*, **10**, 255–266.

Keppler, L.D. and Baker, C.J. (1989) O_2^- initiated lipid peroxidation in a bacteria-induced hypersensitive reaction in tobacco cell suspensions. *Phytopathology*, **79**, 555–562.

Keppler, L.D., Baker, C.J., and Atkinson, M.M. (1989) Active oxygen production during a bacteria-induced hypersensitive reaction in tobacco suspension cells. *Phytopathology*, **79**, 974–978.

Kieffer, F., Simon-Plas, F., Maume, B.F., and Blein, J.P. (1997) Tobacco cells contain a protein, immunologically related to the neutrophil small G protein Rac2 and involved in elicitor-induced oxidative burst. *FEBS Lett.*, **403**, 149–153.

Klotz, M.G. (1993) The importance of bacterial growth phase for in planta virulence and pathogenicity testing: coordinated stress response regulation in fluorescent pseudomonads. *Can. J. Microbiol.*, **39**, 948–957.

Klotz, M.G. and Anderson, A.J. (1994) The role of catalase isozymes in the culturability of the root colonizer *Pseudomonas putida* after exposure to H_2O_2 and antibiotics. *Can. J. Microbiol.*, **40**, 382–387.

Klotz, M.G. and Hutcheson, S.W. (1992) Multiple periplasmic catalases in phytopathogenic strains of *Pseudomonas syringae*. *Appl. Environ. Microbiol.*, **58**, 2468–2473.

Knogge, W. (1996) Fungal infection of plants. *Plant Cell*, **8**, 1711–1722.

Kombrink, E. and Somssich, I.E. (1995) Defense responses of plants to pathogens. *Adv. Bot. Res.*, **21**, 1–34.

Kuchitsu, K., Kosaka, H., Shiga, T., and Shibuya, N. (1995) EPR evidence for generation of hydroxyl radical triggered by N-acetylchitooligosaccharide elicitor and a protein phosphatase inhibitor in suspension-cultured rice cells. *Protoplasma*, **188**, 138–142.

Lamb, C. and Dixon, R.A. (1997) The oxidative burst in plant disease resistance. *Annu. Rev. Plant Physiol. Plant Mol. Biol.*, **48**, 251–275.

Legendre, L., Heinstein, P.F., and Low, P.S. (1992) Evidence for participation of GTP-binding proteins in elicitation of the rapid oxidative burst in cultured soybean cells. *J. Biol. Chem.*, **267**, 20140–20147.

Legendre, L., Rueter, S., Heinstein, P.F., and Low, P.S. (1993a) Characterization of the oligalacturonide-induced oxidative burst in cultured soybean (*Glycine max*) cells. *Plant Physiol.*, **102**, 233–240.

Legendre, L., Yueh, Y.G., Crain, R., Haddock, N., Heinstein, P.F., and Low, P.S. (1993b) Phospholipase C activation during elicitation of the oxidative burst in cultured plant cells. *J. Biol. Chem.*, **268**, 24559–24563.

Levine, A., Pennell, R.I., Alvarez, M.E., Palmer, R., and Lamb, C. (1996) Calcium-mediated apoptosis in a plant hypersensitive disease resistance response. *Curr. Biol.*, **6**, 427–437.

Levine, A., Tenhaken, R., and Dixon, R. (1994) H_2O_2 from the oxidative burst orchestrates the plant hypersensitive disease resistance response. *Cell*, **79**, 583–593.

Ligterink, W., Kroj, T., zur Nieden, U., Hirt, H., and Scheel, D. (1997) Receptor-mediated activation of a MAP kinase in pathogen defense of plants. *Science*, **276**, 2054–2057.

Low, P.S. and Heinstein, P.F. (1986) Elicitor stimulation of the defense response in cultured plant cells monitored by fluorescent dyes. *Arch. Biochem. Biophys.*, **249**, 472–479.

Low, P.S. and Merida, J.R. (1996) The oxidative burst in plant defense: Function and signal transduction. *Physiol. Plant.*, **96**, 533–542.

Ma, M. and Eaton, J.W. (1992) Multicellular oxidant defense in unicellular organisms. *Proc. Natl. Acad. Sci. USA*, **89**, 7924–7928.

May, M.J., Hammond-Kosack, K.E., and Jones, J.D.G. (1996) Involvement of reactive oxygen species, glutathione metabolism, and lipid peroxidation in the *Cf*-gene-dependent defense response of tomato cotyledons induced by race-specific elicitors of *Cladosporium fulvum*. *Plant Physiol.*, **110**, 1367–1379.

Mehdy, M.C., Sharma, Y.K., Sathasivan, K., and Bays, N.W. (1996) The role of activated oxygen species in plant disease resistance. *Physiol. Plant.*, **98**, 365–374.

Minardi, P. and Mazzucchi, U. (1988) No evidence of direct superoxide anion effect in hypersensitive death of *Pseudomonas syringae* van Hall in tobacco leaf tissue. *J. Phytopathol.*, **122**, 351–358.

Mithöfer, A., Daxberger, A., Fromhold-Treu, D., and Ebel, J. (1997) Involvement of an NAD(P)H oxidase in the elicitor-inducible oxidative burst of soybean. *Phytochemistry*, **45**, 1101–1107.

Miura, Y., Yoshioka, H., and Doke, N. (1995) An autophotographic determination of the active oxygen generation in potato tuber discs during hypersensitive response to fungal infection or elicitor. *Plant Sci.*, **105**, 42–52.

Murphy, T.M., Vu, H., and Nguyen, T. (1998) The superoxide synthases of rose cells. *Plant Physiol.*, **117**, 1301–1305.

Naton, B., Hahlbrock, K., and Schmelzer, E. (1996) Correlation of rapid cell death with metabolic changes in fungus-infected, cultured parsley cells. *Plant Physiol.*, **112**, 433–444.

Nürnberger, T., Colling, C., Hahlbrock, K., Jabs, T., Renelt, A., Sacks, W.R., and Scheel, D. (1994a) Perception and transduction of an elicitor signal in cultured parsley cells. *Biochem. Soc. Symp.*, **60**, 173–182.

Nürnberger, T., Nennstiel, D., Jabs, T., Sacks, W.R., Hahlbrock, K., and Scheel, D. (1994b) High affinity binding of a fungal oligopeptide elicitor to parsley plasma membranes triggers multiple defense responses. *Cell*, **78**, 449–460.

Orlandi, E.W., Hutcheson, S.W., and Baker, C.J. (1992) Early physiological responses associated with race-specific recognition in soybean leaf tissue and cell suspensions treated with *Pseudomonas syringae* pv. *glycinea*. *Physiol. Mol. Plant Pathol.*, **40**, 173–180.

Otte, O. and Barz, W. (1996) The elicitor-induced oxidative burst in cultured chickpea cells drives the rapid insolubilization of two cell wall structural proteins. *Planta*, **200**, 238–246.

Peng, M. and Kuc, J. (1992) Peroxidase-generated hydrogen peroxide as a source of antifungal activity in vitro and on tobacco leaf disks. *Phytopathology*, **82**, 696–699.

Piedras, P., Hammond-Kosack, K.E., Harrison, K., and Jones, J.D.G. (1998) Rapid, *Cf-9*- and Avr9-dependent production of active oxygen species in tobacco suspension cultures. *Mol. Plant-Microbe Interact.*, **11**, 1155–1166.

Rustérucci, C., Stallaert, V., Milat, M.-L., Pugin, A., Ricci, P., and Blein, J.-P. (1996) Relationship between active oxygen species, lipid peroxydation, necrosis, and phytoalexin production induced by elicitins in *Nicotiana*. *Plant Physiol.*, **111**, 885–891.

Salzer, P., Hebe, G., Reith, A., Zitterell-Haid, B., Stransky, H., Gaschler, K., and Hager, A. (1996) Rapid reactions of spruce cells to elicitors released from the ectomycorrhizal fungus *Hebeloma crustuliniforme* and inactivation of these elicitors by extracellular spruce cell enzymes. *Planta*, **198**, 118–126.

Sanchez, L.M., Doke, N., and Kawakita, K. (1993) Elicitor-induced chemiluminescence in cell suspension cultures of tomato, sweet pepper and tobacco plants and its inhibition by suppressors from *Phytophthora* spp. *Plant Sci.*, **88**, 141–148.

Scholtens-Toma, I.M.J. and De Wit, P.J.G.M. (1988) Purification and primary structure of a necrosis-inducing peptide from the apoplastic fluids of tomato infected with *Cladosporium fulvum* (syn. *Fulvia fulva*). *Physiol. Mol. Plant Pathol.*, **33**, 59–67.

Schwacke, R. and Hager, A. (1992) Fungal elicitors induce a transient release of active oxygen species from cultured spruce cells that is dependent on Ca^{2+} and protein kinase activity. *Planta*, **187**, 136–141.

Sekizawa, Y., Haga, M., Hirabayashi, E., Takeuchi, N., and Takino, Y. (1987) Dynamic behavior of superoxide generation in rice leaf tissue infected with blast fungus and its regulation by some substances. *Agric. Biol. Chem.*, **51**, 763–770.

Sutherland, M.W. (1991) The generation of oxygen radicals during host plant responses to infection. *Physiol. Mol. Plant Pathol.*, **39**, 79–93.

Svalheim, O. and Robertsen, B. (1993) Elicitation of H_2O_2 production in cucumber hypocotyl segments by oligo-1,4-á-D-galacturonides and an oligo-â-glucan preparation from cell walls of *Phytophthora megasperma* f.sp. *glycinea*. *Physiol. Plant.*, **88**, 675–681.

Takahashi, A., Kawasaki, T., Henmi, K., Shil, K., Kodama, O., Satoh, H., and Shimamoto, K. (1999) Lesion mimic mutants of rice with alterations in early signaling events of defense. *Plant J.*, **17**, 535–545.

Takahashi, H., Chen, Z., Du, H., Liu, Y., and Klessig, D.F. (1997) Development of necrosis and activation of disease resistance in transgenic tobacco plants with severely reduced catalase levels. *Plant J.*, **11**, 993–1005.

Tavernier, E., Wendehenne, D., Blein, J.-P., and Pugin, A. (1995) Involvement of free calcium in action of cryptogein, a proteinaceous elicitor of hypersensitive reaction in tobacco cells. *Plant Physiol.*, **109**, 1025–1031.

Taylor, A.T.S. and Low, P.S. (1997) Phospholipase D involvement in the plant oxidative burst. *Biochem. Biophys. Res. Commun.*, **237**, 10–15.

Tenhaken, R., Levine, A., Brisson, L.F., Dixon, R.A., and Lamb, C. (1995) Function of the oxidative burst in hypersensitive disease resistance. *Proc. Natl. Acad. Sci. USA*, **92**, 4158–4163.

Tenhaken, R. and Rübel, C. (1998) Cloning of putative subunits of the soybean plasma membrane NADPH oxidase involved in the oxidative burst by antibody expression screening. *Protoplasma*, **205**, 21–28.

Thordal-Christensen, H., Zhang, Z., Wei, Y., and Collinge, D.B. (1997) Subcellular localization of H_2O_2 in plants, H_2O_2 accumulation in papillae and hypersensitive response during the barley-powdery mildew interaction. *Plant J.*, **11**, 1187–1194.

Torres, M.A., Onouchi, H., Hamada, S., Machida, C., Hammond-Kosack, K.E., and Jones, J.D.G. (1998) Six *Arabidopsis thaliana* homologues of the human respiratory burst oxidase (*gp91phox*). *Plant J.*, **14**, 365–370.

Tzeng, D.D. and DeVay, J.E. (1993) Role of oxygen radicals in plant disease development. *Adv. Plant Pathol.*, **10**, 1–34.

Van Camp, W., Van Montagu, M., and Inzé, D. (1998) H_2O_2 and NO: redox signals in disease resistance. *Trends Plant Sci.*, **3**, 330–334.

Van den Ackerveken, G.F.J.M., Van Kan, J.A.L., and De Wit, P.J.G.M. (1992) Molecular analysis of the avirulence gene *avr9* of the fungal tomato pathogen *Cladosporium fulvum* fully supports the gene-for-gene hypothesis. *Plant J.*, **2**, 359–366.

Van Gestelen, P., Asard, H., and Caubergs, R.J. (1997) Solubilization and separation of a plant plasma membrane NADPH-O_2^- synthase from other NAD(P)H oxidoreductases. *Plant Physiol.*, **115**, 543–550.

Vera-Estrella, R., Blumwald, E., and Higgins, V.J. (1992) Effect of specific elicitors of *Cladosporium fulvum* on tomato suspension cells. Evidence for the involvement of active oxygen species. *Plant Physiol.*, **99**, 1208–1215.

Vera-Estrella, R., Higgins, V.J., and Blumwald, E. (1994) Plant defense response to fungal pathogens. II. G-protein-mediated changes in host plasma membrane redox reactions. *Plant Physiol.*, **106**, 97–102.

Viard, M.-P., Martin, F., Pugin, A., Ricci, P., and Blein, J.-P. (1994) Protein phosphorylation is induced in tobacco cells by the elicitor cryptogein. *Plant Physiol.*, **104**, 1245–1249.

Wojtaszek, P. (1997) Oxidative burst: an early plant response to pathogen infection. *Biochem. J.*, **322**, 681–692.

Wu, G., Shortt, B.J., Lawrence, E.B., León, J., Fitzsimmons, K.C., Levine, E.B., Raskin, I., and Shah, D.M. (1997) Activation of host defense mechanisms by elevated production of H_2O_2 in transgenic plants. *Plant Physiol.*, **115**, 427–435.

Wu, G., Shortt, B.J., Lawrence, E.B., Levine, E.B., Fitzsimmons, K.C., and Shah, D.M. (1995) Disease resistance conferred by expression of a gene encoding H_2O_2-generating glucose oxidase in transgenic potato plants. *Plant Cell*, **7**, 1357–1368.

Xing, T., Higgins, V.J., and Blumwald, E. (1997) Race-specific elicitors of *Cladosporium fulvum* promote translocation of cytosolic components of NADPH oxidase to the plasma membrane of tomato cells. *Plant Cell*, **9**, 249–259.

Yu, L., Zhen, L., and Dinauer, M.C. (1997) Biosynthesis of the phagocyte NADPH oxidase cytochrome b_{558}. Role of heme incorporation and heterodimer formation in maturation and stability of gp91^{phox} and p22^{phox} subunits. *J. Biol. Chem.*, **272**, 27288–27294.

Zimmermann, S., Nürnberger, T., Frachisse, J.-M., Wirtz, W., Guern, J., Hedrich, R., and Scheel, D. (1997) Receptor-mediated activation of a plant Ca^{2+}-permeable ion channel involved in pathogen defense. *Proc. Natl. Acad. Sci. USA*, **94**, 2751–2755.

6   Photosensitizing Tetrapyrroles Induce Antioxidative and Pathogen Defense Responses in Plants

Hans-Peter Mock, Ulrich Keetman and Bernhard Grimm

INTRODUCTION

Photoreactions generally bear the potential for the generation of reactive oxygen species (ROS). This chapter will provide an introduction into the current understanding of plant defense processes in response to accumulating photosensitizing tetrapyrroles. The photodynamic consequences of chemically induced or hereditary accumulation of photoreactive tetrapyrroles by the modification of their metabolic pathway have previously been described and have stimulated the development of strategies to benefit from the photodestructive capacity of tetrapyrroles in agriculture and plant biotechnology as well as in medicine.

Tetrapyrroles consist of a porphyrin skeleton of four pyrrole molecules linked by methine (ß) groups. They are synthesized in a single, branched multienzymatic pathway that supplies pigments and cofactors for a range of diverse proteins and protein complexes involved e.g. in energy transformation (chlorophyll and heme in photosynthetic and fermentation complexes, respectively), light perception (phytochromobilin), protection against xenobiotics (cytochrome b_{450}), and removal of ROS (CAT, peroxidase). This list of pivotal functions already indicates the outstanding role of tetrapyrroles that is based on their chemical and physical properties and their ubiquitous spread among all organisms. Their photoreactive qualities are the prerequisite for photosynthesis and have also been used in pharmaceutical and agrochemical applications. The following review will focus on the description of cellular defense strategies to circumvent the deleterious action of cyclic tetrapyrroles in plants, in which the tight control of the tetrapyrrolic pathway is modified. These cellular responses against tetrapyrrole-dependent oxidative stress resemble many of the symptoms observed on lesion mimicking mutants and on plants exposed to various environmental stresses (Dietrich *et al.*, 1994; Eckey-Kaltenbach *et al.*, 1994) The analogy of these processes opens new prospects to find consistencies in the signaling and defense pathway upon generation of ROS.

BIOSYNTHESIS OF TETRAPYRROLES

The pathway of tetrapyrrole synthesis is tightly controlled to match the plant's current demands that are mainly defined by environmental and developmental conditions. The coordinated gene expression and enzyme activities in the pathway guarantee a controlled substrate flux to avoid the accumulation of highly photoreactive tetrapyrrole intermediates. In the recent years, scientific interest in tetrapyrrole biosynthesis has focused on genetical and biochemical studies of single enzymatic steps of the pathway and, to an increasing extent, on the elucidation of the regulatory network that matches gene expression and enzyme activity with the demand of the substrate flow.

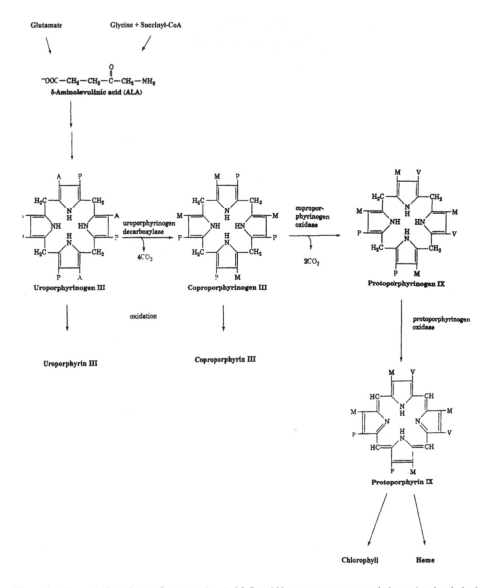

Figure 1. Biosynthetic pathway of tetrapyrroles. A, M, P and V, represent acetate, methyl, propionyl and vinyl residues respectively

The tetrapyrrole biosynthetic pathway is depicted in Figure 1. For a complete description of tetrapyrrole biosynthesis, the reader is referred to recent reviews (Smith and Griffith, 1993; Chadwick and Ackrill, 1994; von Wettstein *et al.*, 1995; Reinbothe and Reinbothe, 1996; Porra, 1997; Grimm, 1998; Beale, 1999). Briefly, the pathway starts with the formation of 5-aminolevulinic acid (ALA). ALA is formed either in a one-step condensation of glycine and succinyl-coenzyme A by ALA synthase (in purple bacteria, fungi, and animals) or from glutamate by the C5 pathway involving three enzymatic steps (in plants, algae, and most bacteria). Two ALA molecules are condensed to the monopyrrole porphobilinogen of which four molecules are sequentially fused to form the first linear tetrapyrrole hydroxymethylbilane that is stabilized as the cyclic molecule uroporphyrinogen III. The decarboxylation of all acetate side chains by uroporphyrinogen decarboxylase (UROD), the oxidative decarboxylation of the propionate side chains of ring A and B by coproporphyrinogen oxidase (CPO), and finally the oxidation by protoporphyrinogen oxidase (PPOX) yield the more hydrophobic molecule protoporphyrin IX that is the common precursor of heme and chlorophyll in which iron and magnesium are chelated, respectively. The intermediates of the oxidative reaction sequence, uroporphyrinogen, coproporphyrinogen, and protoporphyrinogen appear colorless because of the lack of the conjugated ring system. In respect to photodynamic reactions, it is of particular interest that these substrates can circumstantially be non-enzymatically oxidized to yield the phototoxic uroporphyrin, coproporphyrin or protoporphyrin, respectively. In the regular tetrapyrrole pathway, magnesium protoporphyrin is sequentially transformed into chlorophyll *a* by formation of a fifth isocyclic ring, several reduction steps of side chains of the tetrapyrrole and an esterification with a long chain fatty alcohol, mainly phytol or geranylgeraniol.

PHOTOTOXICITY OF TETRAPYRROLES

The photosensitizing capacity of isolated chlorophyll on cells has been described as early as 1908 by the work of Hausmann. Porphyrins and Mg-porphyrins are potent photosensitizers in light-triggered reactions, a property which is lost in the Fe-containing species. The system of conjugated double bonds of the porphyrins determines the (light) energy-absorbing function of tetrapyrroles. The photochemical and photophysical properties of tetrapyrroles have been reviewed by Spikes and Bommer (1991) and Ochsner (1997). In general, the photodynamic action (Figure 2) of an excited photosensitizer is exerted via a radical mechanism involving electron or hydrogen transfer (type I reaction) or transfer of energy to oxygen with the formation of singlet oxygen (type II reaction). A range of free radicals were observed when porphyrins dissolved in organic solution were illuminated (Haseloff *et al.*, 1989). The relative contribution of the type I or II mechanisms to photodynamic damage is dependent on the local oxygen concentration of the tissue and the cellular microenvironment (Ochsner, 1997). No significant difference was obtained in the formation of singlet oxygen after UV-A irradiation of protoporphyrin, coproporphyrin, or hematoporphyrin in acetone solution (Arakane *et al.*, 1996). The quantum yield for singlet oxygen formation was different when coproporphyrin was irradiated in aqueous solution or in micellar distribution (Lambert *et al.*, 1986). The lipophilicity of tetrapyrroles determines their cellular distribution and phototoxic efficacy.

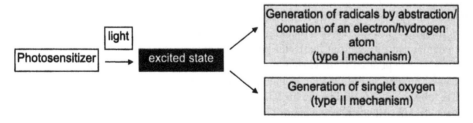

Figure 2. Molecular mechanisms of photosensitization.

Only the lipophilic protoporphyrin showed a phototoxic effect on endothelial cells *in vitro* when the efficacy of uroporphyrin, coproporphyrin, and protoporphyrin for vascular damage was compared (Strauss *et al.*, 1995).

The cellular targets of ROS generated by tetrapyrroles include membranes, DNA, and proteins (Kessel, 1986). *In vitro* experiments showed that the uroporphyrin induced photo-inactivation of heme-synthesizing enzymes of human erythrocyte hemolysates occurred preferentially by a type I mechanism (Afonso *et al.*, 1996). Studies with erythrocyte ghosts and use of cholesterol as an *in situ* probe demonstrated that porphyrin-sensitized lipid peroxidation occurred via a type II mechanism (Bachowski *et al.*, 1988; Girotti, 1993).

In photosynthesis, the energy trapped is transferred towards the reaction centers of the two photosystems. Excess energy will be safely re-emitted by fluorescence or quenched by nonradiative decay, but the energy can also be transferred from the long-lived excited triplet state of the tetrapyrrole to molecular oxygen. This tetrapyrrole-photosensitized process is harmful when protective components, such as ascorbate, carotenoids and tocopherol, cannot sufficiently detoxify the photochemically generated ROS.

In conclusion, in spite of the same mechanisms of ROS generation by accumulated tetrapyrroles, the photodynamic effects in plants and animals will differ in many aspects. This is due to the different subcellular localization of tetrapyrrole biosynthesis. In animals, enzymes of heme formation are located either in mitochondria or in the cytoplasm. In plants, heme and chlorophyll are synthesized in plastids and, in addition, the last steps of heme synthesis are located in mitochondria. Below we will describe the analysis of tetrapyrrole-sensitized plants and compare the results with selected data obtained in studies of human porphyrias or photodynamic therapy of cancer.

TRANSGENIC PLANTS WITH NECROTIC LEAF LESIONS DUE TO THE EXPRESSION OF SENSE OR ANTISENSE RNA OF TARGET ENZYMES IN PORPHYRIN BIOSYNTHESIS

Recently, cloning and characterization of plant genes that encode UROD (EC 4.1.1.37) (Mock *et al.*, 1995) and the consecutive enzyme in the pathway CPO (EC 1.3.3.3) (Madsen *et al.*, 1993; Kruse *et al.*, 1995a) enabled genetic engineering of the tetrapyrrole pathway in plants by introduction of transgenes in sense and antisense orientation.

The deduced amino acid sequence of the open reading frame for tobacco UROD contains 391 residues including an amino-terminal transit peptide that apparently consists of 39 amino acids. The mature protein was shown to be imported *in vitro* into pea chloroplasts and a transit peptide was cleaved off resulting in a 39-kDa polypeptide resident in the stroma fraction. Active enzyme could be overproduced in *Escherichia coli* cultured at 37°C (Mock *et al.*, 1995). Bacterial cells that harbored an expression plasmid for tobacco CPO immediately stopped growth at 37°C upon induction of the synthesis of the recombinant protein, but led to the overproduction of the active plant CPO when the cultures were kept at 20°C. *In vitro* import experiments into green pea plastids resulted in a processed protein of 39 kDa that accumulated in the stroma fraction (Kruse *et al.*, 1995a).

The tobacco cDNA sequences for UROD and CPO were used to design antisense and sense gene constructs under the control of the CaMV 35S promoter by which leaf discs of tobacco plants were transformed (Kruse *et al.*, 1995b; Mock and Grimm, 1997; Mock *et al.*, 1998). Most of the transformants grow significantly more slowly than the wild type and their leaves vary from wrinkled or shriveled to having areas of desiccated brownish dead tissue. In most transformed plants, young leaves show wild type-like pigmentation and acquire necrotic lesions at the stage of cell differentiation, elongation and at chloroplast propagation. Necrosis appears especially along the major and the minor veins of the leaves, although in some transformants necrotic spots are also randomly distributed. With increasing age, the necrotic injuries cover almost the entire leaf. Root formation is also negatively affected and flowering is delayed in parallel with the impaired development of stems and leaves compared to control plants. In addition to the very pronounced leaf necrosis of some transgenic lines, others show gray or dirty greenish leaves. Other plants with a less severe phenotype loose their green pigmentation earlier than the wild type, indicative of premature senescence. Transgenic lines containing the UROD antisense construct are generally characterized by a more patchy distribution of necrosis (Mock and Grimm, 1997) whereas transformants with CPO antisense genes particularly form a net of necrotic lesions along the minor veins (Kruse *et al.*, 1995b)

The content of chlorophyll and heme is only slightly reduced in most of the transformants. Reductions of up to 20% on a fresh weight basis were measured in the most severely affected lines. The chlorophyll *a/b* ratio is also not significantly changed (Kruse *et al.* 1995b; Mock and Grimm 1997). The results indicate that these plants do not suffer from a lack of tetrapyrrole end products.

The necrotic leaf lesions of transgenic lines correlate with decreased steady-state UROD and CPO RNA levels, respectively, which lead to lower contents of the encoded enzyme. The reduced activity of the two target enzymes correlates inversely with the accumulation of the respective substrates uroporphyrinogen and coproporphyrinogen in young leaves of transformants. The more porphyrins initially accumulate in premature leaves, the more pronounced the necrotic phenotype will be in fully developed leaves. Up to 500-fold of the uroporphyrinogen and coproporphyrinogen concentration measured in wild-type leaves was detected in strongly affected UROD and CPO antisense lines, respectively (Kruse *et al.*, 1995b; Mock and Grimm, 1997). The gene expression of other enzymes involved in tetrapyrrole biosynthesis (e.g. glutamyl-tRNA synthetase, glutamate 1-semialdehyde aminotransferase, 5-aminolevulinic acid dehydratase (ALAD), and the subunits of magnesium chelatase) is not affected in transformants by the modification of UROD or CPO synthesis, indicating that the transgenic phenotype can solely be assigned to reduced UROD or CPO activity. Moreover, deregulation of these two enzymes does not lead to feedback-controlled changes of the expression of enzymes situated early in the pathway.

The formation of necrotic leaf lesions is strictly dependent on light intensities and light periods. A few transformants among the UROD and CPO antisense plants could be selected that grow under low light or short light period conditions without visible transgenic phenotype although similar amounts of uroporphyrinogen or coproporphyrinogen are accumulated under these growth conditions compared to high light or long illumination periods. Leaves of these plants showed necrotic lesions within 24–48 hours upon transfer from low to high light conditions (unpublished results). Another example for the light dose-dependent photodynamic destruction was the reduced activity of two other enzymes of the pathway, ALAD and porphobilinogen deaminase, in UROD antisense plants with necrotic lesions. *In vitro* experiments with protein extracts from erythrocytes (Afonso *et al.*, 1996) and with purified tobacco ALAD (H.P. Mock, unpublished results) have also demonstrated the high sensitivity of this enzyme towards photooxidative inactivation after incubation with uroporphyrin. In photosensitized cells of CPO antisense plants, gene expression for the nuclear-encoded early light-inducible protein (ELIP) increased, whereas it decreased for the light-harvesting chlorophyll-binding proteins (LHCP) and for the plastid-encoded D1 protein, which reflects an impaired signaling between plastids and nucleus in response to photodestructive processes in plastids (Kruse *et al.*, 1995b).

Similar results have also been obtained in some transgenic plants harboring a CPO-encoding gene in sense orientation under the control of the CaMV 35S promoter. CPO activity is not enhanced in these transformed lines, reflecting the frequently observed phenomenon of cosuppression. These plants are also characterized by the up to 100-fold accumulation of coproporphyrin(ogen) compared to wild type and by a necrotic phenotype very reminiscent to that of the corresponding CPO antisense plants (Mock *et al.*, 1997). Other transformants with one to several copies of the same CPO-encoding transgenes display higher CPO mRNA, protein and activity levels, but have no effect on the following steps of the pathway, which indicates that CPO is not involved in limiting the flow rate of the metabolites through the pathway.

The mutator tagged maize mutant *Les22* displays small necrotic spots on leaves in a developmentally and light-dependent manner (Hu *et al.*, 1998). This mutant was initially grouped among other maize mutants characterized by the formation of various types of leaf lesions. The *Les22* gene encodes UROD. The deficiency in UROD activity was demonstrated by the accumulation of uroporphyrin(ogen) in mutant leaves compared to wild-type maize, confirming that the null mutation in the *UroD* gene accounted for the lesion phenotype similar to that of UROD antisense plants.

An example of metal-triggered photodynamic lesions as result of impaired chlorophyll biosynthesis has been published by Marschner (1965) and Shalygo *et al.* (1997, 1998). Etiolated leaves or seedlings from barley and tobacco incubated with 15 mM cesium chloride for 8 h, accumulate uroporphyrin(ogen). The UROD activity extracted from the leaves was less than 20% of that of the water-incubated control leaves. Because CsCl in millimolar concentration cannot affect the native and the recombinant plant UROD *in vitro*, we suggest that cesium ions do not directly inhibit UROD but rather disturb the cellular metal homeostasis by inhibition of the potassium uptake into roots, and perhaps into leaf cells and their plastids (Shalygo *et al.*, 1997).

Conclusively, the transgenic plants with deregulated UROD and CPO activity experience multiple molecular disturbances in response to the photodynamic action of accumulated porphyrins. The changes in many molecular processes, such as modified gene expression and enzyme activities, finally result in the development of light dose-dependent necrotic lesions.

NECROTIC LESIONS IN CPO AND UROD ANTISENSE PLANTS ARE CAUSED BY REACTIVE OXYGEN SPECIES GENERATED BY PHOTOOXIDIZED PORPHYRINOGEN

More results were compiled to prove that the macroscopic phenotype provoked by the partial loss of UROD or CPO function in plants is attributed to intrinsically caused oxidative stress. The molecular reactions and the resulting necrotic phenotype of these transformants resemble those of other experimental plant systems suffering from oxidative stress. Chloroplasts usually face high concentrations of excited chlorophyll molecules (triplet state) as result of rapidly changing environmental conditions and of impaired coordination of the energy transfer steps between light trapping in the antenna of photosystems and storage of the energy transformed into redox equivalents (Foyer, 1997). These photosynthetic reactions potentially generate excess amounts of ROS that cause deleterious reactions. The photodynamic processes in response to excessive photoexcitation or porphyrin accumulation should affect the contents and activity of low molecular as well as enzymatic antioxidants. Plants have developed an efficient antioxidative stress defense system to compensate the excessive formation of ROS, such as superoxide and hydrogen peroxide, which are formed as normal side products of redox reactions.

An approximately 2-fold increase in total SOD activity was measured in leaves of a CPO antisense line. Elevated SOD isoform activities could be confirmed in protein extracts of CPO and UROD antisense plants by electrophoretic separation on nondenaturing gels that were stained for SOD activity (Mock *et al.*, 1998). In particular, the cytosolic Cu/ZnSOD and the mitochondrial MnSOD were found to be more active. In addition to higher SOD isoform activity, APX, MDHAR, and GR also display higher enzyme activity, indicating an increased demand for ascorbate recycling. Simultaneously, transcript levels of antioxidant mRNAs that encode SOD, CAT, and GPX increased in selected transformants. However, despite the elevated transcriptional and enzyme activities, the limited capacity of the antioxidative system is apparent from decreased total amounts and altered redox ratios of ascorbate and glutathione. The tocopherol content is also severely reduced in the transgenic tissue. CAT which is known as a photosensitive enzyme (Streb and Feierabend, 1996), displays lower activity, especially in the older and more necrotic leaves of the transformants, in spite of higher transcript levels than in control plants. This observation is interpreted as enhanced turn-over under the stress conditions (Mock *et al.*, 1998). Reduced activity of the heme-dependent CAT in the maize mutant Les22 is attributed to heme deficiency as a result of an insufficient amount of UROD, although the labile state of CAT upon oxidative stress had not been taken into account (Hu *et al.*, 1998).

The chlorophyll biosynthesis including the enzymes used in these studies is exclusively located in plastids (Jacobs and Jacobs 1995). We conclude from the general response of the antioxidative defense system in all cellular compartments that either accumulated photosensitizing porphyrins and/or ROS leak out of the plastids and spread into other cell compartments. Signals for the activation of the local protective systems are assumed to be released in various compartments before the cells are finally intoxicated. The cellular site of the deregulated metabolic pathway is most probably separated from the site at which the initial destruction takes place.

CHEMICALLY INDUCED AND INHERENT ACCUMULATION OF PHOTOSENSITIZING TETRAPYRROLES DUE TO A BLOCK IN THE BIOSYNTHETIC PATHWAY

A few examples of naturally occurring and experimentally caused photosensitization of plants by accumulating porphyrins and Mg-porphyrins are summarized to illustrate the need for a tight regulation of the pathway and for coordinated synthesis with the pigment-binding apoproteins. Barley grains germinated in the dark develop yellow leaves because of the lack of the light-requiring conversion from protochlorophyllide to chlorophyllide. Feeding of ALA to these etiolated seedlings in the dark results in the accumulation of large amounts of protochlorophyllide. Apparently, all enzymes required to convert ALA into protochlorophyllide are present in the plastids of dark-grown leaves. The formation of ALA is thus rate limiting for the entire pathway. ALA treatment bypasses the downregulation of ALA synthesis by which accumulation of protochlorophyllide is normally avoided. The ALA-fed plants are poisoned when they are exposed to light and undergo photodynamic damage caused by the excess of non-photoconvertible proto-chlorophyllide and the formation mainly of singlet oxygen, which leads to desiccated leaf areas within only a few hours (Spikes and Bommer, 1991).

Mutations in four barley genes (*tigrina-b*, *tigrina-d*, *tigrina-n*, and *tigrina-o*) are characterized by the deficiency in the restriction of ALA synthesis during dark growth and by accumulation of 2- to 10-fold the wild-type amount of protochlorophyllide in the dark (Nielsen, 1974; von Wettstein et al., 1995). The shape of the green-white striped seedlings when grown in dark and light cycles resembles a tiger tail. The white, often necrotic, leaf domains are caused by the accumulated protochlorophyllide in the dark, which, in the light, cannot be transformed by the protochlorophyllide oxidoreductase present at only normal amounts in the mutants. The photodynamic damage to the plastids can be prevented and, thus, the *tigrina-d*^{12} mutant be saved when grown under short light pulses that enable the successive conversion of protochlorophyllide into chlorophyll.

The transgenic phenotype of UROD and CPO antisense plants resembles very much the symptoms observed when plants are treated with biphenyl ether-type herbicides. The action of these herbicides is explained by the concept of chemically induced accumulation of porphyrins (Rebeiz et al., 1984; Dodge, 1994; Duke and Rebeiz, 1994). Upon treatment with these chemicals, plants are quickly poisoned and very rapidly develop large necrotic lesions in the light. These herbicides have been found to inhibit PPOX, the enzyme directly upstream of the branch point towards chlorophyll and heme synthesis. The block at this point leads to a transient accumulation of protoporphyrinogen IX. The highly photosensitizing effect caused by the inhibition of PPOX can be explained by the distribution of the enzyme's substrate in the plasma membrane, a site which is probably less protected against oxidative stress (Duke and Kenyon, 1985; Kenyon et al., 1985, 1988; Lehnen et al., 1990). Membrane-bound unspecific peroxidases form protoporphyrin from protoporphyrinogen (Jacobs et al., 1991). Protoporphyrin is highly photosensitizing and gives rise to a chain of deleterious events, including singlet oxygen and radical formation, which leads to severe damage of essential cell constituents.

Lermontova et al. (1997) described the cloning and characterization of plastidal and mitochondrial isoforms of tobacco PPOX. It remains open whether both isoforms are inhibited and equally contribute to the photodynamic susceptibility of plants towards the herbicides. Analysis of transgenic plants that lack the activity of either one of the isoforms are currently done in our laboratory. Protoporphyrinogen needs to be actively targeted to mitochondria. This transfer through the cytoplasm can be accelerated when the plastidal enzyme is inhibited, which might explain the enhanced photosensitive effects of protoporphyrin(ogen) compared to coproporphyrin(ogen) and uroporphyrin-(ogen). Protoporphyrinogen can perhaps be easily released from the plastids *into* the cytoplasm before the plastid membrane looses its entire integrity, whereas the transfer of the other two less hydrophobic precursors is not as efficient. The photoreactive potential of protoporphyrin can also be testified by the fact that its 4-fold accumulation relative to control levels is sufficient to cause the same severity of necrosis than the up to 500-fold accumulation of coproporphyrin(ogen) or uroporphyrin(ogen) in the transgenic plants mentioned above.

Very similar to the cellular reactions observed upon porphyrin-induced oxidative stress in the UROD and CPO antisense plants are the antioxidative effects observed in soybean cultures treated with oxyfluorfen, an herbicide targeted against PPOX (Knörzer *et al.*, 1996). The principal difference between both experimental systems is the capability to maintain a highly reduced ascorbate pool in the soybean cell culture which is considered to be the limiting factor in withstanding the imposed photooxidative stress.

UROD AND CPO ANTISENSE PLANTS SHOW CHARACTERISTIC FEATURES AFTER PATHOGEN ATTACK

To characterize the tetrapyrrole-dependent photooxidized transgenic plants we analyzed phenolic compounds that are formed in response to various stresses (Dixon and Paiva 1995). The transgenic plants with the reduced UROD and CPO activity accumulate substantial amounts of the coumarin, scopolin, in leaves (Mock *et al.*, 1999). Scopoletin, the aglycon of this component, exhibits antimicrobial activity (Jurd *et al.*, 1971; Ahl-Goy *et al.*, 1993; Valle *et al.*, 1997). In several plant-pathogen interactions, a rapid and pronounced synthesis of scopoletin was observed after an incompatible interaction, whereas slower and reduced formation was found in compatible interactions (Tal and Robeson, 1986; El Modafar *et al.*, 1995; Valle *et al.*, 1997). Scopolin also accumulates during the hypersensitive reaction of tobacco leaves infected by tobacco mosaic virus (TMV) (Fritig and Hirth, 1971; Tanguy and Martin, 1972). These findings suggest similarities between the tetrapyrrole-sensitized lesion formation in UROD and CPO antisense plants and the defense responses after pathogen attack.

This conclusion was substantiated by the observed induction of several classes of pathogenesis-related (PR) proteins and by increased levels of free and conjugated salicylic acid in leaves of UROD and CPO antisense plants (Mock *et al.*, 1999). PR protein expression and accumulation of scopolin are mainly found in leaf areas with necrotic spots. These results are consistent with the cellular responses observed after TMV infection (Antoniw and White, 1986; Enyedi *et al.*, 1992; Heitz *et al.*, 1994). The coordinated expression of several PR proteins, increased levels of salicylic acid, and the accumulation of low-molecular weight antimicrobial compounds contribute to pathogen resistance of plants (Hammond-Kosack and Jones, 1996). We therefore tested the physiological significance of tetrapyrrole-induced pathogen defense mechanisms in several lines of UROD and CPO antisense plants by inoculation with TMV and found a clear reduction in the local spread of the virus relative to wild-type controls. The reduction in the amount of TMV RNA correlated with the severity of necrotic symptoms when several transgenic lines were compared (Mock *et al.*, 1999).

Increased pathogen resistance in plants can also be induced by treatment with prooxidant chemicals, such as herbicides (e.g. paraquat, acifluorfen), sodium chlorate, phosphate, and heavy metals (Irving and Kuc, 1990; Lummerzheim *et al.*, 1995; Strobel and Kuc, 1995). The production of ROS is a key step in plant defense responses (Hammond-Kosack and Jones, 1996; Dangl *et al.*,1996; Doke 1997; Alvarez and Lamb, 1997) and was also demonstrated for the incompatible interaction between tobacco and TMV (Doke and Ohashi, 1988). These ROS might (i) exert direct antimicrobial activity; (ii) provide cellular protection by contributing to cell wall rigidification, or (iii) act as signaling molecules triggering cellular defense mechanisms. One of the earliest responses mediated might be the hypersensitive reaction observed in incompatible plant-pathogen interactions. The hypersensitive reaction in tobacco leaves challenged with TMV follows a programmed cell death mechanism (Mittler *et al.*, 1997). In analogy, we propose that ROS generated by photosensitizing tetrapyrroles or other radical species subsequently formed can interact with the signaling pathway, thereby triggering plant defense responses including a hypersensitive-like reaction.

Lesion formation and PR protein cross-induction in UROD and CPO antisense plants share symptoms not only with TMV infected plants, but are also reminiscent of plants with reduced CAT activity (Chamnongpol *et al.*, 1996; Takahashi *et al.*, 1997), lesion mimic mutants (Dietrich *et al.*, 1994; Dangl *et al.*, 1996, Dangl, 1999), or plants suffering from environmental stresses, such as ozone (Eckey-Kaltenbach *et al.*, 1994). Initial experiments indicate common aspects in the cellular reactions of plants exposed to tetrapyrrole-induced oxidative stress and of plants with reduced CAT activity, e.g. accumulation of scopolin (H.P. Mock and W. Van Camp, unpublished results).

TETRAPYRROLE INDUCED OXIDATIVE STRESS IN HUMANS: PORPHYRIA DISEASES AND PHOTODYNAMIC THERAPY

The tetrapyrrole biosynthetic pathway is ubiquitous in all organisms. Consequently, they are subject to the genetically or chemically induced deregulation of this pathway eventually leading to tetrapyrrole-induced oxidative stress. Here, we will discuss some of the cellular responses to photosensitizing tetrapyrroles observed in patients suffering from porphyria diseases or in model systems used to establish photodynamic therapy and we will compare these findings with results described in the previous sections for photosensitized plants.

Porphyrias are inherited diseases that are caused by deficiency in one of the eight enzymes of heme biosynthesis (Doss and Sassa, 1993; Elder, 1993). In addition, porphyrias can be chemically induced (Ockner and Schmid, 1961). Liver and erythropoietic cells have the highest demand for heme in the body and are therefore organs involved in the development of porphyrias (Brun and Sandberg, 1991). Millimolar concentrations of protoporphyrin can accumulate in erythrocytes of patients who suffer from porphyria (Brun *et al.*, 1988). The accumulation of excessive amounts of tetrapyrroles in erythropoietic protoporphyria can lead to the deposition of crystalline pigments exhibiting the typical porphyrin fluorescence (Bruguera *et al.*, 1976). Most porphyrias are accompanied by cutaneous symptoms. Porphyrins accumulated in porphyric patients bind to albumin or lipoproteins or diffuse freely, depending on their lipophilicity. Endothelial cells can take up and accumulate protoporphyrin IX from these molecules leading to the elicitation of cutaneous symptoms (Ochsner, 1997). In plants, high concentrations of harmful tetrapyrroles will rapidly induce the death of the whole organism as evidenced from the treatment with high concentrations of herbicides or ALA. In contrast to human, all organs of the aerial parts of plants are exposed to high light intensities, which render them particularly sensitive.

The photosensitizing effect of porphyrins can intentionally be used for the treatment of certain types of cancer by photodynamic therapy (PDT) (McCaughan, 1999). This treatment can be divided into several stages: (i) generation of the sensitizer; (ii) cellular and subcellular distribution; (iii) light treatment to generate ROS; (iv) cellular reactions, e.g.

oxidative stress defense; and (v) cell death. In analogy to plants, porphyrin accumulation can be induced by the administration of porphyrin derivatives (Ochsner 1997) or of the precursor molecule ALA (Peng *et al.*, 1997; Fritsch *et al.*, 1999) or by inhibition of protoporphyrinogen oxidation (Fingar *et al.*, 1997). Porphyrins preferentially accumulate in metabolically active tissues, such as tumors. The most well-known sensitizer used in clinical trials is Photofrin®, a mixture of different porphyrins. The site of the primary cellular localization of the sensitizer strongly depends on its lipophilicity or hydrophilicity and will determine the observed photodynamic damages (Ochsner, 1997). Quantitative, but not qualitative, differences in the photodynamic effects were found when protoporphyrin was generated in mice by feeding them with the PPOX inhibitor griseofulvin or by externally applying protoporphyrin itself. In the latter case, more protoporphyrin was found in lysosomes, resulting in increased inactivation of lysosomal enzymes. The detrimental effects determined in isolated mitochondria comprised uncoupling and inhibition of oxidative phosphorylation, energy dissipation, inhibition of respiration, and finally swelling and disruption of mitochondria. The phototoxicity of deuteroporphyrin or protoporphyrin was much higher than that of coproporphyrin or uroporphyrin and dependent on oxygen (Sandberg and Ramslo, 1981).

In plants, ALA feeding, inactivation of PPOX by herbicides, or decreased activities of heme and chlorophyll-synthesizing enzymes initially result in the accumulation of tetrapyrroles in plastids and mitochondria. These organelles of chlorophyll and heme biosynthesis therefore constitute important targets for photosensitization. Upon the spread of tetrapyrroles, other cellular sites will become vulnerable, which is indicated e.g. by the decrease of peroxisomal CAT activity in UROD and CPO antisense plants (Mock *et al.*, 1998).

The oxidative stress associated with photosensitization leads to transiently increased transcription of the early response genes *c-fos*, *c-jun*, *c-myc*, and *egr-1* in murine radiation-induced fibrosarcoma cells. Inhibition experiments indicated that the oxidative stress response involved a protein kinase-mediated pathway (Luna *et al.*, 1994). The gene products of these early-response genes act as transcription factors and regulate the expression of a number of genes. Photofrin®-mediated photosensitization of mouse leukemia cells was followed by the enhanced DNA binding of nuclear factor NFκB (Ryter and Gomer, 1993). This transcription factor is activated by ROS (Pahl and Baeuerle, 1994). The involvement of transcription factors and other elements of the early responses in mediating cellular defense reactions to tetrapyrrole-induced oxidative stress is currently also investigated in plants.

CONCLUSIONS AND PERSPECTIVES

Figure 3 summarizes our current model on cellular reactions induced by photosensitizing tetrapyrroles in UROD and CPO antisense plants. The reduced activity of UROD or CPO and the subsequent accumulation of excessive amounts of photosensitizing tetrapyrroles additionally generate ROS. These species, or radicals formed as a consequence, induce the cellular antioxidative defense system. In addition, the increased formation of ROS might also interfere with the signaling mechanisms leading to cellular responses that resemble the hypersensitive reaction found after TMV infection. The cellular site of interference is still unknown. The general response of the antioxidative defense system indicates that the

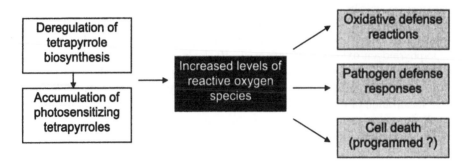

Figure 3. Cellular responses to excessive tetrapyrroles.

generation of ROS by accumulated tetrapyrroles is not restricted to plastids. Future experiments will reveal the subcellular localization of accumulated photosensitizing tetrapyrroles. Molecular and biochemical efforts will be necessary to work out a detailed understanding of the signaling events triggered upon accumulation of photosensitizing tetrapyrroles. It is tempting to speculate that lesion formation in leaves of UROD or CPO antisense plants is exerted via a programmed cell death mechanism as described for TMV infected tobacco (Mittler *et al.*, 1997). However, fast photodynamic damage of cells, which excludes a controlled cellular response, cannot be ruled out in photosensitized transformants. Photosensitizing porphyrins could either induce apoptotic or necrotic cell death in a model system for the photodynamic treatment of cancer (Dellinger 1996). Both, the UROD and the CPO antisense plants develop their necrotic phenotype in a light intensity-dependent manner. Necrosis rapidly develops when low light-grown plants are transferred to high light (H.-P. Mock, unpublished results). These plants will be a valuable tool to investigate the kinetics of the biochemical and molecular responses caused by photodynamically active tetrapyrroles upon transition from the non-necrotic to the necrotic state.

The induction of pathogen defense responses by deregulation of chlorophyll biosynthesis offers the prospect to genetically improve the resistance of plants against invading pathogens. Experiments are necessary to elucidate the spectrum of pathogen resistance induced by this approach. Increased resistance of plants to some pathogens has been provided by expression of the cholera toxin (Beffa *et al.*, 1995), yeast invertase (Herbers *et al.*, 1996), bacterio-opsin (Abad *et al.*, 1997), lipid transfer protein (Molina and Gracia-Olmedo, 1997), glucose oxidase (Wu *et al.*, 1995) and by reduction of CAT activity (Chamnongpol *et al.*, 1996; Takahashi *et al.*, 1997). In this context, it might be favorable for agricultural applications that the inducible defense response against pathogens is primed by permanent stress signals which result from sublethal deregulation of tetrapyrrole biosynthesis. It would be even more advantageous if the accumulation of photosensitizing porphyrins was tightly controlled (i.e. only triggered if needed) by the combination of a pathogen defense-related promoter and antisense constructs of genes involved in tetrapyrrole biosynthetic pathway.

REFERENCES

Abad, M., Hakimi, S.M., Kaniewski, W.K., Tommens, C.M.T., Shulaev, V., Lam, E., and Shah, D.M. (1997) Characterization of acquired resistance in lesion-mimic transgenic potato expressing bacterio-opsin. *Mol. Plant-Microbe Interact.*, **10**, 635–645.

Afonso, S.G., Polo, C.F., De Salamanca, R.F., and Battle, A. (1996) Mechanistic studies on uroporphyrin I-induced photoinactivation of some heme-enzymes. *Int. J. Biochem. Cell Biol.*, **28**, 415–420.

Ahl Goy, P., Signer, P., Aichholz, R., Blum, W., Schmidt, E., and Kessmann, H. (1993) Accumulation of scopoletin is associated with the high disease resistance of the hybrid *Nicotiana glutinosa* x *Nicotiana debneyi*. *Planta*, **191**, 200–206.

Alvarez, M.E. and Lamb, C. (1997) Oxidative burst-mediated defense responses in plant disease resistance. In J.G. Scandalios, (ed.), *Oxidative Stress and the Molecular Biology of Antioxidants, (Monograph Series,* Vol. 34), Cold Spring Harbor Laboratory Press, Cold Spring Harbor, pp. 815–839.

Antoniw, J.F. and White, R.F. (1986) Changes with time in the distribution of virus and PR protein around single local lesions of TMV infected tobacco. *Plant Mol. Biol.*, **6**, 145–149.

Arakane, K., Ryu, A., Hayashi, C., Masunage, T., Shinmoto, K., Mahiko, S., Nagano, T., and Hrobe, M. (1996) Singlet oxygen ($^1\Delta_g$) generation from coproporphyrin in *Propionibacterium acnes* on irradiation. *Biochem. Biophys. Res. Commun.*, **223**, 578–582.

Bachowski, G.J., Thomas J.P., and Girotti, A.W. (1988) Ascorbate-enhanced lipid peroxidation in photooxidized cell membranes: cholesterol product analysis as a probe of reaction mechanism. *Lipids*, **23**, 580–586.

Beale, S. (1999) Enzymes of chlorophyll biosynthesis. *Photosynthesis Res.*, **60**, 43–73.

Beffa, R., Szell, M., Meuwly, P., Pay, A., Vögeli-Lange, R., Metraux, J.P., Neuhaus, G., Meins, F. jr and Nagy, F.U. (1995) Cholera toxin elevates pathogen resistance and induces pathogenesis-related gene expression in tobacco. *EMBO J.*, **14**, 5753–5761.

Bruguera, M., Esquerda, J.E., Mascaró, J.M., and Pinol, J. (1976) Erythropoietic protoporphyria. A light, electron and polarization microscopical study of the liver in three patients. *Arch. Pathol. Lab. Med.*, **100**, 587–589.

Brun, A. and Sandberg, S. (1991) Mechanisms of photosensitivity in porphyric patients with special emphasis on erythropoietic protoporphyria. *J. Photochem. Photobiol.*, **10**, 285–302.

Brun, A., Steen, H.B., and Sandberg, S. (1988) Erythropoietic protoprophyria: a quantitative determination of erythrocyte protoporhyrin in indivudual cells by flow cytometry. *Scand. J. Clin. Lab. Invest.*, **48**, 261–267.

Chadwick, D.J. and Ackrill, K. (1994). *The Biosynthesis of the Tetrapyrrole Pigments*, (The Ciba Foundation Symposium 180), John Wiley and Sons, Chichester.

Chamnongpol, S., Willekens, H., Langebartels, C., Van Montagu, M., Inzé, D., and Van Camp, W. (1996) Transgenic tobacco with a reduced catalase activity developes necrotic lesions and induces pathogenesis-related expression under high light. *Plant J.*, **10**, 491–503.

Dangl, J. (1999) Plant defence. Long view from a high plateau. *Nature* **401**, 543–544

Dangl, J.L., Dietrich, R.A., and Richberg, M.H. (1996) Death don't have no mercy: cell death programs in plant-microbe interactions. *Plant Cell*, **8**, 1793–1807.

Dellinger, M. (1996) Apoptosis or necrosis following photofrin photosensitization: Influence of the incubation protocol. *Photochem. Photobiol.*, **64**, 182–187.

Dietrich, R.A., Delaney, T.P., Uknes, S.J., Ward, E.R., Ryals, J.A., and Dangl, J.L. (1994) Arabidopsis mutants simulating disease resistance response. *Cell*, **77**, 565–578.

Dixon, R.A. and Paiva, N.L. (1995) Stress-induced phenylpropanoid metabolism. *Plant Cell*, **7**, 1085–1097.

Dodge, A.D. (1994) Herbicide action and effects on detoxification processes. In C.H. Foyer, P. Mullineaux, (eds.), *Causes of Photooxidative Stress and Amelioration of Defense Systems in Plants*, CRC Press, Boca Raton, pp. 219–236.

Doke, N. (1997) The oxidative burst: roles in signal transduction and plant stress. In J.G. Scandalios, (ed.), *Oxidative Stress and the Molecular Biology of Antioxidants*, (Monograph Series, Vol. 34), Cold Spring Harbor Laboratory Press, Cold Spring Harbor, pp. 785–813.

Doke, N. and Ohashi, Y. (1988) Involvement of $O_2 \cdot^-$ generating reaction in formation of necrotic lesion in tobacco leaves infected with tobacco mosaic virus. *Physiol. Mol. Plant Pathol.*, **32**, 163–175.

Doss, M.O. and Sassa, S. (1993) The porphyrias. In D.A. Noe and R.C. Rock, (eds.), *Laboratory Medicine: The Selection and Interpretation of Clinical Laboratory Studies*, Williams and Wilkins, Baltimore, pp. 535–553.

Duke, S.O. and Kenyon, W.H. (1987) A non-metabolic model of acifluorfen activity. *Z. Naturforsch. C*, **42**, 813–819.

Duke, S.O. and Rebeiz, C.A. (1994) *Porphyric pesticides. Chemistry, Toxicology and Pharmaceutical Applications*.(ACS Symposium Series 559), American Chemical Society, Washington.

Eckey-Kaltenbach, H., Ernst, D., Heller, W., and Sandermann, H. (1994) Biochemical plant responses to ozone. IV. Cross-induction of defensive pathways in parsley (*Petroselinum crispum* L.) plants. *Plant Physiol.*, **104**, 67–74.

El Modafar, C., Clérivet, A., Vigouroux, A., and Macheix, J.J. (1995) Accumulation of phytoalexins in leaves of plane tree (*Platanus* ssp.) expressing susceptibility or resistance to *Ceratocystis fimbriata* f. sp. *platani*. *Eur. J. Plant Pathol.*, **101**, 503–509.

Elder, G.H. (1993) Molecular genetics of disorders of haem biosynthesis. *J. Clin. Pathol.*, **46**, 977–981.

Enyedi, A.J., Yalpani, N., Silverman, P., and Raskin, I. (1992) Localization, conjugation, and function of salicylic acid in tobacco during the hypersensitive reaction of tobacco mosaic virus. *Proc. Natl. Acad. Sci. USA*, **89**, 2480–2484.

Fingar, V.H., Wieman, T.J., McMahon, K.S., Haydon, P.S., Halling, B.P., Yuhas, D.A., and Winkelman, J.W. (1997) Photodynamic therapy using a protoporphyrinogen oxidase. *Cancer Res.*, **57**, 4551–4556.

Foyer, C.H. (1997) Oxygen metabolism and electron transport in photosynthesis. In J. Scandalios, (ed.), *Oxidative Stress and the Molecular Biology of Antioxidants*, (Monograph Series, Vol. 34), Cold Spring Harbor Laboratory Press, Cold Spring Harbor, pp. 587–621.

Fritig, B. and Hirth, L. (1971) Biosynthesis of phenylpropanoids and coumarins in TMV-infected tobacco leaves and tobacco cultures. *Acta Phytopathol. Acad. Scient. Hungar.*, **6**, 21–29.

Fritsch, C., Lehmann, P., Stahl, W., Schulte, K.W., Blohm, E., Lang, K., Sies, H., Ruzicka, T. (1999) Optimum porphyrin accumulation in epithelial skin tumours and psoriatic lesions after topical application of delta-aminolaevulinic acid. *Br. J. Cancer*, **79**, 1603–1608.

Girotti, A.W. (1993) Photodynamic therapy for neoplastic disease: reaction pathways, cellular targets, and defense mechanisms. In K. Yagi, (ed.), *Active Oxygens, Lipid Peroxidation and Antioxidants*. CRC Press, Boca Raton, pp. 313–318.

Grimm, B. (1998) Novel insights into the control of tetrapyrrole metabolism of higher plants. *Curr. Opin. Plant Biol.*, **1**, 245–250

Hammond-Kosack, K.E., and Jones, J.D.G. (1996) Resistance gene-dependent plant defense responses. *Plant Cell*, **8**, 1773–1791.

Haseloff, R.F., Ebert, B., and Roeder, B. (1989) Generation of free radicals by photoexcitation of pheophorbide a, haematoporphyrine and protoporphyrin. *J. Photochem. Photobiol. B*, **3**, 593–602.

Hausmann, W. (1908) Über die photodynamische Wirkung chlorophyll-haltiger Pflanzenextrakte. *Biochem. Z.*, **12**, 331.

Heitz, T., Fritig, B., and Legrand, M. (1994) Local and systemic accumulation of pathogenesis-related proteins in tobacco plants infected with tobacco mosaic virus. *Mol. Plant-Microbe Interact.*, **7**, 776–779.

Herbers, K., Meuwly, P., Frommer, W.B., Metraux, J.P., and Sonnewald, U. (1996) Systemic acquired resistance mediated by the ectopic expression of invertase: possible hexose sensing in the secretory pathway. *Plant Cell*, **8**, 793–803.

Hu, G., Yalpani N., Briggs, S.P., and Johal, G.S. (1998) A porphyrin pathway impairment is responsible for the phenotype of a dominant disease lesion mimic mutant of maize. *Plant Cell*, **10**, 1095–1105.

Irving, H.R. and Kuc, J. (1990) Local and systemic resistance to anthracnose in cucumber by phosphates. *Phytopathology*, **79**, 176–179.

Jacobs, J.M. and Jacobs, N.J. (1995) Porphyrin accumulation and export by isolated barley (*Hordeum vulgare*) plastids. *Plant Physiol.*, **101**, 1181–1187.

Jacobs, J.M., Jacobs, N.J., Sherman, T.D., and Duke, S.O. (1991) Effect of diphenyl ether herbicides on oxidation of protoporphyrinogen to protoporphyrin in organellar and plasma membrane enriched fractions of barley. *Plant Physiol.*, **97**, 197–203.

Jurd, L., Corse, A.D., and King Jr, A.D. (1971) Antimicrobial properties of 6,7-dihydroxy-, 7,8-dihydroxy-, 6-hydroxy- and 8-hydroxycoumarins. *Phytochemistry*, **10**, 2971–2974.

Kenyon, W.H., Duke, S.O., and Vaughn, K.C. (1985) Sequence of effects of acifluorfen on ultrastructural and physiological parameters in cucumber cotyledon discs. *Pestic. Biochem. Physiol.*, **24**, 240–250.

Kenyon, W.H., Duke, S.O., and Paul, R.N. (1988) Effects of temperature on the activity of the *p*-nitrosubstituted diphenyl ether herbicide acifluorfen in cucumber (*Cucumis sativus* L.). *Pestic. Biochem. Physiol.*, **30**, 57–66.

Kessel, D. (1986) Sites of photosensitization by derivatives of hematoporphyrin. *Photochem. Photobiol.*, **44**, 489–493.

Knörzer, O.C., Durner, J., and Böger, P. (1996) Alterations in the antioxidative system of suspension-cultured soybean cells (*Glycine max*) induced by oxidative stress. *Physiol. Plant*, **97**, 388–396.

Kruse, E., Mock, H.P., and Grimm, B. (1995a) Coproporphyrinogen oxidase from barley and tobacco—sequence analysis and initial expression studies. *Planta*, **196**, 796–803.

Kruse, E., Mock, H.P., and Grimm, B. (1995b) Reduction of coproporphyrinogen oxidase level by antisense RNA synthesis leads to deregulated gene expression of plastid proteins and affects the oxidative defence system. *EMBO J.*, **4**, 3712–3720.

Lambert C.R., Reddi, E., Spikes, J.D. Rodgers, M.A., and Jori, G. (1986) The effects of porphyrin structure and aggregation state on photosensitized processes in aqueous and micellar media. *Photochem. Photobiol.*, **44**, 595–601.

Lehnen, L.P., Sherman, T.D., Becerril J.M. and Duke, S.O. (1990) Tissue and cellular localization of acifluorfen-induced porphyrins in cucumber cotyledons. *Pestic. Biochem. Physiol.*, **37**, 329–248.

Lermontova, I., Kruse, E., Mock, H.-P., and Grimm, B. (1997) Cloning and characterisation of a plastidal and a mitochondrial isoform of tobacco protoporphyrinogen IX oxidase. *Proc. Natl. Acad. Sci. USA*, **94**, 8895–8900.

Lummerzheim, M., Sandroni, M., Castesana, C., De Olivera, D., Van Montagu, M., Roby, D., and Timmerman, B. (1995) Comparative microscopic and enzymatic characterization of the lead necrosis induced in *Arabidopsis thaliana* by lead nitrate and by *Xanthomonas campestris* pv. *campestris* after foliar spray. *Plant Cell Environ.*, **18**, 499–509.

Luna, M.C., Wong, S., and Gomer, C.J. (1994) Photodynamic therapy mediated induction of early response genes. *Cancer Res.*, **54**, 1374–1380.

Madsen, O., Sandal, L., Sandal, N.N., and Marcker, K. (1993) A soybean coproporphyrinogen oxidase gene is highly expressed in root nodules. *Plant Mol. Biol.*, **23**, 35–43.

Marschner, H. (1965) Anreicherung von Porphyrinen und Protochlorophyll in Gerstensprossen unter dem Einfluß von Cäsiumionen. *Flora*, **155**, 558–572.

McCaughan, J.S. Jr (1999) Photodynamic therapy: a review. *Drugs aging*, **15**, 49–68.

Mittler, R., Simon, L., and Lam, E. (1997) Pathogen-induced programmed cell death in tobacco. *J. Cell Sci.*, **110**, 1333–1344.

Mock, H.-P. and Grimm, B. (1997) Reduction of uroporphyrinogen decarboxylase by antisense RNA expression affects activities of other enzymes involved in tetrapyrrole biosynthesis and leads to light-dependent necrosis. *Plant Physiol.*, **113**, 1101–1112.

Mock, H.-P., Heller, W., Molina, A., Neubohn, B., Sandermann, H. Jr, and Grimm, B. (1999) Expression of uroporphyrinogen decarboxylase or coproporphyrinogen oxidase antisense RNA in tobacco induces pathogen defense responses conferring increased resistance to tobacco mosaic virus. *J. Biol. Chem.*, **274**, 4231–4238.

Mock, H.-P., Keetman, U., Kruse, E., Rank, B., and Grimm, B. (1998) Defense responses to tetrapyrrole-induced oxidative stress in transgenic plants with reduced uroporphyrinogen decarboxylase or coproporphyrinogen oxidase activity. *Plant Physiol.*, **116**, 107–116.

Mock, H.-P., Lermontova, I., Keetman, U., and Grimm, B. (1997) Consequences of photo-oxidation in transgenic tobacco with co-suppression of coproporphyrinogen oxidase. *Phyton*, **37**, 169–174.

Mock, H.P., Trainotti, L., Kruse, E., and Grimm, B. (1995) Isolation, sequencing and expression of cDNA sequences encoding uroporphyrinogen decarboxylase from tobacco and barley. *Plant Mol. Biol.*, **28**, 245–256.

Molina, A. and Garcia-Olmedo, F. (1997) Enhanced tolerance to bacterial pathogens caused by the transgenic expression of barley lipid transfer protein LTP2. *Plant J.*, **12**, 669–675.

Nielsen, O.F. (1974) Macromolecular physiology of plastids. *Hereditas*, **76**, 269–304.

Ochsner, M. (1997) Photophysical and photobiological processes in the photodynamic therapy of tumours. *J. Photochem. Photobiol. B*, **39**, 1–18.

Ockner, R.K. and Schmid, R. (1961) Acquired porphyria in man and rat due to hexachlorbenzene intoxication. *Nature*, **189**, 499.

Pahl, H.L. and Baeuerle, PA. (1994) Oxygen and the control of gene expression. *BioEssays*, **16**, 497–502.

Peng, Q., Berg, K., Moan, J., Kongshaug, M., and Nesland, J.M. (1997) 5-Aminolevulinic acid-based photodynamic therapy: principles and experimental research. *Photochem. Photobiol.*, **65**, 235–251.

Porra, R.J. (1997) Recent progress in porphyrin and chlorophyll biosynthesis. *Photochem. Photobiol.*, **65**, 492–516.

Rebeiz, C.A., Montazer-zouhoor, A., Hopen, H.J., and Wu, S.M. (1984) Photodynamic herbicides. I. Concept and phenomenology. *Enzyme Microbiol. Technol.*, **6**, 390–401.

Reinbothe, S. and Reinbothe, C. (1996) The regulation of enzymes involved in chlorophyll biosynthesis. *Eur. J. Biochem.*, **237**, 323–342.

Ryter, S.W. and Gomer, C.J. (1993) Nuclear factor kappa B binding activity in mouse L1210 cells following photofrin II-mediated photosensitization. *Photochem. Photobiol.*, **58**, 753–756.

Sandberg, S. and Romslo, I. (1981) Phototoxicity of protoporphyrin as related to its subcellular localisation in mice livers after short-term feeding with griseofulvin. *Biochem. J.*, **198**, 67–74.

Shalygo, N.V., Averina, N.G., Grimm, B., and Mock, H.-P. (1997) Influence of cesium on tetrapyrrole biosynthesis in etiolated and greening barley leaves. *Physiol. Plant.*, **99**, 160–168.

Shalygo, N.V., Mock, H.P., Averina, N.G., and Grimm, B. (1998) Photodynamic action of uroporphyrin and protochlorophyllide in greening barley leaves treated with cesium chloride. *J. Photochem. Photobiol. B*, **42**, 151–158

Smith, A.G. and Griffith, W.T. (1993) Enzymes of chlorophyll and heme biosynthesis. In P.M. Dey, and J.B. Harborne, (eds.), *Methods in Plant Biochemistry*, Vol. 9., Academic Press, London, pp. 299–343.

Spikes, J.D. and Bommer, J.C. (1991) Chlorophyll and related pigments as photosensitizers in biology and medicine. In H. Scheer, (ed.), *Chlorophylls*, CRC Press, Boca Raton, pp. 1181–1204.

Strauss, W.S.L., Gschwend, M.H., and Sailer, R. (1995) Intracellular fluorescence behavior of *meso*-tetra (4-sulphonatophenyl)porphyrin during photodynamic treatment at various growth phases of cultured cells. *J. Photochem. Photobiol. B*, **28**, 155–161.

Streb, P. and Feierabend, J. (1996) Oxidative stress responses accompanying photoinactivation of catalase in NaCl-treated rye leaves. *Bot. Acta*, **109**, 125–132.

Strobel, N.E. and Kuc, J.A. (1995) Chemical and biological inducers of systemic resistance to pathogens protect cucumber and tobacco plants from damage caused by paraquat and cupric chloride. *Phytopathology*, **85**, 1306–1310.

Takahashi, H., Chen, Z., Du, H., Liu, Y., and Klessig, D.F. (1997) Development of necrosis and activation of disease resistance in transgenic tobacco plants with severely reduced catalase levels. *Plant J.*, **11**, 993–1005.

Tal, B., and Robeson, D.J. (1986) The metabolism of sunflower phytoalexins ayapin and scopoletin. *Plant Physiol.*, **82**, 167–172.

Tanguy, J. and Martin, C. (1972) Phenolic compounds and the hypersensitivity reaction in *Nicotiana tabacum* infected with tobacco mosaic virus. *Phytochemistry*, **11**, 19–28.

Valle, T., López, J.L., Hernández, J.M., and Corchete, P. (1997) Antifungal activity of scopoletin and its differential accumulation in *Ulmus pumila* and *Ulmus campestris* cell suspension cultures infected with *Ophiostoma ulmi* spores. *Plant Sci.*, **125**, 97–101.

von Wettstein, D., Gough, S., and Kannangara, G. (1995) Chlorophyll biosynthesis. *Plant Cell*, **7**, 1039–1057.

Wu, G., Shortt, B.J., Lawrence, E.B., Levine, E.B., Fitzsimmons, K.C. and Shah, D.M. (1995) Disease resistance conferred by expression of a gene encoding H_2O_2-generating glucose oxidase *Plant Cell*, **7**, 1357–1368.

7 Metal Ion-Activated Oxidative Stress and its Control

Jean-François Briat

INTRODUCTION

"Anybody who's able to hold his breath the time required for 1,000 inhalations and 1,000 exhalations will experience immortality." (Koan Zen)

The meaning of such a paradox is clear for a biologist interested in oxidative stress: oxygen, although necessary for all forms of aerobic life, is one of the most highly toxic pollutants on the earth (Balentine, 1982). Beyond this "first order" paradox, a second one, less obvious to comprehend, is almost as important and concerns the heart of cellular metabolism. Most of the basic redox reactions involved in either oxygen production, during photosynthesis for example, or oxygen consumption, during mitochondrial respiration, require metalloproteins that contain copper or iron. Although essential, both these transition metals, as discussed below, are major pro-oxidants because of their ability to activate reduced forms of oxygen, which are natural side-products of the basic reactions mentioned above, and also by their interference with antioxidant mechanisms.

The principal aims of this chapter are: (i) to outline the role that metal ions (mainly copper and iron) play in promoting or aggravating oxidative stress, and (ii) to review our actual knowledge of the various mechanisms that participate in the protection against oxidative stress by controlling free-transition metal concentrations.

CHEMISTRY AND BIOCHEMISTRY OF MOLECULAR OXYGEN REDUCTION AND RELATION TO TRANSITION METALS

Activation of Reduced Forms of Oxygen by Metal Ions: The Haber Weiss Reaction

Oxygen-free radicals are produced when molecular oxygen accepts electrons from other molecules. This property makes oxygen toxic to cells because many intracellular reactions reduce oxygen to superoxide ($O_2^{\bullet -}$) or hydrogen peroxide (H_2O_2). Although these molecules are not very reactive with other molecules, they can form hydroxyl radicals ($OH^{\bullet}$) that are probably responsible for most of the oxidative damage in biological systems (Cadenas, 1989; Halliwell and Gutteridge, 1990). The one-electron reduction of molecular oxygen to the superoxide radical is thermodynamically unfavorable ($E_0(O_2/O_2^{\bullet -}) = -0.33V$) (Illan *et al.*, 1976), but this restriction can be overcome by interaction with another paramagnetic centre (which has $E_0 > 0$). Therefore, transition metals such as iron and copper (M), which frequently have unpaired electrons, are very good catalysts of oxygen reduction, following the reaction:

$$M^n + O_2 \longrightarrow M^{n+1} + O_2^{\bullet -}$$

In aqueous neutral pH solutions, $O_2^{\bullet -}$ can generate H_2O_2, which can subsequently decompose to produce $OH^{\bullet}$ by the Haber-Weiss reaction which involves copper or iron (M) following the reactions

$$M^{n+1} + O_2^{\bullet-} \longrightarrow M^n + O_2$$

$$M^n + H_2O_2 \longrightarrow M^{n+1} + OH^- + OH^\bullet$$

which are usually summarized as

$$O_2^{\bullet-} + H_2O_2 \longrightarrow O_2 + OH^- + OH^\bullet$$

When iron is the transition metal in the Haber-Weiss reaction, it is called the Fenton reaction.

In biological systems, as pointed out by Goldstein and Czapski (1986), transition metal complexes may act as protectors against $O_2^{\bullet-}$ damage, or as sensitizers to $O_2^{\bullet-}$ toxic effects. This role is dependent on the kinetic properties of the metal ligands and on the steady state concentrations of the required O_2 species. Protection can be achieved through disproportionation of $O_2^{\bullet-}$ catalyzed by several metal ligands (L-M) according to a mechanism analogous to a superoxide dismutase reaction (Fridovich 1995)

$$L - M^{n+1} + O_2^{\bullet-} \longrightarrow L - M^n + O_2$$
$$L - M^n + O_2^{\bullet-} + 2H^+ \longrightarrow L - M^{n+1} + H_2O_2$$

Oxidative damage mediated through the Haber-Weiss reaction described above causes alterations to various important biological molecules present in the vicinity of the site of production of hydroxyl radicals. These effects will now be considered.

Deleterious Effects of Free Radicals on Biological Structures

DNA nicking

The hydroxyl radical ($OH^\bullet$) has a high reactivity with organic molecules and can be produced in close proximity to DNA. It is therefore a unique oxyradical that is able to add H atoms to the DNA bases as well as to remove H atoms from the DNA backbone (Pryor, 1988); it could be responsible for 10^4 to 10^5 DNA base modifications per cell per day (Ames *et al.*, 1991). Luo *et al.* (1994) have reported that DNA damage during *in vitro* iron-mediated Fenton reaction might be the result of the production of three types of oxidants. Type I are sensitive to H_2O_2 and moderately resistant to ethanol; the Fe^{2+} ions involved in their production being complexed to a phosphate residue. Type II oxidants, resistant to both ethanol and H_2O_2, are produced from Fe^{2+} coordinated with ring nitrogens and perhaps also with a DNA phosphate group. Type III oxidants are sensitive to H_2O_2 ethanol and *t*-butanol and could be formed by Fe^{2+} ions free in solution. Production of three chemically distinct types of oxidants formed by the iron-mediated Fenton reaction in the presence of DNA has been correlated with the two modes of killing *Escherichia coli* cells with H_2O_2 (Imlay and Linn, 1988). Therefore, particular interactions of DNA with transition metals could explain, in part, the toxicity of H_2O_2, and also some aspects of cell death (Jacobson, 1996). Furthermore, independently of these oxidative processes, iron can relax mitochondrial DNA from the supercoiled to the open circular form, in a dose-dependent manner. This change in DNA topology has been correlated to the formation of iron-colloid on mitochondrial DNA strands (Yaffee *et al.*, 1996).

Amino acids and protein oxidation

The role of metal ions (Cu, Fe, Mn, etc.) in the oxidative modifications of free amino acids and proteins has been reviewed by Stadtman (1993). Rapid oxidation of amino acids by H_2O_2 and Fe^{II} is completely dependent upon the presence of bicarbonate and can be either stimulated or inhibited by metal ion chelators. Multiple pathways, not yet completely elucidated, yield NH_4^+, α-ketoacids, oximes, CO_2 and carboxylic acids. The opposite effects, stimulation versus inhibition, of chelating agents are functions of the Fe salt versus chelate ratio and of the affinity constant of a given chelate for iron (Stadtman and Berlett, 1991). Maximal rates of amino acid oxidation by H_2O_2 require iron in two forms: one complexed to a chelator, such as EDTA for example, and one to an amino acid residue or to bicarbonate. The observation that these rates were not affected by hydroxyl scavengers led to the proposition that metal ion catalyzed oxidation of amino acids is a caged process, in which oxygen radicals are not released into solution, but react directly in a complex, including bicarbonate and the amino acid target (Stadtman and Berlett, 1991).

A number of metal-catalyzed oxidation systems that can to oxidize amino acid residues in proteins, in the presence of Fe^{III} or Cu^{II}, oxygen, and an electron donor have been described (for review, see Stadtman 1993). Histidine, arginine, lysine, proline, methionine and cysteine residues have been shown to be the most common sites of oxidation in proteins, and their major oxidation products have been identified. Frequently, only one amino acid residue in a given protein is modified by oxidation. These modifications correspond to site-specific processes in which amino acid residues at metal-binding sites are specific targets, explaining therefore why metal-catalyzed modification of proteins is relatively insensitive to inhibition by free radical scavengers. Protein oxidation catalyzed by metal ions has been explained by Haber-Weiss reaction at the active sites of proteins because of Fe^{II}/Fe^{III} or Cu^{I}/Cu^{II} cycling in the presence of H_2O_2 (or $O_2^{\bullet-}$), leading to $OH^\bullet$ production (Goldstein and Czapski, 1986, Stadtman, 1993). This caged process at the protein active sites, in which a transition metal ion is coordinated to given amino acid residues can explain the contradictory effects of metal chelators on these reactions; activation versus inhibition would be selected according to the relative strength of the binding constant for the metal ion of the protein-active site, compared with the metal-binding capacity of the chelator. A major consequence of oxygen-free radical damage to proteins is to target them for degradation by common proteases *in vitro* (Rivett, 1985a, Hunt *et al.*, 1988), and *in vivo* by proteases in bacteria (Davies *et al.*, 1987, Roseman and Levine, 1987) mammalian cells (Rivett, 1985b, 1985c; Lee *et al.*, 1988), and very probably in plants (Laulhère *et al.*, 1989, 1990; Lobréaux and Briat, 1991). Another important aspect of oxidation of proteins by reduced forms of oxygen involves the release of Fe^{II} from [4Fe-4S] clusters of some dehydratases such as aconitases (Goldstein and Czapski 1986). This event can promote the Fenton reaction producing $Fe^{II}O$ or $Fe^{III} + OH^\bullet$, and has also been implicated in gene regulation (see below).

Lipid peroxidation

A free-radical chain reaction with initiation, propagation and termination phases is responsible for lipid peroxidation, and biological membranes rich in polyunsaturated fatty acids are extremely susceptible to these reactions (Scholz *et al.*, 1990). Oxygen and transition metals are implicated in this process. Initiation of lipid peroxidation is started by the removal of an H atom by a free-radical initiator, which could require iron to be generated (Halliwell and Gutteridge, 1984, Aruoma *et al.*, 1989). During the propagation phase, alkyl radicals are formed and react with molecular oxygen to form lipid hydroperoxyl radicals (ROO•). Then, a chain reaction occurs in which ROO• reacts with another polyunsaturated fatty acid and produces a lipid hydroperoxide (ROOH) and another alkyl radical (R•). Iron ions are thought to be responsible for the continuation of the peroxidative reaction by interacting with a relatively stable ROOH to produce additional radical species in the following manner (Halliwell and Gutteridge, 1984):

$$ROOH + Fe^{3+} \longrightarrow Fe^{2+} + H^+ + ROO^\bullet$$
$$ROOH + Fe^{2+} \longrightarrow Fe^{3+} + OH^- + RO^\bullet$$

Ascorbic acid is known to quench lipid peroxy and alkoxy radicals in the aqueous phase. However, illustrating again the paradox presented in the introduction, ascorbate can also function as a pro-oxidant by reducing Fe^{3+} or Cu^{2+}, allowing the metal-catalyzed transition of lipid hydroperoxides to radical species.

INTEGRATION OF METAL ION AND OXYGEN METABOLISM IN BACTERIA, YEAST, AND ANIMALS

Integration of metal ion and oxygen metabolism at the level of gene expression regulation has recently received increased attention. In *Escherichia coli*, the Fur protein is a transcriptional repressor of the biosynthetic aerobactin operon involved in iron uptake that is regulated by its interaction with Fe^{II} (Bagg and Neilands, 1987). Fur is also known to repress the *sodA* gene coding for Mn superoxide dismutase (Hassan and Sun, 1992). Recently, Touati *et al.* (1995) have demonstrated that iron overload occurred in *fur* mutants, leading to oxidative stress and DNA damage and causing mutagenic lesions and death. Overproduction of the iron storage ferritin H-like protein in this *fur* background eliminated oxygen toxicity, providing further evidence of the link that exists between iron overload and oxidative stress. Again in *E. coli*, at least two other regulatory proteins have been shown to play a role in O_2 metabolism via an iron cofactor. Firstly, the fumarate nitrate reductase (FNR), a [4Fe-4S] cluster protein, senses O_2 level and binds to promoters of the genes of the anaerobic metabolism in the absence of O_2; upon exposure to oxygen, the iron-sulfur cluster is immediately destroyed and FNR is no longer able to bind to its DNA target (Khoroshilova *et al.*, 1995). Secondly, SoxR regulates the *soxRS* oxidative stress regulon in *E. coli* that is activated by superoxide-generating agents or by nitric oxide. Hidalgo and Demple (1994) have demonstrated that SoxR exists as two forms, apoSoxR and Fe-SoxR, both can bind to the *soxS* promoter target; however, only the oxidized Fe-SoxR can activate transcription. Molecular genetic information is also available on the coordinate regulation of *Bacillus subtilis* peroxide stress genes by H_2O_2 and Mn^{II} (Chen *et al.*, 1995).

In the yeast *Saccharomyces cerevisiae*, several recent reports clearly indicate a link between metal ion homeostasis and oxidative stress, which is illustrated by the following examples. Firstly, the yeast *par1* gene confers resistance to iron chelators, and *par1* null mutants are hypersensitive to oxidative stress mediated by H_2O_2 treatment (Schnell and Entian, 1991). Secondly, MAC1 is a nuclear regulatory protein whose N-terminal region is very similar to the copper and DNA-binding domains of ACE1, the transcriptional *trans*-regulator of the *cup1* metallothionein gene; *MAC1* loss-of-function mutants are defective in plasmalemma Cu^{II} and Fe^{III} reductase activity and are hypersensitive to H_2O_2 exposure (Jungmann *et al.*, 1993). Thirdly, the *ATX1* gene, a suppressor of oxygen toxicity in yeast, which lacks superoxide dismutase, codes for a protein similar to bacterial metal transporters; it is effectively involved in the transport and/or partitioning of copper, and loss-of-function mutants of *ATX1* are hypersensitive to paraquat or H_2O_2 treatments (Lin and Culotta, 1995). Indeed, ATX1 suppression of SOD1 deficiency is dependent of ATX2, a Mn-trafficking protein that localizes to Golgi-like vesicles (Lin and Culotta, 1996). These results are in good agreement with the observations that (i) a defect in Cu/ZnSOD in SOD1 yeast deletion mutants can be overcome by overproduction of yeast or monkey copper metallothionein (Tamai *et al.*, 1993), and (ii) that transcription of the *CUP1* metallothionein gene is activated by menadione, a generator of superoxide anions (Liu and Thiele, 1996). Finally, a mutant in the AFT1 protein ($AFT1-1^{up}$) accumulates toxic levels of iron (Yamaguchi-Iwai *et al.*, 1995). As this protein is responsible for the iron-regulated transcriptional control of the yeast iron uptake system (Yamaguchi-Iwai *et al.*, 1996) such a mutant is a promising tool for developing a genetic approach to the study of the coupling of iron and oxygen toxicity in eukaryotic cells. It has already been demonstrated that iron excess was responsible of yeast cell cycle arrest and inhibition of G1 cyclin translation (Philpott *et al.*, 1998).

In animal and human cell lines, direct evidence has also been reported that links metal ion metabolism and oxygen toxicity. At the chemical level, the redox cycling of the Cu^{II}-1, 10-phenanthroline complex in the presence of oxygen generates hydroxyl radicals that induce internucleosomal DNA fragmentation in human HepG2 cells, revealing that DNA ladders, as an hallmark of apoptosis, are not unique to endonuclease activity (Tsang et al., 1996). Molecular approaches have also yielded information to start unraveling the network linking metal ions and oxidative stress. For example, the two transcription factors AP1 (Jun/Fos) and NF-κB, are both activated in response to metal ion-mediated oxidative stress (Schreck *et al.*, 1991, Meyer *et al.*, 1993). Activation of NF-κB by treating monocytes with H_2O_2 can be decreased by exposure of these cells to iron chelators such as 2,3-dihydroxybenzoic acid and its ethyl ester derivative; addition of ferric ions to these compounds almost completely restores activation of NF-κB (Sappey *et al.*, 1995). Recently, AP1, activated by iron-generated hydroxyl radicals has been implicated in the transcriptional activation of GADD153, a CCAAT-enhancer binding protein responsible for growth of HeLa cells in response to stress signals (Guyton *et al.*, 1996). However, the most documented molecular links between oxidative stress and metal ion concern ferritin (Theil, 1987), a cytosolic iron storage protein that accomodates thousands of iron atoms in its central cavity defined by a coat of 24 subunits, and the transferrin receptor (TfR) responsible in part for iron uptake in animal cells (Müllner and Kühn, 1988, Müllner *et al.*, 1989). Functional evidence that ferritin could play a role in cytoprotection against pro-oxidant treatment was obtained by Balla *et al.* (1992), who have demonstrated that apoferritin added to endothelial cells was taken up by these cells, leading to a protection against oxidant-mediated cytolysis; in contrast, addition of a mutant ferritin protein unable to load iron, failed to protect endothelial cells. Regulation of ferritin and TfR synthesis also appears to be related to oxidative stress. Cairo *et al.* (1995) have reported that treatment of rats with phorone, a glutathione-depleting drug, stimulated ferritin synthesis in the liver. Deregulation of the expression of the transferrin receptor by transfecting Chinese hamster

V79 cells with a human TfR cDNA lacking its 3' untranslated iron regulatory sequence led to constitutive iron uptake from transferrin. As a consequence the transfected cells had an abnormally high intracellular iron content and were more sensitive to DNA damage by H_2O_2 (Nascimento and Meneghini, 1995). The basic molecular framework for the regulation of ferritin and TfR synthesis has been elucidated. Translational control of animal ferritin synthesis and regulation of mRNA stability of TfR in response to iron are mediated through RNA-binding proteins. These Iron Responsive Proteins (IRPs) bind to a specific stem-loop structure, the iron-responsive element (IRE), found in the 5' untranslated region of ferritin mRNA and in the 3' untranslated region of TfR mRNA (Theil, 1994; Klausner *et al.*, 1993; Hentze and Kühn, 1996). One of these *trans*-regulators, IRP1, is similar to the cytosolic aconitase, and the loss of the [4Fe-4S] cluster from this enzyme at low intracellular iron concentrations switches its activity to IRE binding, resulting both in increased TfR mRNA stability and in repression of ferritin mRNA translation. A link also exists at this level between iron metabolism and oxidative stress, because exposure of cells to H_2O_2 leads to reduced ferritin synthesis and to increased TfR synthesis, through induction of a new transduction pathway stimulating the binding of IRP1 to IREs (Martins *et al.*, 1995; Pantopoulous and Hentze, 1995; Rouault and Klausner, 1996).

EVIDENCE FOR OXIDATIVE STRESS IN PLANTS PROMOTED BY METAL IONS

Physiological Evidence

Most of the evidence establishing a link between metal ion homeostasis and oxidative stress in plants comes from physiological studies that have clearly described the role played by these metals in the promotion and/or amplification of oxidative stress. Iron-dependent oxygen radical production *in vivo* has been reported (Caro and Puntarulo, 1996). Also, physiological disorders due to iron toxicity responsible for necrotic spots in leaves are known for a long time (Ponnamperuma *et al.*, 1955). Indeed, simulation of iron toxicity by an artificial iron overload of *Nicotiana plumbaginifolia* revealed an increase in oxidized glutathione and dehydroascorbate levels and a correlated decrease in reduced glutathione and ascorbate concentrations, proving that a link exists between iron toxicity and the plant glutathione/ascorbate antioxidant system (Kampfenkel *et al.*, 1995a). Furthermore, in this system, ascorbate peroxidase (APX) and catalase activities doubled in leaves treated with 100 μM Fe^{III}-EDTA. These data are in perfect agreement with the observation of abnormally high levels of iron and copper in the tomato *chloronerva* mutant, defective in nicotianamine synthesis (Pich *et al.*, 1994); increased production of oxygen radicals and activation of antioxidant enzymes occurs in this mutant (Pich and Scholz, 1993). The combined effects of iron overload and light intensity on the photosynthetic activity of mung bean chloroplasts has been studied (Sook and Jung, 1993). Enhanced incorporation of non-heme iron in the thylakoids acts as a potent sensitizer, increasing the photosensitivity of low-light adapted seedlings and stimulating linearly the generation rate of the singlet oxygen ($^1O_2$). Iron toxicity in plants could, therefore, result from Fe-dependent photosensitization mediated through $^1O_2$; in contrast, toxicity due to excess Cu^{II} or Zn^{II}, known to decrease photosynthetic activity, cannot be explained by an increase in the susceptibility of chloroplasts to photoinhibition. Interestingly, from a chemical point of view, the production of $OH^•$ during the Haber-Weiss reaction has been challenged recently, and it has been proposed that $^1O_2$ could be the *in vivo* reactive species, and not $OH^•$ according to the following equations (Khan and Kasha, 1994):

$$O_2^{•-} + H_2O_2 \longrightarrow {}^1O_2 + OH^• + OH^- \text{ (Haber-Weiss reaction)}$$

$$OH^• + O_2^{•-} \longrightarrow {}^1O_2 + OH^• \text{ (eletron transfer)}$$

$$2H^+ + 2O_2^{•-} \longrightarrow {}^1O_2 + H_2O_2 \text{ (dismutation)}$$

A combined effect of light and metals to promote oxidative stress in leaves is not restricted to iron. Very recently, Gonzalez *et al.* (1998) have studied the effect of light intensity on antioxidants, antioxidant enzymes, and chlorophyll content in common bean exposed to excess manganese. These authors observed that a 10-day-exposure to toxic levels of Mn increased APX and SOD activities in leaves by 78% under low light and by 235% under high light. Ascorbate depletion was also observed prior to the onset of chlorosis, whereas no lipid peroxidation was measured.

Concerning development, transition metals, and more probably iron, have been implicated in the activation of free radical production during nodule senescence (Becana and Klucas, 1992). Chemical activation of superoxide anion production by treatment of detached pea leaves with paraquat is less damaging for biological structures than when leaves are pretreated with desferrioxamine, a strong Fe^{III} chelator, demonstrating the mediatory role of iron in the toxicity of this herbicide (Zer *et al.*, 1994). Other transition metals also participate in oxidative damage to plants. For example, it has been known for a long time that excess copper mediates lipid peroxidation of photosynthetic membranes (Sandmann and Böger, 1980). Also, the integrated uptake of iron and other heavy metals by roots could explain the toxic effects in response to deficiency in one of these elements (Briat *et al.*, 1995a; von Wiren *et al.*, 1996; Cohen *et al.*, 1998). Iron starvation, for instance, leads to metallothionein synthesis in barley and pea, but this is probably the consequence of copper overload (Okumura *et al.*, 1991; Evans *et al.*, 1992). Similarly, zinc deficiency has been reported to increase superoxide radical concentrations in roots, but again this could be due to a correlative iron and/or copper overload (Cakmak and Marschner, 1988). Finally, among the mechanisms of aluminum toxicity that leads to plant root growth inhibition, induction of oxidative stress via lipid peroxidation has been suggested (Kochian, 1995).

Molecular Evidence

Currently, little is known about how the plant genome perceives oxidative damage and responds to it, and even less is known regarding the cellular and molecular networks that are responsible for activation of oxidative stress by metal ions in plants. To understand these mechanisms, it is essential to identify genes that respond to both transition metals and oxidative stress, in order to study their structure, regulation, and expression. Cytosolic APX and Cu/ZnSOD could be good candidates for a molecular study of the cellular network that link transition metal metabolism and oxidative stress, because these proteins accumulate into the tomato *chloronerva* mutant, which contains toxic levels of copper and iron (Herbick *et al.*, 1996), and since their activities increased in response to Mn toxicity

(Gonzalez *et al.*, 1998). Furthermore, APX mRNA has been demonstrated to accumulate in response to iron overload in cotyledons of *Brassica napus* (Vansuyt *et al.*, 1997). Such a result is consistent with the fact that tobacco Fe-superoxide dismutase gene expression is repressed at the transcriptional level in response to iron deficiency (Kurepa *et al.*, 1997). Also, in *Arabidopsis thaliana*, aluminum induces oxidative stress genes such as peroxidase, glutathione *S*-transferase, reticuline/oxygen oxidoreductase, and Cu/ZnSOD. In addition, three transcripts that encode proteins of unknown function and have an increased abundance in response to aluminum treatment, accumulate also in response to ozone treatment, further documenting the link between aluminum toxicity and activation of antioxidant defenses (Richards *et al.*, 1998). Finally, expressed sequence tags from *Arabidopsis thaliana* and *Oryza sativa* have been identified that are homologuous to the *Saccharomyces cerevisiae ATX1* gene, which is involved in the transport and/or partitioning of copper, and that can suppress oxygen toxicity in the yeast *sod1* mutant (Lin and Culotta, 1995). The recent cloning of a functional *ATX1* homolog from *Arabidopis thaliana* constitutes an attractive entry point for studying Cu/O$_2$relationships in plants (Himelblau *et al.*, 1998).

Synthesis of the plant iron storage protein ferritin (for a review see Briat *et al.*, 1995b) can be activated at the transcriptional level by iron overload or H$_2$O$_2$ treatment and antagonized by antioxidant or phosphatase inhibitor treatments (Lescure *et al.*, 1991; Lobréaux *et al.*, 1995; Gaymard *et al.*, 1996; Savino *et al.*, 1997). Ferritin promoters from maize (Fobis-Loisy *et al.*, 1995) and *Arabidopsis thaliana* (O. Van Wuytswinkel, S. Lobréaux, F. Gaymard, J.-F. Briat, unpublished data) are available and their activation by iron, which is under study in our laboratory, should give an insight into the regulatory elements that control gene expression in plants in response to an iron-mediated oxidative stress.

CONTROL OF METAL ION TOXICITY IN PLANTS

Although little is known regarding the control of metal ion toxicity in plants, three major principles are probably be applied: (i) regulation of the acquisition of metal ions at the root level to avoid overloading, (ii) buffering and storing metal ions inside plant cells as non-toxic forms, and (iii) metal-dependent activation of oxygen detoxification mechanisms when metal ions are in excess.

Regulation of Metal Ion Uptake

The root barrier plays a key role at the level of metal ion uptake. The first checkpoint is located at the apoplast level and has been well studied for iron (Bienfait *et al.*, 1985; Zhang *et al.*, 1991). When iron is resupplied to iron-starved plants, the total root iron is 10-fold increased after 48 hours, but 90% of this iron is extracellular and trapped in the apoplastic-free space (Lobréaux *et al.*, 1992). Possible apoplastic iron forms that are deposited in roots, besides ferric ions bound to cation exhange sites, are ferric hydroxide and ferric phosphate. Iron toxicity in flooded anaerobic soils is due to the predominance of ferrous iron that is readily taken up by roots (Ponnamperuma *et al.*, 1955); rapid detoxification of excess ferrous ions by generating an apoplastic ferric deposit has been described. This highlights the role that the root apoplast can play in limiting iron toxicity (Green and Etherington, 1977; Foy *et al.*, 1978; Ando *et al.*, 1983). This apoplastic ferric iron deposit constitutes the iron reserve that is transported through the plasmalemma membrane via various iron deficiency-activated transport systems (for a review, see Briat and Lobréaux, 1997). At the molecular level, a probable ferrous iron transporter has been isolated from *Arabidopsis thaliana* (IRT1) and was identified by functional complementation of a yeast mutant. IRT1 is encoded by a small gene family, is homologous to an expressed sequence tag of rice, and its mRNA is expressed in roots and accumulates in response to iron starvation (Eide *et al.*, 1996).

Unlike iron, copper and zinc are probably not buffered by the apoplast. Their uptake at the root plasmalemma level is, however, tightly linked to iron uptake (for a review, see Briat *et al.*, 1995a). At the molecular level, functional complementation of a yeast mutant has also enabled the identification of a putative copper transporter from *Arabidopsis thaliana* named COPT1, which seems to be mainly expressed in leaves (Kampfenkel *et al.*, 1995b). With regard to zinc, a yeast gene (*ZRT1*) encoding a zinc transporter protein for a high-affinity uptake system induced by zinc limitation has been cloned (Zhao and Eide, 1996). A functional approach based on complementation by *Arabidopsis thaliana* cDNA expression of the *zrt1* mutant strain enabled the characterization of three plant Zn transporters (ZIP1, ZIP2 and ZIP3) (Grotz *et al.*, 1998). ZIP1 and ZIP3 are produced in roots in response to iron deficiency, and ZIP2 production has not been detected yet. A fourth related protein of *Arabidopsis*, ZIP4, has been identified thanks to the genome sequence project; ZIP4 is induced in both shoots and roots of zinc-limited plants, suggesting it could transport zinc intracellularly or between plant tissues.

Intracellular Storage of Metal Ions in a Non-Toxic Form

Although metal ion uptake is strictly controlled as described above, environmental fluctuations experienced by plants can lead to shifts in the internal concentrations of metal ions. The second level of control of metal ion toxicity, therefore, is at the level of mechanisms that maintain intracellular metal ions in non-toxic forms, i.e., in terms of oxidative stress, unable to activate reduced forms of oxygen or to deplete pools of antioxidant molecules. Various subcellular compartments participate in this control by harbouring various molecules more or less specific for their storage and detoxification function of metal ions.

Iron storage and detoxification

It has already been mentioned that the apoplast because of its location at the cell periphery is the first barrier to avoid iron toxicity. However, when high iron concentrations are present in the soil, the amount of iron trapped in the root apoplast can be huge, and this iron can be stored by the plant for use in periods of deprivation. Therefore, the apoplast should also be considered as an iron storage compartment that plays a key role in the buffering of this element. This statement also applies to the leaf apoplast as demonstrated by the tomato *chloronerva* mutation that results in an iron overload of leaf cells, with iron accumulated in both the apoplast and the symplast (Pich and Scholz, 1991). As iron has a tendency to precipitate in the presence of oxygen and to react with reduced forms of oxygen, it is sequestred by a variety of molecules including phenolics, organic acids, amino acids and some of their derivatives, and proteins (for review, see Briat *et al.*, 1995a), of which the specificity of the chelating capacities and iron affinity vary. Among this broad range of compounds, two deserve to be considered in more detail, namely citrate and nicotianamine. Organic acids, and mainly citrate, are used as a shuttle to transport ferric iron to the leaves by the xylem sap (Cataldo *et al.*, 1988). Furthermore, a link between iron metabolism and citrate concentration in plants has been clearly established (Pich *et al.*, 1991). In iron overload conditions, citrate might also be involved in detoxification processes, because it is found in the vacuoles that could serve as iron storage under these conditions (Lescure *et al.*, 1990). The second important compound thought to play an important role in the control of iron trafficking in plants is nicotianamine (Stephan and Scholz, 1993). The importance of nicotianamine in the regulation of iron homeostasis in plants has been demonstrated with the tomato *chloronerva* mutant that is defective in nicotianamine synthesis; as a consequence of this mutation, plants take up and translocate an excess iron and heavy metals, with as a result an increased production of oxygen radicals and the activation of antioxidant enzymes (Pich and Scholz, 1993; Pich *et al.*, 1994). In addition, nicotianamine has been postulated to be an Fe^{II} shuttle in the phloem sap that signals the iron status of the leaves to the roots to modulate the efficiency of iron uptake systems. Besides the low-molecular weight compounds mentioned above, a class of ubiquitous multimeric iron storage proteins, the ferritins, have the ability to sequester several thousands of iron atoms per molecule (Briat *et al.*, 1995b). In plants, these proteins are located in the plastids (Lescure *et al.*, 1991), and it has recently been demonstrated that an iron mediated-oxidative stress is responsible in part for their synthesis (Lobréaux *et al.*, 1995; Gaymard *et al.*, 1996). Plant ferritin emerges, therefore, as an important component of the oxidative stress response in plants and probably participates in the protection of plastids against oxidative stress by storing excess free iron. In addition to an increase in ferritin concentration, iron-mediated oxidative stress also increases the concentration of dehydroascorbate and decreases the ascorbate concentration (Kampfenkel *et al.*, 1995a). This observation is interesting because ascorbate concentrations lower than 2.5 mM favours iron uptake *in vitro* by pea seed ferritin, whereas higher concentrations promote iron release (Laulhère *et al.*, 1990; Laulhère and Briat, 1993). Therefore, an oxidative stress in response to iron toxicity leads to both ferritin accumulation and increased iron uptake by the protein. The intracellular concentrations of transition and heavy metals, other than iron, in particular cadmium, zinc, and copper, are controlled by two main classes of molecules (phytochelatins and metallothioneins).

Phytochelatins

Phytochelatins, also called class III metallothioneins, are peptides with $(\gamma\text{-Glu-Cys})_n(\text{Gly})$ as general structure ($n = 2$ to 7); these peptides are, therefore, not primary gene products (for a review, see Rauser 1990, 1995, Grill *et al.*, 1991). Variants around this general structure have also been identified, such as $\gamma\text{-Glu-Cys-}\beta\text{-Ala}$ in certain legumes, $(\gamma\text{-Glu-Cys})_n$ and $(\gamma\text{-Glu-Cys})_n(\text{Glu})$ in maize, and $(\gamma\text{-Glu-Cys})_n(\text{Ser})$ in some species of the Poaceae family. It is clear from these structures that phytochelatins arise from chain extension of glutathione (or homoglutathione in some legumes). This observation is confirmed by (i) the concomitent disappearance of glutathione and the appearance of phytochelatins, (ii) the absence or reduced phytochelatin synthesis in plant mutants deficient in glutathione synthase or γ-glutamylcysteine synthetase, the two key enzymes of the glutathione biosynthetic pathway, and (iii) inhibition of phytochelatin synthesis by buthionine sulfoximine, an inhibitor of γ-glutamylcysteine synthetase. Phytochelatins synthesis is activated upon treatment of plant cells by various metals (Cd, Ni, Cu, Zn, Ag, Sn, Sb, Te, W, Au, Hg, Pb, and Bi). Although the enzyme involved, phytochelatin synthase (γ-glutamylcysteine dipeptidyl transpeptidase), is constitutively expressed, it is a self-regulating enzyme because the product of the reaction chelates the enzyme-activating metals. At the cellular level, phytochelatins have been shown to be synthesized in the cytosol and transported to the vacuoles via a tonoplast ABC-type transporter (Ortiz *et al.*, 1992, 1995; Salt and Rauser, 1995), indicating a role for phytochelatins in homeostasis by buffering transition metals. This observation has been suggested from *in vitro* studies (for a review, see Rauser 1995). However, the *in vivo* situation is still unclear. For example, the *cad1–3* mutant of *Arabidopsis thaliana*, which has no detectable phytochelatins or phytochelatin synthase activity, grows normally in the presence of Cu and Zn (Howden *et al.* 1995).

Phytochelatin intereactions with transition metals can be viewed as another example of the paradoxal link between metal ion and oxidative stress. Metal chelation protects cells against activation of reduced forms of oxygen, but the concomitent glutathione pool depletion triggers them toward a pro-oxidant state.

Metallothionein

Class II metallothioneins are small proteins with a molecular weight of 6000 to 7000 that have been identified in various organisms. They sequester excess amounts of certain metal ions, most commonly Zn^{2+}, Cu+ and Cd^{2+}, by coordination of these metals with cysteine residues organized as CysXCys or CysXXCys repeats (for a review, see Hamer 1986, Robinson *et al.*, 1993). The synthesis of class II metallothioneins is activated at the transcriptional level by metal ions, and this has been extensively studied in the case of the metallothionein *CUP1* gene of *Saccharomyces cerevisiae* (Fürst *et al.*, 1988). The *in vivo* function of these molecules is still a matter of debate. Some of them may participate in the control of oxidative stress. This hypothesis has been suggested by recent evidence that oxidative stress-induced DNA strand breakage is reduced in the presence of elevated metallothionein levels, and enhanced in Chinese hamster cells that express a metallothionein antisense construct (Chubatsu and Meneghini, 1993).

Class II metallothioneins of plants have received little attention this last decade because intracellular chelation of metals was thought to be carried out exclusively by phytochelatins (Grill *et al.*, 1987). Until recently, the only functional evidence that such proteins do exist in plants came from the characterization of the Ec protein from wheat germ (for a review, see Robinson *et al.*, 1993). The Ec protein is synthesized transiently during early germination from mRNA stored in the dry seeds and contains canonical cysteine residue repeats that associate with Zn^{2+} with an approximate ratio per protein of 5:1. In contrast to class II metallothionein genes from yeast and animals, transcription of *Ec* genes during early embryogenesis is under the control of abscisic acid, and is not activated by metal ions. Furthermore, plant metallothionein-like genes have been identified by sequence comparison, but no functional evidence concerning their ability to chelate metal ions in plants has been reported. However, the pea *PsMTA* gene, when expressed in *Escherichia coli*, produces a protein that can bind copper, zinc, and cadmium (Kille *et al.*, 1991; Tommey *et al.*, 1991).

That plants indeed contain functional homologs of metallothionein has been demonstrated by Zhou and Goldsbrough (1994), who have cloned two cDNAs from *Arabidopsis thaliana*, named MT1 and MT2, which share all the structural features of yeast metallothioneins. Expression of these metallothionein-like cDNAs in the metallothionein-defective mutant *cup1* from *Saccharomyces cerevisiae* restores copper and cadmium resistance to this yeast strain. MT1 and MT2 are developmentally regulated, MT1 mRNA being more abundant in roots and MT2 mRNA in leaves. Furthermore, their accumulation is activated in response to copper and cadmium treatment, but not by zinc treatment. From that study, it was concluded that MT1 and MT2 are functional homologs of yeast metallothionein and that phytochelatins are not the sole molecules involved in heavy metal tolerance in plants. In total, five *Arabidopsis thaliana* metallothionein genes have now been characterized (Zhou and Goldsbrough, 1995): MT1a, MT1c, MT2a, and MT2b are transcribed whereas MT1b is probably a pseudogene. The *Arabidopsis thaliana* class II metallothionein genes may also provide a promising system for the study of the mechanisms of metal-regulated gene expression in plants. In addition, the probable role played by copper excess in activating oxidative stress can now be addressed.

Metal Ion Activation of Enzymes Involved in Oxygen Detoxification

Metal ion overload can lead to intracellular concentrations beyond the storage and detoxification capacities of plant cells, and such a displacement of the balance towards a pro-oxidant state should normally activate anti-oxidant defenses. This direct link between metal ion toxicity and activation of responses against oxidative stress is poorly documented. As already mentioned, artificial iron overload of *Nicotiana plumbaginifolia* doubled APX and catalase activities in the leaves (Kampfenkel *et al.*, 1995a). Also, characterization of the tomato *chloronerva* mutant revealed abnormally high levels of iron and copper in leaves, as well as an increase in oxygen radical production, and activation of antioxidants enzymes (Pich and Scholz, 1993; Pich *et al.*, 1994). Among these enzymes, cytosolic APX and Cu/ZnSOD are more abundant in this mutant (Herbick *et al.*, 1996). Furthermore, it has now been demonstrated that APX mRNA abundance increases in response to iron overload in cotyledons of *Brassica napus* (Vansuyt *et al.*, 1997). Induction of plant FeSOD and MnSOD by different metals has also been reported (del Río *et al.*, 1991). In *Arabidopsis thaliana*, aluminum induces oxidative stress genes such as peroxidase, glutathione *S*-transferase, reticuline/oxygen oxidoreductase, and Cu/ZnSOD; in addition, three transcripts that code for proteins of unknown function and that are more abundant in response to aluminum treatment, accumulate also in response to ozone treatment. These observations document the link between aluminum toxicity and activation of antioxidant defenses (Richards *et al.*, 1998). Finally, an increased transcription of the genes for gluthatione synthesis, γ-glutamylcysteine synthethase and glutathione synthetase, as well as gluthathione reductase were activated in *Arabidopsis thaliana* in response to copper and cadmium treatment (Xiang and Oliver, 1998). This response was specific for those metals, and was mimicked by jasmonic acid treatment but not by H_2O_2 treatment, or oxidized or reduced gluthathione treatment.

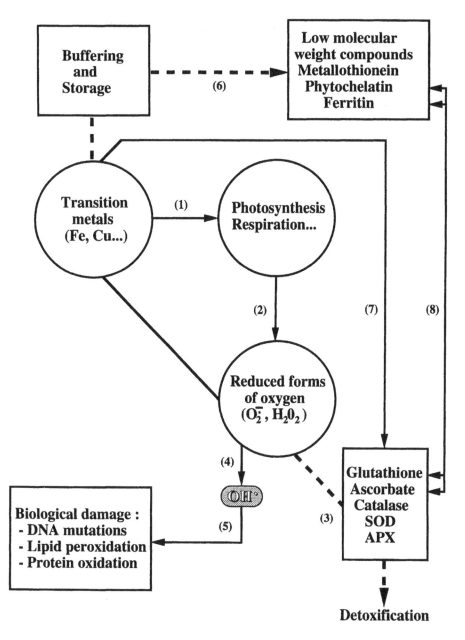

Figure 1. Schematic representation of the network linking oxidative stress and metal ion metabolism. Transition metals are necessary for many cellular reactions (1) that are essential to produce reduced forms of oxygen as secondary products (2). These molecules are detoxified by various systems including superoxide dismutases (SOD), catalases, ascorbate peroxidases (APX), reduced glutathione and ascorbate (3), but they can also react with transition metals to produce radical species (4), which are highly deleterious for biological structures (5). Therefore, buffering and storage of excess-free transition metals by various molecules, such as metallothioneins, phytochelatins ferritins, and low-molecular weight compounds appear to be important for plants to protect themselves against oxidative stress (6). Further links between transition metal and oxygen metabolism may occur for these metals through regulation of detoxification systems of reduced oxygen forms (7) and the requirement for glutathione in phytochelatin synthesis and ascorbate for iron loading of ferritin (8).

CONCLUDING REMARKS

Although a major checkpoint for metal ion homeostasis occurs at the uptake level through modulation of transporters, the intracellular concentrations of metal ions also need to be tightly controlled to avoid metal ion-activated oxidative stress (Figure 1). Only a limited number of the genes for transporters or involved in the processes described in Figure 1 are actually available. In addition, information concerning regulation of the expression of those that are actually cloned is still limited. Characterization of further genes involved in metal ion metabolism, specifically those activated in response to oxidative stress, is required to obtain a deeper insight into the mechanisms. Furthermore, molecular physiological approaches with transgenic plants thar are altered in the various pathways schematized in Figure 1, or characterization of mutants affected in such functions, should lead to a better understanding of metal ion-activated oxidative stress in plants.

ACKNOWLEDGEMENTS

Drs. Helen Logan and Stephane Lobréaux are acknowledged for critical reading of this manuscript, and for helpful discussions. J.-.F.B.'s work quoted in this review was supported by the Centre National de la Recherche Scientifique (ATIPE Grant 93N60/0563 and Programme "Biologie Cellulaire: du Normal au Pathologique", Grant 960003).

REFERENCES

Ames, B.A., Shingenaga, M.K., and Park, E.M. (1991) DNA oxidative damage. In K.J.A. Davies, (ed.), *Oxidative Damage and Repair: Chemical, Biological and Medical Aspects*, Pergamon Press, New York, pp. 181–187.

Ando, T., Yoshida, S., and Nishiyama, I. (1983) Nature of oxidizing power of rice roots. *Plant Soil*, **72**, 57–71.

Aruoma, O.I., Halliwell, B., Laughton, M.L., Quinlan, G.J., and Gutteridge, J.M. (1989) The mechanism of initiation of lipid peroxidation. Evidence against a requirement for an iron(II)-iron(III) complex. *Biochem. J.*, **258**, 617–620.

Bagg, A. and Neilands, J.B. (1987) Molecular mechanism of regulation of siderophore-mediated iron assimilation. *Microbiol. Rev.*, **51**, 509–518.

Balentine, J.D. (1982) *Pathology of oxygen toxicity*, Academic Press, New York.

Balla, G., Jacob, H.S., Balla, J., Rosenberg, M., Nath, K., Apple, F., Eaton, J.W., and Vercellotti, G.M. (1992) Ferritin: a cytoprotective antioxidant stratagem of endothelium. *J. Biol. Chem.* **267**, 18148–18153.

Becana, M. and Klucas, R.V. (1992) Transition metals in legume root nodules: Iron-dependent free radical production increases during nodule senescence. *Proc. Natl. Acad. Sci. USA*, **89**, 8958–8962.

Bienfait, H.F., Van Den Briel, W., and Mesland-Mul, N.T. (1985) Free space iron pools in roots. Generation and mobilisation. *Plant Physiol.*, **78**, 596–600.

Briat, J.F. and Lobréaux, S. (1997) Iron transport and storage in plants. *Trends Plant Sci.*, **2**, 187–193.

Briat, J.F., Fobis-Loisy, I., Grignon, N., Lobréaux, S., Pascal, N., Savino, G., Thoiron, S., von Wirén, N., and Van Wuytswinkel, O. (1995a) Cellular and molecular aspects of iron metabolism in plants. *Biol. Cell*, **84**, 69–81.

Briat, J.F., Labouré, A.M., Laulhère, J.P., Lescure, A.M, Lobréaux, S., Pesey, H., Proudhon, D., and Van Wuytswinkel, O. (1995b) Molecular and cellular biology of plant ferritins. In J. Abadia, (ed.), *Iron in Plants and Soils*, Kluwer Academic Publishers, Dordrecht, pp. 265–276.

Cadenas, E. (1989) Biochemistry of oxygen toxicity. *Annu. Rev. Biochem.*, **58**, 79–110.

Cairo, G., Tacchini, L., Pogliaghi, G., Anzon, E., Tomasi, A., and Bernelli-Zazzera, A. (1995) Induction of ferritin synthesis by oxidative stress. Transcriptional and post-transcriptional regulation by expansion of the free iron pool. *J. Biol. Chem.*, **270**, 700–703.

Cakmak I., and Marschner, H. (1988) Enhanced superoxide radical production in roots of zinc-deficient plants. *J. Exp. Bot.*, **207**, 1449–1460.

Caro, A., and Puntarulo, S. (1996) Effect of *in vivo* iron supplementation on oxygen radical production by soybean roots. *Biochim. Biophys. Acta*, **1291**, 245–251.

Chen, L., Keramati, L., and Helmann, J.D. (1995) Coordinate regulation of *Bacillus subtilis* peroxide stress genes by hydrogen peroxide and metal ions. *Proc. Natl. Acad. Sci. USA*, **92**, 8190–8194.

Cataldo, D.A., Mc Fadden, K.M., Garland, T.R., and Wildung R.E. (1988) Organic constituents and complexation of nickel (II), iron (III), cadmium (II) and plutonium (IV) in soybean xylem exudates. *Plant Physiol.*, **86**, 734–739.

Chubatsu, L.S. and Meneghini, R. (1993) Metallothionein protects DNA from oxidative damage. *Biochem. J.*, **291**, 193–198.

Cohen, C.K., Fox, T.C., Garvin, D.F. and Kochian L.V. (1998) The role of iron-deficiency stress responses in stimulating heavy-metal transport in plants. *Plant Physiol.*, **116**, 1063–1072.

Davies, K.J.A., Lin, S.W., and Pacifici, R.E. (1987) Protein damage and degradation by oxygen radicals. IV. Degradation of denatured protein. *J. Biol. Chem.*, **262**, 9914–9920.

del Río, L.A., Sevilla, F., Sandalio, L.M., and Palma J.M. (1991) Nutritional effect and expression of SODs: induction and gene expression; diagnostics; prospective protection against oxygen toxicity. *Free Rad. Res. Commun.*, **12–13**, 819–827.

Eide, D., Broderius, M., Fett, J., and Guerinot, M.L. (1996) A novel iron-regulated metal transporter from plants identified by functional expression in yeast. *Proc. Natl. Acad. Sci. USA*, **93**, 5624–5628.

Evans, E.M., Gatehouse, J.A., Lindsay, W.P., Shi, J., Tommey, A.M., and Robinson, N.J. (1992) Expression of the pea metallothionein-like gene *PsMT$_A$* in *Escherichia coli* and *Arabidopsis thaliana* and analysis of trace metal ion accumulation: implications for *PsMT$_A$* functions. *Plant Mol. Biol.*, **20**, 1019–1028.

Fobis-Loisy, I., Loridon, K., Lobréaux, S., Lebrun, M., and Briat, J.F. (1995) Structure and differential expression in response to iron and abscisic acid of two maize ferritin genes. *Eur. J. Biochem.*, **231**, 609–619.

Foy, C.D., Chaney, R.L., and White, M.C. (1978) The physiology of metal toxicity in plants. *Annu. Rev. Plant Physiol.*, **29**, 511–566.

Fridovich, I. (1995) Superoxide radical and superoxide dismutases. *Annu. Rev. Biochem.*, **64**, 97–112.

Fürst, P, Hu, S., Hackett, R., and Hamer, D.H. (1988) Copper activates metallothionein gene transcription by altering the conformation of a specific DNA binding protein. *Cell*, **55**, 705–717.

Gaymard, F., Boucherez, J., and Briat, J.F. (1996) Characterization of a ferritin mRNA from *Arabidopsis thaliana* accumulated in response to iron through an oxidative pathway independant of abscissic acid. *Biochem. J.*, **318**, 67–73.

Goldstein, S. and Czapski, G. (1986) The role and mechanism of metal ions and their complexes in enhancing damage in biological systems or in protecting these systems from the toxicity of O_2^{*p0-}. *Free Rad. Biol. Med.*, **2**, 3–11.

Gonzalez, A., Steffen, K.L., and Lynch, J.P. (1998) Light and excess manganese. Implications for oxidative stress in common bean. *Plant Physiol.*, **118**, 493–504.

Green, M.S. and Etherington, J.R. (1977) Oxidation of ferrous iron by rice (*Oryza sativa*) roots: a mechanism of waterlogging tolerance? *J. Exp. Bot.*, **28**, 678–690.

Grill, E., Winnacker, E.L., and Zenk, M.H. (1987) Phytochelatins, a class of heavy-metal-binding peptides from plants, are functionally analogous to metallothioneins. *Proc. Natl. Acad. Sci. USA*, **84**, 439–443.

Grill, E., Winnacker, E.L., and Zenk, M.H. (1991) Phytochelatins. In J.F. Riordan, and B.L. Vallee, (eds.), *Metallobiochemistry, Part B: Metallothionein and Related Molecules* (Methods in Enzymology, Vol. 205), Academic Press, New York, pp. 333–345.

Grotz, N., Fox, T., Connolly, E., Park, W., Guerinot, M.L., and Eide D. (1998) Identification of a family of zinc transporter genes from *Arabidopsis* that respond to zinc deficiency. *Proc. Natl. Acad. Sci. USA*, **95**, 7220–7224.

Guyton, K.Z., Xu, Q., and Holbrook, N.J. (1996) Induction in the mammalian stress response gene GADD153 by oxidative stress: role of AP-1 element. *Biochem J.*, **314**, 547–554.

Halliwell, B., and Gutteridge, J.M.C. (1984) Oxygen toxicity, oxygen radicals, transition metals and disease. *Biochem. J.*, **219**, 1–14.

Halliwell, B., and Gutteridge, J.M.C. (1990) Role of free radicals and catalytic metal ions in human disease: an overview. In L. Packer, and A.N. Glazer, (eds.), *Oxygen Radicals in Biological Systems, Part B: Oxygen Radicals and Antioxidants* (Methods in Enzymology, Vol. 186), Academic Press, New York, pp. 1–85.

Hamer D.H. (1986) Metallothionein. *Annu. Rev. Biochem.*, **55**, 913–951.

Hassan, H.M. and Sun, H.C.H. (1992) Regulatory roles of Fnr, Fur, and Arc in expression of manganese-containing superoxide dismutase in *Escherichia coli*. *Proc. Natl. Acad. Sci. USA*, **89**, 3217–3221.

Hentze, M.W. and Kühn, L.C. (1996). Molecular control of vertebrate iron metabolism: mRNA based regulatory circuits operated by iron, nitric oxide, and oxidative stress. *Proc. Natl. Acad. Sci. USA*, **93**, 8175–8182.

Herbick, A., Giritch, A., Hortsmann, C., Becker, R., Balzer, H.J., Bäumlein, H., and Stephan, U.W. (1996) Iron and copper nutrition-dependent changes in protein expression in a tomato wild type and the nicotianamine-free mutant *chloronerva*. *Plant Physiol.*, **111**, 533–540.

Hidalgo, E. and Demple, B. (1994) An iron-sulfur center essential for transcriptional activation by the redox-sensing SoxR protein. *EMBO J.*, **13**, 138–146.

Himelblau, E., Mira, H., Lin, S.J., Culotta V.C., Penarrubia, L. and Amasino R.M. (1998) Identification of a functional homolog of the yeast copper homeostasis gene *ATX1* from *Arabidopsis*. *Plant Physiol.*, **117**, 1227–1234.

Howden, R., Goldsbrough, P.B, Andersen, C.R., and Cobbett, C.S. (1995) Cadmium-sensitive *cad1* mutants of *Arabidopsis thaliana* are phytochelatin deficient. *Plant Physiol.*, **107**, 1059–1066.

Hunt, J.V., Simpson, J.A., and Dean, R.T. (1988) Hydroperoxide-mediated fragmentation of proteins. *Biochem. J.*, **250**, 87–93.

Illan, Y.A., Czapski, G., and Meisel, D. (1976) The one-electron transfer redox potentials of free radicals. I. The oxygen/superoxide system. *Biochim. Biophys. Acta*, **430**, 209–224.

Imlay, J.A. and Linn, S. (1988) DNA damage and oxygen radical toxicity. *Science*, **240**, 1302–1309.

Jacobson, M.D. (1996) Reactive oxygen species and programmed cell death. *Trends Biochem. Sci.*, **21**, 83–86.

Jungmann, J., Reins, H.A., Lee, J., Romeo, A., Hassett, R., Kosman D., and Jentsch, S. (1993) MAC1, a nuclear regulatory protein related to Cu-dependent transcription factors is involved in Cu/Fe utilization and stress resistance in yeast. *EMBO J.*, **12**, 5051–5056.

Kampfenkel, K., Van Montagu, M., and Inzé, D. (1995a) Effects of iron excess on *Nicotiana plumbaginifolia* plants. Implications to oxidative stress. *Plant Physiol.*,**107**, 725–735.

Kampfenkel, K., Kushnir, S., Babiychuk, E., Inzé D., and Van Montagu, M. (1995b) Molecular characterization of a putative *Arabidopsis thaliana* copper transporter and its yeast homologue. *J. Biol. Chem.*, **270**, 28479–28486.

Khan, A.U. and Kasha M. (1994) Singlet molecular oxygen in the Haber-Weiss reaction. *Proc. Natl. Acad. Sci. USA*, **91**, 12365–12367.

Khoroshilova, N., Beinert, H., and Kiley, P.J. (1995) Association of a polynuclear iron-sulfur center with a mutant FNR protein enhances DNA binding. *Proc. Natl. Acad. Sci. USA*, **92**, 2499–2503.

Kille, P., Winge, D.R., Harwood, J.L., and Kay, J. (1991) A plant metallothionein produced in *E. coli. FEBS Lett.*, **295**, 171–175.

Klausner, R.D., Rouault, T.A., and Harford, J.B. (1993) Regulating the fate of mRNA: the control of cellular iron metabolism. *Cell*, **72**, 19–28.

Kochian, L.V. (1995) Cellular mechanisms of aluminium toxicity and resistance in plants. *Annu. Rev. Plant Physiol. Plant Mol. Biol.*, **46**, 237–260.

Kurepa, J., Bueno, P., Kampfenkel, K., Van Montagu, M., Van den Bulcke, M., and Inzé, D. (1997). Effects of iron deficiency on iron superoxide dismutase expression in *Nicotiana tabacum*. *Plant Physiol. Biochem.*, **35**, 467–474.

Laulhère, J.P. and Briat, J.F. (1993) Iron release and uptake by plant ferritin as affected by pH, reduction and chelation. *Biochem. J.*, **290**, 693–699.

Laulhère, J.P., Labouré, A.M., and Briat, J.F. (1989) Mechanism of the transition from plant ferritin to phytosiderin. *J. Biol. Chem.*, **264**, 3629–3635.

Laulhère, J.P., Labouré, A.M., and Briat, J.F. (1990) Photoreduction and incorporation of iron into ferritins. *Biochem J.*, **269**, 79–84.

Lee, Y.S., Park, S.C., Goldberg, A.L., and Chung C.H. (1988) Protease So from *Escherichia coli* preferentially degrades oxidatively damaged glutamine synthetase. *J. Biol. Chem.*, **263**, 6643–6646.

Lescure, A.M., Massenet, O., and Briat, J.F. (1990) Purification and characterization of an iron-induced ferritin from soybean (*Glycine max*) cell suspensions. *Biochem. J.*, **272**, 147–150.

Lescure, A.M., Proudhon, D., Pesey, H., Ragland, M., Theil, E.C., and Briat, J.F. (1991) Ferritin gene transcription is regulated by iron in soybean cell cultures. *Proc. Natl. Acad. Sci. USA*, **88**, 8222–8226.

Lin, S.J. and Culotta, V.C. (1995) The *ATX1* gene of *Saccharomyces cerevisiae* encodes a small metal homeostasis factor that protects cells against reactive oxygen toxicity. *Proc. Natl. Acad. Sci. USA*, **92**, 3784–3788.

Lin, S.J. and Culotta, V.C. (1996) Suppression of oxidative damage by *Saccharomyces cerevisiae* ATX2, which encodes a manganese-trafficking protein that localizes to Golgi-like vesicles. *Mol. Cell. Biol.*, **16**, 6303–6312.

Liu, X.D. and Thiele, D.J. (1996) Oxidative stress induced heat shock factor phosphorylation and HSF-dependent activation of yeast metallothionein gene transcription. *Genes Dev.*, **10**, 590–603.

Lobréaux, S. and Briat, J.F. (1991) Ferritin accumulation and degradation in different organs of pea during development. *Biochem. J.*, **274**, 601–606.

Lobréaux, S., Massenet, O., and Briat J.F. (1992) Iron induces ferritin synthesis in maize plantlets. *Plant Mol. Biol.*, **19**, 563–575.

Lobréaux, S., Thoiron, S., and Briat, J.F. (1995) Induction of ferritin synthesis in maize leaves by an iron-mediated oxidative stress. *Plant J.*, **8**, 443–449.

Luo, Y., Han, Z., Chin, S.M., and Linn, S. (1994) Three chemically distinct types of oxidants formed by iron-mediated Fenton reactions in the presence of DNA. *Proc. Natl. Acad. Sci. USA*, **91**, 12438–12442.

Martins, E.A., Robalinho, R.L., and Meneghini, R. (1995) Oxidative stress induces activation of a cytosolic protein responsible for control of iron uptake. *Arch. Biochem. Biophys.*, **316**, 128–134.

Meyer, M., Schreck, R., and Bauerle, P.A. (1993) H_2O_2 and antioxidants have opposite effects on activation of NF-KB and AP-1 in intact cells: AP-1 as secondary antioxidant-responsive factor. *EMBO J.*, **12**, 2005–2015.

Müllner, E.W. and Kühn, L.C. (1988) A stem-loop in the 3' untranslated region mediates iron-dependent regulation of transferrin receptor mRNA stability in the cytoplasm. *Cell*, **53**, 815–825.

Müllner, E.W., Neupert, B., and Kühn, L.C. (1989) A specific mRNA binding factor regulates the iron dependent stability of cytoplasmic transferrin receptor mRNA. *Cell*, **58**, 373–382.

Nascimento, A.L. and Meneghini, R. (1995) Cells transfected with transferrin receptor cDNA lacking the iron regulatory domain become more sensitive to the DNA-damaging action of oxidative stress. *Carcinogenesis*, **16**, 1335–1338.

Okumura, N., Nishizawa, N.K., Umehara, Y., and Mori, S. (1991) An iron deficiency-specific cDNA from barley roots having two homologous cystein-rich MT domains. *Plant Mol. Biol.*, **17**, 531–533.

Ortiz, D.F., Kreppel, L., Speiser, D.M., Scheel, G., McDonald, G., and Ow, D.W. (1992) Heavy metal tolerance in the fission yeast requires an ATP-binding cassette-type vacuolar membrane transporter. *EMBO J.*, **11**, **3491–3499.**

Ortiz, D.F., Ruscitti, T., McCue, K.F., and Ow, D.W. (1995) Transport of metal-binding peptides by HMT1, a fission yeast ABC-type vacuolar membrane protein. *J. Biol. Chem.*, **270**, 4721–4728.

Pantopoulous, K. and Hentze, M.W. (1995) Rapid responses to oxidative stress mediated by iron regulatory protein. *EMBO J.*, **14**, 2917–2924.

Philpott, C.C., Rashford, J., Yamaguchi-Iwai, Y., Rouault, T.A., Dancis, A., and Klausner, R.D. (1998) Cell-cycle arrest and inhibition of G1 cyclin translation by iron in *AFT1-1^{up}* yeast. *EMBO J.*, **17**, 5026–5036.

Pich, A. and Scholz, G. (1991) Nicotianamine and the distribution of iron into apoplast and symplast of tomato (*Lycopersicon esculentum* Mill.). II Uptake of iron by protoplasts from the variety Bonner Beste and its nicotinanamine-less mutant *chloronerva* and the compartimentation of iron in leaves. *J. Exp. Bot.*, **42**, 1517–1523.

Pich, A. and Scholz, G. (1993) The relationship between the activity of various iron-containing and iron-free enzymes and the presence of nicotianamine in tomato seedlings. *Physiol. Plant.*, **88**, 172–178.

Pich, A., Scholz, G., and Seifert, K. (1991) Effect of nicotianamine on iron uptake and citrate accumulation in two genotypes of tomato, *Lycopersicon esculentum* Mill. *J. Plant Physiol.*, **137**, 323–326.

Pich, A., Scholz, G., and Stephan, U.W. (1994) Iron dependent changes of heavy metals, nicotianamine and citrate in different plant organs and in the xylem exudate of two tomato genotypes. Nicotianamine as possible copper translocator. *Plant Soil*, **165**, 189–196.

Ponnamperuma, F.N., Bradfield, R., and Peech, M. (1955) Physiological disease of rice attributable to iron toxicity. *Nature*, **175**, 275.

Pryor, W.A. (1988) Why is the hydroxyl radical the only radical that commonly adds to DNA? Hypothesis: it is a rare combination of high electrophilicity, high thermochemical reactivity, and a mode of production that occur near DNA. *Free Rad. Biol. Med.*, **4**, 219–223.

Rauser, W.E. (1990) Phytochelatins. *Annu. Rev. Biochem.*, **59**, 61–86.

Rauser, W.E. (1995) Phytochelatins and related peptides. Structure, biosynthesis, and function. *Plant Physiol.*, **109**, 1141–1149.

Richards, K.D., Schott, E.J., Sharma, Y.K., Davis, K.R., and Gardner, R.C. (1998) Aluminum induces oxidative stress genes in *Arabidopsis thaliana*. *Plant Physiol.*, **116**, 409–418.

Rivett, A.J. (1985a) The effect of mixed-function oxidation of enzymes on their susceptibility to degradation by nonlysosomal cysteine protease. *Arch. Biochem. Biophys.*, **243**, 624–632.

Rivett, A.J. (1985b) Preferential degradation of the oxidatively modified form of glutamine synthetase by intracellular mammalian proteases. *J. Biol. Chem.*, **260**, 300–305.

Rivett, A.J. (1985c) Purification of a liver alkaline protease which degrades oxidatively modified glutamine synthetase. Characterization as a high molecular weight cysteine proteinase. *J. Biol. Chem.*, **260**, 12600–12606.

Robinson, N.J., Tommey, A.M., Kuske, C., and Jackson, P.J. (1993) Plant metallothioneins. *Biochem. J.*, **295**, 1–10.

Roseman, J.E. and Levine, R.L. (1987) Purification of a protease from *Escherichia coli* with specificity for oxidized glutamine synthetase. *J. Biol. Chem.*, **262**, 2101–2110.

Rouault, T.A. and Klausner, R.D. (1996) Iron-sulfur clusters as biosensors of oxidants and iron. *Trends Biochem. Sci.*, **21**, 174–177.

Salt, D.E. and Rauser, W.E. (1995) MgATP-dependent transport of phytochelatins across the tonoplast of oat roots. *Plant Physiol.*, **107**, 1293–1301.

Sandmann, G. and Böger, P. (1980) Copper-mediated lipid peroxidation processes in photosynthetic membranes. *Plant Physiol.*, **66**, 797–800.

Sappey, C., Boelaert, J.R., Legrand-Poels, S., Grady R.W., and Piette, J. (1995) NF-κB transcription factor activation by hydrogen peroxide can be decreased by 2,3-dihydrobenzoic acid and its ethyl ester derivative. *Arch. Biochem. Biophys.*, **321**, 263–270.

Savino, G., Briat, J.F., and Lobréaux, S. (1997). Inhibition of the iron-induced *ZmFer1* maize ferritin gene expression by antioxidants and serine/threonine phosphatase inhibitors. *J. Biol. Chem.*, **272**, 33319–33326.

Schnell, N. and Entian, K.D. (1991) Identification and characterization of a *Saccharomyces cerevisiae* gene (*par1*) conferring resistance to iron chelators. *Eur. J. Biochem.*, **200**, 487–493.

Scholz, R.W, Graham, K.S., and Wynn, M.K. (1990) Interaction of glutathione and α-tocopherol in the inhibition of lipid peroxidation of rat liver microsomes. In C.C. Reddy, G.A. Hamilton, and K.M. Madyastha, (eds.), *Biological Oxidation Systems, Volume 2*, Academic Press, San Diego, pp. 841–867.

Schreck, R., Rieber, P., and Bauerle, P.A. (1991) Reactive oxygen intermediate as apparently widely used messengers in the activation of the NF-κB transcription factor and HIV-1. *EMBO J.*, **10**, 2247–2258.

Sook, C. and Jung, J. (1993) The susceptibility of mung bean chloroplasts to photoinhibition is increased by an excess supply of iron to plants: a photobiological aspect of iron toxicity in plant leaves. *Photochem. Photobiol.*, **58**, 120–126.

Stadtman, E.R. (1993) Oxidation of free amino acids and amino acid residues in proteins by radiolysis and by metal-catalysed reactions. *Annu. Rev. Biochem.*, **62**, 797–821.

Stadtman, E.R. and Berlett, B.S. (1991) Fenton Chemistry: Amino acid oxidation. *J. Biol. Chem.*, **266**, 17201–17211.

Stephan, U.W. and Scholz, G. (1993). Nicotianamine: mediator of transport of iron and heavy metals in the phloem? *Physiol. Plant.*, **88**, 522–529.

Tamai, K.T., Gralla, E.B., Ellerby, L.M., Valentine, J.S., and Thiele, D.J. (1993) Yeast and mammalian metallothioneins functionally substitute for yeast copper-zinc superoxide dismutase. *Proc. Natl. Acad. Sci. USA*, **90**, 8013–8017.

Theil, E.C. (1987) Ferritin: structure, gene regulation, and cellular function in animals, plants and microorganisms. *Annu. Rev. Biochem.*, **56**, 289–315.

Theil, E.C. (1994) Iron regulatory elements (IREs): a family of mRNA non-coding sequences. *Biochem. J.*, **304**, 1–11.

Tommey, A.M., Shi, J., Lindsay, W.P., Urwin, P.E., and Robinson, N.J. (1991) Expression of the pea gene *PsMTA* in *E. coli*: metal-binding properties of the expressed protein. *FEBS Lett.*, **292**, 48–52.

Touati, D., Jacques, M., Tardat, B., Bouchard, L., and Despied, S. (1995) Lethal oxidative damage and mutagenesis are generated by iron in Δfur mutants of *Escherichia coli*: protective role of superoxide dismutase. *J. Bacteriol.*, **177**, 2305–2314.

Tsang, S.Y., Tam, S.C., Bremmer, I., and Burkitt, M.J. (1996) Copper-1,10-phenantroline induces internucleosomal DNA fragmentation in HepG2 cells, resulting from direct oxidation by the hydroxyl radical. *Biochem. J.*, **317**, 13–16.

Vansuyt, G., Lopez, F., Inzé, D., Briat, J.-F., and Fourcroy, P. (1997) Iron triggers a rapid induction of ascorbate peroxidase gene expression in *Brassica napus*. *FEBS Lett.*, **410**, 195–200.

von Wiren, N., Marschner, H., and Röhmeld, V. (1996) Roots of iron-efficient maize also absorb phytosiderophore-chelated zinc. *Plant Physiol.*, **111**, 1119–1125.

Xiang, C. and Oliver, D.J. (1998) Glutathione metabolic genes coordinately respond to heavy metals and jasmonic acid in *Arabidopsis. Plant Cell*, **10**, 1539–1550.

Yaffee, M., Walter, P., Richter, C., and Muller, M. (1996) Direct observation of iron-induced conformational changes of mitochondrial DNA by high-resolution field-emission in-lens scanning electron microscopy. *Proc. Natl. Acad. Sci. USA*, **93**, 5341–5346.

Yamaguchi-Iwai, Y., Dancis, A., and Klausner, R.D. (1995) AFT1: a mediator of iron regulated transcriptional control in *Saccharomyces cerevisiae. EMBO J.*, **14**, 1231–1239.

Yamaguchi-Iwai, Y., Stearman, R., Dancis, A., and Klausner, R.D. (1996) Iron-regulated DNA binding by the AFT1 protein controls the iron regulon in yeast. *EMBO J.*, **15**, 3377–3384.

Zer, H., Peleg, I., and Chevion, M. (1994) The protective effect of desferrioxamine on paraquat treated pea (*Pisum sativum*). *Physiol. Plant.*, **92**, 437–442.

Zhang, F.S., Röhmeld, V., and Marschner, H. (1991) Role of the root apoplast for iron acquisition by wheat plants. *Plant Physiol.*, **97**, 1302–1305.

Zhao, H. and Eide, D. (1996) The yeast *ZRT1* gene encodes the zinc transporter protein of a high-affinity uptake system induced by zinc limitation. *Proc. Natl. Acad. Sci. USA*, **93**, 2454–2458.

Zhou, J. and Goldsbrough, P.B. (1994) Functional homologs of fungal metallothionein genes from *Arabidopsis. Plant Cell*, **6**, 875–884.

Zhou, J. and Goldsbrough, P.B. (1995) Structure, organization and expression of the metallothionein gene family in *Arabidopsis. Mol. Gen. Genet.*, **248**, 318–328.

8 Engineering Stress Tolerance in Maize

Frank Van Breusegem, Marc Van Montagu and Dirk Inzé

INTRODUCTION

Chilling Stress in Maize

Plants that grow in cold climates are well adapted to the ambient chilling temperatures, but the adaptation to low temperatures in these plants is directed more towards survival rather than to performance (harvestable yield). For crop plants such as maize, which are grown outside their natural environment, performance is the only valuable criterion. Greaves (1996) defined a suboptimal stress for maize as any reduction in growth or induced metabolic, cellular, or tissue injury that results in limitations to the genetically determined yield potential, caused as a result of exposure to temperatures below the thermal thresholds for optimal biochemical and biophysical activity or morphological development. Optimal growth of maize crop occurs in climates with midsummer temperatures between 21°C and 27°C and the optimal temperature for maximum grain yields lies around 25°C. Maize plants that are subjected to temperatures below 20°C are believed to undergo already physiological and biochemical changes. Of course, the damage will increase according to the duration and severity of the chilling conditions. For example, maize plants grown continuously at 17/15°C (day/night temperatures) are seriously retarded in growth. The stressed plants reach the same developmental stage 10 days after nonstressed plants. Besides the disruption of plasma membranes and decreased activities of metabolic enzymes, photooxidative stress is considered the main cause of damage during chilling stress in chilling-sensitive species such as maize.

Photooxidative stress

The inhibition of photosynthesis is an early event during chilling stress. Under optimal growth conditions, light energy absorbed by leaves is used primarily for the assimilation of carbon in the process of photosynthesis. When plants experience suboptimal growth temperatures in the field, as is almost always the case during the early growing season in Northern Europe, light absorbed by the leaves cannot be used efficiently for photosynthesis and becomes potentially damaging because the excess electrons react with the abundantly present oxygen. Together with the electron transport chain of mitochondria, chloroplasts are the main site of active oxygen species (AOS) production (for a schematic overview, see Figure 1). Electrons passing through the transport chain in the photosystems and mitochondria can react with oxygen to form superoxide radicals ($O_2^{\cdot-}$) and hydrogen peroxide (H_2O_2). Other sources of AOS production are the microbodies. In the glyoxisomes, H_2O_2 is produced during fatty acid degradation in the glyoxylate cycle. In the peroxisomes, photorespiration is responsible for H_2O_2 production. Although neither $O_2^{\cdot-}$ nor H_2O_2 seem particularly harmful at physiological concentrations, their toxicity *in vivo* is enhanced by a metal ion-dependent conversion into hydroxyl radicals ($OH^{\cdot}$), one of the most reactive oxygen species known in chemistry.

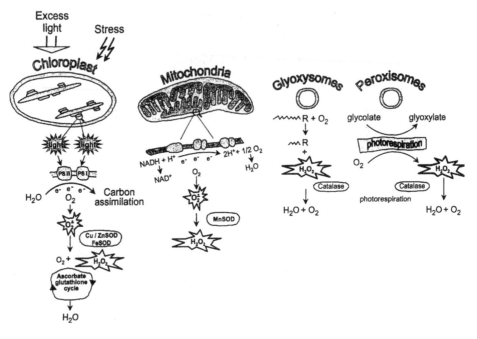

Figure 1. Subcellular production and scavenging of AOS in plants. The electron transport chains of mitochondria and chloroplasts are the main sites of AOS production. In the peroxisomes, H_2O_2 is mainly produced during photorespiration. Fatty acid degradation in the glyoxylate cycle in the glyoxisomes also generates H_2O_2. Abbreviations: PS, photosystem; SOD, superoxide dismutase; $\wedge \wedge \wedge$ R, β-fatty acids

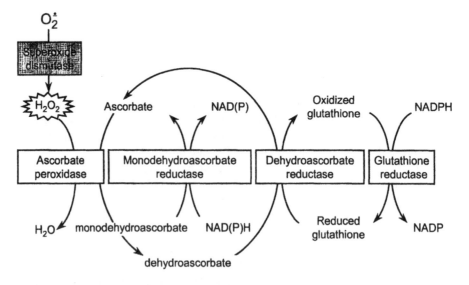

Figure 2. The ascorbate–glutathione cycle. Superoxide radicals are dismutated by superoxide dismutases. H_2O_2 is removed by APX and ascorbate is regenerated by the ascorbate–glutathione cycle. Ascorbate is oxidized to monodehydroascorbate reductase. Monodehydroascorbate can spontaneously disproportionate into ascorbate and dehydroascorbate. Dehydroascorbate reductase recycles ascorbate using reduced glutathione. Oxidized glutathione is regenerated by glutathione reductase in a NADPH-dependent reaction.

$$O_2^{\bullet-} + H_2O_2 \rightarrow O_2 + 2\, OH^{\bullet} \qquad \text{(Haber-Weiss reaction)}$$

Hydroxyl radicals are capable of reacting indiscriminately causing lipid peroxidation, denaturation of proteins, and mutation of DNA (Breen and Murphy, 1995; Desimone *et al.*, 1996). Fortunately, plants have the capacity to cope with these aggressive agents by eliminating them with an AOS-scavenging system. Under moderate stress conditions, the produced radicals can be efficiently scavenged, but during periods of more severe stress, the scavenging systems become saturated with the increased rate of radical production. The presence of excessive AOS results in damage to the photosynthetic apparatus (photoinhibition), bleaching of the leaves by oxidation of the pigments, and, hence, severe yield losses. AOS are produced in a wide variety of stresses (drought, heat, salt, pollution, chilling, ozone, etc.). These, at first sight, unrelated stresses have all in common the electron leakage from electron transport chains in chloroplasts and mitochondria.

In certain situations, the plant uses AOS in a beneficial way. AOS play an important role in the induction of protection mechanisms during biotic and abiotic stresses. The best known example is their role in the activation of resistance responses during incompatible plant-pathogen interactions. Upon infection, an NADPH oxidase of the plasma membrane is activated, producing superoxide radicals (Desikan *et al.*, 1996). Superoxide radicals are, via spontaneous dismutation or via superoxide dismutase (SOD) activity converted to hydrogen peroxide (H_2O_2). The defensive properties of H_2O_2 are situated at several levels. (i) High levels of H_2O_2 are toxic for both pathogen and plant cells. Killing the plant cells surrounding the infection site inhibits spreading of a (biotrophic) pathogen (Levine *et al.*, 1996). (ii) Hydrogen peroxide can serve as a substrate in peroxidative cross-linking reactions of lignin precursors (Dixon *et al.*, 1996) and induces cross-linking of cell wall proteins (Brisson *et al.*, 1994). A reinforced plant cell wall slows down the spreading of the pathogen and makes new infections more difficult. (iii) Because H_2O_2 is relatively stable and diffusible (in contrast with superoxide) through membranes, it is a perfect candidate to act as a signal molecule during stress responses. H_2O_2 induces several pathogen defense genes (coding for pathogenesis-related proteins, and phytoalexins) and a programmed cell death pathway (Desikan *et al.*, 1998). These different effects of H_2O_2 are thought to be regulated not only through the level and timing of H_2O_2 induction, but also through interactions with other potential signals, such as salicylic acid, superoxide and nitric oxide (Alvarez *et al.*, 1998; Delledonne *et al.*, 1998; Van Camp *et al.*, 1998). The role of H_2O_2 as a molecular signal for the induction of gene expression may not be limited to plant-pathogen interactions. Studies on maize hypocotyls and on potato and mustard seedlings have shown that H_2O_2 mediates a subsequent cold and heat tolerance, respectively (Prasad *et al.*, 1994a, 1994b; Dat *et al.*, 1998; Lopez-Delgado *et al.*, 1998). In barley, H_2O_2 treatment broke dormancy of seeds (Fontaine *et al.*, 1994). AOS production is hence vital to plant development and in defense against pathogens, but it needs to be tightly controlled in order to avoid cellular damage.

To limit cellular damage caused by excess AOS levels, plants have evolved a broad variety of nonenzymatic and enzymatic protection mechanisms that efficiently scavenge AOS (Inzé and Van Montagu, 1995). The best-known nonenzymatic antioxidants are ascorbate, glutathione, α-tocopherol and carotenoids. They are present in relatively high concentrations within plant cells. For a detailed overview on these components, the reader is referred to Alscher and Hess (1993).

Since hydroxyl radical are too reactive to be directly controlled, aerobic organisms prefer to eliminate the less reactive precursor forms superoxide and H_2O_2 and hence prevent the formation of hydroxyl radicals. Superoxide dismutases scavenge superoxide radicals while catalases and peroxidases remove hydrogen peroxide. Catalases consume the bulk of H_2O_2 and (ascorbate) peroxidases — because of their higher affinity and their diverse subcellular locations — remove H_2O_2 that is not accessible to catalase. Other enzymes that are involved in the removal of AOS are: monodehydroascorbate reductase, dehydroascorbate reductase, glutathione reductase and glutathione peroxidase. However these enzymes will not be described in detail in this review. For a detailed overview the reader is referred to Noctor and Foyer (1998).

The following section gives a general overview on the (molecular) biology of the AOS defense system in plants. Because of the abundancy of sequence and expression data, only an examplary selection is presented, with emphasis on monocotyledonous species. In a separate section the literature on the AOS defense system in maize (and its involvement in environmental stress) is reviewed in more detail. To end, an overview is given of the use of antioxidant genes in generating transgenic plants with enhanced oxidative stress tolerance capacities.

AOS SCAVENGING SYSTEMS IN PLANTS

Superoxide Dismutases

Superoxide dismutases (SOD; superoxide:superoxide oxidoreductase; EC 1.15.1.1) can be considered key enzymes of the antioxidative stress defense mechanism (Van Camp et al., 1994a). They directly determine the cellular concentrations of $O_2^{\bullet-}$ and H_2O_2 since they dismutate superoxide into O_2 and H_2O_2.

$$2H^+ + 2O_2^{\bullet-} \xrightarrow{\text{SOD}} H_2O_2 + O_2$$

Apart from a few exceptions, SODs are present in all aerobic organisms, and in all subcellular compartments that have to deal with AOS. They are classified according to their metal cofactor as isozymes containing copper/zinc (Cu/Zn), iron (Fe), and manganese (Mn). FeSOD and MnSOD proteins are structurally similar (Stallings et al., 1984), while the Cu/ZnSOD family is structurally unrelated (Bowler et al., 1994). SODs catalyze the disproportionation of superoxide through an oxidation-reduction cycle of the prosthetic Cu, Mn or Fe cofactor. Experimentally the three SOD types can easily be distinguished via *in situ* gel staining (Beauchamp and Fridovich, 1971). Incubating the gels with KCN or H_2O_2 allows the discrimination between the different classes. Cu/ZnSOD is characterized as being sensitive to both H_2O_2 and KCN; FeSOD is sensitive only to H_2O_2 while MnSOD is resistant to both inhibitors. However, in a recent article, an FeSOD from rice was shown to be resistant to both inhibitors (Kaminaka et al., 1999). Besides their differential sensitivity towards inhibitors, the subcellular distribution of these isozymes is also distinctive. The amount and relative abundance of SOD isozymes varies within each organism. Developmental control and environmental stresses that generate AOS (UV, ozone, air pollutants, low temperatures, salt stress, drought, heat shock, pathogen infections, etc.) have been shown to induce plant SOD activities (Van Camp et al., 1994a).

MnSOD is found in the mitochondria of all eukaryotic cells, including plants; Cu/ZnSODs are found in the cytosol, peroxisomes and chloroplasts of higher plants. Multiple Cu/ZnSOD isoforms are also present in the extracellular fluids of Scots pine (Streller and Wingsle, 1994; Schinkel *et al.*, 1998). Until now, FeSODs were only found in prokaryotes and in the chloroplasts of plants.

Cytosolic and chloroplastic Cu/ZnSOD

The Cu/ZnSODs are dimeric proteins that contain one Cu and one Zn molecule per subunit. The enzyme is localized in the chloroplast (stromal and thylakoid associated), the cytosol and the peroxisomes (Bueno *et al.*, 1995; Ogawa *et al.*, 1995). In higher plants, cDNAs and genes encoding cytosolic Cu/ZnSODs have been isolated from a wide variety of species: tomato (Perl-Treves *et al.*, 1988), pea (White and Zilinskas, 1991), poplar (Akkapeddi *et al.*, 1994), tall goldenrod (Murai and Murai, 1996), spinach (Sakamoto *et al.*, 1990), *Nicotiana plumbaginifolia* (Tsang *et al.*, 1991), *Arabidopsis thaliana* (Kliebenstein *et al.*, 1998), Scots pine (Karpinski *et al.*, 1992), rice (Sakamoto *et al.*, 1995a, 1995b) and maize (Cannon and Scandalios, 1989). Cytosolic Cu/ZnSOD genes are strongly expressed in most plant species and their expression is enhanced by several stresses (methyl viologen, heat shock, chilling, ozone and salt stress, etc.). The rice Cu/ZnSOD genes are thought to be stress responsive as they contain multiple heat shock elements and an element responsive to the phytohormone abscisic acid (ABA) (Sakamoto *et al.*, 1995a; Tanaka, 1998).

The chloroplastic isoforms are nuclear encoded and targeted to the chloroplast by an amino-terminal transit peptide. Chloroplastic Cu/ZnSOD cDNAs have been isolated among others, from *N. plumbaginifolia* (Kurepa *et al.*, 1997), tomato (Perl-Treves *et al.*, 1988), pea (Reed Scioli and Zilinskas, 1988), wheat (EMBL U69632), and rice (Kaminaka *et al.*, 1997). In tomato, tobacco, and pea chloroplastic Cu/ZnSOD is developmentally and stress regulated (Kurepa *et al.*, 1997; Perl-Treves and Galun, 1991).

In maize, ten different SOD isozymes are known: four cytosolic Cu/ZnSODs, a chloroplast-associated Cu/ZnSOD, four mitochondria-associated MnSODs, and a novel type of chloroplastic FeSOD (see also Table 1). Cu/ZnSOD is the most abundant isoform

Table 1. Genes or cDNAs, coding for maize AOS defense enzymes

Enzyme	Gene	Reference
MnSOD	*sod3.1*	Zhu and Scandalios (1993)
	sod3.2	
	sod3.3	
	sod3.4	
FeSOD		F. Van Breusegem (unpublished data)
Cu/ZnSOD (cytosolic)	*sod2*	Cannon *et al.* (1987)
	sod4	Kernodle and Scandalios (1996)
	sod4A	
Catalase	*cat1*	Guan and Scandalios (1996) (and references therein)
	cat2	
	cat3	
cAPX		Van Breusegem *et al.* (1995)
GR		F. Van Breusegem (unpublished data)

in maize leaves. Four different cytosolic and one chloroplastic Cu/ZnSODs are identified. The cytoplasmic SOD-2, SOD-4, SOD-4A, and SOD-5 are dimeric proteins consisting of subunits of approximately 17 kDa (Baum and Scandalios, 1981; Cannon and Scandalios, 1989). The maize Cu/ZnSOD (SOD-5) and the plastidic Cu/ZnSOD (SOD-1) have not been purified nor their DNA sequences cloned.

Sod2 is the first maize Cu/ZnSOD cDNA that was isolated. Besides its presence at high levels in scutella, little is known on the regulation of *Sod2*. *Sod4* and *Sod4a* are 95% identical in the coding region and hence code for two very similar proteins. The deduced amino acid sequences only differ at two positions and both isozymes are biochemically indistinguishable (Cannon and Scandalios, 1989). The coding regions share a great degree of homology with *Sod2*. The promoters of both genes contain several well-known stress-related regulatory elements (such as ABA-responsive, heat shock, and cold stress-responsive elements). The 3'untranslated part of the cDNAs are completely different (White *et al.*, 1990; Acevedo and Scandalios, 1992). The *Sod4* and *Sod4a* transcripts are found in most tissues of the maize plant. Both genes are differentially induced by various stress conditions. By a treatment of seedlings with methyl viologen, there is an opposite expression profile. The *Sod4* transcripts are induced at high levels while *Sod4a* transcripts are absent (Kernodle and Scandalios, 1996). In leaves *Sod4* and *Sod4a* also respond to ABA and osmotic stress via alternate pathways (Guan and Scandalios, 1998a). Such opposite expression patterns of antioxidant genes show that the different SOD isoforms have different tasks and are, in addition, spatially and temporally distributed within the stress defense system, enabling the plant to respond to various oxidative stress situations at different developmental time points and at different subcellular sites.

Mitochondrial MnSOD

MnSOD is a homo-tetrameric mitochondrial enzyme of approximately 85 kDa. Plant *Mnsod* sequences have been isolated from *N. plumbaginifolia* (Bowler *et al.*, 1989), pea (Wong-Vega *et al.*, 1991), *Arabidopsis* (Kliebenstein *et al.*, 1998), rubber tree (Miao and Gaynor, 1993), wheat (EMBL U72212) and rice (EMBL L34039; Sakamoto *et al.*, 1993; Chen *et al.*, 1997). The MnSODs are nuclear encoded and imported in the mitochondria via an amino-terminal transit targeting peptide. In maize, MnSOD (SOD-3) is a tetrameric protein of approximately 90 kDa (Baum and Scandalios, 1981) and is located in the mitochondrial matrix (White and Scandalios, 1987; Zhu and Scandalios, 1995). Unlike the single *Mnsod* gene reported in animal and most other plants, the maize MnSOD is encoded by a multigene family: *Sod3.1*, *Sod3.2*, *Sod3.3*, and *Sod3.4* (Zhu and Scandalios, 1993). All four deduced amino acid sequences have a mitochondrial transit peptide (31 residues) from which the first nine amino acids (the matrix targeting sequence) are identical. This suggests that all maize MnSOD proteins are mitochondrial enzymes. The four genes are highly similar in the coding region, but vary in the 5' and 3' noncoding regions. The nucleotide sequence of *Sod3.3* and *Sod3.4* are nearly identical in both coding and non-coding regions. Promoter regions of the respective genes are not cloned yet. The different *Mnsod* transcripts are developmentally and spatially regulated, except for *Sod3.1* which is expressed in all examined tissue (from embryo to leaf). ABA induces three of the four maize MnSOD genes but total SOD-3 protein and enzymatic activity remain constant. The specific mechanism whereby ABA affects expression of the MnSOD genes is not known.

A plausible explanation is that ABA induces major metabolic changes, which, in turn regulate expression of MnSOD genes (Zhu and Scandalios, 1994). The expression pattern of *Sod3.3* coincides with the onset of dehydration and can be considered a "low temperature-responsive" gene because it is also induced by chilling stress. Interestingly, expression patterns of *Sod3.3* follow the increase of mitochondrial respiratory activities, meaning that the expression of nuclear-encoded genes can be influenced by mitochondrial activities. As described above, AOS themselves are good candidates to act here as signal molecules between mitochondria and the nucleus and influence gene expression levels.

Chloroplastic FeSOD

MnSOD and FeSOD genes probably have a common ancestor and can be considered as structural homologues. The metal cofactor of FeSODs and MnSODs is generally specific but is interchangeable within a few prokaryotic organisms (Meier *et al.*, 1982). FeSOD is a mono-dimeric enzyme, consisting of 24 kDa subunits that are nuclear encoded. FeSOD is thought to be the oldest SOD because anaerobic prokaryotes have a FeSOD, but no MnSOD. FeSOD was identified in prokaryotes, eukaryotic algae, protozoa, mosses, dicotyledonous seed plants and recently in some monocots. Because of their absence in fungi and animals and because of the specific sequence similarities between plant and cyanobacterial FeSODs, it is thought that the *Fesod* gene was acquired by endosymbiotic uptake from the chloroplast ancestor.

Single plant cDNA clones are isolated from *Nicotiana plumbaginifolia*, soybean, rice and maize (Van Camp *et al.*, 1990; Crowell and Amasino, 1991; Kaminaka *et al.*, 1999; unpublished results). *Arabidopsis thaliana* has at least three different *Fesod* genes that are independently regulated, raising the possibility that they are responsible for protecting the chloroplast against different forms of oxidative stress (Kliebenstein *et al.*, 1998). The rice FeSOD is strongly expressed in stems of young tissues, etiolated seedlings and embryogenic calli, but was only minimally expressed in the leaves and roots. As in tobacco and barley, rice FeSOD transcripts are induced by light. In rice seedlings, the level of FeSOD mRNA was increased significantly 60 minutes after the onset of the light induction and reached a maximum level after 4 hours (Kaminaka *et al.*, 1999). The inducibility of FeSOD during stress conditions seems to be dependent on the developmental status of the stressed leaf. In barley light stress accumulates *Fesod* transcripts in young leaves, but not in mature-senescent leaves (Casano *et al.*, 1994). In tobacco FeSOD transcripts are more abundant in young compared to old leaves (Tsang *et al.*, 1991). In maize leaves, a similar phenomenon is observed: FeSOD transcripts are induced upon growth at low temperatures (after one day) and the induction is most pronounced in younger and healthy leaves. This observation is in contrast with the expression levels of a cytosolic Cu/ZnSOD (*Sod4*) during the same stress regime. *Sod4* was only induced in older, bleached leaves after several days of chilling stress. In these leaves, FeSOD mRNA levels dropped back to undetectable levels (F. Van Breusegem, unpublished results). *Sod4* transcript levels also increased dramatically in seedlings that were treated with high concentrations of etephon, which promotes early senescence and wilting of the leaves (Kernodle and Scandalios, 1996), suggesting that FeSOD is responding to an early oxidative stress, whereas Cu/ZnSOD transcripts only accumulate in irreversibly damaged maize leaves.

ASCORBATE PEROXIDASE

Peroxidases are ubiquitous enzymes found in plants. Besides the peroxidases, of which the oxidation products play mainly physiological roles (lignification, cross-linking of cell wall matrices), there is a second class that forms part of the AOS defense system. Ascorbate peroxidases (APX; EC 1.11.1.11) destroy harmful H_2O_2 via the ascorbate-glutathione pathway in chloroplasts and cytosol of plants, algae, and some cyanobacteria (Figure 2). Recently APX activity was also identified in insects and purified from bovine eye tissue (Mathews *et al.*, 1997; Wada *et al.*, 1998). The ascorbate-glutathione pathway provides protection against oxidative stress by a series of coupled redox reactions, particularly in photosynthetic tissues (Foyer and Halliwell, 1976), but also in mitochondria and peroxisomes (Jiménez *et al.*, 1997). APX eliminates H_2O_2 by using ascorbate as an electron donor in an oxidation-reduction reaction. Ascorbate is then oxidized to monodehydroascorbate (MDA).

$$2 \text{ ascorbate} + H_2O_2 \xrightarrow{\text{APX}} 2 \text{ monodehydroascorbate} + 2H_2O$$

Ascorbate is regenerated from MDA in the chloroplast membrane by ferredoxin (Miyake and Asada, 1994) or in the stroma by monodehydroascorbate reductase (MDHAR) at the expense of NADPH. MDA can also spontaneously dissociate into ascorbate and dehydroascorbate (DHA). DHA is re-reduced by dehydroascorbate reductase (DHAR). DHAR uses reduced glutathione as an electron donor. Oxidized glutathione is then recycled by NADPH-consuming glutathione reductase (Figure 2). APX is directly involved in the defense response against oxidative stress. In pea and radish, APX activities are induced by salt and drought stress (Hernández *et al.*, 1995; Lopez *et al.*, 1996). Winter acclimated pine needles contain up to 65-fold more APX activity than summer needles, whereas in maize activities (together with SOD) are constitutively higher in a chilling-resistant line than in a sensitive one (Ievinsh *et al.*, 1995; Hodges *et al.*, 1996). In wheat roots and rice seedlings, APX activities rise after anoxic stress (Biemelt *et al.*, 1998; Ushimaru *et al.*, 1997). In wheat a correct temporal expression of APX is an important factor for efficient seed germination. APX activities increase during germination in parallel with the rise of ascorbate levels (De Gara *et al.*, 1997).

Based on the available sequence data, seven different APXs are distinguished in plants: two soluble cytosolic forms, three types of cytosol membrane-bound, including a glyoxisome-bound form, one chloroplastic stromal, and one thylakoid membrane-bound APX (Jespersen *et al.*, 1997). The various isoforms are different in several molecular and enzymatic properties, such as molecular weight, electron donor specificity, lability in the absence of ascorbate, pH optimum, and ascorbate and H_2O_2 affinity (Creissen *et al.*, 1994). In general, the chloroplastic isoforms are very specific for ascorbate as electron donor, whereas the cytosolic APX can also oxidize pyrogallol (Koshiba, 1993).

Cytosolic APX

Cytosolic APX (cAPX) cDNAs have been reported from several plant species (Jespersen *et al.*, 1997). Translation products of the cAPX cDNAs have in general an overall sequence similarity of 90%. Transcript levels are induced by drought, sulfur dioxide, ozone fumigation, iron stress, and excess light (Mittler and Zilinskas, 1994; Conklin and Las, 1995; Kubo *et al.*, 1995; Karpinski *et al.*, 1997; Örvar and Ellis, 1997a; Vansuyt *et al.*, 1997). In *Arabidopsis*, the two cAPX genes are regulated by the redox status of the plastoquinone pool (Karpinski *et al.*, 1997, 1999). In several cases, APX activity is posttranscriptionally regulated. The levels of APX transcript in radish were almost identical in salt-treated and control plants in both leaves and roots. However, in leaves and in roots, a 2-fold and a 7-fold increase in APX activity was observed, respectively (Lopez *et al.*, 1996). Additionally, during recovery of drought stress in pea, APX activities are posttranslationnally regulated: levels of cAPx protein remain identical, while the activity rises (Mittler and Zilinskas, 1994). Upon pathogen attack, a posttranscriptional suppression of cAPX enables the accumulation of H_2O_2 and the acceleration of programmed cell death pathways (Mittler *et al.*, 1998).

Chloroplastic APX (Stromal and Thylakoid bound)

Stromal APX (sAPX) cDNAs were isolated from *A. thaliana*, spinach, and pumpkin (Ishikawa *et al.*, 1996; Jespersen *et al.*, 1997). The deduced amino acid sequences are very homologous to the thylakoid-bound isoforms: only one residue is different in the N-terminal 364 amino acids in the spinach isoform and 84% identity in the *A. thaliana* deduced protein sequences. The only apparent difference is the absence of the C-terminal extension that contains the hydrophobic membrane-spanning domain. The cDNAs of sAPX and thylakoid-bound APX (tAPX) originate from a single-copy gene and both APXs are produced by alternative splicing (Ishikawa *et al.*, 1997; Mano *et al.*, 1997).

In *A. thaliana*, spinach, and pumpkin, cDNAs that encode a tAPX have been isolated (Ishikawa *et al.*, 1996; Yamaguchi *et al.*, 1996; Jespersen *et al.*, 1997). Their mature coding regions are more or less 50% similar with their cytosolic homologues. The tAPX isoforms have an amino-terminal signal peptide, which shares the characteristics of a chloroplast transit peptide. As in all other chloroplast transit peptides, there is no sequence homology among the transit peptides of the different species. Besides the chloroplast transit peptide, all three have also a carboxyl-terminal extension of 90 residues. The amino acid sequences are highly similar (up to 65% identical) to each other and they all possess a hydrophobic region, which is a putative membrane-spanning domain. The tAPXs are probably anchored in the thylakoid membrane via their C-terminal membrane-spanning domain. This observation is in contrast with other known thylakoid proteins, which possess their membrane anchor domain at the N-terminus or within the mature protein. If these C-terminal domains are actually responsible for the thylakoidic localization, it would be the first class of chloroplast-targeting peptides that share sequence homology. For engineering stress tolerance in plants, tAPXs could be interesting target enzymes. The conspicuous presence of anchored H_2O_2 scavengers in the immediate presence of AOS production in the thylakoid membranes would diminish leakage of chloroplastic H_2O_2 in the cytoplasm and prevent further damage in the cytosol.

APX in Microbodies and Mitochondria

In mitochondria and peroxisomes of pea leaves, APX activity was demonstrated, as well as the occurrence of all other enzymatic activities of the ascorbate-glutathione cycle. Intact mitochondria and microbodies had no latent APX activity, which is an indication that the active site of the APX is exposed to the cytosol and scavenges H_2O_2 leaking from microbodies and mitochondria (Jiménez *et al.*, 1997). APX activity was also found in the glyoxisomal membranes of pumpkin, cotton, and spinach. The corresponding cDNAs code for 32 kDa proteins with a single, putative membrane-spanning domain near the carboxyl-terminal end (Yamaguchi *et al.*, 1995; Bunkelmann and Trelease, 1996, 1997; Ishikawa *et al.*, 1998; Zhang *et al.*, 1998).

In maize, a monomeric 28 kDa APX isozyme was purified and partially sequenced. It is abundantly present in the cytosol of young tissues of maize, such as coleoptiles, mesocotyls, young leaves, and roots. In mature green leaves, small amounts of the enzyme are distributed in the vascular system (Koshiba, 1993). The N-terminal amino acid sequence of this protein is different from the translation product of a cDNA sequence obtained by screening a maize seedling cDNA library (Van Breusegem *et al.*, 1995). No detailed expression data is currently available. Although APX activity was reported in maize chloroplasts, no corresponding sequences are published.

CATALASES

Plants, unlike animals, have multiple forms of catalase (H_2O_2:H_2O_2 oxidoreductase; EC 1.11.1.6) that are mainly found in peroxisomes and glyoxisomes. Catalase activity was also found in the mitochondria in maize (Willekens *et al.*, 1994a; Guan and Scandalios, 1996). Catalases directly consume H_2O_2 or oxidize substrates (R), such as methanol, ethanol, formaldehyde and formic acid.

$$2H_2O_2 \xrightarrow{\text{catalase}} 2H_2O + O_2$$

$$H_2O_2 + RH_2 \xrightarrow{\text{catalase}} 2H_2O + R$$

The organization of the catalase gene family in different species has some striking similarities. Our laboratory showed that catalases can be divided into three classes according to their expression (Willekens *et al.*, 1995). The transition from glyoxisomes to leaf peroxisomes during seedling development is associated with the disappearance of class III catalases and the induction of class I catalases. In maize, however, both class I and class III are expressed in seeds, indicating that the class I catalase has a dual function in maize. Class I is most prominent in photosynthetic tissues, where they are involved in the removal of photorespiratory H_2O_2. Class II catalases are highly expressed in vascular tissues where they might play a role in lignification, but their exact biological role remains unknown. As mentionned above, class III is only abundant in seeds and young seedlings and its activity is linked to the removal of excessive H_2O_2 that is produced during fatty acid degradation in the glyoxylate cycle in the glyoxisomes.

Since catalase isozymes are rapidly induced by UV-B, ozone, and also chilling, they may play a direct role in stress protection (Willekens *et al.*, 1994b; Auh and Scandalios, 1997). In *A. thaliana*, *N. plumbaginifolia*, rice, and maize sequences were isolated for the three different classes (Mori *et al.*, 1992; Morita *et al.*, 1994; Willekens *et al.*, 1994b; Frugoli *et al.*, 1996; Guan and Scandalios, 1996; Higo and Higo, 1996).

In maize, the three isozymes are temporally and spatially regulated, but they are also influenced by environmental factors. Heat stress, fungal infections, high osmoticum, and cold upregulate catalase transcript and protein (activity) levels (Auh and Scandalios, 1997; Guan and Scandalios, 1998a, 1998b; for a detailed review, see Scandalios, 1994). All three maize catalases consist of four monomers with a molecular mass of 60 kDa and are structurally similar to those found in other organisms. They are approximately 65% similar to each other on the amino acid level and are capable of interacting to form functional heterotetramers. The maize class I catalase (nominated CAT-2) is highly abundant in glyoxisomes and peroxisomes of seeds and bundle sheath cells of young plants. The maize class II (CAT-3) isoform is biochemically different from other plant catalases: it has an enhanced peroxidatic activity and it is localized within the mitochondria of bundle sheath cells. CAT3 has a 70-fold higher peroxidatic activity, suggesting that CAT3 has other metabolic functions besides H_2O_2 scavenging.

ENVIRONMENTAL STRESSES AFFECT THE AOS-SCAVENGING MACHINERY IN MAIZE

In maize, correlative studies have been reported between environmental stress tolerance and enhanced levels of AOS-scavenging enzymes. Stress-resistant maize genotypes often have higher antioxidant capacities than sensitive ones. In combination with carbohydrate concentrations, the activities of catalase, MDHAR, and APX serve as indicators of chilling stress in a rapid screening technique for the chilling sensitivity of maize (Hodges *et al.*, 1997).

The specific activities of SOD, glutathione reductase (GR), APX, and DHAR at 5°C are higher in the cold-tolerant *Zea diploperennis* than in the cold-sensitive *Zea mays*. In addition, total SOD and APX activities are twofold higher in the resistant line, both at normal and chilling temperatures. These differences might be related to the differential resistance to chilling-dependent photoinhibition (Jahnke *et al.*, 1991). Chilling tolerance can be induced in sensitive plants by chemical treatments. Paclobutrazol, a member of the growth regulator family of triazole, induces a variety of morphological and biochemical responses in plants, including retarded shoot elongation, stimulated rooting, and protection from various environmental stresses. Triazole-treated plants exhibit increased tolerance to SO_2, heat stress, methyl viologen, and chilling (Davis and Curry, 1991; Kraus *et al.*, 1995). The mechanism for this induced stress tolerance is not clear, but it is related with increased antioxidant levels. In maize, the paclobutrazol-mediated chilling tolerance correlates with an induction of APX and SOD activities in leaves (Pinhero *et al.*, 1997). When seedlings and callus tissues of drought-resistant and drought-sensitive maize cultivars were treated with brassinolide, uniconazole, and methyl jasmonate the activities of SOD, catalase and APX increased and remained higher in the resistant cultivar, whereas the levels remained unchanged in the sensitive cultivar.

The changes in AOS-scavenging activities are often very specific and limited within certain parts of the plant. In chilled maize seedlings, none of the antioxidant enzymes are significantly affected in the coleoptile, leaf, or root. However, in the mesocotyl, which is most affected during chilling stress, CAT3 is elevated and prevents the accumulation of deleterious levels of H_2O_2. Acclimation of young maize seedlings can be obtained by pre-exposing them for a short period to chilling temperatures. The mild oxidative stress situation that is then created induces the CAT3 isoform, which protects the seedlings from a subsequent more severe oxidative stress. More than 60% of the acclimated seedlings survived an extended chilling stress, whereas all non-acclimated plants died. The direct contribution of CAT3 to chilling tolerance was shown by treating the seedlings with aminotriazole. This herbicide inhibits specifically CAT3. Aminotriazole-treated seedlings could not be acclimated against a subsequent chilling stress (Prasad *et al.*, 1994a; Prasad, 1996).

Interestingly, Doulis *et al.* (1997) reported on the differential localization of antioxidants in maize. Total SOD and APX activity in maize can only be detected in bundle sheath cells, whereas GR and DHAR are almost exclusively located in the mesophyll tissue. The capacity of each cell type to destroy AOS may be important for the tolerance of maize to stresses, such as chilling. Hence, this partitioning of antioxidants probably contributes seriously to the chilling sensitivity of maize. The absence of SOD and APX in the mesophyll cells is a disadvantage during stress conditions because mesophyll cells contain chloroplasts with both PSI and PSII, in contrast with the bundle sheath cells that are deficient in PSII and, hence, exhibit little net O_2 evolution. Consequently, the majority of AOS in maize leaves is formed in the mesophyll cells.

TRANSGENIC PLANTS WITH MODIFIED ANTIOXIDANT ENZYME LEVELS

Because of the involvement of AOS in a wide variety of environmental stresses, antioxidant enzymes are interesting molecular targets for the production of new plant varieties that can cope with these stresses. Several antioxidative stress enzymes have been genetically engineered into plants to assess their potential capacity for enhancing oxidative stress tolerance (Table 2). The beneficial effects observed in some of these transgenic plants can lead to interesting agronomic applications.

The tolerance towards oxidative stress is often tested *in vitro* with the use of the light-activated herbicide methyl viologen (MV). In the light, MV becomes an electron acceptor from PSI, subsequently reducing dioxygen to $O_2^{\cdot -}$. In this way, the herbicide strongly enhances the formation of superoxide radicals and is hence a fairly good mimic of the process of superoxide formation as it occurs *in vivo* in illuminated chloroplasts (Bowler *et al.*, 1991). MV accepts also electrons from the respiratory electron transport chain in the mitochondria and forms superoxide radicals in the dark as well. *In vitro* tolerance towards MV correlates with enhanced stress tolerance in controlled conditions and in the field. Overproduction of *Nicotiana plumbaginifolia* MnSOD in alfalfa confers resistance to freezing stress, whereas tobacco plants overproducing the same construct show visible ozone tolerance (Van Camp *et al.*, 1994b; McKersie *et al.*, 1996, 1999). A correlation between MV tolerance and chilling tolerance was also found in transgenic tobacco plants that produce a chloroplast-localized Cu/ZnSOD from pea (Sen Gupta *et al.*, 1993a, 1993b).

A 30-fold increase in chloroplastic Cu/ZnSOD activity in transgenic tobacco did however not confer resistance towards different concentrations of MV (0.1–100 μM) (Tepperman and Dunsmuir, 1990). Similarly, transgenic tomato plants, with 4-fold increases in Cu/ZnSOD activity did not exhibit tolerance to low temperatures or low CO_2 concentrations, in combination with high light intensities (Tepperman and Dunsmuir, 1990). This lack of induced tolerance was explained by the inability of the plants to cope with the elevated levels of H_2O_2 that were produced by the enhanced dismutation of superoxide radicals. Pitcher *et al.* (1991) also found that the same plants were not protected

Table 2. Stress-tolerant transgenic plants overproducing or underproducing AOS-scavenging enzymes

Transgenic plant	Target enzyme	Ectopic location	Stress assessment	Tolerance	Reference
A. With enhanced AOS-scavenging capacities					
Tobacco	Petunia chloroplastic Cu/ZnSOD	Chloroplast	Ozone	No	Pitcher et al. (1991)
			Paraquat	No	Tepperman and Dunsmuir (1990)
	N. plumbaginifolia MnSOD	Chloroplast	Methyl viologen	Less sensitive (light and dark)	Bowler et al. (1991)
		Mitochondria	Ozone	Reduced visible injury	Van Camp et al. (1994b)
	A. thaliana FeSOD	Chloroplast	Methyl viologen	Increased tolerance	Van Camp et al. (1996)
	Pea glutathione reductase	Cytosol	Methyl viologen	Increased tolerance	Broadbent et al. (1995)
		Chloroplast	Methyl viologen	Increased tolerance	
		Cytosol and chloroplasts	Ozone	Less sensitive	
	Spinach chloroplastic Cu/ZnSOD	Chloroplast	Methyl viologen	Increased tolerance	Hironori (1997)
	Pea chloroplastic Cu/ZnSOD	Chloroplast	Chilling	Higher photosynthetic rates	Sen Gupta et al. (1993a)
	Pea cytosolic Cu,ZnSOD	Cytosol	Methyl viologen	Increased resistance	Pitcher and Zilinskas (1996)
			Ozone	Partial resistance to foliar necrosis	
	A. thaliana APX	Chloroplast	Aminotriazole	Complete protection	L. Slooten (unpublished data)
			Ozone	No protection	Orvar and Ellis (1997b)
	Pea APX	Chloroplast	Ozone	No protection	Torsethaugen et al. (1997)
	E. coli GR × rice Cu/ZnSOD	Cytosol	Methyl viologen	Increased resistance	Aono et al. (1995a)
	Tobacco GST/GPX	Cytosol	Chilling/salt stress	Enhanced growth	Roxas et al. (1997)
Tomato	Petunia Cu/ZnSOD	Chloroplast	Chilling	No	Tepperman and Dunsmuir (1990)
Potato	E. coli GR	Chloroplast	Low CO_2	No	Brüggemann et al. (1999)
			Chilling	No	
	Tomato cytosolic and chloroplastic Cu/ZnSOD	Cytosol and chloroplasts	Methyl viologen	Less chlorosis and wilting	Perl et al. (1993)
				Enhanced root growth	

Table 2. (continued)

Transgenic plant	Target enzyme	Ectopic location	Stress assessment	Tolerance	Reference
Alfalfa	N. plumbaginifolia MnSOD	Chloroplast and Mitochondria	Acifluorfen	Increased tolerance	McKersie et al. (1993, 1996)
			Freezing	Faster regrowth	
			Winter survival	Yes	McKersie et al. (1999)
Poplar	E. coli GR	Chloroplast	Methyl viologen	Increased anti-oxidant capacity	Foyer et al. (1995)
Maize	A. thaliana FeSOD	Chloroplast	Methyl viologen	Yes	Arisi et al. (1998)
	N. plumbaginifolia MnSOD	Chloroplast	Paraquat	Increased tolerance	Van Breusegem et al. (1999a)
			Chilling	Tendencies for better growth	
	A. thaliana FeSOD	Chloroplast	Paraquat	Increased tolerance	Van Breusegem et al. (1999b)
			Chilling	Better growth	
Cotton	N. plumbaginifolia MnSOD	Chloroplast	Chilling/high light	No	Payton et al. (1997)

B. With lower scavenging capacities

		Effect			
Tobacco	APX	Higher susceptibility for ozone injury			Örvar and Ellis (1997a)
Tobacco	GR	Increased paraquat sensitivity			Aono et al. (1995b)
Tobacco	CAT	Development of necrosis			Takahashi et al. (1997), Chamnongpol et al. (1996), Willekens et al. (1997)

against ozone stress. In contrast, transgenic potato plants producing tomato cytosolic or chloroplastic Cu/ZnSOD lacked a chlorotic and wilting phenotype upon MV treatment. In medium with 10 μM MV, root cultures from the same transgenic potatoes grew at rates similar to those in control medium (Perl *et al.*, 1993). Transgenic tobacco plants overproducing spinach chloroplastic Cu/ZnSOD had 2- to 3-fold SOD activity and also enhanced endogenous APX activity (3–4 fold). These plants were more tolerant to 10 μM MV (Hironori, 1997). Overproduction of pea cytosolic Cu/ZnSOD in transgenic tobacco confers partial resistance to ozone-induced foliar necrosis (Pitcher and Zilinskas, 1996). Up to six-fold enhancement of SOD activity led to 30% to 50% less visible foliar damage. The beneficial effect of the transgene could however, only be detected in older leaves (leaf 3 to 5). In the youngest leaves (5 and 6), no difference between transgenic and wild-type plants could be observed. Since Cu/ZnSODs are sensitive to H_2O_2, it is possible that some of the introduced Cu/ZnSOD activity is inhibited by its own end product. In this way, MnSOD, which is insensitive to H_2O_2, could be a better candidate for engineering stress tolerance.

MnSOD from *Nicotiana plumbaginifolia* has often been reported to confer resistance towards oxidative stress. Transgenic tobacco plants overproducing MnSOD in mitochondria or chloroplasts were less sensitive to MV in both dark and light conditions (Bowler *et al.*, 1991). When the same plants were subjected to ozone fumigation, enhanced MnSOD activity in the mitochondria had only a minor effect on ozone tolerance. However, overproduction of MnSOD in the chloroplasts resulted in a 3- to 4-fold reduction of visible ozone injury (Van Camp *et al.*, 1994b). Transgenic alfalfa plants with twice the amount of total SOD activity by overproducing *Nicotiana plumbaginifolia* MnSOD in the chloroplast had increased tolerance against acifluorfen. Acifluorfen is a photobleaching, *p*-nitrodiphenylether herbicide that promotes accumulation of the chlorophyll precursor protoporphyrin; in the light it generates singlet oxygen that causes peroxidation in the tonoplast, plasmalemma, and chloroplast envelope. After freezing stress, transgenic alfalfa plants had also a more rapid regrowth (McKersie *et al.*, 1993, 1996, 1999).

Overproduction of an *Arabidopsis thaliana* APX in tobacco chloroplasts provided an almost complete protection of the PSII reaction center against aminotriazole. Aminotriazole inhibits catalase and in this way provokes accumulation of H_2O_2. Tolerance to MV was slightly better but tolerance to eosin (a singlet oxygen generator) or chilling-induced photoinhibition was not enhanced (L. Slooten, personal communication). A 50% increase in APX activity in an ozone-sensitive transgenic tobacco did not decrease ozone injury (Örvar and Ellis, 1997b).

In transgenic tomatoes with approximately 60-fold increased GR activity, the chilling sensitivity of the photosynthetic apparatus was identical to that of wild-type plants (Brüggemann *et al.*, 1999). In contrast, the enhancement of the pool of GSSG by overproducing GR increased the total antioxidant capacity of the leaves and in two cases led to increased resistance towards oxidative stress (Broadbent *et al.*, 1995; Foyer *et al.*, 1995). The above described experiments indicate that overproduction of AOS scavengers by genetic engineering is a valuable strategy to make plants less sensitive to several environmental stress situations.

By combining superoxide scavengers and H_2O_2 scavengers even better results can be envisaged, the most spectacular results having been reported in insects. In *Drosophila melanogaster*, the overproduction of SOD and catalase resulted in a delay of aging and a greater longevity whereas overproduction of either of the two enzymes alone had only minor effects (Orr and Sohal, 1994; Sohal *et al.*, 1995). Transgenic tobacco overproducing an *Escherichia coli* GR or a rice Cu/ZnSOD were 5-fold more resistant towards ion leakage caused by MV treatments. Crossings between both transgenics were highly tolerant to MV concentrations of 50 μM, whereas control plants were sensitive to concentrations as low as 1 μM MV (Aono *et al.*, 1995a).

Underproduction of antioxidative stress enzymes often leads to increased sensitivity towards the experienced stress. Tobacco plants with decreased APX or GR activity, obtained via antisense technology have enhanced susceptibility to ozone stress and MV, respectively (Aono *et al.*, 1995b; Örvar and Ellis, 1997a). Transgenic plants deficient in catalase (class I) can only be grown under low light conditions (< 100 μmol m^{-2} s^{-1} photosynthetic photon fluence rate). When exposed to higher light intensities, these plants developped white necrotic lesions on the leaves after 1 to 2 days. Lesion formation was induced by photorespiration, because damage was prevented under elevated CO_2. Stress analysis revealed that Cat1-deficient plants were more sensitive to paraquat, salt, and ozone stress, indicating that Cat1 is a key component of several stress defenses (Chamnongpol *et al.*, 1996; Willekens *et al.*, 1997).

WHY ENHANCE CHILLING TOLERANCE IN MAIZE?

Maize can be considered the most important agricultural crop in Europe and Northern America. Estimated production worldwide of maize (grain) in 1997 was approximately 565 million tons. In 2005, the production will have increased by 10%, giving a total of approximately 670 million tons (www.europa.eu.int/comm/dg06/). Within the European Union, 24% of the cereals used for animal feeding is maize. In 1995, maize was grown on 7.5×10^6 hectares. Being such an economically important crop, maize has been the subject of extensive research. Research programs focus on the improvement of maize (grain quality and nutritional value) and on resistance to pests and diseases. Over the past decades, these research programs mostly used conventional breeding techniques to improve quality of maize. Because maize originates from subtropical regions, it is not surprising that it is very sensitive to environmental stresses, such as chilling and freezing. After expansion of maize growth areas towards Northern climates (Northern Europe and North America), acclimation tolerance to chilling conditions became a major research target. Optimal growth conditions for maize are between 20–30°C. However, in Northern Europe, temperatures of 4°C to 15°C are not rare in the early growing season. Moreover, the combination of high light intensities and low temperatures, such as those experienced on cold but sunny mornings in spring, can cause dramatic damage to young maize seedlings. Low temperature stress is an important determinant for a total area of 1.3×10^6 hectares of maize grown in Northern parts of Europe. The average yearly losses have been estimated to 18 million Euro. Significant economic losses due to low-temperature stress are encountered in Northern France, Belgium, the United Kingdom, Ireland, Germany, Denmark, and The Netherlands. Conventional breeding programs in conjunction with transformation technologies are now being employed to overcome problems in those areas with suboptimal growing temperatures. Four of the major advantages to produce chilling-tolerant maize lines are: (i) production of the same quantity on fewer hectares, thereby reducing production costs; (ii) growth in regions where, until now, no or limited areas of maize could be cultivated; (iii) higher resistance to parasites, thus considerably reducing chemical treatments; and (iv) the ability of sowing maize in early spring would avoid hot dry summer periods during pollination and fertilization.

As many agriculturally important traits, chilling resistance is genetically complex and polygenic. In addition, a limitation of conventional breeding approaches for the improvement of chilling tolerance in commercial hybrids is the use of yield (also a genetically complex trait) as the main selection index and it takes 10 to 15 years for one typical crop improvement cycle (Pauls *et al.*, 1995). Genetic engineering can greatly contribute to crop improvement programs and will accelerate the production of chilling-tolerant lines. The identification of specific genes involved in chilling tolerance, the availability of various methods for the production of transgenic maize, and the development of *in vitro* assays for the rapid measurement of tolerance will save considerable time on the itinerary towards production of stress-tolerant maize.

ENGINEERING STRESS TOLERANCE IN MAIZE

One of the research projects within our laboratory focuses on the role of SODs and APX in maize during chilling stress. The working hypothesis is that maize lines with an improved scavenging system for oxygen radical will be better protected against chilling stress. To this end, the fundamental insights into the molecular biology of the oxidative stress defense systems explored in model plants, such as tobacco and *Arabidopsis* (see previous sections) are applied to a more economically important crop, maize. Target enzymes for transformation were MnSOD (*N. plumbaginifolia*), FeSOD (*A. thaliana* and *Z. mays*), and APX (*Z. mays*). Transgenic maize has been generated with particle bombardment that overproduce a *N. plumbaginifolia* MnSOD cDNA fused to a chloroplast transit peptide from pea under control of the cauliflower mosaic virus 35S (CaMV35S) promoter. The recombinant MnSOD was correctly targeted to chloroplasts and its enzymatic activity could be distinguished on SOD activity gels. One transgenic line showed enhanced tolerance to MV. The growth characteristics of transgenic maize lines were followed during growth at ambient (22–25°C) and chilling temperatures (15–17°C). Although the transgenic lines in all experiments had a growth advantage compared to the wild-type lines, no statistically significant increase in growth could be observed (Van Breusegem *et al.*, 1999a). To extrapolate the *in vitro* oxidative stress tolerance to improved growth effects during environmental stress conditions, such as chilling, we initiated field trials at different places in Europe in order to scale up the experiment and to check the behavior of the transgenic plants under natural environmental stress conditions.

Immunolocalization experiments revealed that in transgenic maize, the recombinant MnSOD is mainly (although not exclusively) located in the chloroplasts of the bundle sheath cells (Van Breusegem *et al.*, 1998). The low levels of MnSOD in the mesophyll cells could be attributed to a different expression capacity of the CaMV35S promoter, but posttranscriptional or posttranslational regulation cannot be excluded. Wilson *et al.* (1995) have also shown the vascular-specific activity of β-glucuronidase (GUS) in leaves of maize plants transformed with a CaMV35S-GUS fusion. Since endogenous SOD and APX activities are restricted to the bundle sheath cells, whereas GR and DHAR activities could only be detected in the mesophyll tissue in maize, a clear partitioning of the AOS defense system between the mesophyll and bundle sheath cells must be present (Doulis *et al.*, 1997). This differential localization correlates with the need for NADPH of the respective enzymes. Because NADPH is limited in the bundle sheath cells, GR and DHAR activities are rate limited in this compartment. This differential distribution of antioxidants is of course crucial for maize plants to deal with oxidative stress. The overproduction of AOS scavengers within each cell type could restore this natural imbalance and, hence, confer a higher tolerance against chilling-associated oxidative stress to maize plants. Exclusive presence of the transgenic MnSOD in the bundle sheath cells would probably have influenced the expected protective effect of MnSOD in the transgenic maize plants during chilling stress.

Transgenic maize lines overproducing an *Arabidopsis thaliana* FeSOD in the chloroplasts suffered also less from paraquat damage than controls as indicated by decreased membrane leakiness and by higher photosynthetic activity. In contrast with the MnSOD lines, the transgenic FeSOD maize plants exhibited also a significantly increased growth rate at low temperatures (as estimated from fresh weight and summed leaf length determinations) (Van Breusegem *et al.*, 1999b). These and previous results in tobacco suggest that FeSOD is a better candidate enzyme for protection of plants against oxidative stress. The reason for this fact could be the difference in suborganellar location between the overproduced MnSOD and FeSOD. Because of their chloroplastic and mitochondrial origin, respectively, FeSOD and MnSOD might have different properties. Van Camp *et al.* (1996) showed that in tobacco the transgenic FeSOD is at least partially bound to the chloroplast membrane, whereas transgenic MnSOD behaves more like a stromal enzyme. This differential subcellular location might provoke different protective effects against oxygen radicals. Transgenic FeSOD might be able to bind electrostatically to the chloroplast membrane in the vicinity of the site of radical production, resulting in increased protective properties.

PERSPECTIVES

The production of crop plants that can cope with adverse environmental conditions is a very important research objective within the agro-industry. Improved production rates of crops during stress situations, such as drought and chilling or the resistance against pathogen attacks will certainly improve the life quality of the ever-growing world population in the next century. The rationale for the production of chilling-tolerant maize lines is to reduce the yearly losses during the early growing season and to expand the growing range within Northern Europe. In the production process of chilling-tolerant maize, two main routes can be followed. The enhancement of lipid desaturation leads to a higher fluidity of the plant cell membranes during chilling stress. Ishizaki-Nishizawi *et al.* (1996) have shown that by overproducing an acyl lipid desaturase from *Anacystis nidulans* transgenic tobacco plants were better protected against chilling stress.

The second route is the study (and eventually modification) of the antioxidants in maize. Besides the production and evaluation of transgenic lines, we are now focusing on a detailed characterization of the AOS-scavenging machinery in maize. The apparent difference in SOD activity between bundle sheath and mesophyll cells could be a main cause for the chilling susceptibility of maize. Enhancing the levels of SOD in the mesophyll cells could lead to chilling-tolerant maize lines. Recent evidence indicates a dual role for H_2O_2. It can either serve as a signal molecule to induce AOS-scavenging enzymes during chilling acclimation or, when present at higher levels, act as a destructive molecule again. As in animal systems, H_2O_2 seems to play an important role in the signal transduction cascade of stress responses. Modulation of H_2O_2 levels is, hence, an interesting route to unravel and improve the oxidative stress signal transduction pathway(s).

The isolation of novel isoforms of AOS enzymes in maize will also lead to better insights into its defense mechanisms. The molecular characterization of several chloroplastic APX in *Arabidopsis thaliana* clearly shows that the AOS-scavenging enzyme families in plants are larger than previously thought (Jespersen *et al.*, 1997). We also characterized a novel chloroplastic FeSOD in maize. This isozyme is located in the chloroplast and is transcriptionally induced during growth at chilling conditions. Until now, it was thought that FeSODs were limited to dicotyledonous plants. With the help of rapid screening methods, such as different display techniques or through the outcome of several genome and cDNA sequence initiatives, new isoforms of the different AOS-scavenging enzyme families will certainly be discovered. Together with the characterization of the signal transduction pathways involved in oxidative stress, these novel isozymes will open new opportunities for the engineering of stress tolerance in plants.

REFERENCES

Acevedo, A. and Scandalios, J.G. (1992) Differential expression of the catalase and superoxide dismutase genes in maize ear shoot tissues. *Plant Cell Physiol.*, **33**, 1079–1088.

Akkapeddi, A.S., Shin, D.I., Stanek, M.T., Karnosky, D.F., and Podila, G.K. (1994) cDNA and derived amino acid sequence of the chloroplastic copper/zinc-superoxide dismutase from aspen (*Populus tremuloides*). *Plant Physiol.*, **106**, 1231–1232.

Alscher, R.G. and Hess, J.L. (1993) *Antioxidants in Higher Plants*, CRC Press, Boca Raton.

Alvarez, M.E., Pennell, R.I., Meijer, P.-J., Ishikawa, A., Dixon, R.A., and Lamb, C. (1998) Reactive oxygen intermediates mediate a systemic signal network in the establishment of plant immunity. *Cell*, **92**, 773–784.

Anonymous (1996) La reprise annoncée du commerce mondial. *AGPM info*, **285**, 2–3.

Aono, M., Saji, H., Sakamoto, A., Tanaka, K., Kondo, N., and Tanaka, K. (1995a) Paraquat tolerance of transgenic *Nicotiana tabacum* with enhanced activities of glutathione reductase and superoxide dismutase. *Plant Cell Physiol.*, **36**, 1687–1691.

Aono, M., Saji, H., Fujiyama, K., Sugita, M., Kondo, N., and Tanaka, K. (1995b) Decrease in activity of glutathione reductase enhances paraquat sensitivity in transgenic *Nicotiana tabacum*. *Plant Physiol.*, **107**, 645–648.

Arisi, A.-C.M., Noctor, G., Foyer, C.H., and Jouanin, L. (1997) Modification of thiol contents in poplars (*Populus tremula* x *P. alba*) overexpressing enzymes involved in glutathione synthesis. *Planta*, **203**, 362–372.

Auh, C.-K. and Scandalios, J.G. (1997) Spatial and temporal responses of the maize catalases to low temperature. *Physiol. Plant.*, **101**, 149–156.

Baum, J.A. and Scandalios, J.G. (1981) Isolation and characterization of the cytosolic and mitochondrial superoxide dismutases of maize. *Arch. Biochem. Biophys.*, **206**, 249–264.

Beauchamp, C.O. and Fridovich, I. (1971) Superoxide dismutase: improved assays and an assay applicable to acrylamide gels. *Anal. Biochem.*, **44**, 276–287.

Biemelt, S., Keetman, U., and Albrecht, G. (1998) Re-aeration following hypoxia or anoxia leads to activation of the antioxidative defense system in roots of wheat seedlings. *Plant Physiol.*, **116**, 651–658.

Bowler, C., Alliotte, T., De Loose, M., Van Montagu, M., and Inzé, D. (1989) The induction of manganese superoxide dismutase in response to stress in *Nicotiana plumbaginifolia*. *EMBO J.*, **8**, 31–38.

Bowler, C., Slooten, L., Vandenbranden, S., De Rycke, R., Botterman, J., Sybesma, C., Van Montagu, M., and Inzé, D. (1991) Manganese superoxide dismutase can reduce cellular damage mediated by oxygen radicals in transgenic plants. *EMBO J.*, **10**, 1723–1732.

Bowler, C., Van Camp, W., Van Montagu, M., and Inzé, D. (1994) Superoxide dismutase in plants. *CRC Crit. ev. Plant Sci.*, **13**, 199–218.

Breen, A.P. and Murphy, J.A. (1995) Reactions of oxyl radicals with DNA. *Free Rad. Biol. Med.*, **18**, 1033–1077.

Brisson, L.F., Tenhaken, R., and Lamb, C. (1994) Function of oxidative cross-linking of cell wall structural proteins in plant disease resistance. *Plant Cell*, **6**, 1703–1712.

Broadbent, P., Creissen, G.P., Kular, B., Wellburn, A.R., and Mullineaux, P.M. (1995) Oxidative stress responses in transgenic tobacco containing altered levels of glutathione reductase activity. *Plant J.*, **8**, 247–255.

Brüggemann, W., Beyel, V., Brodka, M., Poth, H., Weil, M., and Stockhaus, J. (1999) Antioxidants and antioxidative enzymes in wild-type and transgenic *Lycopersicon* genotypes of different chilling tolerance. *Plant Sci.*, **140**, 145–154.

Bueno, P., Varela, J., Giménez-Gallego, G., and del Río, L.A. (1995) Peroxisomal copper,zinc superoxide dismutase: characterization of the isoenzyme from watermelon cotyledons. *Plant Physiol.*, **108**, 1151–1160.

Bunkelmann, J.R. and Trelease, R.N. (1996) Ascorbate peroxidase. A prominent membrane protein in oilseed glyoxysomes. *Plant Physiol.*, **110**, 589–598.

Bunkelmann, J.R. and Trelease, R.N. (1997) Expression of glyoxysomal ascorbate peroxidase in cotton seedlings during postgerminative growth. *Plant Sci.*, **122**, 209–216.

Cannon, R.E. and Scandalios, J.G. (1989) Two cDNAs encode two nearly identical Cu/Zn superoxide dismutase proteins in maize. *Mol. Gen. Genet.*, **219**, 1–8.

Cannon, R.E., White, J.A., and Scandalios, J.G. (1987) Cloning of cDNA for maize superoxide dismutase 2 (SOD2). *Proc. Natl. Acad. Sci. USA*, **84**, 179–183.

Casano, L.M., Martín, M., and Sabater, B. (1994) Sensitivity of superoxide dismutase transcript levels and activities to oxidative stress is lower in mature-senescent than in young barley leaves. *Plant Physiol.*, **106**, 1033–1039.

Chamnongpol, S., Willekens, H., Langebartels, C., Van Montagu, M., Inzé, D., and Van Camp, W. (1996) Transgenic tobacco with a reduced catalase activity develops necrotic lesions and induces pathogenesis-related expression under high light. *Plant J.*, **10**, 491–503.

Chen, J.C., Wei, D.S., and Pan, S.M. (1997) Cloning and characterization of rice manganese superoxide dismutases. *Taiwania*, **42**, 53–62.

Conklin, P.L. and Last, R.L. (1995) Differential accumulation of antioxidant mRNAs in *Arabidopsis thaliana* exposed to ozone. *Plant Physiol.*, **109**, 203–212.

Creissen, G.P., Edwards, E.A., and Mullineaux, P.M. (1994) Glutathione reductase and ascorbate peroxidase. In C.H. Foyer and P.M. Mullineaux, (eds.), *Causes of Photooxidative Stress and Amelioration of Defense Systems in Plants*, CRC Press, Boca Raton, pp. 343–364.

Crowell, D.N. and Amasino, R.M. (1991) Nucleotide sequence of an iron superoxide dismutase complementary DNA from soybean. *Plant Physiol.*, **96**, 1393–1394.

Dat, J.F., Lopez-Delgado, H., Foyer, C.H., and Scott, I.M. (1998) Parallel changes in H_2O_2 and catalase during thermotolerance induced by salicylic acid or heat acclimation in mustard seedlings. *Plant Physiol.*, **116**, 1351–1357.

Davis, T.D. and Curry, E.A. (1991) Chemical regulation of vegetative growth. *Crit. Rev. Plant Sci.*, **10**, 151–188.

De Gara, L., de Pinto, M.C., and Arrigoni, O. (1997) Ascorbate synthesis and ascorbate peroxidase activity during the early stage of wheat germination. *Physiol. Plant.*, **100**, 894–900.

Delledonne, M., Xia, Y., Dixon, R.A., and Lamb, C. (1998) Nitric oxide functions as a signal in plant disease resistance. *Nature*, **394**, 585–588.

Desikan, R., Hancock, J.T., Coffey, M.J., and Neill, N.J. (1996) Generation of active oxygen in elicited cells of *Arabidopsis thaliana* is mediated by a NADPH oxidase-like enzyme. *FEBS Lett.*, **382**, 213–217.

Desikan, R., Reynolds, A., Hancock, J.T., and Neill, S.J. (1998) Harpin and hydrogen peroxide both initiate programmed cell death but have differential effects on defence gene expression in *Arabidopsis* suspension cultures. *Biochem. J.*, **330**, 115–120.

Desimone, M., Henke, A., and Wagner, E. (1996) Oxidative Stress induces partial degradation of the large subunit of ribulose-1,5-bisphosphate carboxylase/oxygenase in isolated chloroplasts of barley. *Plant Physiol.*, **111**, 789–796.

Dixon, R.A., Lamb, C.J., Masoud, S., Sewalt, V.J., and Paiva, N.L. (1996) Metabolic engineering: prospects for crop improvement through the genetic manipulation of phenylpropanoid biosynthesis and defense responses—a review. *Gene*, **179**, 61–71.

Doulis, A.G., Debian, N., Kingston-Smith A.H., and Foyer, C.H. (1997) Differential localization of antioxidants in maize leaves. *Plant Physiol.*, **114**, 1031–1037.

Fontaine, O., Huault, C., Pavis, N., and Billard, J.-P. (1994) Dormancy breakage of *Hordeum vulgare* seeds: effects of hydrogen peroxide and scarification on glutathione level and glutathione reductase activity. *Plant Physiol. Biochem.*, **32**, 677–683.

Foyer, C.H. and Halliwell, B. (1976) The presence of glutathione and glutathione reductase in chloroplasts: a proposed role in ascorbic acid metabolism. *Planta*, **133**, 21–25.

Foyer, C.H., Souriau, N., Perret, S., Lelandais, M., Kunert, K.-J., Pruvost, C., and Jouanin, L. (1995) Overexpression of glutathione reductase but not glutathione synthetase leads to increases in antioxidant capacity and resistance to photoinhibition in poplar trees. *Plant Physiol.*, **109**, 1047–1057.

Frugoli, J.A., Zhong, H.H., Nuccio, M.L., McCourt, P., McPeek, M.A., Thomas, T.L. and McClung, C.R. (1996) Catalase is encoded by a multigene family in *Arabidopsis thaliana* (L.) Heynh. *Plant Physiol.*, **112**, 327–336.

Greaves, J.A. (1996) Improving suboptimal temperature tolerance in maize—the search for variation. *J. Exp. Bot.*, **47**, 307–323.

Guan, L.Q. and Scandalios, J.G. (1996) Molecular evolution of maize catalases and their relationship to other eukaryotic and prokaryotic catalases. *J. Mol. Evol.*, **42**, 570–579.

Guan, L. and Scandalios, J.G. (1998a) Two structurally similar maize cytosolic superoxide dismutase genes, *Sod4* and *Sod4A*, respond differentially to abscisic acid and high osmoticum. *Plant Physiol.*, **117**, 217–224.

Guan, L. and Scandalios, J.G. (1998b) Effects of the plant growth regulator abscisic acid and high osmoticum on the developmental expression of the maize catalase genes. *Physiol. Plant.*, **104**, 413–422.

Hernández, J.A., Olmos, E., Corpas, F.J., Sevilla, F., and Del Río, L.A. (1995) Salt-induced oxidative stress in chloroplasts of pea plants. *Plant Sci.*, **105**, 151–167.

Higo, K. and Higo, H. (1996) Cloning and characterization of the rice *CatA* catalase gene, a homologue of the maize *Cat3* gene. *Plant Mol. Biol.*, **30**, 505–521.

Hironori, K. (1997) Responses of rice superoxide dismutase genes to oxidative stresses and tolerance of superoxide dismutase overproducing transgenic plants to paraquat treatment. *Plant Physiol.*, Supp. **114**, 102 (Abstract #438).

Hodges, D.M., Andrews, C.J., Johnson, D.A., and Hamilton, R.I. (1996) Antioxidant compound responses to chilling stress in differentially sensitive inbred maize lines. *Physiol. Plant.*, **98**, 685–692.

Hodges, D.M., Andrews, C.J., Johnson, D.A., and Hamilton, R.I. (1997) Antioxidant enzyme and compound responses to chilling stress and their combining abilities in differentially sensitive maize hybrids. *Crop Sci.*, **37**, 857–863.

Ievinsh, G., Valcina, A., and Ozola, D. (1995) Induction of ascorbate peroxidase activity in stressed pine (*Pinus sylvestris* L.) needles: a putative role for ethylene. *Plant Sci.*, **112**, 167–173.

Inzé, D., and Van Montagu, M. (1995) Oxidative stress in plants. *Curr. Opin. Biotechnol.*, **6**, 153–158.

Ishikawa, T., Sakai, K., Yoshimura, K., Takeda, T., and Shigeoka, S. (1996) cDNAs encoding spinach stromal and thylakoid-bound ascorbate peroxidase, differing in the presence or absence of their 3'-coding regions. *FEBS Lett.*, **384**, 289–293.

Ishikawa, T., Yoshimura, K., Tamoi, M., Takeda, T., and Shigeoka, S. (1997) Alternative mRNA splicing of 3'-terminal exons generates ascorbate peroxidase isoenzymes in spinach (*Spinacia oleracea*) chloroplasts. *Biochem. J.*, **328**, 795–800.

Ishikawa, T., Yoshimura, K., Sakai, K., Tamoi, M., Takeda, T., and Shigeoka, S. (1998) Molecular characterization and physiological role of a glyoxysome-bound ascorbate peroxidase from spinach. *Plant Cell Physiol.*, **39**, 23–34.

Ishizaki-Nishizawa, O., Fujii, T., Azuma, M., Sekiguchi, K., Murata, N., Ohtani, T., and Toguri, T. (1996) Low-temperature resistance of higher plants is significantly enhanced by a nonspecific cyanobacterial desaturase. *Nature Biotechnol.*, **14**, 1003–1006.

Jahnke, L.S., Hull, M.R., and Long, S.P. (1991) Chilling stress and oxygen metabolizing enzymes in *Zea mays* and *Zea diploperennis*. *Plant Cell Environm.*, **14**, 97–104.

Jespersen, H.M., Kjærsgård, I.V.H., Østergaard, L., and Welinder, K.G. (1997) From sequence analysis of three novel ascorbate peroxidases from *Arabidopsis thaliana* to structure, function and evolution of seven types of ascorbate peroxidase. *Biochem. J.*, **326**, 305–310.

Jiménez, A., Hernández, J.A., del Río, L.A., and Sevilla, F. (1997) Evidence for the presence of the ascorbate-glutathione cycle in mitochondria and peroxisomes of pea leaves. *Plant Physiol.*, **114**, 275–284.

Kaminaka, H., Morita, S., Yokoi, H., Masumura, T., and Tanaka, K. (1997) Molecular cloning and characterization of a cDNA for plastidic copper/zinc-superoxide dismutase in rice (*Oryza sativa* L.). *Plant Cell Physiol.*, **38**, 65–69.

Kaminaka, H., Morita, S., Tokumoto, M., Yokoyama, H., Masumura, T., and Tanaka, K. (1999) Molecular cloning and characterization of a cDNA for an iron-superoxide dismutase in rice (*Oryza sativa* L.). *Biosci. Biotechnol. Biochem.*, **63**, 302–308.

Karpinski, S., Wingsle, G., Olsson, O., and Hällgren, J.-E. (1992) Characterization of cDNAs encoding CuZn-superoxide dismutases in Scots pine. *Plant Mol. Biol.*, **18**, 545–555.

Karpinski, S., Escobar, C., Karpinska, B., Creissen, G., and Mullineaux, P.M. (1997) Photosynthetic electron transport regulates the expression of cytosolic ascorbate peroxidase genes in Arabidopsis during excess light stress. *Plant Cell*, **9**, 627–640.

Karpinski, S., Reynolds, H., Karpinska, B., Wingsle, G., Creissen, G., and Mullineaux, P. (1999) Systemic signaling and acclimation in response to excess excitation energy in *Arabidopsis*. *Science*, **284**, 654–657.

Kernodle, S.P. and Scandalios, J.G. (1996) A comparison of the structure and function of the highly homologous maize antioxidant Cu/Zn superoxide dismutase genes, *Sod4* and *Sod4A*. *Genetics*, **144**, 317–328.

Kliebenstein, D.J., Monde, R.-A., and Last, R.L. (1998) Superoxide dismutase in Arabidopsis: an eclectic enzyme family with disparate regulation and protein localization. *Plant Physiol.*, **118**, 637–650.

Koshiba, T. (1993) Cytosolic ascorbate peroxidase in seedlings and leaves of maize (*Zea mays*). *Plant Cell Physiol.*, **34**, 713–721.

Kraus, T.E., McKersie, B.D., and Fletcher, R.A. (1995) Paclobutrazol-induced tolerance of wheat leaves to paraquat may involve increased antioxidant enzyme activity. *J. Plant Physiol.*, **145**, 570–576.

Kubo, A., Saji, H., Tanaka, K., and Kondo, N. (1995) Expression of *arabidopsis* cytosolic ascorbate peroxidase gene in response to ozone or sulfur dioxide. *Plant Mol. Biol.*, **29**, 479–489.

Kurepa, J., Hérouart, D., Van Montagu, M., and Inzé, D. (1997) Differential expression of CuZn- and Fe-superoxide dismutase genes of tobacco during development, oxidative stress and hormonal treatments. *Plant Cell Physiol.*, **38**, 463–470.

Levine, A., Pennell, R.I., Alvarez, M.E., Palmer, R., and Lamb, C. (1996) Calcium-mediated apoptosis in a plant hypersensitive disease resistance response. *Curr. Biol.*, **6**, 427–437.

Lopez-Delgado, H., Dat, J.F., Foyer, C.H., and Scott, I.M. (1998) Induction of thermotolerance in potato microplants by acetylsalicylic acid and H_2O_2. *J. Exp. Bot.*, **49**, 713–720.

Lopez, F., Vansuyt, G., Casse-Delbart, F., and Fourcroy, P. (1996) Ascorbate peroxidase activity, not the mRNA level, is enhanced in salt-stressed *Raphanus sativus* plants. *Physiol. Plant.*, **97**, 13–20.

Mano, S., Yamaguchi, K., Hayashi, M., and Nishimura, M. (1997) Stromal and thylakoid-bound ascorbate peroxidases are produced by alternative splicing in pumpkin. *FEBS Lett.*, **413**, 21–26.

Mathews, M.C., Summers, C.B., and Felton, G.W. (1997) Ascorbate peroxidase: a novel antioxidant enzyme in insects. *Arch. Insect Biochem. Physiol.*, **34**, 57–68.

McKersie, B.D., Chen, Y., de Beus, M., Bowley, S.R., Bowler, C., Inzé, D., D'Halluin, K., and Botterman, J. (1993) Superoxide dismutase enhances tolerance of freezing stress in transgenic alfalfa (*Medicago sativa* L.). *Plant Physiol.*, **103**, 1155–1163.

McKersie, B.D., Bowley, S.R., Harjanto, E., and Leprince, O. (1996) Water-deficit tolerance and field performance of transgenic alfalfa overexpressing superoxide dismutase. *Plant Physiol.*, **111**, 1177–1181.

McKersie, B.D., Bowley, S.R., and Jones, K.S. (1999) Winter survival of transgenic alfalfa overexpressing superoxide dismutase. *Plant Physiol.*, **119**, 839–847.

Meier, B., Barra, D., Bossa, F., Calabrese, L., and Rotilio, G. (1982) Synthesis of either Fe- or Mn-superoxide dismutase with an apparently identical protein moiety by an anaerobic bacterium dependent on the metal supplied. *J. Biol. Chem.*, **257**, 13977–13980.

Miao, Z. and Gaynor, J.J. (1993) Molecular cloning, characterization and expression of Mn-superoxide dismutase from the rubber tree (*Hevea brasiliensis*). *Plant Mol. Biol.*, **23**, 267–277.

Mittler, R. and Zilinskas, B.A. (1994) Regulation of pea cytosolic ascorbate peroxidase and other antioxidant enzymes during the progression of drought stress and following recovery from drought. *Plant J.*, **5**, 397–405.

Mittler, R., Feng, X., and Cohen, M. (1998) Post-transcriptional suppression of cytosolic ascorbate peroxidase expression during pathogen-induced programmed cell death in tobacco. *Plant Cell*, **10**, 461–473.

Miyake, C. and Asada, K. (1994) Ferredoxin-dependent photoreduction of the monodehydroascorbate radical in spinach thylakoids. *Plant Cell Physiol.*, **35**, 539–549.

Mori, H., Higo, K.-i., Higo, H., Minobe, Y., Matsui, H., and Chiba, S. (1992) Nucleotide and derived amino acid sequence of a catalase cDNA isolated from rice immature seeds. *Plant Mol. Biol.*, **18**, 973–976.

Morita, S., Tasaka, M., Fujisawa, H., Ushimaru, T., and Tsuji, H. (1994) A cDNA clone encoding a rice catalase isozyme. *Plant Physiol.*, **105**, 1015–1016.

Murai, R. and Murai, K. (1996) Different transcriptional regulation of cytosolic and plastidic Cu/Zn-superoxide dismutase genes in *Solidago altissima* (Asteraceae). *Plant Sci.*, **120**, 71–79.

Noctor, G. and Foyer, C.H. (1998) Ascorbate and glutathione: keeping active oxygen under control. *Annu. Rev. Plant Physiol. Plant Mol. Biol.*, **49**, 249–279.

Ogawa, K., Kanematsu, S., Takabe, K., and Asada, K. (1995) Attachment of CuZn-superoxide dismutase to thylakoid membranes at the site of superoxide generation (PSI) in spinach chloroplasts: detection by immuno-gold labeling after rapid freezing and substitution method. *Plant Cell Physiol.*, **36**, 565–573.

Orr, W.C. and Sohal, R.S. (1994) Extension of life-span by overexpression of superoxide dismutase and catalase in *Drosophila melanogaster. Science*, **263**, 1128–1130.

Örvar, B.L. and Ellis, B.E. (1995) Isolation of a cDNA encoding cytosolic ascorbate peroxidase in tobacco. *Plant Physiol.*, **108**, 839–840.

Örvar, B.L. and Ellis, B.E. (1997a) Transgenic tobacco plants expressing antisense RNA for cytosolic ascorbate peroxidase show increased susceptibility to ozone injury. *Plant J.*, **11**, 1297–1305.

Orvar, B.L. and Ellis, B.E. (1997b) The effect of overexpression of ascorbate peroxidase on ozone-induced necrosis in the ozone-sensitive tobacco cultivar Bel-W3. *Plant Physiol.*, Supp., **114**, 103 (Abstract #440).

Pauls, K.P. (1995) Plant biotechnology for crop improvement. *Biotechnol. Adv.*, **13**, 673–693.

Payton, P., Allen, R.D., Trolinder, N., and Holaday, A.S. (1997) Over-expression of chloroplast-targeted Mn superoxide dismutase in cotton (*Gossypium hirsutum* L., cv. Coker 312) does not alter the reduction of photosynthesis after short exposure to low temperature and high light intensity. *Photosynthesis Res.*, **52**, 233–244.

Perl, A., Perl-Treves, R., Galili, S., Aviv, D., Shalgi, E., Malkin, S., and Galun, E. (1993) Enhanced oxidative-stress defense in transgenic potato expressing tomato Cu,Zn superoxide dismutases. *Theor. Appl. Genet.*, **85**, 568–576.

Perl-Treves, R. and Galun, E. (1991) The tomato Cu,Zn superoxide dismutase genes are developmentally regulated and respond to light and stress. *Plant Mol. Biol.*, **17**, 745–760.

Perl-Treves, R., Nacmias, B., Aviv, D., Zeelon, E.P., and Galun, E. (1988) Isolation of two cDNA clones from tomato containing two different superoxide dismutase sequences. *Plant Mol. Biol.*, **11**, 609–623.

Pinhero, R.G., Rao, M.V., Paliyath, G., Murr, D.P., and Fletcher, R.A. (1997) Changes in activities of antioxidant enzymes and their relationship to genetic and paclobutrazol-induced chilling tolerance of maize seedlings. *Plant Physiol.*, **114**, 695–704.

Pitcher, L.H., Brennan, E., Hurley, A., Dunsmuir, P., Tepperman, J.M., and Zilinskas, B.A. (1991) Overproduction of petunia chloroplastic copper/zinc superoxide dismutase does not confer ozone tolerance in transgenic tobacco. *Plant Physiol.*, **97**, 452–455.

Pitcher, L.H. and Zilinskas, B.A. (1996) Overexpression of copper/zinc superoxide dismutase in the cytosol of transgenic tobacco confers partial resistance to ozone-induced foliar necrosis. *Plant Physiol.*, **110**, 583–588.

Prasad, T.K. (1996) Mechanisms of chilling-induced oxidative stress injury and tolerance in developing maize seedlings: changes in antioxidant system, oxidation of proteins and lipids, and protease activities. *Plant J.*, **10**, 1017–1026.

Prasad, T.K., Anderson, M.D., Martin, B.A., and Stewart, C.R. (1994a) Evidence for chilling-induced oxidative stress in maize seedlings and a regulatory role for hydrogen peroxide. *Plant Cell*, **6**, 65–74.

Prasad, T.K., Anderson, M.D., and Stewart, C.R. (1994b) Acclimation, hydrogen peroxide, and abscisic acid protect mitochondria against irreversible chilling injury in maize seedlings. *Plant Physiol.*, **105**, 619–627.

Reed Scioli, J. and Zilinskas, B.A. (1988) Cloning and characterization of a cDNA encoding the chloroplastic copper/zinc-superoxide dismutase from pea. *Proc. Natl. Acad . Sci. USA*, **85**, 7661–7665.

Roxas, V.P., Smith, R.K. Jr, Allen, E.R., and Allen, R.D. (1997) Overexpression of glutathione S-transferase/ glutathione peroxidase enhances the growth of transgenic tobacco seedlings during stress. *Nature Biotechnol.*, **15**, 988–991.

Sakamoto, A., Ohsuga, H., Wakaura, M., Mitsukawa, N., Hibino, T., Masumura, T., Sasaki, Y., and Tanaka, K. (1990) Nucleotide sequence of cDNA for the cytosolic Cu/Zn-superoxide dismutase from spinach (*Spinacia oleracea* L.). *Nucleic Acids Res.*, **18**, 4923.

Sakamoto, A., Nosaka, Y., and Tanaka, K. (1993) Cloning and sequencing analysis of a complementary DNA for manganese-superoxide dismutase from rice (*Oryza sativa* L.). *Plant Physiol.*, **103**, 1477–1478.

Sakamoto, A., Okumura, T., Kaminaka, H., Sumi, K., and Tanaka, K. (1995a) Structure and differential response to abscisic acid of two promoters for the cytosolic copper/zinc-superoxide dismutase genes, *SodCc1*, and *SodCc2*, in rice protoplasts. *FEBS Lett.*, **358**, 62–66.

Sakamoto, A., Okumura, T., Kaminaka, H., and Tanaka, K. (1995b) Molecular cloning of the gene (*Sodcc1*) that encodes a cytosolic copper/zinc-superoxide dismutase from rice (*Oryza sativa* L.). *Plant Physiol.*, **107**, 651–652.

Scandalios, J.G. (1994) Regulation and properties of plant catalases. In C.H. Foyer and P.M. Mullineaux, (eds.), *Causes of Photooxidative Stress and Amelioration of Defense Systems in Plants*, CRC Press, Boca Raton, pp. 275–315.

Schinkel, H., Streller, S., and Wingsle, G. (1998) Multiple forms of extracellular superoxide dismutase in needles, stem tissues and seedlings of Scots pine. *J. Exp. Bot.*, **49**, 931–936.

Sen Gupta, A., Heinen, J.L., Holaday, A.S., Burke, J.J., and Allen, R.D. (1993a) Increased resistance to oxidative stress in transgenic plants that overexpress chloroplastic Cu/Zn superoxide dismutase. *Proc. Natl. Acad. Sci. USA*, **90**, 1629–1633.

Sen Gupta, A., Webb, R.P., Holaday, A.S., and Allen, R.D. (1993b) Overexpression of superoxide dismutase protects plants from oxidative stress. *Plant Physiol.*, **103**, 1067–1073.

Sohal, R.S., Agarwal, A., Agarwal, S., and Orr, W.C. (1995) Simultaneous overexpression of copper- and zinc-containing superoxide dismutase and catalase retards age-related oxidative damage and increases metabolic potential in *Drosophila melanogaster*. *J. Biol. Chem.*, **270**, 15671–15674.

Stallings, W.C., Pattridge, K.A., Strong, R.K., and Ludwig, M.L. (1984) Manganese and iron superoxide dismutases are structural homologs. *J. Biol. Chem.*, **259**, 10695–10699.

Streller, S. and Wingsle, G. (1994) *Pinus sylvestris* L. needles contain extracellular CuZn superoxide dismutase. *Planta*, **192**, 195–201.

Takahashi, H., Chen, Z., Du, H., Liu, Y., and Klessig, D.F. (1997) Development of necrosis and activation of disease resistance in transgenic tobacco plants with severely reduced catalase levels. *Plant J.*, **11**, 993–1005.

Tanaka, K. (1998) Gene structures and expression control of active oxygen scavenging enzymes in rice. In K. Satoh and N. Murata, (eds.), *Stress Responses of Photosynthetic Organisms: Molecular Mechanisms and Molecular Regulations*, Elsevier, Amsterdam, pp. 53–68.

Tepperman, J.M. and Dunsmuir, P. (1990) Transformed plants with elevated levels of chloroplastic SOD are not more resistant to superoxide toxicity. *Plant Mol. Biol.*, **14**, 501–511.

Torsethaugen, G., Pitcher, L.H., Zilinskas, B.A., and Pell, E.J. (1997) Overproduction of ascorbate peroxidase in the tobacco chloroplast does not provide protection against ozone. *Plant Physiol.*, **114**, 529–537.

Tsang, E.W.T., Bowler, C., Hérouart, D., Van Camp, W., Villarroel, R., Genetello, C., Van Montagu, M., and Inzé, D. (1991) Differential regulation of superoxide dismutases in plants exposed to environmental stress. *Plant Cell*, **3**, 783–792.

Ushimaru, T., Maki, Y., Sano, S., Koshiba, K., Asada, K., and Tsuji, H. (1997) Induction of enzymes involved in the ascorbate-dependent antioxidative system, namely, ascorbate peroxidase, monodehydroascorbate reductase and dehydroascorbate reductase, after exposure to air of rice (*Oryza sativa*) seedlings germinated under water. *Plant Cell Physiol.*, **38**, 541–549.

Van Breusegem, F., Villarroel, R., Van Montagu, M., and Inzé, D. (1995) Ascorbate peroxidase cDNA from maize. *Plant Physiol.*, **107**, 649–650.

Van Breusegem, F., Kushnir, S., Slooten, L., Bauw, G., Botterman, J., Van Montagu, M., and Inzé, D. (1998) Processing of a chimeric protein in chloroplasts is different in transgenic maize and tobacco plants. *Plant Mol. Biol.*, **38**, 491–496.

Van Breusegem, F., Slooten, L., Stassart, J.-M., Botterman, J., Moens, T., Van Montagu, M., and Inzé, D. (1999a) Effects of overexpression of tobacco MnSOD in maize chloroplasts on foliar tolerance to cold and oxidative stress. *J. Exp. Bot.*, **50**, 71–78.

Van Breusegem, F., Slooten, L., Stassart, J.-M., Moens, T., Botterman, J., Van Montagu, M., and Inzé, D. (1999b) Overproduction of *Arabidopsis thaliana* FeSOD confers oxidative stress tolerance to transgenic maize. *Plant Cell Physiol.*, **40**, 515–523.

Van Camp, W., Bowler, C., Villarroel, R., Tsang, E.W.T., Van Montagu, M., and Inzé, D. (1990) Characterization of iron superoxide dismutase cDNAs from plants obtained by genetic complementation in *Escherichia coli. Proc. Natl. Acad. Sci. USA*, **87**, 9903–9907.

Van Camp, W., Van Montagu, M., and Inzé, D. (1994a) Superoxide dismutases. In C.H. Foyer and P.M. Mullineaux, (eds.), *Causes of Photooxidative Stress and Amelioration of Defense Systems in Plants*, CRC Press, Boca Raton, pp. 317–341.

Van Camp, W., Willekens, H., Bowler, C., Van Montagu, M., Inzé, D., Reupold-Popp, P., Sandermann Jr, H., and Langebartels, C. (1994b) Elevated levels of superoxide dismutase protect transgenic plants against ozone damage. *Bio/technology*, **12**, 165–168.

Van Camp, W., Capiau, K., Van Montagu, M., Inzé, D., and Slooten, L. (1996) Enhancement of oxidative stress tolerance in transgenic tobacco plants overexpressing Fe-superoxide dismutase in chloroplasts. *Plant Physiol.*, **112**, 1703–1714.

Van Camp, W., Van Montagu, M., and Inzé, D. (1998) H_2O_2 and NO: redox signals in disease resistance. *Trends Plant Sci.*, **3**, 330–334.

Vansuyt, G., Lopez, F., Inzé, D., Briat, J.F., and Fourcroy, P. (1997) Iron triggers a rapid induction of ascorbate peroxidase gene expression in *Brassica napus*. *FEBS Lett.*, **410**, 195–200.

Wada, N., Kinoshita, S., Matsuo, M., Amako, K., Miyake, C., and Asada, K. (1998) Purification and molecular properties of ascorbate peroxidase from bovine eye. *Biochem. Biophys. Res. Commun.*, **242**, 256–261.

White, D.A. and Zilinskas, B.A. (1991) Nucleotide sequence of a complementary DNA encoding pea cytosolic copper/zinc superoxide dismutase. *Plant Physiol.*, **96**, 1391–1392.

White, J.A. and Scandalios, J. G. (1987) In vitro synthesis, importation and processing of Mn-superoxide dismutase (SOD-3) into maize mitochondria. *Biochim. Biophys. Acta*, **926**, 16–25.

White, J.A., Plant, S., Cannon, R.E., Wadsworth, G.J., and Scandalios, J.G. (1990) Developmental analysis of steady-state levels of Cu/Zn and Mn superoxide dismutase mRNAs in maize tissues. *Plant Cell Physiol.*, **31**, 1163–1167.

Willekens, H., Villarroel, R., Van Montagu, M., Inzé, D., and Van Camp, W. (1994a) Molecular identification of catalases from *Nicotiana plumbaginifolia* (L.). *FEBS Lett.*, **352**, 79–83.

Willekens, H., Van Camp, W., Van Montagu, M., Inzé, D., Sandermann Jr, H., and Langebartels, C. (1994b) Ozone, sulfur dioxide, and ultraviolet B have similar effects on mRNA accumulation of antioxidant genes in *Nicotiana plumbaginifolia* (L.). *Plant Physiol.*, **106**, 1007–1014.

Willekens, H., Inzé, D., Van Montagu, M., and Van Camp, W. (1995) Catalases in plants. *Mol. Breeding*, **1**, 207–228.

Willekens, H., Chamnongpol, S., Davey, M., Schraudner, M., Langebartels, C., Van Montagu, M., Inzé, D., and Van Camp, W. (1997) Catalase is a sink for H_2O_2 and is indispensable for stress defence in C_3 plants. *EMBO J.*, **16**, 4806–4816.

Wilson, H.M., Bullock, W.P., Dunwell, J.M., Ellis, J.R., Frame, B., Register III, J., and Thompson, J.A. (1995) Maize. In K. Wang, A. Herrera-Estrella, and M. Van Montagu, (eds.), *Transformation of Plants and Soil Microorganisms*, (Plant and Microbial Biotechnology Research Series, Vol. 3), Cambridge University Press, Cambridge, pp. 65–80.

Wong-Vega, L., Burke, J.J., and Allen, R.D. (1991) Isolation and sequence analysis of a cDNA that encodes pea manganese superoxide dismutase. *Plant Mol. Biol.*, **17**, 1271–1274.

Yamaguchi, K., Hayashi, M., and Nishimura, M. (1996) cDNA cloning of thylakoid-bound ascorbate peroxidase in pumpkin and its characterization. *Plant Cell Physiol.*, **37**, 405–409.

Yamaguchi, K., Mori, H., and Nishimura, M. (1995) A novel isoenzyme of ascorbate peroxidase localized on glyoxysomal and leaf peroxisomal membranes in pumpkin. *Plant Cell Physiol.*, **36**, 1157–1162.

Zhang, H., Wang, J., Nickel, U., Allen, R.D., and Goodman, H.M. (1997) Cloning and expression of an *Arabidopsis* gene encoding a putative peroxisomal ascorbate peroxidase. *Plant Mol. Biol.*, **34**, 967–971.

Zhu, D. and Scandalios, J.G. (1993) Maize mitochondrial manganese superoxide dismutases are encoded by a differentially expressed multigene family. *Proc. Natl. Acad. Sci. USA*, **90**, 9310–9314.

Zhu, D. and Scandalios, J.G. (1994) Differential accumulation of manganese-superoxide dismutase transcripts in maize in response to abscisic acid and high osmoticum. *Plant Physiol.*, **106**, 173–178.

Zhu, D. and Scandalios, J.G. (1995) The maize mitochondrial MnSODs encoded by multiple genes are localized in the mitochondrial matrix of transformed yeast cells. *Free Rad. Biol. Med.*, **18**, 179–183.

9 Early Events in Environmental Stresses in Plants— Induction Mechanisms of Oxidative Stress

Jun'ichi Mano

INTRODUCTION

Environmental stress is defined as a set of physical and chemical factors of the environment that are unfavorable to the growth of a plant species, such as high or low temperatures, water deficit, ultraviolet radiation, and pollutant gases. The study of plant environmental stress is important, because (i) agricultural productivity worldwide has been greatly restricted by environmental stresses, and (ii) a high capacity to tolerate environmental stresses is the prerequisite to the immotile life of plants, which is affected greatly by ever-fluctuating environments.

Under stressful conditions, the stress factor or the toxic molecules derived from the stress factor attack the most sensitive molecules (primary targets) in cells to impair their functions (Figure 1). Cells are protected by the endogenous molecular systems that mitigate the stress. The damaged targets are recovered either by repair or by replacement via *de novo* biosynthesis. When the stress is too intense and severely damages target molecules, catastrophic cascades of events set in leading to cell death.

Upon stresses, cells have to tolerate them until the new metabolism is established, which may take several hours to days. The destiny of the cells is determined by their pre-existing protective capacity and the intensity and duration of the stress. Investigation of the cellular responses in early stages of environmental stresses will reveal which endogenous and exogenous factors determine the stress tolerance of plants.

Reactive oxygen species (ROS) play crucial roles in the cellular damage resulting from environmental stresses. This is based on the facts that upon stresses (i) the production of ROS and of oxidized target molecules increases; (ii) the antioxidant levels or contents decrease; (iii) the expression of the genes for antioxidative functions increases; and (iv) the scavenging capacity for ROS and the tolerance against stresses are positively correlated. This chapter aims at providing an overview of the action of ROS in environmentally stressed plants, discriminating the occurrence of ROS as causes and as results. Discussion will be focused on the initial targets in early stages of stresses, for which the basic biochemistry of ROS, i.e. their production, reactions, and scavenging will be introduced. Stresses caused by biological interactions will not be treated in this chapter.

PRODUCTION OF REACTIVE OXYGEN SPECIES

ROS are produced in almost every cell compartment during normal metabolism, e.g. for biosynthesis, cell defense, and intra- and inter-cellular signaling. Thus, ROS act as indispensable as well as toxic factors to life. The chloroplasts and leaf tissues, rich in pigments and evolving O_2 under light, are the major source of ROS and are the primary sites of the stress-induced damages in plants under light (Asada and Takahashi, 1987). The

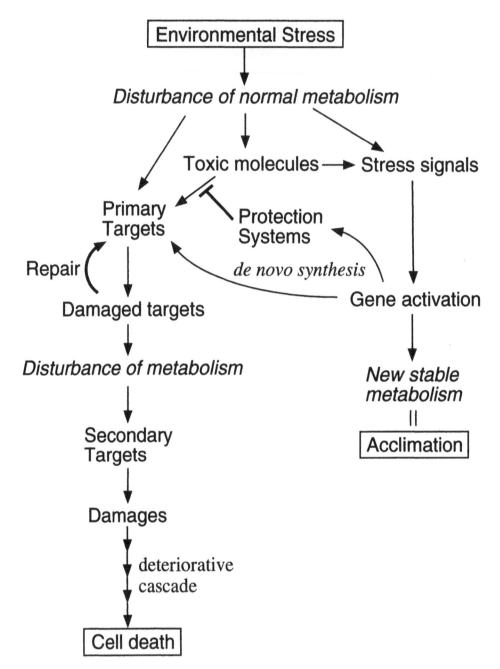

Figure 1. Response of cells to an environmental stress. An environmental stress disturbs the normal metabolism of cells. As a consequence, vulnerable molecules (primary targets) are damaged directly by the stress factor or indirectly by the toxic molecules that are produced due to the stress. The initial disturbance of metabolism or the derived toxic molecules act as the signals to activate a set of stress-responsive genes that enforce the primary targets and the protection systems and achieve acclimation. Unless the primary targets are protected from the damages or repaired within the time of the genetic response to establish a new metabolism to survive the stress conditions, further disturbance of metabolism will enhance the production of toxic molecules and cause the cascade of deteriorative events, leading to cell death.

primary stimuli may arise from extrachloroplastic sites; for example, ozone that penetrates into leaf tissues first interacts with apoplastic components, for which reason apoplastic antioxidant capacity becomes important. Biotic stresses, such as bacterial infection and grazing, also arise from the periphery of the cells. In case of drought and salt stress, stimuli are sensed primarily by roots as well as by leaves. In this section, the production of ROS associated with photosynthetic electron transport in the chloroplast will be introduced first. Their production in other compartments will be briefly discussed later.

Production of ROS in Chloroplasts

Production, reactions, and scavenging of ROS in chloroplasts are summarized in Figure 2. When the chlorophyll (Chl) molecule absorbs light energy at the photochemical reaction center in thylakoid membranes, a high-potential oxidative power (a positive charge) and a low-potential reducing power (a negative charge) are generated. On the oxidative side of photosystem II (PSII), which is the oxidative terminus of the photosynthetic electron transport chain, water is oxidized to O_2. On the reducing side of photosystem I (PSI), the opposite terminus, the iron-sulfur protein ferredoxin (Fd) is reduced. The reduced Fd provides electrons for CO_2 fixation and other reactions in chloroplasts. There are two potential production sites for ROS: one is the reducing side of PSI, and the other is PSII.

The redox potential of the FeS centers at the terminus of PSI, -0.4 V, is low enough to reduce O_2 univalently to produce superoxide radical ($O_2^{\bullet-}$):

$$O_2 + PSI \text{ (reduced) } O_2^{\bullet-} \rightarrow + PSI \text{ (oxidized)}$$

The photoreduction of O_2 to $O_2^{\bullet-}$ by PSI (Asada and Kiso 1973b) occurs inevitably and uses 10–20% of the photosynthetic electron flux even under conditions where CO_2 supply is saturating (Asada and Takahashi, 1987). $O_2^{\bullet-}$ is disproportionated to H_2O_2 and O_2 via catalysis by superoxide dismutase (SOD) that resides in the stroma (Asada *et al.*, 1973).

$$2\,O_2^{\bullet-} + 2\,H^+ \rightarrow H_2O_2 + O_2$$

These reactions account for most of the photoproduction of H_2O_2 in chloroplasts (Mehler reaction; Mehler 1951).

H_2O_2 is produced via non-enzymic reduction of $O_2^{\bullet-}$ with ascorbate (L-AA) or glutathione (GSH):

$$O_2^{\bullet-} + AH \rightarrow H_2O_2 + A\cdot$$

where AH and A· represent a reductant, either L-AA or GSH, and its radical, respectively. Under normal physiological conditions, however, this mechanism is negligible because the produced $O_2^{\bullet-}$ is immediately disproportionated with SOD, which is located near the production site of $O_2^{\bullet-}$ (Ogawa *et al.*, 1995).

O_2 that is photoproduced from H_2O in PSII is finally reduced to H_2O in PSI with catalysis by SOD and ascorbate peroxidase (APX) to form a cycle of electron flow (water-water cycle) (Asada *et al.*, 1998; Miyake *et al.*, 1998; Asada, 1990). As far as the produced ROS are scavenged *in situ* by the enzymes of the water-water cycle, the photoreduction of O_2 to $O_2^{\bullet-}$ is not detrimental, but rather indispensable for preventing photoinhibition of chloroplasts, acting as a "safety valve" that dissipates excessive excitation energy as heat

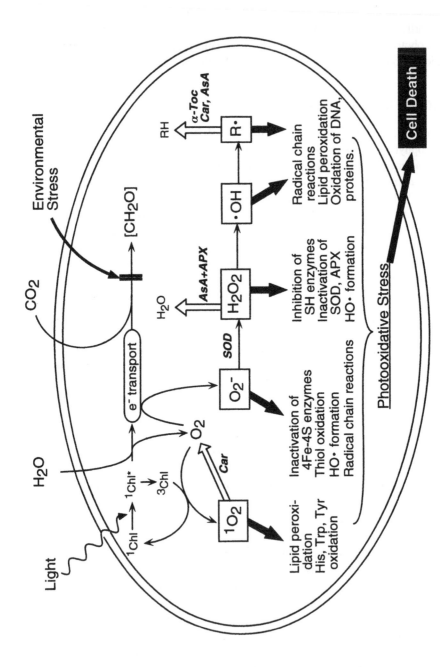

Figure 2. Production, reactions, and scavenging of reactive oxygen species in chloroplasts. Light energy absorbed by Chl is converted to electrochemical potential, which oxidizes H_2O to O_2 and generate the electrons for reducing CO_2 to sugar (represented as $[CH_2O]$). Production of O_2^- and $^1O_2$ are possible events and inevitable. Inhibition of CO_2 reduction by environmental stresses increases the possibilities of the production of these reactive oxygen species. O_2^- is converted to H_2O_2, which, if not scavenged, produces $HO^•$. $HO^•$ oxidizes organic compounds to their radicals ($R^•$). These reactive oxygen species react with targets (thick closed arrows) specifically or nonspecifically to bring about damages, or are scavenged (open arrows) enzymically or non-enzymically with small molecules, *i.e.* carotenoids (Car), AsA, glutathione (GSH), and α-Toc.

(Schreiber and Neubauer, 1990; Neubauer and Yamamoto, 1992; Osmond and Grace, 1995; Laisk and Edwards, 1998). Even at 1.1% CO_2, which saturates photoreduction of CO_2 in chloroplasts, the electron flow to O_2 prevents photoinhibition, despite the production of $O_2^{\bullet -}$ (Park et al., 1996). This efficient scavenging of $O_2^{\bullet -}$ and H_2O_2 is ensured by the high molecular activities and the intraorganellar microlocalization of the enzymes in the water-water cycle (Asada et al., 1998). The chloroplastic flavoenzyme monodehydroascorbate reductase (MDAR) has recently been suggested to catalyze the photoproduction of $O_2^{\bullet -}$ at PSI (Miyake et al., 1998).

H_2O_2 is also produced outside of chloroplasts not only via the disproportionation of $O_2^{\bullet -}$, but also via the divalent reduction of O_2 catalyzed by various oxidases through divalent oxidation (mainly located in peroxisomes; see below).

When reductants and an appropriate catalyst (e.g. transition metal ions, quinones, and Fd; Jakob and Heber, 1996) are provided, H_2O_2 is reduced to form a highly toxic hydroxyl radical (HO$^{\bullet}$):

$$H_2O_2 + AH \rightarrow HO^{\bullet} + OH^- + A \qquad \text{(Haber-Weiss reaction)}$$

L-AA, GSH, and $O_2^{\bullet -}$ can be reductants for this reaction. As catalysts, the FeS centers in the PSI reaction center complex (Sonoike 1996b) and in Fd (Jakob and Heber, 1996) might produce HO$^{\bullet}$ *in situ*. Transition metal ions, e.g. Fe, Cu, and Mn, are also effective catalysts, in case they are released from metalloenzymes for some reasons. Cd from the environment also catalyzes the Haber-Weiss reaction. HO$^{\bullet}$ production is implied in the oxidative stress caused by excess Fe in tobacco (Kampfenkel et al., 1995). HO$^{\bullet}$ is also detected on the donor side of PSII that is impaired by UV-B (Hideg and Vass, 1996), although the source and the reaction to produce this radical have not been identified yet. HO$^{\bullet}$ is highly oxidative (redox potential of HO$^{\bullet}$/H_2O is +2.3 V) and oxidizes organic molecules at a rate constant of 10^9 M^{-1} s^{-1}, hence is the most toxic molecule among reactive oxygen species (Halliwell and Gutteridge, 1989, also see below), as well as singlet oxygen is.

On the other end of the electron transport chain, when the separated charges at the Chl dimer of reaction center recombine, the triplet state of Chl (^{3}Chl) is formed and reacts rapidly with the ground state oxygen ($^3O_2$) to form singlet oxygen ($^1O_2$):

$$^3\text{Chl} + {}^3O_2 \rightarrow {}^1\text{Chl} + {}^1O_2$$

$^1O_2$ is also produced via a similar photodynamic reaction with heme groups in proteins and with flavins, and via various reactions from $O_2^{\bullet -}$ and H_2O_2 (Halliwell and Gutteridge, 1999). In PSII reaction centers, $^1O_2$ is produced when the primary acceptor quinone Q_A is fully reduced (Vass and Styring, 1993). The photoproduction of $^1O_2$ in PSII has been observed *in vitro* (Macpherson et al., 1993) and *in vivo* (Hideg et al., 1998). $^1O_2$ is highly reactive with organic molecules and, hence, is highly toxic (see below). The oxidative potential generated in the PSII reaction center, which is required for the oxidation of water to oxygen, is potentially toxic to the PSII complex itself (Anderson et al., 1998). The Mn cluster of water oxidase accumulates four positive charges to oxidize water to oxygen, and does not release the possibly generated intermediates of water oxidation, HO$^{\bullet}$, H_2O_2, and $O_2^{\bullet -}$. When water oxidase is destroyed for some reasons, e.g. UV-B or heat, the photogenerated oxidative power, either as P680$^+$ or TyrZ$^+$, may oxidize the surrounding protein matrix or neighboring molecules to inactivate PSII complex (donor side-induced photoinhibition; Blubaugh et al., 1991; Aro et al., 1993). In addition, ROS that could be produced via photooxidation of water may be released (Ananyev et al., 1992; Fine and Frasch, 1992; Hideg et al., 1994).

Production of ROS in Non-Chloroplastic Compartments

In extrachloroplastic compartments the major production reactions for ROS are not only the univalent reduction of O_2 to O_2 but the divalent reduction of O_2 to H_2O_2. Peroxisomes contain divalent reaction oxidases and produce H_2O_2 in association with oxidative metabolisms, such as photorespiration and β-oxidation of lipids. In C3 plants a substantial amount of H_2O_2 is produced accompanying photorespiration via the peroxisomal glycolate oxidase:

$$O_2 + \text{glycolate} \rightarrow H_2O_2 + \text{glyoxylate}$$

Acyl-CoA oxidase in peroxisome catalyzes divalent oxidation of acyl-CoA to trans-2,3-dehydroacyl-CoA by O_2 in the β-oxidation of lipids and produces H_2O_2.

$O_2^{\bullet -}$ is produced in mitochondria. In mammalian mitochondria, $O_2^{\bullet -}$ production due to electron leakage from the electron transport to O_2 accounts for 1–2% of total electron flux through the chain (Chance et al., 1979) and is increased several fold by the inhibitors of electron transport, uncouplers, and other agents to disrupt mitochondrial functions (Richter and Schweizer, 1997). The production of $O_2^{\bullet -}$ in submitochondrial particles from pea leaves has been demonstrated (Hernández et al., 1993). Presumably mitochondria are the major production sites of $O_2^{\bullet -}$ in nonphotosynthetic cells. Whether the production of $O_2^{\bullet -}$ in mitochondria has a physiologically positive significance as that in chloroplasts is not yet elucidated.

$O_2^{\bullet -}$ is also produced in peroxisomes and plasma membranes. In plant peroxisomes, $O_2^{\bullet -}$ is produced via xanthine oxidase and at least three distinct NAD(P)H oxidases (del Río et al., 1998). Peroxisomal $O_2^{\bullet -}$ production is increased during senescence and the derived ROS decompose cellular components (Brennan and Frenkel, 1977, del Río et al., 1998). Participation in the production of $O_2^{\bullet -}$ of a mammalian-like NADPH-oxidase on the plasma membranes in plant cells has been demonstrated upon extracellular stimuli (Auh and Murphy, 1995, Allan and Fluhr, 1997) and during lignification (Ogawa et al., 1997).

REACTIVITY, TARGETS, AND SCAVENGING OF ROS

Understanding the reaction mechanism of ROS with both the target biomolecules and the scavengers provides a deep insight into stress physiology in plant cells. Details of the reaction mechanism of each species have been summarized by Halliwell and Gutteridge (1999), and of the scavenging mechanisms by Asada (1996) and Polle (1997), therefore only specific topics are being reviewed here.

Superoxide Radical

$O_2^{\bullet -}$ is a relatively stable or unreactive molecule among the ROS. However, the protonated form HO_2 (pKa = 4.8) is much more reactive and can initiate lipid peroxidation, whereas $O_2^{\bullet -}$ cannot. In addition, HO_2 can pass across lipid bilayers, but $O_2^{\bullet -}$ cannot. In an aqueous solution, $O_2^{\bullet -}$ is spontaneously disproportionated to form H_2O_2 and O_2:

$$O_2^{\bullet -} + O_2^{\bullet -} + 2H^+ \rightarrow H_2O_2 + O_2$$

At lower pH, the following reactions can occur:

$$O_2^{\bullet -} + HO_2 + H^+ \rightarrow H_2O_2 + O_2$$
$$HO_2 + HO_2 \rightarrow H_2O_2 + O_2$$

The second-order rate constants for these reactions are < 0.35 M^{-1} s^{-1}, 1.02 × 10^7 M^{-1} s^{-1}, and 8.60 × 10^5 M^{-1} s^{-1}, respectively. Since the reaction rate constant is largest for the second reaction, the disproportionation is fastest at pH 4.8. The apparent second-order rate constant for the disproportionation of $O_2^{\bullet -}$, 5 × 10^5 M^{-1} s^{-1} at pH 7.0, thus decreases by 10-fold per each pH unit increase in the range over pH 5 (Bielski, 1978).

$O_2^{\bullet -}$ is a reductant of the transition metal ions in the Haber-Weiss reaction to produce $HO^\bullet$ from H_2O_2. $O_2^{\bullet -}$ also propagates radical chain reactions especially in the presence of quinones. When quinones are univalently reduced to the semiquinones ($QH^\bullet$) with quinone reductases which abundantly occur in plant cells, part of $QH^\bullet$ reduces dioxygen to produce $O_2^{\bullet -}$. $O_2^{\bullet -}$ oxidizes the quinols that have been produced via the disproportionation of $QH^\bullet$, to reproduce $QH^\bullet$. This chain reaction is effectively terminated by SOD (Cadenas *et al.*, 1992; Maro *et al.*, 2000).

It is important to note that $O_2^{\bullet -}$ is highly reactive with reduced sulfur compounds, such as thiols and FeS clusters. $O_2^{\bullet -}$ oxidizes thiols to produce thiyl radicals, which may initiate radical chain reactions (see *Thiyl radicals* subsection below). $O_2^{\bullet -}$ also oxidizes the 4Fe-4S cluster of aconitase in mammalian mitochondria or in bacteria at the order of 10^6–10^7 M^{-1} s^{-1} to the inactive 3Fe-4S form. The Fe^{2+} ion released as a consequence is a potent catalyst for the Haber-Weiss reaction.

In plant cells, the major SOD isozymes are located in the chloroplasts (Cu/ZnSOD and FeSOD in some species), cytoplasm (Cu/ZnSOD), and mitochondria (MnSOD). The occurrence of Cu/ZnSOD in the apoplast and nucleus has been confirmed by immunoelectron microscopy (Ogawa *et al.*, 1995). The presence of SOD implies the *in situ* production of $O_2^{\bullet -}$. Cu/ZnSOD and FeSOD are sensitive to H_2O_2. These SODs are potential targets when H_2O_2-scavenging systems do not operate properly.

Hydrogen Peroxide

H_2O_2 is a non-radical neutral molecule below pH 10, and can diffuse across biomembranes as does water. The function of H_2O_2 as a stress signal (Doke, 1997; Karpinski *et al.*, 1999) is partly based on its intra- and intercellular diffusibility. H_2O_2 is a relatively weak oxidant; the oxidative potential of the H_2O_2/ H_2O pair is +320 mV. However, metalloenzymes in general are sensitive targets of H_2O_2. Heme proteins can catalyze the Haber-Weiss reaction and can be degraded by the resulting $HO^\bullet$ (Puppo and Halliwell, 1988). Chloroplastic APX isozymes are inactivated by H_2O_2 in the absence of electron donors (Hossain and Asada, 1984), because compound I is irreversibly oxidized by H_2O_2 (Miyake and Asada, 1996). Cu/ZnSOD is inactivated by H_2O_2 (Bray et al., 1974) via the reduction of the Cu^{2+} ion at the reaction center to Cu^+ and the subsequent production of $HO^•$ (Hodgson and Fridovich, 1975). Cu/ZnSOD in isolated chloroplasts of wheat leaves are inactivated by weak light, most probably because of photoproduced H_2O_2 (Casano et al., 1997). FeSOD is also inactivated by H_2O_2 (Beyer and Fridovich, 1987). The inactivation of these enzymes has been observed in vitro at the micromolar to sub-millimolar range of H_2O_2, which can be reached in vivo as well when H_2O_2-scavenging systems do not operate efficiently. H_2O_2 oxidizes thiols to the sulfenic acids, which react with thiols to form disulfides. The reaction between H_2O_2 and cysteine is slow (the apparent second-order rate constant being $1 \ M^{-1} \ s^{-1}$), but on the surface of proteins the reaction may be largely accelerated by the presence of basic residues, such as Lys and Arg, when they neighbor the thiol groups (Armstrong and Buchanan, 1978). H_2O_2 at micromolar concentrations in darkness inhibit CO_2 fixation in chloroplasts by 50% in 10 min (Kaiser, 1979). This inhibition is due to the oxidation of the active-site thiols to disulfide in the Calvin cycle enzymes, fructose-1,6-bisphosphatase, NADP-glyceraldehyde-3-phosphate dehydrogenase, and ribulose-5-phosphate kinase. The inhibition of these enzymes are reversibly released by the reduction with reduced thioredoxin (Wolosiuk and Buchanan, 1977). However, when cessation of CO_2 fixation continues under light, it will lead to the situation with excess of light energy, in which the production of ROS increases.

H_2O_2 is scavenged by catalase or peroxidase. Catalase scavenges H_2O_2 via the disproportionation of H_2O_2 to O_2 and H_2O, corresponding to a turnover rate of about $10^7 \ min^{-1}$ (Scandalios et al., 1997):

$$2H_2O_2 \ \rightarrow \ O_2 + 2H_2O.$$

Plants have several catalase isozymes, the expression of which is regulated stage and tissue specifically (Scandalios et al., 1997). Catalase is mainly localized in peroxisomes and responsible for scavenging the H_2O_2 produced in photorespiration and β-oxidation of lipids.

Peroxidases scavenge H_2O_2 via reduction to H_2O using an electron donor (AH):

$$H_2O_2 + 2 \ AH \ \rightarrow \ 2H_2O + 2 \ A$$

The major peroxidase in chloroplasts is ascorbate peroxidase (APX), which specifically utilizes L-AA as the electron donor and produces monodehydroascorbate radical ($MDHA^•$):

$$H_2O_2 + 2L\text{-}AA \ \rightarrow \ 2H_2O + 2MDHA^•$$

The soluble and the thylakoid-bound isoforms of APX have been identified in chloroplasts (Miyake and Asada, 1992; Asada, 1997). The L-AA concentration in chloroplasts (approximately 10 mM) is much higher than the K_m value of APX for L-AA (0.5 mM), so the H_2O_2-scavenging capacity in chloroplasts is governed by the APX content. Because of the APX contained at 70 μM (soluble and thylakoid-bound isoforms) in chloroplasts, the steady-state concentration of H_2O_2 under illumination is kept as low as 0.5 μM (Asada, 1994). Once APX is inactivated, the photoproduced H_2O_2 is not scavenged efficiently and diffuses out of chloroplasts (Nakano and Asada, 1981). APX occurs in the cytosol, peroxisomes (Yamaguchi *et al.*, 1995), and mitochondria (Jiménez *et al.*, 1997). Some cyanobacteria scavenge H_2O_2 with peroxidase using photoreducing equivalents as the electron donors, whereas others adopt catalase (Miyake *et al.*, 1991). Peroxiredoxin scavenges H_2O_2 and lipid peroxides in the cyanobacteria (Yamamoto *et al.*, 1998), as confirmed in *Escherichia coli* and yeast. In higher plants the peroxide detoxication by peroxiredoxin has not been demonstrated clearly, although homologous genes are expressed, for instance in *Arabidopsis thaliana* (Baier and Dietz, 1997).

The guaiacol peroxidases (the peroxidases represented by the "horseradish peroxidase" that prefer phenolics such as guaiacol and pyrogallol to ascorbate) in the apoplasts and vacuoles may participate in the scavenging of H_2O_2, via the redox mediation by phenolics (Takahama and Oniki, 1992; Otter and Polle, 1994; Mehlhorn *et al.*, 1996) or flavonoids (Yamasaki *et al.*, 1997), using L-AA as the electron donor (Mehlhorn *et al.*, 1996; Polle, 1997).

Hydroxyl Radical

Because of its high oxidative power (+2.3 V), $HO^{\bullet}$ react with organic and inorganic molecules at the order of 10^7–10^{10} M^{-1} s^{-1}, via hydrogen abstraction, hydroxylation, or electron transfer (Halliwell and Gutteridge, 1999). The reaction products are also radicals, which are usually less reactive hence live longer than $HO^{\bullet}$. When such long-life radicals react with O_2, radical chain reactions are propagated (see *Lipid peroxides* subsection below).

$HO^{\bullet}$ reacts nonselectively with most molecules nearby its production site, and hence does not diffuse for a long distance. In other words, there are no specific scavengers for $HO^{\bullet}$. Effective protection of cells from the toxicity of $HO^{\bullet}$ is therefore achieved indirectly by keeping the concentrations of precursors ($O_2^{\bullet-}$ and H_2O_2) and of catalysts (transition metal ions) at low level. Once produced in cells, $HO^{\bullet}$ is scavenged by the mass reaction. L-AA and GSH, the most abundant reductants in every cell compartment, can act as the first defenses when they are present at the production site of $HO^{\bullet}$. It should also be noted that some compatible solutes accumulated during water stress, such as mannitol and trehalose, can scavenge $HO^{\bullet}$ (Smirnoff and Cumbes, 1989) and may contribute to the stress tolerance (Smirnoff, 1993; Shen *et al.*, 1997).

Singlet Oxygen ($^1O_2$)

The reactivity of $^1O_2$ with organic molecules is much higher than that of $^3O_2$, because of its high energy (higher than $^3O_2$ by 22.5 kcal mol^{-1}) and spin state. $^1O_2$ is nucleophilic, oxidizes other molecules often divalently, adds to double bonds to form hydroperoxides or endoperoxides, and damages proteins by oxidizing His, Trp, and Met residues at the order of 3.5×10^7 M^{-1} s^{-1} (Halliwell and Gutteridge, 1999). Since $^1O_2$ is not a radical, these reactions do not produce radicals and do not directly initiate chain reactions. However, when the resulting peroxides are reduced, highly reactive alkoxyl radicals are formed (see below). $^1O_2$ also oxidizes a low-potential compounds to produce $O_2^{\bullet-}$ and a radical (Saito *et al.*, 1983):

$$^1O_2 + A \rightarrow O_2^{\bullet-} + A^{\bullet+}$$

The most effective endogenous scavenger for $^1O_2$ is β-carotene. It directly reacts with $^1O_2$ at the order of 3×10^{10} M^{-1} s^{-1} and dissipates the excitation energy as heat. β-Carotene also reacts with ^3Chl and convert it to ^1Chl, thereby preventing the formation of $^1O_2$. The $^1O_2$ produced via the photosensitization of Chl is primarily quenched by β-carotene molecules that are localized close to the Chl molecules in the Chl-binding proteins and quench the *in situ* produced ^3Chl and $^1O_2$ (Telfer *et al.*, 1994; Yamamoto and Bassi, 1996). α-Tocopherol (α-Toc) in the lipid bilayer of thylakoid membranes also quenches $^1O_2$. In aqueous solution, $^1O_2$ is quenched by L-AA and GSH if they are present at millimolar levels. A close relation between the $^1O_2$ production and the photoinhibition of PSII reaction center has been demonstrated *in vitro* and *in vivo* (Hideg *et al.*, 1994, 1998).

Secondary Products of ROS

Lipid peroxides

Lipid peroxides, including lipid hydroperoxides and peroxyl radicals, are produced via the chain reactions starting with $HO^\bullet$ (Halliwell and Gutteridge, 1989). Initiation reaction is the hydrogen abstraction from the methylene ($-CH_2-$) part of lipids by $HO^\bullet$ and thiyl radicals (see below), resulting in the formation of lipid radical ($L^\bullet$). The formation of $L^\bullet$ in biomembranes readily leads to the chain reaction, because of high local concentration of lipid molecules in the membranes and high solubility of O_2 in the hydrophobic environment of the membranes. O_2 is added to $L^\bullet$, forming lipid peroxyl radical ($LOO^\bullet$). $LOO^\bullet$ abstracts hydrogen from another lipid molecule (LH) to produce $L^\bullet$ and lipid hydroperoxide (LOOH). LOOH is formed also via the addition of $^1O_2$ to unsaturated lipids (Takahama, 1979). LOOH itself is not highly reactive, like H_2O_2, but in the presence of reductants and catalysts it is reduced to form alkoxyl radical ($LO^\bullet$). $LO^\bullet$ proceeds to β-scission to produce carbonyl compounds, including aldehydes, and alkyl radicals, which again propagate chain reactions. The resulting reaction intermediates and aldehydes react with thiol groups and amino groups on proteins and on DNA molecules and modify them covalently.

   Lipid radicals in biomembranes are primarily scavenged by α-Toc that reacts with $HO^\bullet$, $LO^\bullet$, and $LOO^\bullet$ at 10^{10}, 10^8, and 10^6 M^{-1} s^{-1}, respectively. The reactions of $HO^\bullet$ and $LO^\bullet$ with target lipid molecules are so fast that α-Toc cannot compete with the lipid oxidation by $HO^\bullet$ and $LO^\bullet$ effectively. Therefore, α-Toc terminates the chain reactions mainly via the reaction with $LOO^\bullet$ (Niki *et al.*, 1995). α-Toc reduces $LOO^\bullet$ to lipid alcohol (LOH) and is oxidized to its chromanoxyl radical (α-$Toc^\bullet$). α-$Toc^\bullet$ is reduced to α-Toc by L-AA (Packer *et al.*, 1979):

$$\alpha\text{-Toc} + LOO^\bullet \rightarrow \alpha\text{-}Toc^\bullet + LOH$$

$$\alpha\text{-}Toc^\bullet + L - AA \rightarrow \alpha\text{-Toc} + MDHA^\bullet$$

In mammals, lipid peroxides are scavenged specifically by glutathione peroxidase (GPX). From cultured citrus cells an enzyme homologous to the mammalian GPX was purified (Beeor-Tzahar *et al.*, 1995). The purified protein catalyzes the reduction of phospholipid hydroperoxide and is distinct from glutathione-S-transferase, which also exhibits a peroxidase activity. The induction of this phospholipid hydroperoxide GPX by salt stress suggests that this enzyme is responsible for scavenging of lipid peroxides that could accumulate under salt stress (Gueta-Dahan *et al.*, 1997). Peroxiredoxin, a chloroplast protein, may also scavenge lipid peroxides; overproduction of this protein from *A. thaliana* confers a tolerance toward alkyl hydroperoxide to *E. coli* (Baier and Dietz, 1997).

Direct detection of lipid radicals in vivo has not been available so far, but there are many reports to indicate their production by the detection of endproducts of lipid oxidation. Accumulation of the endproduct malondialdehyde, observed in various stressed plants, would reflect the inhibition of antioxidant enzymes and shortage of antioxidants in cells, which are terminal symptoms in the damage process.

Thiyl radicals

Thiyl radicals are formed via the univalent oxidation of thiols, such as GSH and cysteine. $O_2^{\bullet-}$ oxidizes thiols (Asada and Kanematsu, 1976), and the highly oxidizing thiyl radicals are formed as a result (Nishimura *et al.*, 1996, Winterbourn and Metodiewa, 1999). Because of their reactivity with thiols, most thiyl radicals would be scavenged to finally form disulfides. Thiyl radicals react quickly with polyunsaturated fatty acids at the order of 10^7 M^{-1} s^{-1} and initiate lipid peroxidation (Schöneich *et al.*, 1992). Guaiacol peroxidase (Harman *et al.*, 1984) and APX (Chen and Asada, 1992) catalyze the univalent oxidation of thiols to the thiyl radicals by H_2O_2, whereas no thiyl radicals are formed in the reaction catalyzed by mammalian GPX (Harman *et al.*, 1986). *In situ* production of the radical leads to the inactivation of APX (Chen and Asada, 1992). It is also possible that Cys residues on protein molecules are oxidized to Cys thiyl radicals, which may lead to protein cleavage or polymerization (Nishimura *et al.*, 1996). The production of the thiyl radicals in living cells has not been demonstrated, but the potential toxicity of the radicals may not be negligible, especially when SOD does not function in the production sites of $O_2^{\bullet-}$, while thiols are present.

Peroxiredoxin, known as thiol-specific antioxidant enzyme or thioredoxin reductase, shows a thiyl radical reductase activity in *E. coli* (Yim *et al.*, 1994). For plants, Baier and Dietz (1996) have cloned several homologous genes from an *A. thaliana* cDNA library. Whether the gene products show thiyl radical reductase activity has not been elucidated yet.

Monodehydroascorbate Radical

L-AA functions as the "terminal antioxidant" in cells, because the redox potential of the L-AA/MDHA• pair (+280 mV) is lower than that of most bioradicals (Buettner and Jurkiewicz, 1993). When L-AA reduces ROS in non-enzymatic and enzymatic reactions, MDHA• is formed. The major reaction to produce MDHA• in chloroplast stroma is the APX reaction. In the thylakoid lumen, MDHA• is produced via the photooxidation on the donor side of PSII (Mano *et al.*, 1997) when water oxidase is inactivated, and on the donor side of PSI when PSII are suppressed (J. Mano, unpublished data). The redox potential of MDHA• is moderate and the spontaneous disproportionation rate is high (5×10^5 M^{-1} s^{-1} at pH 7). MDHA• is relatively inert to cellular components, ensuring the "safe" scavenging of ROS by L-AA.

MDHA•, however, may be cytotoxic in some cases. MDHA• oxidizes the compound I of catalase to compound II to inactivate it (Davison *et al.*, 1986). The lymphocyte microsomal 3-hydroxy-3-methylglutaryl CoA reductase is inhibited by 50% with MDHA[•] at 10 μM (Harwood *et al.*, 1986). MDHA[•] also propagates radical chain reactions mediated by redox-active quinones and counteracts the chain-terminating action of SOD (Jarabak *et al.*, 1997).

Several specific systems for scavenging MDHA[•] have been known. MDHA[•] photoproduced at PSI is reduced to L-AA by reduced Fd, preferentially to the reduction of NADP[+] (Miyake and Asada, 1994). Monodehydroascorbate reductase (MDHAR) in chloroplast stroma (Marrè and Arrigoni, 1958, Hossain *et al.*, 1984) and cytosol is an FAD enzyme that catalyzes the univalent reduction of MDHA[•] to L-AA by NAD(P)H (Hossain and Asada, 1985). MDAR catalyzes the reaction at diffusion-controlled rates (Sano *et al.*, 1995). There are soluble MDAR in soybean root nodules (Dalton *et al.*, 1992), in mitochondria of potato tubers (Leonardis *et al.*, 1995), and membrane-bound MDAR in mitochondria (Jiménez *et al.*, 1997), peroxisomes (Bowditch and Donaldson, 1990; Bunkelmann and Trelease, 1996; Jiménez *et al.*, 1997), and plasma membranes (Bérczi and Møller, 1998). The very high molecular activity of MDAR and its ubiquitous occurrence in plant cells might result from the necessity to suppress detrimental effects of MDHA[•] as described above.

INVOLVEMENT OF ROS IN ABIOTIC STRESSES

Environmental stresses cause oxidative stress in cells directly or indirectly. Pollutant gases such as SO_2, O_3, and NO_2 and the herbicide methyl viologen (MV) initiate or propagate radical chain reactions, thereby accelerating the production of ROS. Alternatively, low or high temperature conditions or enhanced production of $O_2^{\bullet -}$ because of MV may inhibit the enzymes and antioxidants for scavenging ROS. Even when the protection systems are not affected directly, severe stress in general distorts the cellular metabolism for energy transduction and maintenance and subsequently increases the production or decreases the scavenging of ROS, or both. These mechanisms sometimes overlap; once ROS occur as the primary cause, they will impair the protection system against them, and as a consequence, the steady-state ROS levels will rise, and the damages will further propagate to other part of cells and tissues.

High Irradiance

Excessiveness of light energy is determined by the balance between light intensity and other environmental and endogenous factors. The more plants are capable of using light energy for assimilation and dissipation, the more tolerant they are to light. A good example is the high-light induced suppression of growth of the cyanobacterium *Synechocystis* 6803 under low CO_2 and its recovery in high CO_2 (Asada, 1996). Excessive light energy causes overreduction of the photosynthetic electron transport chain, leading to both increased production of $O_2^{\bullet -}$ on the acceptor side of PSI and the increased production of $^1O_2$, unless the water-water cycle operates properly (Asada *et al.*, 1998). Oxidative stress due to high irradiance is often demonstrated in combination with other unfavorable conditions such as low temperature, low CO_2 or water shortage. A very high irradiance increases the production of MDHA radicals in darkness and under light (Heber *et al.*, 1996). The plants acclimated to high irradiance show higher contents of water-water cycle enzymes (Grace and Logan, 1996). These indicate that the production of $O_2^{\bullet-}$ at PSI increases under high irradiance. Indeed, high irradiance enhances the electron partitioning to O_2 at PSI (Miyake and Vokota, 2000). The Golden Leaf variety of the tropical fig *Ficus microcarpa* L.f. lacks heat-stable dehydroascorbate reductase (DHAR) activity in leaves and shows a hypersensitivity to strong light (Yamasaki *et al.*, 1999), suggesting a crucial role of the L-AA regeneration system for the tolerance against high irradiance.

Photorespiration supplies electron acceptors to PSI. Kozaki and Takeba (1996) have demonstrated the protective role of photorespiration against the damage due to strong illumination. The transgenic tobacco by overexpressing plastidic glutamine synthetase exhibited higher capacity of photorespiration, and tolerated a 24-hour illumination at 2,000 μmol m^{-2} s^{-1}, which caused complete bleaching of Chl in the leaves of control plants. The Chl bleaching under high light, a terminal symptom of oxidative stress, could be caused by H_2O_2 or its derivative HO$^{\bullet}$, because it is enhanced in tobacco that is deficient in peroxisomal catalase (Willekens *et al.*, 1997) and is prevented by exogenous introduction of catalase (Chamnongpol *et al.*, 1996).

Catalase is another possible target of strong irradiation. Grotjohann *et al.*, (1997) have observed photoinactivation of catalase isozymes from sunflower cotyledons, accompanied by the degradation of the hemes. In rye leaves catalase is turned over under light at a relatively high rate ($t_{1/2}$: 3-4 h), indicating its photoinactivation *in vivo* (Hertwig *et al.*, 1992).

The D1 protein, which is the key component of the PSII reaction center complex, is one of the best studied targets of high irradiance stress. The D1 protein turns over most rapidly among thylakoidal proteins even under medium light (Matoo *et al.*, 1984). The modification of D1 under light, accompanied by the presumably enzymatic degradation, is an inevitable reaction due to the intrinsically produced $^{1}O_2$ or P680^{+} (Anderson *et al.*, 1998). Under normal physiological conditions the rates of degradation and *de novo* synthesis of D1 proteins are balanced, so that the photoinhibition of PSII is not apparent. This is a very intriguing example of homeostasis that maintains chloroplast functions. For the mechanisms of PSII photoinhibition, see reviews by Aro *et al.* (1993) and Andersson and Barber (1996).

With respect to oxidative stress in chloroplasts, the PSII complex is the ultimate source of electrons for the generation of $O_2^{\bullet-}$. In addition, PSII reaction center complex may produce strong oxidants when water oxidase activity is inhibited, as described previously. Under various environmental stresses where the supply of electron acceptors is lowered, electron supply from PSII is down-regulated via the xanthophyll cycle (Demmig-Adams and Adams 1992). The degradation of D1 proteins, can be regarded as another molecular mechanism that attenuates electron supply, thereby preventing the conversion of excess excitation energy to toxic molecules. The photoproduction of $^{1}O_2$ in PSII is greater when the turnover of D1 proteins is inhibited by lincomycin (Hideg *et al.*, 1998), which implies that damaged PSII center is more harmful than the native one. If D1 proteins were not degraded and turned over, the production of ROS would not be suppressed and the chloroplast would be more readily burned out under illumination.

Methyl Viologen and Other Herbicides

The herbicide methyl viologen (MV), or paraquat, exerts its toxicity by promoting photoproduction of $O_2^{\bullet-}$ in chloroplasts, hence has been frequently used as a stressor to experimentally induce oxidative stress in plants. MV, a bis-cation in aqueous solution, is readily photoreduced to its monocation radical on the acceptor side of PSI. MV radical reduces dioxygen at a very high rate to produce $O_2^{\bullet -}$ (Farrington et al., 1973). Since most electrons from photosynthetic electron transport are trapped by MV, the photoreduction of MDHA$^\bullet$ radical, which is indispensable to maintain the L-AA content in chloroplasts, is inhibited. When MV-administered chloroplasts are illuminated, APX activity is lost in several minutes. The loss of the H_2O_2-scavenging capacity in leaves, as determined by the photoproduction of MDHA$^\bullet$, is attributable to the loss of APX activity in chloroplasts (Mano et al., 1998), because of the oxidative decomposition of the compound I under a high H_2O_2/L-AA ratio (Miyake and Asada, 1996). Then the inactivation of other ROS-scavenging enzymes in chloroplasts (Iturbe-Ormaetxe et al., 1998) and dissociation of ferredoxin-NADP oxidoreductase from thylakoid membranes (Palatnik et al., 1997) will follow. In wheat leaves, the treatment with MV plus medium light (200 μmol m^{-2} s^{-1}) first inactivates APX and DHAR in 2.5 h, and then SOD and glutathione reductase (GR) in 5–25 h (Kraus and Fletcher 1994). Enhanced production of H_2O_2 and $O_2^{\bullet -}$ leads to the production of hydroxyl radical via the reductive fission of H_2O_2, catalyzed by transition metal ions or quinones. These data indicate that the toxicity of MV is attributable to both H_2O_2 and $O_2^{\bullet -}$. The suppression of MV toxicity by metal chelators in rice leaves (Chang and Kao, 1997) implies the participation of HO$^\bullet$ in the MV-induced damage, but the production of HO$^\bullet$ would be a secondary effect of the increase in free metal ions, most probably released from metalloproteins because of the damage by H_2O_2 and $O_2^{\bullet -}$. Guard cells of stomata from Commelina communis exposed to MV at more than 10 μM are irreversibly injured in 2–3 h (McAinsh et al., 1996), suggesting that these cells are also primary targets of MV.

Carotenoid biosynthesis inhibitors, such as norflurazone and fluridone, prevents chloroplast biogenesis and inhibit the plant's growth, because Chl without an association of carotenoids cannot escape from the attack by $^1O_2$, which is produced in situ via the photosensitization of Chl molecules themselves. Herbicides of another class, chlorophyll biosynthesis inhibitors, such as oxyfluorfen and thidiazimin, cause accumulation of protochlorophyllides and enhance the production of $^1O_2$ by photosensitizing proto-chlorphyllides (Böger and Sandmann, 1998).

Pollutant Gases

Sulfur dioxide

Sulfur dioxide (SO_2) enters leaf cells through stomata to form sulfite ion (HSO_3^-/SO_3^{2-}). Under illumination, sulfite reacts with the photoproduced $O_2^{\bullet -}$ to form HSO_3^- and HO$^\bullet$ to propagate radical chain reactions in illuminated chloroplasts (Asada and Kiso, 1973a). The resulting enhancement of oxidative stress in leaves is detected as an increase in the photoproduction of MDHA radical (Veljovic-Jovanovic et al., 1998). The SO_2-induced oxidative damages in leaves, specifically pigment bleaching and lipid peroxidation, are thus light dependent (Shimazaki et al., 1980; Peiser et al., 1982). The H_2O_2 produced via the SO_2-initiated chain reactions inhibits thiol enzymes in the Calvin cycle (Tanaka et al., 1982b). Although the inhibition of these enzymes is reversible, long photoproduction of $O_2^{\bullet -}$ and inhibition of CO_2 assimilation would result in irreversible damages of biomolecules, such as inactivation of APX, catalase, GR (Tanaka et al.,

1982a) and SOD, chlorophyll destruction (Shimazaki *et al.*, 1980), and lipid peroxidation (Peiser *et al.*, 1982). Inhibition of catalase would facilitate the chemical decomposition of glyoxylate and hydroxypyruvate in the peroxisomes (Zelitch, 1973), leading to a decrease in photorespiration and a relative increase in the photoproduction of $O_2^{\bullet-}$ at PSI. Combined with the accumulation of H_2O_2, this situation causes further oxidative stresses.

Toxicity of sulfite in plants is exhibited even in the dark. Shimazaki *et al.* (1980) observed a 60% decrease in SOD activity in spinach leaves that were fumigated with 2 ppm SO_2 for 2 hours. Catalase in spinach leaves is sensitive to sulfite and infiltration of 0.5 mM sulfite results in a 70% inhibition of catalase, while other heme-containing enzymes, APX and guaiacol-peroxidase are insensitive to the treatment (Veljovic-Jovanovic *et al.*, 1998). PSII is inhibited by 20 mM HSO_3^- in bean leaves, which precedes lipid peroxidation of thylakoid membranes, suggesting that PSII is a primary target site (Covello *et al.*, 1989). It is unclear whether ROS participate in the inhibition.

Ozone

Ozone (O_3) also causes oxidative stress in leaves, but the initial reactions and the target molecules are not well elucidated. Sensitivity to O_3 is different among species and cultivars (Runeckles and Krupa, 1994). O_3 enters plant tissues through the stomata and contacts first the apoplastic fluid. In aqueous solutions, O_3 produces H_2O_2 and subsequently $HO^\bullet$, via the Haber-Weiss reaction. Alternatively, $HO^\bullet$ is directly produced from O_3 via reduction (Byvoet *et al.*, 1995). L-AA in the apoplasts can be a reductant for the $HO^\bullet$ production. Apoplastic L-AA at the same time acts as a first defense against O_3 (Luwe *et al.*, 1993), probably by scavenging $HO^\bullet$. Light-dependent production of radicals in leaves, as detected by electron spin resonance spectrometry, is enhanced by ozone, and the L-AA infiltrated into the apoplasts suppresses the radical production (Runeckles and Vaatnou, 1997). The critical importance of L-AA for the O_3 tolerance is demonstrated by Conklin *et al.* (1996), who showed that an *Arabidopsis* mutant exhibiting lower ascorbate content was more susceptible to ozone. The mechanism and the rate-limiting steps of L-AA regeneration in the apoplasts is of primary importance for the protection, although they have not been well elucidated (Noctor and Foyer, 1998).

Ultraviolet B Region (UV-B)

For recent reviews and a monograph on the effects of UV-B in plants, see Runeckles and Krupa (1994), Teramura and Sullivan (1994), Vass (1996), and Lusmunden (1997). UV light is divided into three regions: 400–320 nm (UV-A), 320–280 nm (UV-B), and 280–200 nm (UV-C). The UV-B is the most interesting because of its biologically detrimental effects and the increasing irradiation on the earth surface due to the breakdown of the stratospheric ozone layer. UV-A is less affecting plants, whereas UV-C, with energy high enough to break covalent bonds in organic molecules, is completely absorbed by the atmosphere and unimportant in the biological processes on the earth surface. The primary endogenous targets and sensitivity to UV-B are quite variable among plant species, partly because of the difference in the action spectra of the UV-B sources used in the experiments

(Runeckles and Krupa, 1994) and the difference among plants in their growth history and contents of UV-B-absorbing protectants such as flavonoids in the epidermal cells (Teramura and Sullivan, 1994; Jansen et al., 1996; Bornman et al., 1997). The UV-B-induced damages have been observed under "unrealistic" conditions in most plants, specifically, with extremely high UV-B irradiance + weak visible and UV-A light (Fiscus and Booker, 1995). Results of experiments conducted under such extreme conditions are described below. Careful treatment of these results is therefore necessary when applying them to the real environmental situations.

Various biological molecules are the target of UV-B: DNA, amino acids and proteins, lipids, quinones, and pigments (Vass, 1996; Lusmunden, 1997). In sensitive plants, the primary damaged target in the leaf cells is proclaimed to be the permeability to ions of thylakoid membranes (Strid et al., 1994), then ribulose-1,5-bisphosphate carboxylase/oxidase (Baker et al., 1997), and the PSII reaction center (Kulandavelu and Noorudeen, 1983). In PSII, the water-oxidase complex, TyrZ on the donor side, and the quinones on the acceptor side are inhibited, among which water-oxidase complex is the most sensitive (Renger et al., 1989). Although the molecular mechanism of the first attack has not been unraveled, it is conceivable that the inactivation of water-oxidase complex brings about further detrimental effects on the PSII complex (Vass et al., 1996). The PSII reaction center generates strong oxidants at +2.0 V under illumination. The breakdown of the water oxidase would lead to donor-side-induced photoinhibition (see above). Hideg et al. (1997) observed increased photoproduction of MDHA• in UV-B-irradiated broad bean leaves, which was at least partly attributed to the photooxidation of L-AA on the donor side of PSII. In Tris-treated thylakoids, in which water oxidase is inactivated, L-AA was photooxidized to MDHA• in the thylakoid lumen (Mano et al., 1997). Ascorbate concentration in the lumen (approximately 4 mM; Foyer and Lelandais, 1996) can support substantial rates of electron transport, when water oxidase is inactivated (Katoh and San Pietro, 1967; Yamashita and Butler, 1968). Since the oxidized L-AA is reduced to L-AA only by DHAR, and not by MDAR nor by ferredoxin in the stroma, the DHAR-dependent system for regenerating L-AA is indispensable for the maintenance of L-AA contents in chloroplasts (Mano et al., 1997).

Oxidative stress in UV-B-irradiated leaves is demonstrated by the increased production of ROS and organic radicals (Hideg and Vass 1996; Dai et al., 1997), lipid peroxidation (Dai et al., 1997), and by the induction of SOD, GR (Jansen et al., 1996), and APX (Rao et al., 1996). Except for the radical production on the donor side of PSII, the sources of the ROS in these UV-B-treated leaves have not been well identified. Nevertheless, the critical importance of the scavenging of ROS for protection from UV-B damage is obvious from the observation that the L-AA-deficient mutant of *A. thaliana* exhibited hypersensitivity to UV-B (Conklin et al., 1996).

Temperature

One of the major questions in the study of the temperature stress is which endogenous factors determine temperature sensitivity/tolerance. Whether ROS and the scavenging systems are included in the determinant factors for this temperature sensitivity, is discussed below.

Chilling temperature

At low temperatures above freezing points (4–7°C), chilling damages are induced. Membrane fluidity, diffusion rates of molecules, and chemical and enzymic reaction rates are decreased by low temperatures. Chloroplast functions are more sensitive to chilling-induced photoinhibition than those of mitochondria (Wise and Naylor, 1987). Disintegration of the ultrastructure of chloroplasts is caused by combination of low temperature and high intensity of light (1,000 μmol m^{-2} s^{-1}) in 6–12 hours, depending on the plant species (Wise and Naylor, 1987). PSII photoinhibition is also observed in chilling-sensitive and chilling-tolerant species after longer illumination (Hetherington *et al.*, 1989). However, the primary site of the chilling-induced photoinhibition is not PSII, but the capacity to accept electrons from PSII governs the light sensitivity of PSII (Öquist and Huner, 1993). Resistance to short-term (hours) photoinhibition in winter cereals is found to be a reflection of the increased capacity to keep Q_A oxidized under high irradiance and low temperatures (Huner *et al.*, 1993). Havaux and Davaud (1994) and Terashima *et al.* (1994) have independently reported that PSI was preferentially inhibited in cold-sensitive plants under weak light (< 300 μmol m^{-2} s^{-1}). The light sensitivity of PSI at chilling temperatures is apparent in cold-sensitive plants such as cucumber, pumpkin and maize, but is not conspicuous in cold-tolerant species (Tjus *et al.*, 1998).

The inhibition of PSI requires both O_2 and electrons from PSII, suggesting that the inhibitory species are the ROS photoproduced on the reducing side of PSI, namely, $O_2^{\bullet -}$ or HO$^{\bullet}$. The FeS centers in the PSI complex are primarily destroyed (Sonoike *et al.*, 1995) and then the damage propagates to the degradation of the PsaB protein (Sonoike, 1996a). The damaging species is presumed to be HO$^{\bullet}$, because the degradation of the PsaB protein is blocked by spin-trapping reagents or the radical scavenger *n*-propyl gallate (Sonoike, 1996a). Terashima *et al.* (1998) observed an accumulation of H_2O_2 in cucumber leaves illuminated at 5°C because of decreased APX activity at that temperature, and attributed the PSI photoinhibition to this suppression of APX activity.

Susceptibility of PSI reaction center in chilling-sensitive plants to low temperature can be explained in several ways: (i) the PSI reaction center is more labile, (ii) the production of ROS is increased to a higher extent, or (iii) ROS scavenging is more inhibited in the sensitive species than in the tolerant species. So far, no reports that compare the temperature dependence of these factors between chilling-sensitive and chilling-tolerant plants are available. Although a facilitated O_2 consumption via the Mehler reaction at low temperatures has been suggested in maize due to a suppression of CO_2 assimilation (Fryer *et al.*, 1998), it is still unclear whether the Calvin cycle enzymes in chilling-sensitive plants are more sensitive to low temperatures than those in tolerant plants. At low temperatures, the water-water cycle enzymes that protect the PSI reaction center (Asada *et al.*, 1998) would be less active and, as a result, the production of ROS may surpass the scavenging capacity. The ascorbate regeneration capacity can be primarily affected by low temperature in maize; in maize leaves, MDAR, DHAR, and GR were more temperature dependent than SOD and APX (Jahnke *et al.*, 1991). On the other hand, Terashima *et al.* (1998) showed that APX activity decreases at low temperature. SOD may be another target in early stages of chilling stress. In a tomato mutant, SOD activities in leaves and stems are reduced by 50% in 2 hours 5°C under mild light (Gianinetti *et al.*, 1993), which is similar to the time at which PSI photoinhibition is observed. The observation that the overproduction of Cu/ZnSOD in chloroplasts conferred chilling tolerance to tobacco (Sen Gupta *et al.*, 1993) may support this possibility. Havaux and Davaud (1994) could mimic the chilling-induced and PSI-preferred photoinhibition in potato leaves by giving a Cu/ZnSOD inhibitor to leaves at 23°C. Whether the inhibition or inactivation of SOD precedes the photoinhibition of PSI has not been investigated so far. Determinant factors for temperature sensitivity thus may vary depending on species and experimental conditions. It is important to examine whether the temperature dependence of these antioxidant enzymes and Calvin cycle enzymes differ or not between cold-sensitive and cold-tolerant plants.

Production of ROS at low temperatures does not always associate with light. In maize leaves grown in the dark, accumulation of H_2O_2 is increased at low temperatures, which is followed by an increase in mitochondrial catalase (Prasad *et al.*, 1994). *Saintpaulia*, a tropical species, is highly sensitive to cold stress. A 3-second contact with water at 5°C brings rapid injury to the leaves, including decreases in chlorophyll fluorescence and PSI activity, and disintegration of thylakoid ultrastructure within seconds to minutes. A concomitant burst of oxygen uptake implies the participation of ROS (Yun *et al.*, 1997). Since light is not necessary for causing the injury, extrachloroplastic origins of the damaging molecules are suggested. Chilling temperatures affect not only leaves but also other parts of plants. Higher activity of antioxidant enzymes in roots may also contribute to the chilling-tolerance of maize (Pinhero *et al.*, 1997).

Heat

Heat stress becomes apparent in most plants over 40°C. It causes uncoupling of electron transport with ATP synthesis, inhibition of PSII water oxidase (Katoh and San Pietro, 1967, Mohanty *et al.*, 1987) and electron transport chain, lipid peroxidation (Mishra and Singhal, 1992), and disruption of ultrastructure (Motowska, 1996). Interestingly, inhibition of PSII induced by heat (40°C) in pea leaves is blocked by weak illumination during the heat treatment (Havaux *et al.*, 1991). Heat treatment (50°C for 2.5 h) induces oxidative stress in wheat seedlings that was mitigated by pretreatment of the plants with paclobutrazol, an artificial inducer of antioxidant enzymes in plants (Kraus and Fletcher, 1994). H_2O_2 production is induced by heat treatment in tobacco (Foyer *et al.*, 1997) and in mustard seedlings (Dat *et al.*, 1998). However, the source of the H_2O_2 and the initial target of heat stress are not well elucidated yet.

Water Stress

Water deficit caused by drought or high salinity of soil is the most serious environmental stress that limits agricultural production in many regions of the world. Plants experience water stress either when water supply to roots becomes difficult or when the transpiration rate becomes very high. These two conditions often coincide under warm and arid climates. Plants in a moderate climate are not the exception: they are exposed to water stress due to dryness for several weeks to months, which is not rare. Tolerance against water stress is seen in every plant species, but its extent is variable from species to species. In contrast to other environmental stresses, the primary target sites of water stress have not been narrowed down to specific tissue or organelles (Alscher *et al.*, 1997).

Effects of water stress on photosynthesis have been extensively studied. The extent of inhibition and the target sites vary, depending on the intensities of water deficit and irradiation (Kaiser, 1987). In isolated mesophyll cells from *Xanthium strumarium*, osmotic stress inhibits the regeneration activity of ribulose 1,5-bisphosphate and uncoupling of the electron transport chain from ATP synthesis in chloroplasts, but electron transport activity is insensitive to low water potential down to -4 MPa (Sharkey and Badger, 1982). From an O_2 exhange study using $^{18}O_2$, the major limiting factor of CO_2 assimilation under drought conditions is CO_2 supply to ribulose-1,5-bisphosphate carboxylase (Tourneux and Peltier, 1995). Under such conditions, the absorbed light energy overflows for photorespiration (Tourneux and Peltier, 1995) and the Mehler reaction (Biehler and Fock, 1996; Cornic and Massaci, 1996).

Induction of oxidative stress in water-stressed plants has been suggested, though evidence for it is rather indirect. In pea plants subjected to slow water stress, the amounts of metal ions available for the Haber-Weiss reaction and oxidative damage of lipids, proteins, and DNA increase, accompanying decreases in catalase, DHAR, and GR and increases in peroxidase and SOD activities (Moran *et al.*, 1994). Salt stress enhances SOD contents in the halophyte *Mesembryanthemum crystallimum* (Miszalski *et al.*, 1998). Tolerance against MV and drought parallels among drought-sensitive and drought-tolerant maize inbred lines (Malan *et al.*, 1990). MnSOD introduced into chloroplasts in alfalfa improves drought tolerance (McKersie *et al.*, 1996). NaCl-tolerant cotton cell lines had significantly higher activities of scavenging enzymes, such as catalase, guaiacol peroxidase, GR, and APX (Gossett *et al.*, 1996) and exhibited lower production of lipid peroxides (Gossett *et al.*, 1994). Most of these oxidative damages and responses in leaves appear to be accounted for as results of the preceding primary distortion of metabolism, e.g. a loss of SOD activity (Hernández *et al.*, 1993) and accumulation of iron (Price and Hendry, 1991).

Hernández *et al.* (1993) have observed increases in $O_2^{\bullet-}$ production and lipid peroxidation due to NaCl stress in mitochondria from pea leaves and ascribed them to the inactivation of the mitochondrial MnSOD. Domae *et al.* (1998) observed an increase in the MV-dependent photoproduction of MDHA$^{\bullet}$ in lettuce leaves due to water stress with polyethyleneglycol or sorbitol for 2–4 hours, preceding a decrease in photosynthetic electron transport. While APX and MDAR were unaffected, chloroplastic Cu/ZnSOD was inactivated by the stress treatment, accompanying the decrease in leaf water potential (Y. Domae, S. Kanematsu, E. Nawata, and J. Mano, unpublished data). The inactivation of SOD in either case may be a cause or a result of oxidative stress, but these results suggest that SOD isozymes may be early targets of water stress.

In considering tolerance against water stress, the responses of root cells should not be neglected in comparison with the insensitivity of leaf osmotic potential to osmotic stresses (Karakas *et al.*, 1997; Schmidhalter *et al.*, 1998). Upon water stress, the energy requirement in root epidermal cells increases for pumping water in, salts out, and to synthesize compatible solutes, so that the mitochondrial respiration will be stimulated. As a consequence, the production of ROS is expected to increase.

Apart from drought and high salinity, water logging is another form of water stress. Plants on flood plains experience abrupt submergence in water and subsequent re-emergence into the air because of water retreat and growth. Cellular metabolism drastically changes on this anaerobiosis-aerobiosis transition (Rumpho and Kennedy, 1991). Rice seeds germinating in water are exposed to air as they grow. Changes have been observed at this transition stage in the levels of antioxidants and of SOD expression in chloroplasts and mitochondria (Ushimaru *et al.*, 1994, 1995).

CONCLUSIONS

ROS are constantly produced in chloroplasts, mitochondria, and other compartments of plant cells. Under normal metabolism, rates of their production are low as compared with photosynthetic O_2 evolution and respiratory O_2 consumption. Steady-state concentrations of the primarily produced species, i.e. $O_2^{\bullet-}$, H_2O_2, and $^1O_2$, are suppressed to low levels by the small antioxidant molecules and by the enzymes that scavenge these species. The chloroplast is the major source of ROS in illuminated plant cells, and hence is potentially the most susceptible target organelle. In the peroxisome, H_2O_2 is produced via photorespiration and β-oxidation. $O_2^{\bullet-}$ is produced also in peroxisomal and plasma membranes.

Environmental stresses enhance oxidative stresses in plants, especially under sunlight (Figure 3) because of the imbalance between the production and scavenging of ROS in chloroplasts. Specific mechanisms are (i) direct generation of ROS (O_3), (ii) accelerated production of $O_2^{\bullet-}$ (MV, SO_2), (iii) inhibition of the scavenging enzymes (low temperature, water deficit), (iv) suppression of antioxidant capacity (carotenoid synthesis inhibitors), and, less directly, (v) acceleration of O_2 photoreduction because of overreduction of the photosynthetic electron transport chain, caused by inhibition of photosynthetic CO_2 assimilation (low temperature, water deficit, and UV-B).

$O_2^{\bullet-}$ and H_2O_2 are toxic enough to bring about initial damages especially to metalloproteins. Cu/ZnSOD, FeSOD, and APX, the major scavenging enzymes, are irreversibly inactivated by H_2O_2. Enzymes containing [4Fe-4S] centers are sensitive to

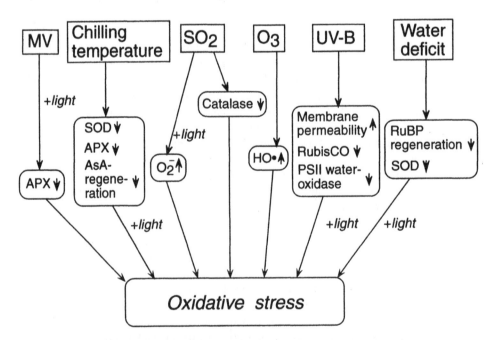

Figure 3. Induction mechanisms of oxidative stress by various environmental stresses. Short arrows beside targets indicate inhibition or inactivation (down) or increases (up). Light is required in some mechanisms. Note that key enzymes for scavenging reactive oxygen species (SOD, APX, and catalase) are vulnerable to environmental stresses, hence, are primary targets.

$O_2^{\bullet-}$. These enzymes are primary targets of $O_2^{\bullet-}$ and H_2O_2. Metal ions released from these proteins can catalyze the Haber-Weiss reaction to produce highly reactive $HO^{\bullet}$. When the decreased capacity to scavenge $O_2^{\bullet-}$ and H_2O_2 are not immediately recovered by repair and *de novo* synthesis, further radical chain reactions propagate and more extensive damages will be brought to lipids, proteins and nucleic acids leading to cell death.

Such imbalance between the production and scavenging of ROS is probably induced by various environmental stresses not only in chloroplasts but also in extrachloroplastic compartments and in nonphotosynthetic tissues. Induction of oxidative stress and the tolerance against it in root cells, as the first sensing tissue of water stress, are important issues for further investigation.

Acknowledgment

The author would like to express his sincere gratitude to Prof. K. Asada for his critical reading of the manuscript.

REFERENCES

Allan, A.C. and Fluhr, R. (1997) Two distinct sources of elicited reactive oxygen species in tobacco epidermal cells. *Plant Cell*, **9**, 1559–1572.

Alscher, R.G., Donahue, J.L., and Cramer, C.L. (1997) Reactive oxygen species and antioxidants: Relationships in green cells. *Physiol. Plant.*, **100**, 224–233.

Ananyev, G., Wydrzynski, T., Renger, G., and Klimov, V. (1992) Transient peroxide formation by the manganese-containing, redox-active donor side of photosystem II upon inhibition of O_2 evolution with lauroylchlorine chloride. *Biochim. Biophys. Acta*, **1100**, 303–311.

Anderson, J.M., Park, Y.I., and Chow, W.S. (1998) Unifying model for the photoinactivation of photosystem II in vivo under steady-state photosynthesis. *Photosynth. Res.*, **56**, 1–13.

Andersson, B. and Barber, J. (1996) Mechanisms of photodamage and protein degradation during photoinhibition of photosystem II. In N.R. Baker, (ed.), *Photosynthesis and the Environment*, Kluwer Academic Publishers, Dordrecht, pp. 101–121.

Armstrong, D.A. and Buchanan, J.D. (1978) Reactions of O_2^-, H_2O_2 and other oxidants with sulfhydryl enzymes. *Photochem. Photobiol.*, **28**, 743–755.

Aro, E.-M., Virgin, I., and Andersson, B. (1993) Photoinhibition of photosystem II. Inactivation, protein damage and turnover. *Biochim. Biophys. Acta*, **1143**, 113–134.

Asada, K. and Kiso, K. (1973a) Initiation of aerobic oxidation of sulfite by illuminated spinach chloroplasts. *Eur. J. Biochem.*, **33**, 253–257.

Asada, K. and Kiso, K. (1973b) The photooxidation of epinephrine by spinach chloroplasts and its inhibition by superoxide dismutase: Evidence for the formation of superoxide radicals in chloroplasts. *Agric. Biol. Chem.*, **37**, 453–454.

Asada, K., Urano, M., and Takahashi, M. (1973) Subcellular location of superoxide dismutase in spinach leaves and preparation and properties of crystalline spinach superoxide dismutase. *Eur. J. Biochem.*, **36**, 257–266.

Asada, K. and Kanematsu, S. (1976) Reactivity of thiols with superoxide radicals. *Agric. Biol. Chem.*, **40**, 867–872.

Asada, K. and Takahashi, M. (1987) Production and scavenging of active oxygen in photosynthesis. In D.J. Kyle, C.B. Osmond and C.J. Arntzen, (eds.), *Photoinhibition*, Elsevier, Amsterdam, pp. 227–287.

Asada, K. (1994) Production and action of active oxygen species in photosynthetic tissues. In C.H. Foyer and P.M. Mullineaux, (eds.), *Causes of Photooxidative Stress and Amelioration of Defense Systems in Plants*, CRC Press, Boca Raton, pp. 77–104.

Asada, K. (1996) Radical production and scavenging in chloroplasts. In N.R. Baker, (ed.), *Photosynthesis and the Environments*, Kluwer Academic Publishers, Dordrecht, pp. 123–150.

Asada, K. (1997) The role of ascorbate peroxidase and monodehydroascorbate reductase in H_2O_2 scavenging in plants. In J.G. Scandalios, (ed.), *Oxidative Stress and the Molecular Biology of Antioxidant Defenses*, (Monograph Series, Vol. 34), Cold Spring Harbor Laboratory Press, Cold Spring Harbor, pp. 715–735.

Asada, K., Endo, T., Mano, J., and Miyake, C. (1998) Molecular mechanism for relaxation of and protection from light stress. In N. Murata and K. Satoh, (eds.), *Stress Responses of Photosynthetic Organisms*, Elsevier, Amsterdam, pp. 37–52.

Asada, K. (1999) The water-water cycle in chloroplasts: scavenging of active oxygens and dissipation of excess photons. *Ann. Rev. Plant Physiol. Plant Mol. Biol.*, **50**, 601–639.

Auh, C.-K. and Murphy, T.M. (1995) Plasma membrane redox enzyme is involved in the synthesis of O_2^- and H_2O_2 by Phytophthora elicitor-stimulated rose cells. *Plant Physiol.*, **107**, 1241–1247.

Baier, M. and Dietz, K.-J. (1996) 2-Cys peroxiredoxin bas1 from *Arabidopsis thaliana* (Accession No. X94218) (PGR 96-031). *Plant Physiol.*, **111**, 651.

Baier, M. and Dietz, K.-J. (1997) The plant 2-Cys peroxiredoxin BAS1 is a nuclear-encoded chloroplast protein: its expressional regulation, phylogenic origin, and implications for its specific physiological function in plants. *Plant J.*, **12**, 179–190.

Baker, N.R., Nogués, S., and Allen, D.J. (1997) Photosynthesis and photoinhibition. In P. Lusmunden, (ed.), *Plants and UV-B*, Cambridge University Press, Cambridge, pp. 95–111.

Beeor-Tzahar, T., Ben-Hayyim, G., Holland, D., Faltin, Z., and Eshdat, Y. (1995) A stress-associated citrus protein is a distinct plant phospholipid hydroperoxide glutathione peroxidase. *FEBS Lett.*, **366**, 151–155.

Bérczi, A. and Møller, I.M. (1998) NADH-monodehydroascorbate oxidoreductase is one of the redox enzymes in spinach leaf plasma membranes. *Plant Physiol.*, **116**, 1029–1036.

Beyer, W.F. and Fridovich, I. (1987) Effect of hydrogen peroxide on the iron-containing superoxide dismutase of *Escherichia coli. Biochemistry*, **26**, 1251–1257.

Biehler, K. and Fock, H. (1996) Evidence for the contribution of the Mehler-peroxidase reaction in dissipating excess electrons in drought-stressed wheat. *Plant Physiol.*, **112**, 265–272.

Bielski, B.H.J. (1978) Reevaluation of the spectral and kinetic properties of HO_2 and O_2^- free radicals. *Photochem. Photobiol.*, **28**, 645–649.

Blubaugh, D.J., Atamian, M., Babcock, G.T., Golbeck, J.H., and Cheniae, G.M. (1991) Photoinhibition of hydroxylamine-extracted photosystem II membranes: Identification of the sites of photodamage. *Biochemistry*, **30**, 7586–7597.

Bornman, J.F., Reuber, S., Cen, Y.-P., and Weissenböck, G. (1997) Ultraviolet radiation as a stress factor and the role of protective pigments. In P. Lusmunden (ed.), *Plants and UV-B*, Cambridge University Press, Cambridge, pp. 157–168.

Bowditch, M.I. and Donaldson, R.P. (1990) Ascorbate free-radical reduction by glyoxysomal membranes. *Plant Physiol.*, **94**, 531–537.

Böger, P. and Sandmann, G. (1998) Action of modern herbicides. In A.S. Raghavendra, (ed.), *Photosynthesis: A Comprehensive Treatise*, Cambridge University Press, Cambridge, pp. 337–351.

Bray, R.C., Cockle, S.A., Fielden, E.M., Roberts, P.B., Rotilio, G., and Calabrese, L. (1974) Reduction and inactivation of superoxide dismutase by hydrogen peroxide. *Biochem. J.*, **139**, 43–48.

Brennan, T. and Frenkel, C. (1977) Involvement of hydrogen peroxide in the regulation of senescence in pear. *Plant Physiol.*, **59**, 411–416.

Bunkelmann, J.R. and Trelease, R.N. (1996) Ascorbate peroxidase. A prominent membrane protein in oilseed glyoxysomes. *Plant Physiol.*, **110**, 589–598.

Buettner, G.R. and Jurkiewicz, B.A. (1993) Ascorbate free radical as a marker of oxidative stress: an EPR study. *Free Rad. Biol. Med.*, **14**, 49–55.

Byvoet, P., Balis, J.U., Shelley, S.A., Montgomery, M.R., and Barber, M.J. (1995) Detection of hydroxyl radicals upon interaction of ozone with aqueous media or extracellular surfactant: The role of trace iron. *Arch. Biochem. Biophys.*, **319**, 464–469.

Cadenas, E., Hochsterin, P., and Erster, L. (1992) Pro- and antioxidant functions of quinones and quinone reductases in mammalian cells. *Advances in Enzymol.*, **65**, 97–146.

Casano, L.M., Gómez, L.D., Lascano, H.R., González, C.A., and Trippi, V.S. (1997) Inactivation and degradation of CuZn-SOD by active oxygen species in wheat chloroplasts exposed to photooxidative stress. *Plant Cell Physiol.*, **38**, 433–440.

Chamnongpol, S., Willekins, H., Langesbartels, C., Van Montagu, M., Inzé, D., and Van Camp, W. (1996) Transgenic tobacco with a reduced catalase activity develops necrotic lesions and induces pathogenesis-related expression under high light. *Plant J.*, **10**: 491–503.

Chance, B., Sies, H., and Boveris, A. (1979) Hydroperoxide metabolism in mammalian organs. *Physiol. Rev.*, **59**, 527–605.

Chang, C.J. and Kao, C.H. (1997) Paraquat toxicity is reduced by metal chelators in rice plants. *Physiol. Plant.*, **101**, 471–476.

Chen, G.-X. and Asada, K. (1992) Inactivation of ascorbate peroxidase by thiols requires hydrogen peroxide. *Plant Cell Physiol.*, **33**, 117–123.

Conklin, P.L., Williams, E.H., and Last, R.L. (1996) Environmental stress sensitivity of an ascorbic acid-deficient *Arabidopsis* mutant. *Proc. Natl. Acad. Sci. USA*, **93**, 9970–9974.

Cornic, G. and Massacci, A. (1996) Leaf photosynthesis under drought stress. In N.R. Baker, (ed.), *Photosynthesis and the Environments*, Kluwer Academic Publishers, Dordrecht, pp. 347–366.

Covello, P.S., Chang, A., Dumbroff, E.B., and Thompson, J.E. (1989) Inhibition of photosystem II precedes thylakoid membrane lipid peroxidation in bisulfite-treated leaves of *Phaseolus vulgaris*. *Plant Physiol.*, **90**, 1492–1497.

Dai, W., Yan, B., Huang, S., Liu, X., Peng, S., Miranda, M.L.L., Chavez, A.Q., Vergara, B.S., and Olszyk, D.M. (1997) Response of oxidative stress defense systems in rice (*Oryza sativa*) leaves with supplemental UV-B radiation. *Physiol. Plant.*, **101**, 301–308.

Dalton, D.A., Langeberg, L., and Robbins, M. (1992) Purification and characterization of monodehydroascorbate reductase from soybean root nodules. *Arch. Biochem. Biophys.*, **292**, 281–286.

Dat, J.F., Lopez-Delgado, H., Foyer, C.H., and Scott, I.M. (1998) Parallel changes in H_2O_2 and catalase during thermotolerance induced by salicylic acid or heat acclimation in mustard seedlings. *Plant Physiol.*, **116**, 1351–1357.

Davison, A.J., Kettle, A.J., and Fatur, D.J. (1986) Mechanism of the inhibition of catalase by ascorbate. Roles of active oxygen species, copper and semidehydroascorbate. *J. Biol. Chem.*, **261**, 1193–1200.

del Río, L.A., Pastori, G.M., Palma, J.M., Sandalio, L.M., Sevilla, F., Corpas, F.J., Jiménez, A., López-Huertas, E., and Hernández, J.A. (1998) The activated oxygen role of peroxisomes in senescence. *Plant Physiol.*, **116**, 1195–1200.

Demmig-Adams, B. and Adams, W.W. III (1992) Photoprotection and other responses of plants to high light stress. *Annu. Rev. Plant Physiol. Plant Mol. Biol.*, **43**, 599–626.

Doke, N. (1997) The oxidative burst: roles in signal transduction and plant stress. In J.G. Scandalios, (ed.), *Oxidative Stress and the Molecular Biology of Antioxidant Defences*, (Monograph Series, Vol. 34), Cold Spring Harbor Laboratory Press, Cold Spring Harbor, pp. 785–813.

Domae, Y., Nawata, E. Sakuratani, T., and Mano, J. (1998) Photooxidative stress is induced in early stage of water stress in lettuce leaves. *Jpn. J. Crop Sci.*, **67** (Extra issue 2), 408–409.

Farrington, J.A., Ebert, M., Land, E.J., and Fletcher, K. (1973) Bipyridium quarternary salts and related compounds. V. Pulse radiolysis studies of the reaction of paraquat radical with oxygen. Implication for the mode of action of bipyridium herbicides. *Biochim. Biophys. Acta*, **314**, 372–381.

Fine, P.L. and Frasch, W.D. (1992) The oxygen-evolving complex requires chloride to prevent hydrogen peroxide formation. *Biochemistry*, **31**, 12204–12210.

Fiscus, E.L. and Booker, F.L. (1995) Is increased UV-B a threat to crop photosynthesis and productivity? *Photosynth. Res.*, **43**, 81–92.

Foyer, C.H. and Lelandais, M. (1996) A comparison of the relative rates of transport of ascorbate and glucose across the thylakoid, chloroplast and plasma membranes of pea leaf mesophyll cells. *J. Plant Physiol.*, **148**, 391–398.

Foyer, C.H., Lopez-Delgado, H., Dat, J.F., and Scott, I.M. (1997) Hydrogen peroxide- and glutathione-associated mechanisms of acclimatory stress tolerance and signalling. *Physiol. Plant.*, **100**, 241–254.

Fryer, M.J., Andrews, J.R., Oxborough, K., Blowers, D.A., and Baker, N.R. (1998) Relationship between CO_2 assimilation, photosynthetic electron transport, and active O_2 metabolism in leaves of maize in the field during periods of low temperature. *Plant Physiol.*, **116**, 571–580.

Gianinetti, A., Lorenzoni, C., and Marocco, A. (1993) Changes in superoxide dismutase and catalase activities in response to low temperature in tomato mutants. *J. Genet. Breed.*, **47**, 353–356.

Gossett, D.R., Millhollon, E.P., and Lucas, M.C. (1994) Antioxidant response to NaCl stress in salt-tolerant and salt-sensitive cultivars of cotton. *Crop Sci.*, **34**, 706–714.

Gossett, D.R., Banks, S.W., Millhollon, E.P., and Lucas, M.C. (1996) Antioxidant response to NaCl stress in a control and an NaCl-tolerant cotton cell line grown in the presence of paraquat, buthionine sulfoximine, and exogenous glutathione. *Plant Physiol.*, **112**, 803–809.

Grace, S.C. and Logan, B.A. (1996) Acclimation of foliar antioxidant system to growth irradiance in three broad-leaved evergreen species. *Plant Physiol.*, **112**, 1631–1640.

Grotjohann, N., Janning, A., and Eising, R. (1997) In vitro photoinactivation of catalase isoforms from cotyledons of sunflower (*Helianthus annuus* L.). *Arch. Biochem. Biophys.*, **346**, 208–218.

Gueta-Dahan, Y., Yaniv, Z., Zilinskas, B., and Ben-Hayyim, G. (1997) Salt and oxidative stress: similar and specific responses and their relation to salt tolerance in Citrus. *Planta*, **203**, 460–469.

Halliwell, B. and Gutteridge, J.M.C. (1999) *Free Radicals in Biology and Medicine*, 3rd ed., Oxford University Press, Oxford.

Harman, L.S., Mottley, C., and Mason, R. (1984) Free radical metabolites of L-cysteine oxidation. *J. Biol. Chem.*, **259**, 5606–5611.

Harman, L.S., Carver, D., Schreiber, J., and Mason, R. (1986) One- and two-electron oxidation of reduced glutathione by peroxidases. *J. Biol. Chem.*, **261**, 1642–1648.

Harwood, H.J. Jr, Greene, Y.J., and Stacpoole, P.W. (1986) Inhibition of human leukocyte 3-hydroxy-3-methylglutaryl coenzyme A reductase activity by ascorbic acid. *J. Biol. Chem.*, **261**, 7127–7135.

Havaux, M., Greppin, H., and Strasser, R.J. (1991) Functioning of photosystems I and II in pea leaves exposed to heat stress in the presence or absence of light. *Planta*, **186**, 88–98.

Havaux, M. and Davaud, A. (1994) Photoinhibition of photosynthesis in chilled potato leaves is not correlated with a loss of photosystem-II activity. Preferential inactivation of photosystem I. *Photosynthesis Res.*, **40**: 75–92.

Heber, U., Miyake, C., Mano, J., Ohno, C., and Asada, K. (1996) Monodehydroascorbate radical detected by electron paramagnetic resonance spectrometry is a sensitive probe of oxidative stress in intact leaves. *Plant Cell Physiol.*, **37**, 1066–1072.

Hernández, J.A., Corpas, F.J., Gómez, M., del Río, L.A., and Sevilla, F. (1993) Salt-induced oxidative stress mediated by activated oxygen species in pea leaf mitochondria. *Plant Physiol.*, **89**, 103–110.

Hernández, J.A., Olmos, E., Corpas, F.J., Sevilla, F., and del Río, L.A. (1995) Salt-induced oxidative stress in chloroplasts of pea plants. *Plant Science*, **105**, 151–167.

Hertwig, B., Streb, P., and Feierabend, J. (1992) Light dependence of catalase synthesis and degradation in leaves and the influence of interfering stress conditions. *Plant Physiol.*, **100**, 1547–1553.

Hetherington, S.E., He, J., and Smillie, R.M. (1989) Photoinhibition at low temperature in chilling-sensitive and -resistant plants. *Plant Physiol.*, **90**, 1609–1615.

Hideg, É., Spetea, C., and Vass, I. (1994) Singlet oxygen and free radical production during acceptor- and donor-side-induced photoinhibition. Studies with spin trapping EPR spectroscopy. *Biochim. Biophys. Acta*, **1186**, 143–152.

Hideg, É. and Vass, I. (1996) UV-B induced free radical production on plant leaves and isolated thylakoid membranes. *Plant Sci.*, **115**, 251–260.

Hideg, É., Mano, J., Ohno, C., and Asada, K. (1997) Increased levels of monodehydroascorbate radical in UV-B-irradiated broad bean leaves. *Plant Cell Physiol.*, **38**, 684–690.

Hideg, É., Kálai, T., Hideg, K., and Vass, I. (1998) Photoinhibition of photosynthesis in vivo results in singlet oxygen production detection via nitroxide-induced fluorescence quenching in broad bean leaves. *Biochemistry*, **37**, 11405–11411.

Hodgson, E.K. and Fridovich, I. (1975) The interaction of bovine erythrocyte superoxide dismutase with hydrogen peroxide: Inactivation of the enzyme. *Biochemistry*, **14**, 5294–5299.

Hossain, M.A. and Asada, K. (1984) Inactivation of ascorbate peroxidase in spinach chloroplasts on dark addition of hydrogen peroxide: Its protection by ascorbate. *Plant Cell Physiol.*, **25**, 1285–1295.

Hossain, M.A., Nakano, Y., and Asada, K. (1984) Monodehydroascorbate reductase in spinach chloroplasts and its participation in regeneration of ascorbate for scavenging hydrogen peroxide. *Plant Cell Physiol.*, **25**, 385–395.

Hossain, M.A. and Asada, K. (1985) Monodehydroascorbate reductase from cucumber is a flavin adenine dinucleotide enzyme. *J. Biol. Chem.*, **260**, 12920–12926.

Huner, N.P.A., Öquist, G., Hurry, V.M., Krol, M., Falk, S., and Grifith, M. (1993) Photosynthesis, photoinhibition and low temperature acclimation in cold tolerant plants. *Photosynthesis Res.*, **37**, 19–39.

Iturbe-Ormaetxe, I., Escuredo, P.R., Arrese-Igor, C., and Becana, M. (1998) Oxidative damage in pea plants exposed to water deficit or paraquat. *Plant Physiol.*, **116**, 173–181.

Jakob, B. and Heber, U. (1996) Photoproduction and detoxification of hydroxyl radicals in chloroplasts and leaves and relation to photoinactivation of photosystems I and II. *Plant Cell Physiol.*, **37**, 629–635.

Jahnke, L.S., Hull, M.R., and Long, S.P. (1991) Chilling stress and oxygen metabolizing enzymes in *Zea mays* and *Zea diploperennis*. *Plant Cell Environ.*, **14**, 97–104.

Jansen, M.A.K., Babu, T.S., Heller, D., Gaba, V., Mattoo, A.K., and Edenman, M. (1996) Ultraviolet-B effects on *Spirodela oligorrhiza*: Induction of different protection mechanisms. *Plant Sci.*, **115**, 217–223.

Jarabak, R., Harvey, R.G., and Jarabak, J. (1997) Redox cycling of polycyclic aromatic hydrocarbon *o*-quinones: Reversal of superoxide dismutase inhibition by ascorbate. *Arch. Biochem. Biophys.*, **339**, 92–98.

Jiménez, A., Hernández, J.A., Pastori, G.M., del Río, L.A., and Sevilla, F. (1997) Evidence for the presence of the ascorbate-glutathione cycle in mitochondria and peroxisomes of pea leaves. *Plant Physiol.*, **114**, 275–284.

Kaiser, W. (1979) Reversible inhibition of the Calvin cycle and activation of oxidative pentose phosphate cycle in isolated intact chloroplasts by hydrogen peroxide. *Planta*, **145**, 377–382.

Kaiser, W.M. (1987) Effects of water deficit on photosynthetic capacity. *Physiol. Plant.*, **71**, 142–149.

Kampfenkel, K., Van Montagu, M., and Inzé, D. (1995) Effects of iron on *Nicotiana plumbaginifolia* plants. Implications to oxidative stress. *Plant Physiol.*, **107**, 725–735.

Karakas, B., Ozias-Akins, P., Stushnoff, C., Suefferheld, M., and Rieger, M. (1997) Salinity and drought tolerance of mannitol-accumulating transgenic tobacco. *Plant Cell Environ.*, **20**, 609–616.

Karpinski, S., Reynolds, H., Karpinska, B., Wingsle, G., Creissen, G., and Mullineaux, P. (1999) Systemic signaling and acclimation in response to excess excitation energy in *Arabidopsis*. *Science*, **284**, 654–657.

Katoh, S. and San Pietro, A. (1967) Ascorbate-supported NADP photoreduction by heated *Euglena* chloroplasts. *Arch. Biochem. Biophys.*, **122**, 144–152.

Kozaki, A. and Takeba, G. (1996) Photorespiration protects C_3 plants from photoinhibition. *Nature*, **384**, 557–560.

Kraus, T.E. and Fletcher, R.A. (1994) Paclobutrazol protects wheat seedlings from heat and paraquat injury. Is detoxification of active oxygen involved? *Plant Cell Physiol.*, **35**, 45–52.

Kulandaivelu, G. and Noorudeen, A.M. (1983) Comparative study of the action of ultraviolet-C and ultraviolet-B radiation on photosynthetic electron transport. *Physiol. Plant.*, **58**, 389–394.

Laisk, A. and Edwards, G.E. (1998) Oxygen and electron flow in C_4 photosynthesis: Mehler reaction, photorespiration and CO_2 concentration in the bundle sheath. *Planta*, **205**, 632–645.

Leonardis, S.D., Lorenzo, G.D., Borracino, G., and Dippiero, S. (1995) A specific ascorbate free radical reductase isozyme participates in the regeneration of ascorbate for scavenging toxic oxygen species in potato tuber mitochondria. *Plant Physiol.*, **109**, 847–851.

Lusmunden, P. (1997) *Plants and UV-B: Responses to Environmental Change*, Cambridge University Press, Cambridge.

Luwe, M., Takahama, U., and Heber, U. (1993) Role of ascorbate in detoxifying ozone in the apoplast of spinach (*Spinacia oleracea* L.) leaves. *Plant Physiol.*, **101**, 969–976.

Macpherson, A.N., Telfer A., Barber J., and Truscott, T.G. (1993) Direct detection of singlet oxygen from isolated photosystem II reaction centers. *Biochim. Biophys. Acta*, **1143**, 301–309.

Malan, C., Greylin, M.M., and Gressel, J. (1990) Correlation between CuZn superoxide dismutase and glutathione reductase, and environmental and xenobiotic stress tolerance in maize inbreds. *Plant Sci.*, **69**, 157–166.

Mano, J., Ushimaru, T., and Asada, K. (1997) Ascorbate in thylakoid lumen as an endogenous electron donor to photosystem II: Protection of thylakoids from photoinhibition and regeneration of ascorbate in stroma by dehydroascorbate reductase. *Photosynthesis Res.*, **53**, 197–204.

Mano, J., Ohno, C., and Asada, K. (1998) Loss of H_2O_2-scavenging capacity due to inactivation of ascorbate peroxidase in methylviologen-fed leaves, an estimation by electron spin resonance spectrometry. In G. Garab, (ed.), *Photosynthesis, Mechanisms and Effects*, Vol. V, Kluwer Academic Publishers, Dordrecht, pp. 3909–3912.

Mano, J., Babiychuk, E., Belles-Boix, E., Hiratake, J., Kimura, A., Inzé, D., Kushnir, S., and Asada, K. (2000) A novel NADPH: diamide oxidoreductase activity in *Arabidopsis thaliana* P1 ζ-crystallin. *Eur.J. Biochem.*, **267**, 3661–3671.

Marré, E. and Arrigoni, O. (1958) Ascorbic acid and photosynthesis. I. "Monodehydroascorbic acid" reductase of chloroplasts. *Biochim. Biophys. Acta*, **30**, 453–457.

Matoo, A.K., Hoffman-Falk, H., Marder, J.B., and Edelman, M. (1984) Regulation of protein metabolism; coupling of photosynthetic electron transport to in vivo degradation of the rapidly metabolized 32-kDa protein of chloroplast membranes. *Proc. Natl. Acad. Sci. USA*, **81**, 1380–1384.

McAinsh, M.R., Clayton, H., Mansfield, T.A., and Hetherington, A.M. (1996) Changes in stomatal behavior and guard cell cytosolic free calcium in response to oxidative stress. *Plant Physiol.*, **111**, 1031–1042.

McKersie, B.D., Bowley, S.R., Harjanto, E., and Leprince, O. (1996) Water-deficit tolerance and field performance of transgenic alfalfa overexpressing superoxide dismutase. *Plant Physiol.*, **111**, 1177–1181.

Mehler, A.H. (1951) Studies on reactivities of illuminated chloroplasts. I. Mechanism of the reduction of oxygen and other Hill reagents. *Arch. Biochem. Biophys.*, **33**, 65–77.

Mehlhorn, H., Lalendais, M., Korth, H.G., and Foyer, C.H. (1996) Ascorbate is the natural substrate for plant peroxidases. *FEBS Lett.*, **378**, 203–206.

Mishra, R.K. and Singhal, G.S. (1992) Function of photosynthetic apparatus of intact wheat leaves under high light and heat stress and its relationship with peroxidation of thylakoid lipids. *Plant Physiol.*, **98**, 1–6.

Miszalski, Z., G[u]llesak, I., Niewiadomska, E., Baczek-Kwinta, R., Lüttige, U., and Ratajczak, R. (1998) Subcellular localization and stress responses of superoxide dismutase from leaves in the C_3-CAM intermediate halophyte *Msembryanthemum crystallinum* L. *Plant Cell Environ.*, **21**, 169–179.

Miyake, C., Michihata, F., and Asada, K. (1991) Scavenging of hydrogen peroxide in prokaryotic and eukaryotic algae: acquisition of ascorbate peroxidase during the evolution of cyanobacteria. *Plant Cell Physiol.*, **32**, 33–43.

Miyake, C. and Asada, K. (1992) Thylakoid-bound ascorbate peroxidase in spinach chloroplasts and photoreduction of its primary oxidation product monodehydroascorbate radicals in thylakoids. *Plant Cell Physiol.*, **33**, 541–553.

Miyake, C. and Asada, K. (1994) Ferredoxin-dependent photoreduction of the monodehydroascorbate radicals in spinach thylakoids. *Plant Cell Physiol.*, **35**, 539–549.

Miyake, C. and Asada, K. (1996) Inactivation mechanism of ascorbate peroxidase at low concentrations of ascorbate; Hydrogen peroxide decomposes compound I of ascorbate peroxidase. *Plant Cell Physiol.*, **37**, 423–430.

Miyake, C., Schreiber, U., Hormann, H., Sano, S., and Asada, K. (1998) The FAD-enzyme monodehydroascorbate radical reductase mediates photoproduction of superoxide radicals in spinach thylakoid membranes. *Plant Cell Physiol.*, **39**, 821–829.

Miyake, C., and Yokota, A. (2000) Determination of the rate of photoreduction of O_2 in the water-water cycle in watermelon leaves and enhancement of the rate by limitation of photosynthesis. *Plant Cell Physiol.*, **41**, 335–343.

Mohanty, N., Murthy, S.D.S., and Mohanty, P. (1987) Reversal of heat-induced alterations in photochemical activities in wheat primary leaves. *Photosynthesis Res.*, **14**, 259–267.

Moran, J.F., Becana, M., Iturbe-Ormaetxe, I., Frechilla, S., Klucas, R.V., and Aparicio-Tejo, P. (1994) Drought induces oxidative stress in pea plants. *Planta*, **194**, 346–352.

Motowska, A. (1996) Environmental factors affecting chloroplasts. In M. Passarakli, (ed.), *Handbook of Photosynthesis*, Marcel Dekker, New York, pp. 407–426.

Nakano, Y. and Asada, K. (1981) Spinach chloroplasts scavenge hydrogen peroxide on illumination. *Plant Cell Physiol.*, **21**, 1295–1307.

Neubauer, C. and Yamamoto, H. Y. (1992) Mehler-peroxidase reaction mediates zeaxanthin-related fluorescence quenching in intact chloroplasts. *Plant Physiol.*, **99**, 1354–1361.

Niki, E. Noguchi, N., Tsuchihashi, H., and Goto, N. (1995) Interaction among vitamin C, vitamin E, and β-carotene. *Am. J. Clin. Nutr.*, **62**, S1322-S1326.

Nishimura, K., Goto, M., and Mano, J. (1996) Participation of the superoxide radical in the beneficial effect of ascorbic acid on heat-induced fish meat gel (Kamaboko). *Biochem. Biotech. Biosci.*, **60**, 1966–1970.

Noctor, G. and Foyer, C.H. (1998) Ascorbate and glutathione: Keeping active oxygen under control. *Annu. Rev. Plant Physiol. Plant Mol. Biol.*, **49**, 249–279.

Ogawa, K. Kanematsu, S., Takabe, K., and Asada, K. (1995) Attachment of CuZn-superoxide dismutase to thylakoid membranes at the site of superoxide generation (PSI) in spinach chloroplasts: detection by immuno-gold labeling after rapid freezing and substitution method. *Plant Cell Physiol.*, **36**, 565–573.

Ogawa, K., Kanematsu, S., and Asada, K. (1997) Generation of superoxide anion and localization of CuZn-superoxide dismutase in the vascular tissue of spinach hypocotyls: Their association with lignification. *Plant Cell Physiol.*, **38**, 1118–1126.

Osmond, C.B. and Grace, S.C. (1995) Perspectives on photoinhibition and photorespiration in the field: Quintessential inefficiencies of the light and dark reactions of photosynthesis? *J. Exp. Bot.*, **46**, 1351–1362.

Otter, T. and Polle, A. (1994) The influence of apoplastic ascorbate on cell wall-associated peroxidase and NADH oxidase activities in Norway spruce (*Picea abies* L.) needles. *Plant Cell Physiol.*, **35**, 1231–1238.

Öquist, G. and Huner, N. P.A. (1993) Cold-hardening-induced resistance to photoinhibition of photosynthesis in winter rye is dependent upon an increased capacity for photosynthesis. *Planta*, **189**, 150–156.

Packer, J.E., Slater, T.F., and Willson, R.L. (1979) Direct observation of a free radical interaction between vitamin E and vitamin C. *Nature*, **278**, 737–738.

Palatnik, J.F., Valle, E.M., and Carillo, N. (1997) Oxidative stress causes ferredoxin-NADP$^+$ reductase solubilization from the thylakoid membranes in methyl viologen-treated plants. *Plant Physiol.*, **115**, 1721–1727.

Park, Y.-I., Chow, W.S., Osmond, C.B., and Anderson, J.M. (1996) Electron transport to oxygen mitigates against the photoinactivation of photosystem II in vivo. *Photosynthesis Res.*, **50**, 23–32.

Peiser, G.D., Lizada, M.C.C., and Yang, S.F. (1982) Sulfite-induced lipid peroxidation in chloroplasts as determined by ethane production. *Plant Physiol.*, **70**, 994–998.

Pinhero, R.G., Rao, M.V., Paliyath, G., Murr, D.P., and Fletcher, R.A. (1997) Changes in activities of antioxidant enzymes and their relationship to genetic and paclobutrazol-induced chilling tolerance of maize seedlings. *Plant Physiol.*, **114**, 695–704.

Polle, A. (1997) Defense against photooxidative damage in plants. In J.G. Scandalios, (ed.), *Oxidative Stress and the Molecular Biology of Antioxidant Defences*, (Monograph Series, Vol. 34), Cold Spring Harbor Laboratory Press, Cold Spring Harbor, pp. 623–666.

Prasad, T.K., Anderson, M.D., Martin, B.A., and Stewart, C.R. (1994) Evidence for chilling-induced oxidative stress in maize seedlings and a regulatory role for hydrogen peroxide. *Plant Cell*, **6**, 65–74.

Price, A.H. and Hendry, G.A.F. (1991) Ion-catalyzed oxygen radical formation and its possible contribution to drought damage in nine native grasses and three cereals. *Plant Cell Environ.*, **14**, 477–484.

Puppo, A. and Halliwell, B. (1988) Formation of hydroxyl radicals from hydrogen peroxide in the presence of iron. *Biochem. J.*, **249**, 185–190.

Rao, M.V., Paliyath, G., and Ormrod, D.P. (1996) Ultraviolet-B- and ozone-induced biochemical changes in antioxidant enzymes of *Arabidopsis thaliana*. *Plant Physiol.*, **110**, 125–136.

Renger, G., Völker, H., Eckert, H.J., Fromme, R., Hohm-Veit, S., and Gräber, P. (1989) On the mechanism of photosystem II deterioration by UV-B irradiation. *Photochem. Photobiol.*, **49**, 97–105.

Richter, C. and Schweizer, M. (1997) Oxidative stress in mitochondria. In J.G. Scandalios, (ed.), *Oxidative Stress and the Molecular Biology of Antioxidant Defences*, (Monograph Series, Vol. 34), Cold Spring Harbor Laboratory Press, Cold Spring Harbor, pp. 169–200.

Rumpho, M.E. and Kennedy, R.A. (1991) Anaerobic metabolism in germinating seeds of *Echinochloa crusgalli* var. *oryzicola* (barnyard grass): Metabolite and enzyme studies. *Plant Physiol.*, **68**, 165–168.

Runeckles, V.C. and Krupa, S.V. (1994) The impact of UV-B radiation and ozone on terrestrial vegetation. *Environ. Pollut.*, **83**, 191–213.

Runeckles, V.C. and Vaatnou, M. (1997) EPR evidence for superoxide anion formation in leaves during exposure to low levels of ozone. *Plant Cell Environ.*, **20**, 306–314.

Saito, I., Matsuura, T., and Inoue, K. (1983) Formation of superoxide ion via one-electron transfer from electron donors to singlet oxygen. *J. Am. Chem. Soc.*, **105**, 3200–3206.

Sano, S., Miyake, C., Mikami, B., and Asada, K. (1995) Molecular characterization of monodehydroascorbate radical reductase from cucumber overproduced in *Escherichia coli*. *J. Biol. Chem.*, **270**, 21354–21361.

Scandalios, J.G., Guan, L., and Polidoros, A.N. (1997) Catalases in plants: Gene structure, properties, regulation, and expression. In J.G. Scandalios, (ed.), *Oxidative Stress and the Molecular Biology of Antioxidant Defenses*, (Monograph Series, Vol. 34), Cold Spring Harbor Laboratory Press, Cold Spring Harbor, pp. 343–406.

Schmithalter, U., Burucs, Z., and Camp, K.H. (1998) Sensitivity of root and leaf water status in maize (*Zea mays*) subjected to mild soil dryness. *Aust. J. Plant Physiol.*, **25**, 307–316.

Schöneich, C., Asmus, K.-D., Dillinger, U., and von Bruchbausen, F. (1992) Oxidation of polyunsaturated fatty acids and lipids through thiyl and sulfonyl radicals: reaction kinetics, and influence of oxygen and structure of thiyl radicals. *Arch. Biochem. Biophys.*, **292**, 456–467.

Schreiber, U. and Neubauer, C. (1990) O_2-dependent electron flow, membrane energization and the mechanism of non-photochemical quenching. *Photosynthesis Res.*, **25**, 279–293.

Sen Gupta, A., Heinen, J.L., Holaday, A.S., Burke, J.J., and Allen, R.D. (1993) Increased resistance to oxidative stress in transgenic plants that overexpress chloroplastic Cu/Zn superoxide dismutase. *Proc. Natl. Acad. Sci. USA*, **90**, 1629–1633.

Sharkey, T.D. and Badger, M.R. (1982) Effects of water stress on photosynthetic electron transport, photophosphorylation, and metabolite levels of *Xanthium strumarium* mesophyll cells. *Planta*, **156**, 199–206.

Shen, B., Jensen, R.G., and Bohnert, H.J. (1997) Increased resistance to oxidative stress in transgenic plants by targeting mannitol biosynthesis to chloroplasts. *Plant Physiol.*, **113**, 1177–1183.

Shimazaki, K., Sakaki, T., Kondo, N., and Sugahara, K. (1980) Active oxygen participation in chlorophyll destruction and lipid peroxidation in SO_2-fumigated leaves of spinach. *Plant Cell Physiol.*, **21**, 1193–1204.

Smirnoff, N. and Cumbes, Q.J. (1989) Hydroxyl radical scavenging activity of compatible solutes. *Phytochemistry*, **28**, 1057–1060.

Smirnoff, N. (1993) The role of active oxygen in the response of plants to water deficit and desiccation. *New Phytol.*, **125**, 27–58.

Sonoike, K., Terashima, I., Iwaki, M., and Itoh, S. (1995) Destruction of photosystem I iron-sulfur centers in leaves of *Cucumis sativus* L. by weak illumination at chilling temperatures. *FEBS Lett.*, **362**, 235–238.

Sonoike, K. (1996a) Degradation of psaB gene product, the reaction center subunit of photosystem I, is caused during photoinhibition of photosystem I: Possible involvement of active oxygen species. *Plant Sci.*, **115**, 157–164.

Sonoike, K. (1996b) Photoinhibition of photosystem I: Its physiological significance in the chilling sensitivity of plants. *Plant Cell Physiol.*, **37**, 239–247.

Strid, Å., Chow, W.S., and Anderson, J.M. (1994) UV-B damage and protection at the molecular level in plants. *Photosynthesis Res.*, **39**, 475–489.

Takahama, U. (1979) Stimulation of lipid peroxidation and carotenoid bleaching by deuterium oxide in illuminated chloroplast fragments: Participation of singlet molecular oxygen in the reactions. *Plant Cell Physiol.*, **20**, 213–218.

Takahama, U. and Oniki, T. (1992) Regulation of peroxidase dependent oxidation of phenolics in the apoplast of spinach leaves by ascorbate. *Plant Cell Physiol.*, **33**, 379–387.

Tanaka, K., Kondo, N., and Sugahara, K. (1982a) Accumulation of hydrogen peroxide in chloroplasts of SO_2-fumigated spinach leaves. *Plant Cell Physiol.*, **23**, 999–1007.

Tanaka, K., Otsubo, T., and Kondo, N. (1982b) Participation of hydrogen peroxide in the inactivation of Calvin-cycle SH enzymes in SO_2-fumigated spinach leaves. *Plant Cell Physiol.*, **23**, 1009–1018.

Telfer, A., Dhami, S., Bishop, S.M., Philips, D., and Barber, J. (1994) β-Carotene quenches singlet oxygen formed by isolated photosystem II reaction centers. *Biochemistry*, **33**, 14468–14474.

Teramura, A.H. and Sullivan, J.H. (1994) Effects of UV-B radiation on photosynthesis and growth of terrestrial plants. *Photosynth. Res.*, **39**, 463–469.

Terashima, I., Funayama, S., and Sonoike, K. (1994) The site of photoinhibition in leaves of *Cucumis sativus* L. at low temperatures is photosystem I, not photosystem II. *Planta*, **193**, 300–306.

Terashima, I., Noguchi, K., Itoh-Nemoto, T., Park, Y.-M., Kubo, A., and Tanaka, K. (1998) The cause of PSI photoinhibition at low temperature in leaves of *Cucumis sativus*, a chilling-sensitive plant. *Physiol. Plant.*, **103**, 295–303.

Tjus, S.E., Møller, B.L., and Scheller, H.V. (1998) Photosystem I is an early target of photoinhibition in barley illuminated at chilling temperatures. *Plant Physiol.*, **116**, 755–764.

Tourneux, C. and Peltier, G. (1995) Effect of water deficit on photosynthetic oxygen exchange measured using $^{18}O_2$ and mass spectrometry in *Solanum tuberosum* L. leaf discs. *Planta*, **195**, 570–577.

Ushimaru, T., Shibasaka, M., and Tsuji, H. (1994) Resistance to oxidative injury in submerged rice seedlings after exposure to air. *Plant Cell Physiol.*, **35**, 211–218.

Ushimaru, T. Ogawa, K., Ishida, N., Shibasaka, M., Kanematsu, S., Asada, K., and Tsuji, H. (1995) Changes in organelle superoxide dismutase isoenzymes dusing air adaptation of submerged rice seedlings: Differential behavior of isoenzymes in plastids and mitochondria. *Planta*, **196**, 606–613.

Vass, I. and Styring, S. (1993) Characterization of chlorophyll triplet promoting states in photosystem II sequentially induced during photoinhibition. *Biochemistry*, **32**, 3334–3341.

Vass, I. (1996) Adverse effects of UV-B light on the structure and function of the photosynthetic apparatus. In M. Passarakli, (ed.), *Handbook of Photosynthesis*, Marcel Dekker, New York, pp. 931–949.

Vass, I., Sass, L., Spetea, C., Bakou, A., Ghanotakis, D., and Petrouleas, V. (1996) UV-B-induced inhibition of photosystem II electron transport studied by EPR and chlorophyll fluorescence. Impairment of donor and acceptor side components. *Biochemistry*, **35**, 8964–8973.

Veljovic-Jovanovic, S., Oniki, T., and Takahama, U. (1998) Detection of monodehydroascorbic acid radical in sulfite-treated leaves and mechanism of its formation. *Plant Cell Physiol.*, **39**, 1203–1208.

Willekens, H., Chamnongpol, S., Davey, M., Schraudner, M., Langebartels, C., Van Montagu, M., Inzé, D., and Van Camp, W. (1997) Catalase is a sink for H_2O_2 and is indispensable for stress defense in C_3 plants. *EMBO J.*, **16**, 4806–4816.

Winterbourn, C.C. and Metodiewa, D. (1999) Reactivity of biologically important thiol compounds with superoxide and hydrogen peroxide. *Free Radic. Biol. Med.* **27**, 322–328.

Wise, R.P. and Naylor, A.W. (1987) Chilling-enhanced photooxidation. The peroxidative destruction of lipids during chilling injury to photosynthesis and ultrastructure. *Plant Physiol.*, **83**, 272–277.

Wolosiuk, R.A. and Buchanan, B.B. (1977) Thioredoxin and glutathione regulate photosynthesis in chloroplasts. *Nature*, **266**, 565–567.

Yamaguchi, K., Mori, H., and Nishimura, M. (1995) A novel isoenzyme of ascorbate peroxidase localized on glyoxysomal and leaf peroxysomal membranes in pumpkin. *Plant Cell Physiol.*, **36**, 1157–1162.

Yamamoto, H.Y. and Bassi, R. (1996) Carotenoids: Localization and function. In D.R. Ort and C.F. Yocum, (eds.), *Oxygenic Photosynthesis: The Light Reactions*, Kluwer Academic Publishers, Dordrecht, pp. 539–563.

Yamamoto, H., Miyake, C., Dietz, K.-J., Tomizawa, K., Yokota, A., and Murata, N. (1998) Thioredoxin peroxidase in cyanobacteria, *Synechocystis* sp. PCC 6803 and *Synechococcus* sp. PCC 7942. *Plant Cell Physiol.*, **39**, s18.

Yamasaki, H., Takahashi, S., and Hesiki, R. (1999) The tropical fig *Ficus microcarpa* L. f. cv. Golden Leaves lacks heat-stable dehydroascorbate reductase activity. *Plant Cell Physiol.*, **40**, 640–646.

Yamasaki, H., Sakihama, Y., and Ikehara, N. (1997) Flavonoid-peroxidase reaction as a detoxification mechanism of plant cells against H_2O_2. *Plant Physiol.*, **115**, 1405–1412.

Yamashita, T. and Butler, W. (1968) Photoreduction and photophosphorylation with Tris-washed chloroplasts. *Plant Physiol.*, **43**, 1978–1986.

Yim, M.B., Chae, H.Z., Rhee, S.G., Chock, P.B., and Stadtman, E.R. (1994) On the protective mechanism of the thiol-specific antioxidant enzyme against the oxidative damage of biomacromolecules. *J. Biol. Chem.*, **269**, 1621–1626.

Yun, J., Hayashi, T., Yazawa, S., Yasuda, Y., and Katoh, T. (1997) Degradation of photosynthetic activity of *saintpaulia* leaf by sudden temperature drop. *Plant Sci.*, **127**, 25–38.

Zelitch, I. (1973) Plant productivity and the control of photorespiration. *Proc. Natl. Acad. Sci. USA*, **70**, 579–584.

10 The Molecular Biology of the Ascorbate-Glutathione Cycle in Higher Plants

Gary P. Creissen and Philip M. Mullineaux

INTRODUCTION

Reactive oxygen intermediates (ROIs), such as hydrogen peroxide and superoxide, are inevitable byproducts of metabolism in all aerobic organisms. Failure to reduce (scavenge) ROIs results in oxidative damage to cellular macromolecules and membranes either directly or via the products of lipid peroxidation. Thus, oxidative stress occurs when the production of ROIs exceeds the capacity of the antioxidative systems to remove them. For plants the danger of oxidative stress is particularly acute because as well as carrying out the reduction of oxygen in processes such as respiration, they also engage in high rates of oxygen metabolism in the chloroplast during photosynthesis and in allied activities such as photorespiration and the Mehler reaction, all of which occurs in the vicinity of highly energised pigment beds.

Illuminated chloroplasts can scavenge large amounts of hydrogen peroxide, the most stable ROI (Asada, 1994). However, chloroplasts lack catalase activity, which is responsible for the reduction of hydrogen peroxide to water in other cellular compartments, and therefore an alternative mechanism for removing H_2O_2 was postulated to exist in the chloroplast. Foyer and Halliwell (1976) proposed that successive oxidations and reductions of ascorbate, glutathione and NADPH catalyzed by ascorbate peroxidase, dehydroascorbate reductase, glutathione reductase and NADP-ferredoxin oxidoreductase would perform this function. These reactions are collectively referred to as the ascorbate-glutathione cycle (Figure 1). Since the original scheme was proposed, components of this

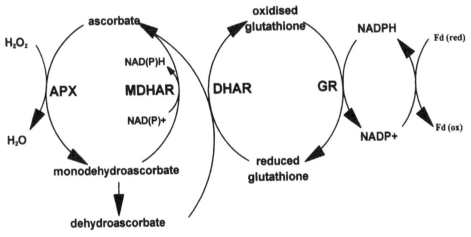

Figure 1. The ascorbate-glutathione cycle. APX, ascorbate peroxidase (EC 1.11.1.11); DHAR, dehydroascorbate reductase (EC 1.8.5.1); GR, glutathione reductase (EC 1.6.4.2); MDHAR, monodehydroascorbate reductase (EC 1.6.5.4).

pathway have been identified in other subcellular compartments such as mitochondria and peroxisomes, as well as in non-photosynthetic tissues, including roots, endosperm, mesocotyl, root nodules and petals (Bielawski and Joy, 1986; Klapheck *et al.*, 1990; Edwards *et al.*, 1990; Mullineaux *et al.*, 1996; Jimenéz *et al.*, 1997). In addition, a further modification has been made to include monodehydroascorbate-free radical reductase, which catalyzes the reduction of monodehydroascorbate-free radical to ascorbate (Figure 1; Asada 1994).

This chapter will focus on the characterization of cDNAs and genes that encode the enzyme components of the ascorbate-glutathione cycle. The enzymes to be considered are ascorbate peroxidase (APX), monodehydroascorbate reductase (MDHAR), dehydroascorbate reductase (DHAR) and glutathione reductase (GR). By considering the types of cDNAs and genes that have been isolated, the complexity of the gene families, the expression of these genes at the transcriptional level both under "normal" and stress conditions and the consequences of engineering their expression in transgenic plants, we shall be able to make an assessment of the extent to which the molecular biology of the ascorbate-glutathione cycle supports its proposed role in determining the plant's ability to withstand oxidative stress.

IDENTIFICATION OF CDNAS AND GENES ENCODING ASCORBATE-GLUTATHIONE CYCLE ENZYMES

Ascorbate Peroxidase

The cytosolic ascorbate peroxidase gene family: identification of cDNAs and genes

Cytosolic ascorbate peroxidase (cAPX) isoforms are encoded by members of a family of related but distinct genes. The first of the APX cDNAs to be identified was that encoding a cytosolic APX isoform from pea. This cDNA (APX1) was isolated by immunoscreening of a bacteriophage λ expression library with a polyclonal antiserum raised against the purified enzyme from pea leaves (Mittler and Zilinskas, 1991a). This example is therefore particularly important because it represents a clear progression from isolation of the protein and obtaining amino acid sequence information, to the subsequent proof of the identity of the cDNA by comparison of the deduced amino acid sequence with the known sequence of the purified protein. Identification of further cAPX cDNAs and genes relies largely, if not entirely, upon their homology to this first cDNA clone.

Most of the APX clones obtained to date encode isozymes that show high degrees of identity at the amino acid level and that possess essentially similar features to APX1 of pea. These key features include the presence of proximal and distal histidine residues (His^{52} and His^{175} in the pea sequence) required for ligation of the heme moiety, and conserved arginine (Arg^{38}) and tryptophan (Trp^{41}) residues located in the distal pocket (Mittler and Zilinskas, 1991a; Asada, 1997).

An exception to this high degree of homology is that of a novel cAPX from spinach that was isolated from a cDNA expression library by immunoscreening with monoclonal antibodies raised against *Euglena* APX (Ishikawa *et al.*, 1995, 1996a; see below). This clone (SAP1) encoded a protein that shared approximately 50% identity with other cytosolic APXs but the conserved Trp was replaced by a Phe next to the distal His. In addition the deduced peptide encoded by SAP1 contained a hydrophobic C-terminal extension compared with other cAPXs, suggesting a possible interaction with membranes. Despite these differences, this clone was confirmed as encoding an APX by production of the recombinant protein in *Escherichia coli*. The recombinant protein showed high specificity for ascorbate and little or no activity with glutathione, NADPH, NADH, or guaiacol as electron donors.

In addition to cDNAs, genes for cytosolic APX have been cloned from pea (Mittler and Zilinskas, 1992) and *Arabidopsis* (Kubo *et al.*, 1993; Santos *et al.*, 1996; see below). A perusal of the sequence databases reveals many APX sequences. These cDNAs and genes have been recovered by several procedures: by homology searches in the *Arabidopsis* expressed sequence tag (EST) databases, as differentially expressed cDNAs under certain stress conditions, by immunoscreening of cDNA libraries, as a prominent unidentified protein in subcellular fractions, which upon characterization turned out to be an APX, or by probing at low stringency with a related member of the gene family.

Plastidial ascorbate peroxidases

The first clones encoding plastidial isoforms of APX were isolated from a spinach leaf cDNA library by screening with monoclonal antibodies raised against *Euglena* APX that recognized cytosolic and plastidial isoforms of spinach APX and had previously been used to isolate novel cAPX cDNAs (Ishikawa *et al.*, 1996b; see below). Analysis of two cDNA clones revealed that the sequences diverge almost precisely at the end of one coding sequence giving rise to a 3′ extension in one of the two classes. The extended cDNA sequence encodes a C-terminal extension to the APX polypeptide of 50 residues with characteristics of a transmembrane domain. The shorter and longer cDNAs code for stromal APX (sAPX) and thylakoid-bound APX (tAPX), respectively, and both isoforms are encoded by one gene giving rise to two different mature transcripts by alternative processing of a single nascent APX transcript (Ishikawa *et al.*, 1996b).

A plastidial APX cDNA isolated from pumpkin has a similar putative C-terminal transmembrane domain and an antibody raised against a C-terminal peptide of 82 residues expressed in *E. coli* from the 3′ region of the cDNA specifically recognized an APX in a thylakoid preparation from isolated pumpkin chloroplasts (Yamaguchi *et al.*, 1996). Therefore, tAPX may be anchored to thylakoid membranes in such a way as to permit the enzyme to catalyze reactions in the stroma, a situation not previously described for any other thylakoid-associated protein (Miyake *et al.*, 1993; Yamaguchi *et al.*, 1996). Analysis of *Arabidopsis* ESTs that represent the sAPX and tAPX isoforms revealed that only the tAPX cDNA contained a predicted C-terminal transmembrane domain (Jespersen *et al.*, 1997). Work in the authors' laboratory has recently confirmed that separate genes that map to different *Arabidopsis* linkage groups (C. Escobar, G. Creissen and P. Mullineaux, unpublished data) encode the sAPX and tAPX transcripts. Therefore, spinach and *Arabidopsis* appear to have adopted different strategies for synthesizing their chloroplast thylakoid-bound and stromal APX isoforms. Alternatively, it may be that there are two thylakoid isoforms: one encoded by alternative splicing of the sAPX mRNA and one encoded by a separate *tAPX* gene. Like the cAPX cDNA reported from pea (Mittler and Zilinskas, 1991a), the identity of the APX cDNAs and the subcellular localization of the particular APX isoform encoded by their respective mRNA was confirmed by comparing the amino acid sequences deduced by cDNA sequencing with the N-terminal amino acid sequences derived by direct sequencing of the purified proteins (Chen and Asada, 1989; Miyake *et al.*, 1993; Ishikawa *et al.*, 1996b; Yamaguchi *et al.*, 1996).

The reliance upon antibodies as the only proof of subcellular localization can go wrong: a cDNA for APX from *Arabidopsis thaliana* obtained by immunoscreening of an expression library with a monoclonal antibody raised against the spinach chloroplast isoform (Kubo *et al.*, 1992) was claimed to represent the chloroplast isoform of APX in this species in spite of the lack of a transit peptide presequence that would allow the import of the mature polypeptide into the plastids. Similarly, the use of an antiserum raised against tea plastidial APX, which apparently recognized bands only in chloroplast extracts, resulted in the recovery of cDNAs encoding only cytosolic isoforms of the enzyme (Santos *et al.*, 1996). Recently, it has been shown that antibodies raised against recombinant spinach sAPX and cAPX isozymes efficiently cross-react with cAPX, sAPX and tAPX (Yoshimura *et al.*, 1998). In addition to highlighting the complexity of the APX gene family in several plant species, all of this cDNA and gene cloning activity has also raised questions regarding the expression of specific genes and the nature of the APX isoforms they encode.

Microbody-associated ascorbate peroxidases

Putative membrane anchoring domains in the C-termini of derived sequences of APX polypeptides, otherwise displaying a high degree of identity to cAPX sequences, have recently been detected. The best evidence, linked to direct biochemical data, comes from an APX that is a readily detectable constituent of glyoxysome membrane fractions from cotton cotyledons (Bunkelmann and Trelease, 1996). In the EST database, a similar class of APX cDNA has been identified that possesses a hydrophobic membrane-anchoring domain displaying high homology to the cotton glyoxysomal APX domain (Santos, 1995; Karpinski *et al.*, 1997) (Figure 2).

APX3 mRNA was detected in all organs of mature plants, but transcript levels were highest in green tissues and paralleled the synthesis of the peroxisomal enzyme, glycollate oxidase (Escobar, 1998). Thus, APX3 production appears to be associated with peroxisomes as well as glyoxysomes. Glyoxysomes and peroxisomes have equivalent structures but different functions that relate to their association with catabolism of seed/ cotyledon storage reserves during post-germinative growth of seedlings and photosynth-esis in leaves, respectively (Mullen and Trelease, 1996). The protein sequences of these two closely related APX cDNAs cannot give any information about the precise location of this enzyme in the membrane of glyoxysomes/peroxisomes. Current opinion is divided between a location of APX on the matrix side or the cytosol surface of the peroxisome/ glyoxysome (Yamaguchi *et al.*, 1995a, 1995b; Bunkelmann and Trelease, 1996; Mullen and Trelease, 1996; Jimenéz *et al.*, 1997). An additional class of cAPX from spinach, termed SAP1, also contains a C-terminal extension with a high degree of hydrophobicity, which has been suggested to reflect a membrane-bound localization for this APX isoform that is quite distinct from a more typical cAPX recovered from spinach in the same study (Ishikawa *et al.*, 1995). There is however, no homology of this C-terminal region to those of the peroxisomal/glyoxysomal class of APXs detected in cotton and *Arabidopsis* (Figure 2). A broader view of the interrelationships of APX sequences within and between organisms is shown graphically in Figure 3 in the form of a dendrogram.

```
             1                                                    50
APX1      .......... .......... .......... .......... ..........
APX2      .......... .......... .......... .......... ..........
APX3      .......... .......... .......... .......... ..........
Thyarab1  .......... .......... .......... .......... ..........
Stroarab  KSITIIPLRK MAERVSLTLN VTLLSPPPTT TTTTMSSSLR STTAASLLLR

             51                                                   100
APX1      .......... .......... .......... .......... ..........
APX2      .......... .......... .......... .......... ..........
APX3      .......... .......... .......... .......... ..........
Thyarab1  .......... .......... .......... .......... ..........
Stroarab  SSSSSSRSTL TLSASSSLSF VRSLVSSPRL SSSSPLSQKK CRIASVNRSF

             101                                                  150
APX1      .MTKNYPTVS EDYKKAVEKC RRKLRGLIAE KNCAPIMVRL AWHSAGTFDC
APX2      MVKKSYPEVK EEYKKAVQRC KRKLRGLIAE KHCAPIVLRL AWHSAGTFDV
APX3      ...MAAPIVD AEYLKEITKA RRELRSLIAN KNCAPIMLRL AWHDAGTYDA
Thyarab1  .......... ........SA KEDIKVLLRT KFCHPILVRL GWHDAGTYNK
Stroarab  NSTTAATKSS SSDPDQLKNA REDIKELLST KFCHPILVRL GWHDAGTYNK

             151                                                  200
APX1      QSRT....GG PFGTMRFDAE QAHGANSGIH IALRLLDPIR EQFPTISFAD
APX2      KTKT....GG PFGTIRHPQE LAHDANNGLD IAVRLLDPIK ELFPILSYAD
APX3      QSKT....GG PNGSIRNEEE HTHGANSGLK IALDLCEGVK AKHPKITYAD
Thyarab1  NIEEWPLRGG ANGSLRFEAE LKHAANAGLL NALKLIQPLK DKYPNISYAD
Stroarab  NIKEWPQRGG ANGSLRFDIE LKHAANAGLV NALNLIKDIK EKYSGISYAD
                                        *
             201                                                  250
APX1      FHQLAGVVAV EVTGGPDIPF HPGREDKPQP ...PPEGRLP DATKGC..DH
APX2      FYQLAGVVAV EITGGPEIPF HPGRLDKVEP ...PPEGRLP QATKGV..DH
APX3      LYQLAGVVAV EVTGGPDIVF VPGRKDSNVC ...PKEGRLP DAKQGF..QH
Thyarab1  LFQLASATAI EEAGGPDIPM KYGRVDVVAP EQCPEEGRLP DAGPPSPADH
Stroarab  LFQLASATAI EEAGGPKIPM KYGRVDASGP EDCPEEGRLP DAGPPSPATH

             251                                                  300
APX1      LRDVFAKQMG LSDKDIVALS GAHTLGRCHK DRSGFEGA.. ..........
APX2      LRDVFGR.MG LNDKDIVALS GGHTLGRCHK ERSGFEGA.. ..........
APX3      LRDVFYR.MG LSDKDIVALS GGHTLGRCHK ERSGFDGP.. ..........
Thyarab1  LRDVFYR.MG LDDKEIVALS GAHTLGRARP DRSGWGKPET KYTKTGPGEA
Stroarab  LREVFYR.MG LDDKDIVALS GAHTLGRSRP ERSGWGKPET KYTKEGPGAP
                                        *
             301                                                  350
APX1      ....WTSNPL IFDNSYFKEL LSGEKEGLLQ LVSDKALLDD PVFRPLVEKY
APX2      ....WTPNPL IFDNSYFKEI LSGEKEGLLG LPTDKALLDD PLFLPFVEKY
APX3      ....WTQEPL KFDNSYFVEL LKGESEGLLK LPTDKTLLED PEFRRLVELY
Thyarab1  GGQSWTVKWL KFDNSYFKDI KEKRDDDLLV LPTDAALFED PSFKNYAEKY
Stroarab  GGQSWTPEWL KFDNSYFKEI KEKRDEDLLV LPTDAAIFED SSFKVYAEKY

             351                                                  400
APX1      AADEDAFFAD YAEAHMKLSE LG..FADA.. .......... ..........
APX2      AADEDASFED YTEAHLKLSE LG..FADKE. .......... ..........
APX3      AKDEDAFFRD YAESHKKLSE LG..FNPNSS AGKAVADSTI LAQSAFGVAV
Thyarab1  AEDVAAFFKD YAEAHAKLSN LGAKFDPPEG IVIENVPEKF VAAKYSTGKK
Stroarab  AADQDAFFKD YAVAHAKLSN LGAEFNPPEG III....... ..........

             401                                                  450
APX1      .......... .......... .......... .......... ..........
APX2      .......... .......... .......... .......... ..........
APX3      AAAVVAFGYF YEIRKRMK.. .......... .......... ..........
Thyarab1  ELSDSMKKKI RAEYEAIGGS PDKPLPTNYF LNIIIAIGVL VLLSTLFGGN
Stroarab  .......... .......... .......... .......... ..........

             451
APX1      .......
APX2      .......
APX3      .......
```

Figure 2. Alignment of the deduced amino acid sequences for the five APX isoforms identified so far in *Arabidopsis thaliana*. The highly conserved proximal and distal histidine residues are indicated by asterisks. The boxed region in APX3 shows high homology to the putative membrane spanning region of the glyoxysomal APX from cotton.

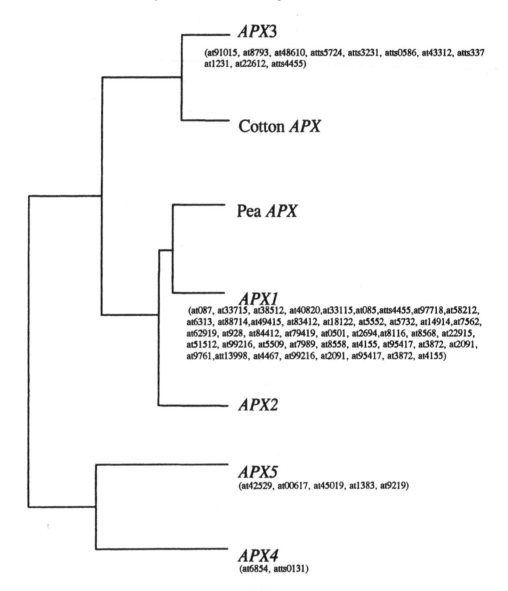

Figure 3. Similarity dendrogram showing the relationships between *APX* sequences from different species and from different compartments. The two plastidial forms (*APX4* and *APX5*) are clearly separated from the cytosolic (*APX1, APX2*) and peroxisomal/glyoxysomal isoform (*APX3*). Accession numbers are shown in small print. References to other sequences are given in the text.

Regulation of ascorbate peroxidase gene expression

It was initially thought that in pea and in *Arabidopsis*, cAPX was encoded by a single gene (*APX1*; Mittler and Zilinskas, 1992; Kubo *et al.*, 1993). However, screening of a database of *Arabidopsis* ESTs revealed a total of 62 cDNAs with homology to core APX amino acid residues all of which could be resolved into one of four classes of *APX* genes, namely *APX1, APX3,* and the genes encoding the tAPX and sAPX isoforms. This screening suggests that in a number of different tissues of *Arabidopsis* under non-stress conditions four *APX* genes are expressed. An *Arabidopsis* gene encoding a second cAPX (termed *APX2*), which is not expressed under non-stress conditions, has been found by low stringency probing with *APX1* sequences (Santos *et al.*, 1996).

This gene is not represented in any EST database and therefore would be missed if this was the only screening method employed. However, *APX2* is expressed under conditions of photoinhibitory stress (see below) and cloning of a cDNA from RNA prepared from stressed leaf tissue has been achieved (Karpinski *et al.*, 1997). Analysis of the derived amino acid sequence of this cAPX isoform indicates that it shares as similar a degree of homology to the *APX1*-encoded isoform as both *Arabidopsis* cAPX isoforms share with pea cAPX (Santos *et al.*, 1996; Karpinski *et al.*, 1997). These data suggest that the two cAPX isoforms in *Arabidopsis* ought to have different properties, but biochemical confirmation of this is not forthcoming at present. Furthermore, the presence of at least one *APX* gene in one plant species, which is expressed only under tightly defined stress conditions, strongly argues that other equivalent genes remain to be discovered in other plant species.

Data on the regulation of expression of the different members of the APX gene family under non-stress conditions are rather limited. *APX1* has been shown to be expressed in leaves, roots, stems and flowers in *Arabidopsis* (Santos *et al.*, 1996). Similarly, *APX3* transcripts have been found in all organs of *Arabidopsis* (see above). A combination of RNA gel blot analysis and *in situ* hybridization has provided more detail on *APX3* expression and has revealed that regulation of *APX3* gene expression is complex. Light and phytochrome positively regulate APX3 and transcripts accumulate during senescence. Data from *in situ* hybridization to flower sections revealed that *APX3* mRNA was limited to tapetal tissue and developing microspores—tissues in which there is enhanced lipid catabolism, resulting in increased H_2O_2 production (C. Escobar, G. Creissen, and P. Mullineaux, manuscript in preparation). This parallels the induction of *APX3* in germinating cotton seed that is associated with glyoxisomal H_2O_2 production (Bunkelmann and Trelease 1997). There are currently no data on the regulation of *APX4* and *APX5* expression.

Glutathione Reductase: cDNAs, Genes and Isoforms

Plastidial and mitochondrial glutathione reductases

Glutathione reductase (GR) activity is distributed between at least two subcellular compartments in all higher plants (plastids and cytosol) and, at least in pea leaves, there is a small but significant mitochondrial fraction (for a review, see Mullineaux and Creissen, 1997). Despite the considerable effort that has been applied to the purification of GR from several diverse plant species and the large numbers of studies correlating enhanced GR activities with the response of plants to environmental stress as well as the increasing information available on the molecular biology of mammalian (human and mouse), yeast and prokaryotic GRs (for a review, see Mullineaux and Creissen, 1997), it was not until 1992 that the first cDNA for GR from a higher plant was published (Creissen *et al.*, 1992). This clone was obtained by immunoscreening of a λgt11 cDNA expression library made from pea leaf mRNA with an antiserum raised against purified GR from pea leaves (Edwards *et al.*, 1990). This cDNA represents the *GOR1* gene of pea (Mullineaux *et al.*, 1996) and was found to encode a polypeptide that clearly showed a high degree of homology at the amino acid level to all other known GRs, either from peptide sequencing or from deduced amino acid sequences derived from analysis of cDNAs or genes. However, the pea cDNA also encoded a N-terminal extension, which, at the time, was thought to be a chloroplast transit peptide sequence. A comparison of the deduced polypeptide encoded by *GOR1* with non-plant GR sequences revealed the presence of a C-terminal extension of 22 residues of unknown function. Subsequently, cDNAs encoding *GOR1* have been cloned from *Arabidopsis* leaves (Kubo *et al.*, 1993), soybean root nodules (Tang and Webb, 1994), tobacco leaves (Creissen and Mullineaux, 1995), maize (EMBL accession number AJ006055), and *Vitis vinifera* (EMBL accession number AF019907).

Each of these cDNAs encodes a single, presumed plastidial, GR isoform and consequently cannot explain the multiple isoforms found in pea and tobacco (Edwards *et al.*, 1990; Foyer *et al.*, 1991; Madamanchi *et al.*, 1992), although multiple isoforms could not be found in *Arabidopsis* (Kubo *et al.*, 1993). It is possible that isoforms of GR could be generated in tobacco by a partially spliced class of GR mRNA that generates a different C-terminal extension, but the low abundance of this transcript make this less likely (Creissen and Mullineaux, 1995).

Expression of the full-length pea *GOR1* cDNA, including its own presequence, in transgenic tobacco (*Nicotiana tabacum*) revealed that the pea *GOR1*-encoded protein can be targeted to chloroplasts and mitochondria simultaneously (Creissen *et al.*, 1995). This cotargeting property (as it is called) of pea GR was revealed because of the need to examine GR activity in the two major sites of oxygen metabolism (i.e. chloroplast and mitochondrion) prior to beginning stress tolerance experiments with these plants (see below). The cotargeting property of the pea GR preprotein has been localized to its targeting sequence (Creissen *et al.*, 1995). These data confirm the presence of GR in pea leaf mitochondria and show that at least two subcellular compartments contain GR isoforms arising from one gene. The biological significance of this observation is still not clear and may not be universal for all plant species, since wild-type tobacco leaves do not contain any detectable mitochondrial GR (Creissen *et al.*, 1995).

Cytosolic isoforms of GR have been detected in tissues of several plant species (Mullineaux and Creissen, 1997). Analysis of the sequence of the pea *GOR1* gene did not reveal any possible mechanism (for example alternative splicing) that could give rise to a cytosolic isoform. It was predicted therefore that, at least in pea, there had to be an additional *GOR* gene that encodes a cytosolic isoform (Mullineaux *et al.*, 1996). Routine screening of the EMBL database for *GOR* sequences revealed one partial cDNA sequence from spinach that, at 53% identity at the DNA level with the *GOR1* sequence from pea, appeared to be distinct from other *GOR* sequences encoding chloroplastic/mitochondrial forms of this enzyme. Polymerase chain reaction (PCR) amplification of spinach cDNA was used to generate a specific probe for this *GOR* sequence that was used in turn, to screen a pea leaf cDNA library. From the cDNA library, two *GOR* cDNAs were recovered that were identical, apart from one being 143 bp shorter than the other, and distinct from the previously characterized chloroplast/mitochondrial *GOR* cDNA class (*GOR1* and see above; Creissen *et al.*, 1992); both cDNAs mapped to linkage group 6 in pea, in contrast to the linkage group 2 location of *GOR1*. A complete sequence of this new *GOR* class (termed *GOR2*) was compiled from the cDNA clones and 5′ rapid amplification of cDNA ends that were needed to obtain the 5' end of the sequence (Stevens *et al.*, 1997). The derived amino acid sequence of *GOR2* contained all of the amino acid residues that have been implicated in the binding of substrates and in the formation of the redox-active site and that are highly conserved between GRs from a wide diversity of organisms (Figure 4; Mullineaux and Creissen, 1997). The GOR2 protein is active when produced in *E. coli* and it will weakly cross-react with GOR1 antibodies, further confirming that this protein is a new and distinct class of higher plant GR. The GOR2 sequence does not contain a putative leader sequence for organellar targeting of the mature protein and, therefore, most probably encodes the cytosolic isoform of the enzyme. *GOR2* genes or cDNAs have also been cloned from *Arabidopsis* (Kubo *et al.*, 1998), rice (Kaminaka *et al.*, 1998), and *Brassica campestris* (Lee *et al.*, 1998).

In summary, the main compartment-specific isoforms of GR have been assigned to their corresponding genes in a few plant species, but the number of genes so far isolated does not explain the apparent multiplicity of isoforms encountered in some species (Edwards *et al.*, 1990; Foyer *et al.*, 1991; Madamanchi *et al.*, 1992; Hausladen and Alscher, 1994). One must also bear in mind that, like APX, there may be GR genes that are expressed only under a tightly defined set of stress conditions.

Monodehydroascorbate-Free Radical Reductase cDNAs

Isolation of cytosolic MDHAR cDNAs by immunoscreening

MDHAR cDNAs were first cloned from cucumber seedlings (Sano and Asada, 1994). This cloning was achieved by screening a λgt22A cDNA library constructed from RNA isolated from the cotyledons of 4-day-old cucumber seedlings and screened with an antiserum raised against purified MDHAR from cucumber fruit. Subsequently, *MDHAR* cDNAs have been cloned from pea and tomato (Murthy and Zilinskas, 1994; Grantz *et al.*, 1995). The cDNAs from the three different plant species all encode a similarly sized polypeptide (approximately 47 kDa), which in all cases is most probably the cytosolic isoform of MDHAR. The three MDHAR peptide sequences show a considerable degree of conservation (approximately 80% identity overall; Figure 5). DNA gel blot analysis of the *MDHAR* genes in tomato suggested a single copy gene for this cytosolic isoform (Grantz *et al.*, 1995). However, the detection of other isoforms of MDHAR in other species and in different subcellular compartments strongly implies that other *MDHAR* genes exist (Jimenéz *et al.*, 1997). Low-stringency probing with a cDNA that encodes a cytosolic isoform of MDHAR would probably not detect any other class of cDNA (Sano and Asada, 1994; Grantz *et al.*, 1995).

Dehydroascorbate Reductase (DHAR) cDNAs

Identification of a cDNAs for dehydroascorbate reductase

DHARs have been purified from a number of species, including spinach (Foyer and Halliwell, 1976), potato (Dipierro and Borraccino, 1991), and rice (Kato *et al.*, 1997). However, the presence of a number of other proteins in plant tissues that have DHAR activity, such as glutaredoxins, protein disulphide isomerases and a Kunitz-type trypsin

Gary P. Creissen and Philip M. Mullineaux

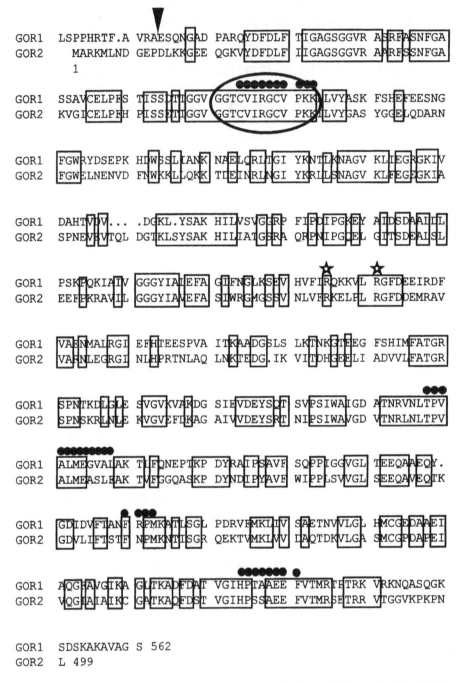

Figure 4. Alignment of the deduced polypeptide sequences encoded by plastidial/mitochondrial (*GOR1*) and cytosolic (*GOR2*) glutathione reductase cDNAs from pea. The arrow indicates the cleavage site in *GOR1* of the mature polypeptide from the presequence. Dots above the alignment show residues important for GSSG binding. The residues contained within the ellipse form the redox-active disulphide domain. The argine residues marked with stars are required for NADP binding. Identical residues between the two deduced sequences are boxed.

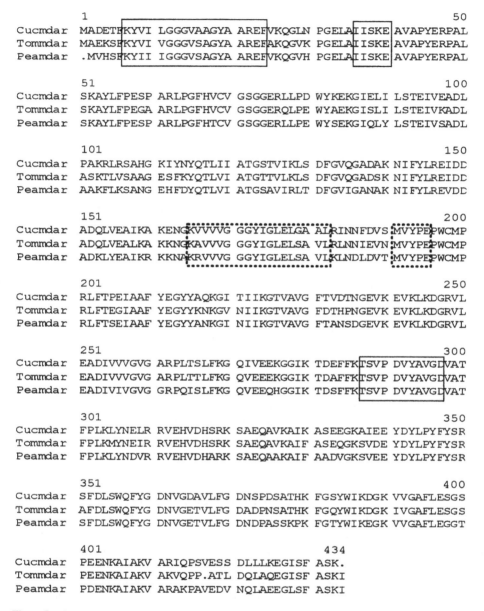

```
          1                                                    50
Cucmdar   MADETHKYVI LGGGVAAGYA AREFVKQGLN PGELAIISKE AVAPYERPAL
Tommdar   MAEKSHKYVI VGGGVSAGYA AREFAKQGVK PGELAIISKE AVAPYERPAL
Peamdar   .MVHSHKYII IGGGVSAGYA AREFVKQGVH PGELAIISKE AVAPYERPAL

          51                                                   100
Cucmdar   SKAYLFPESP ARLPGFHVCV GSGGERLLPD WYKEKGIELI LSTEIVEADL
Tommdar   SKAYLFPEGA ARLPGFHVCV GSGGERQLPE WYAEKGISLI LSTEIVKADL
Peamdar   SKAYLFPESP ARLPGFHTCV GSGGERLLPE WYSEKGIQLY LSTEIVSADL

          101                                                  150
Cucmdar   PAKRLRSAHG KIYNYQTLII ATGSTVIKLS DFGVQGADAK NIFYLREIDD
Tommdar   ASKTLVSAAG ESFKYQTLVI ATGTTVLKLS DFGVQGADSK NIFYLREIDD
Peamdar   AAKFLKSANG EHFDYQTLVI ATGSAVIRLT DFGVIGANAK NIFYLREVDD

          151                                                  200
Cucmdar   ADQLVEAIKA KENGKVVVVG GGYIGLELGA ALRINNFDVS MVYPEPWCMP
Tommdar   ADQLVEALKA KKNGKAVVVG GGYIGLELSA VLRLNNIEVN MVYPEPWCMP
Peamdar   ADKLYEAIKR KKNAKRVVVG GGYIGLELSA VLKLNDLDVT MVYPEPWCMP

          201                                                  250
Cucmdar   RLFTPEIAAF YEGYYAQKGI TIIKGTVAVG FTVDTNGEVK EVKLKDGRVL
Tommdar   RLFTEGIAAF YEGYYKNKGV NIIKGTVAVG FDTHPNGEVK EVKLKDGRVL
Peamdar   RLFTSEIAAF YEGYYANKGI NIIKGTVAVG FTANSDGEVK EVKLKDGRVL

          251                                                  300
Cucmdar   EADIVVVGVG ARPLTSLFKG QIVEEKGGIK TDEFFKTSVP DVYAVGDVAT
Tommdar   EADIVVVGVG ARPLTTLFKG QVEEEKGGIK TDAFFKTSVP DVYAVGDVAT
Peamdar   EADIVIVGVG GRPQISLFKG QVEEQHGGIK TDSFFKTSVP DVYAVGDVAT

          301                                                  350
Cucmdar   FPLKLYNELR RVEHVDHSRK SAEQAVKAIK ASEEGKAIEE YDYLPYFYSR
Tommdar   FPLKMYNEIR RVEHVDHSRK SAEQAVKAIF ASEQGKSVDE YDYLPYFYSR
Peamdar   FPLKLYNDVR RVEHVDHARK SAEQAAKAIF AADVGKSVEE YDYLPYFYSR

          351                                                  400
Cucmdar   SFDLSWQFYG DNVGDAVLFG DNSPDSATHK FGSYWIKDGK VVGAFLESGS
Tommdar   AFDLSWQFYG DNVGETVLFG DADPNSATHK FGQYWIKDGK IVGAFLESGS
Peamdar   SFDLSWQFYG DNVGETVLFG DNDPASSKPK FGTYWIKEGK VVGAFLEGGT

          401                                434
Cucmdar   PEENKAIAKV ARIQPSVESS DLLLKEGISF ASK.
Tommdar   PEENKAIAKV AKVQPP.ATL DQLAQEGISF ASKI
Peamdar   PDENKAIAKV ARAKPAVEDV NQLAEEGLSF ASKI
```

Figure 5.   Comparison of the deduced amino acid sequences for cytosolic MDHARs from cucumber (Cucmdar), tomato (Tommdar), and pea (Peamdar). The residues boxed with solid lines and dotted lines are those thought to be responsible for FAD-binding and NAD(P)H-binding, respectively.

inhibitor (Wells *et al.*, 1990; Trumper *et al.*, 1994) has caused some confusion and has even led to the suggestion that DHA detected in chloroplasts is merely an artefact (Morell *et al.*, 1997). The purification of rice DHAR to homogeneity and determination of its amino-terminal sequence (Kato *et al*, 1997) has enabled us to identify expressed sequence tag cDNAs in the Arabidopsis database that show high homology to the rice protein and little or no homology to other proteins with known DHAR activity (Foyer and Mullineaux,

1998). Furthermore, expression of one of the EST cDNAs in *E. coli* allowed us to confirm that the encoded protein has DHAR activity (G. Creissen, A. Jiménéz, and P. Mullineaux, unpublished data).

EXPRESSION OF ASCORBATE-GLUTATHIONE CYCLE GENES UNDER STRESS CONDITIONS

For two of the four enzymes under consideration here, namely APX and GR, there is a wealth of literature on their activities in different organs of different plant species subjected to a wide range of environmental stresses (Creissen *et al.*, 1994; Mullineaux and Creissen, 1997). It is not within the remit of this chapter to review this literature except to state that most studies have revealed rises (usually in the order of 2-fold) in the extractable activity of these enzymes from stressed compared with unstressed material. In contrast, studies that have been extended to examine synthesis of ascorbate-glutathione cycle enzymes are relatively few and in most cases do not reveal a simple relationship between increases in enzyme activities and the corresponding levels of enzyme protein or its mRNA(s). This will be demonstrated below.

Ascorbate Peroxidase mRNA and Protein Levels Under Stress Conditions

Effects of gaseous pollutants

Oxidative stress brought about by the exposure of *Arabidopsis* and *Nicotiana* sp. to the gaseous pollutants ozone and sulphur dioxide causes an increase in the levels of *APX1* transcripts (homologues of *APX1* in peas; Mittler and Zilinskas, 1992). The timing and amounts of ozone delivered in these experiments were different: from single high doses of 250 nl.l^{-1} for 8 hours to episodic exposure of lower doses for several days (for example 120 nl.l^{-1} for 14 h per day for 6 days). In plants subjected to ozone fumigation for all or part of the day over several days, the consistent observation in both *Arabidopsis* and *Nicotiana* is that the levels of *APX1* mRNA were elevated up to 4-fold compared with control untreated plants (Willekens *et al.*, 1994; Conklin and Last, 1995; Kubo *et al.*, 1995). In *Arabidopsis*, elevated *APX1* transcript was associated with a concomitant increase in total APX activity and was sustained throughout the photoperiod during which levels of *APX1* transcript displayed a diurnal rhythm with maxima at the beginning and end of the photoperiod (Kubo *et al.*, 1995). In variety PBD6 of tobacco exposed to ozone over several days, the rise in *APX1* mRNA was associated with the onset of visible foliar damage. In contrast, the same fumigation regime imposed on *Nicotiana plumbaginifolia*, which did not produce visible symptoms, did not induce any rise in *APX1* transcript level (Willekens *et al.*, 1994). In single-dose experiments on *Arabidopsis*, *APX1* transcript levels rose up to 7-fold within 2 hours of the start of ozone fumigation, but had declined back to pre-exposure levels by the end of the fumigation period of 8 hours (Conklin and Last, 1995). In the same study, episodic exposure to ozone over a 4–day period resulted in a sustained increase in *APX1* transcript levels, although the increase was highest on day 1 and lowest on day 4. Continuous exposure of *Arabidopsis* to ozone resulted in the maintenance of 4-fold elevated levels throughout the period of the experiment (7 days) (Kubo *et al.*, 1995).

In contrast to ozone, fumigation with sulphur dioxide induces only a small increase (1.6 to 2-fold) in *APX1* transcripts in *Arabidopsis* and no increase in *N. plumbaginifolia* (Willekens *et al.*, 1994; Kubo *et al.*, 1995). In *Arabidopsis*, the small rise in transcript was associated with a similar rise in APX activity (Kubo *et al.*, 1995).

The fumigation experiments suggest that there is a relatively simple relationship between elevated levels of *APX1* transcript and APX activity. This relationship is also apparent in transgenic tobacco containing elevated levels of Cu/Zn superoxide dismutase that was associated with an increase in the levels of endogenous *APX1* transcript and APX activity (Sen Gupta *et al.*, 1993). However, this relationship is not apparent in some cases.

Effects of drought and salt stress

A comprehensive study of the regulation of APX synthesis in drought-stressed peas and during the recovery from this stress was carried out by Mittler and Zilinskas (1994). Withdrawal of water for 3 days did not cause any significant increase in immunodetectable APX protein or activity, but 10 hours after resuming watering, both APX activity and protein had increased by approximately 50%. In contrast to these small increases, *APX1* transcript levels increased by approximately 3-fold during the drought period and by 15-fold 10 hours after rewatering. However, the amount of *APX1* mRNA associated with polysomes and, therefore, able to participate in APX protein synthesis, was found to have increased by only 1.8-fold in the rewatering period, which is more in line with the increase in APX activity and immunodetectable protein. Therefore, only a small proportion of the increased amount of *APX1* mRNA was available for protein synthesis, suggesting that a major point of control of *APX* expression in peas is posttranscriptional, at the level of accumulation of steady-state transcripts and protein synthesis (Mittler and Zilinskas, 1994). The transcription rate of *APX1* after 3 days of drought and 10 hours after rewatering was also investigated using nuclear run-on transcription assays. These studies indicated that *APX1* gene transcription increased only 1.5 fold after 3 days of drought and was not significantly higher in the post-drought phase. This increase in transcription rates may have been enough to sustain the observed increase in *APX1* mRNA or possibly that the harvesting of nuclei for this assay should have been done at times earlier than the time of the observed increase in transcript levels (Mittler and Zilinskas, 1994).

Similarly APX1, but not sAPX mRNA levels increased in a salt-tolerant variety of pea exposed to salt stress for 15 days. Interestingly, in a salt-sensitive variety such an increase did not occur (J. Hernandez, A. Jiménez, P. Mullineaux, and F. Sevilla, manuscript in preparation).

This complex relationship between transcription of *APX* genes and enzyme and protein levels may also occur during salt stress in radish. The leaves and roots of radish plants grown on medium with 100 mM NaCl showed up to 2-fold and 7-fold higher levels of APX activity, respectively, but no detectable increase in the levels of *APX1* transcript (Lopez *et al.*, 1996).

It should be borne in mind that most of the stress studies described above have used probes corresponding to a single class of *APX* genes that encode the cytosolic isoform of the enzyme. It is clear that there are other classes of *APX* genes (see above) and therefore the studies on transcript levels in stressed plants and the relationship with increases in APX activity might not be giving the complete picture.

Biotic stress and ascorbate peroxidase expression

The effects of pathogen attack on expression of specific members of the *APX* gene family have not been studied in detail. Mittler *et al.* (1996) found that the hypersensitive response (HR) induced by infecting tobacco plants with tobacco mosaic virus (TMV) was accompanied by steady-state increases in *APX1* transcripts. However, the level of the APX1 protein declined to about 50% of the control level, indicating that post-transcriptional suppression of *APX1* occurred in this treatment (Mittler *et al.*, 1998). A similar effect on *APX1* expression was seen following induction of HR by infiltration of leaves with *Pseudomonas syringae*. H_2O_2 production in response to pathogen infection is thought to be one of the mediators of programmed cell death (PCD) during the HR response (Levine *et al.*, 1994) and thus the suppression of *APX1* expression was suggested to play an important role in preventing the scavenging of H_2O_2 in infected cells that could otherwise have inhibited PCD (Mittler *et al.*, 1998).

Ascorbate peroxidase expression and photo-oxidative stress

That differences in expression patterns of *APX* genes may occur is illustrated by the study of Karpinski *et al.* (1997). When *Arabidopsis* plants, grown under low light (LL) conditions (200 μmol m^{-2} s^{-1}), are exposed suddenly to a photoinhibitory 10-fold excess light (EL) stress, for up to 60 min, photooxidative stress ensues. Under these conditions, the synthesis of *APX2* mRNA, which is not present under non-stress conditions (Santos *et al.*, 1996), is detectable within 7 min and increases a further 40-fold after this initial time point. *APX1* mRNA is induced within 15 min of the onset of the stress and is induced by 18-fold, whereas *APX3* mRNA levels are not altered by this treatment. Equally, the decline in the levels of the *APX1* and *APX2* mRNAs show marked differences: 2 hours after the stress, *APX2* mRNA levels are virtually undetectable and are completely absent by 24 hours after stress. In contrast, *APX1* mRNA levels are still higher 24 hours after stress than in untreated control plants.

The induction in expression of these *APX* genes requires changes in electron transport through photosystem II (PSII) and is mediated in some way by foliar antioxidant/prooxidant balance. More recent work (S. Karpinski, H. Reynolds, B. Karpinska, G. Wingsle, G. Creissen, and P. Mullineaux, manuscript submitted) has revealed a central role for H_2O_2, in conjunction with PSII, in inducing *APX2* expression both locally and systemically, in regions of the plant remote from EL exposure. Pre-treatment of leaves with enzymic or non-enzymic antioxidants inhibits the EL-mediated induction of *APX2* expression. This systemic signal can be associated with an acclimatory response in such remote leaves, what renders them more tolerant to further episodes of EL. *APX2* induction is at least an indicator of this phenomenon, which has been termed systemic acquired acclimation.

Expression of Glutathione Reductase Genes During Stress

The most striking feature concerning the levels of the *GOR1* transcript is its constancy in a wide range of species subjected to diverse stress conditions. Pea plants subjected to paraquat treatment, chilling, long term cold treatment, drought, exposure of etiolated seedlings to strong light and ozone fumigation did not exhibit any change in the levels of

GOR1 mRNA (Edwards *et al.*,1994; Stevens *et al.*, 1997). Only in UV-B irradiated peas has a 3-fold increase in *GOR1* transcript been noted (Strid, 1993). Equally, by using a PCR-based technique, the *GOR1* gene appears to be expressed in every organ of the pea plant (leaf, root, bract, petiole, stem, flower, pod, and developing seed) (Mullineaux *et al.*, 1996), although *GOR1* transcript may vary up to 5-fold, depending upon the age of the leaf (Strid, 1993). Similarly in *Arabidopsis*, the *GOR1* transcript level remains constant during an exposure to the high light stress that causes the levels of *APX* transcripts to be raised (see above; Karpinski *et al.*, 1997). A decline in the steady-state level of *GOR1* mRNA has been observed in ozone-fumigated *Arabidopsis* plants (Conklin and Last, 1995). Transcripts that encode cytosolic and plastidial isoforms of superoxide dismutase, but not those encoding plastidial GR, are elevated in Scots pine needles that suffer oxidative stress brought on by high light intensities in chilling conditions or by treatment of shoots with oxidized glutathione (GSSG) (Karpinski *et al.*, 1993; Wingsle and Karpinski, 1996).

In contrast to the constancy of *GOR1* transcript levels during stress, preliminary experiments with both drought-stressed and chilled peas and with *Arabidopsis* treated with high light, showed a small increase (2–3 fold) in the amount of *GOR2* transcript during the stress period and a more dramatic increase in the post-stress periods. After stress, the levels of GOR2 mRNA were up to 10-fold higher in the case of re-watered peas and 4-fold higher following 24 hours recovery from high light stress by returning *Arabidopsis* to LL conditions (Figure 6) (Karpinski *et al.*, 1997; Stevens *et al.*, 1997). In the stressed peas, the changes in *GOR2* transcript levels occurred against a background of no changes in total GR activity. This may not be too surprising, because cytosolic GR activity may be no more than 20% of the total activity of this enzyme (Edwards *et al.*, 1990). We can speculate that, if during oxidative stress it is the cytosolic pool of glutathione which becomes more oxidized, then an increase in the level of cytosolic GR may be part of the process of restoring the redox state of glutathione to its predominantly reduced form.

MDHAR Gene Expression in Response to Stress

There are very few data on the behaviour of *MDHAR* transcripts during stress. The high-light stress applied to *Arabidopsis* (see above) does not bring about a change in cytosolic *MDHAR* mRNA levels (Karpinski *et al.*, 1997). *MDHAR* mRNA is most abundant in the roots of tomato compared with stem and leaf tissue. Interestingly this is the inverse of the ascorbate levels in these organs, i.e. highest in leaves and lowest in roots, leading to the suggestion that *MDHAR* gene expression may be sensitive to prevailing ascorbate levels. In developing and ripening tomato fruit, *MDHAR* mRNA shows a steady rise from immature green to the red ripe stage. Ascorbate levels remained constant throughout the fruit developmental phase but showed a slight decline on ripening, coincident with the highest fruit *MDHAR* mRNA levels (Grantz *et al.*, 1995). Wounding applied to leaves, roots, stem or fruit at the different stages of development, caused a 3- to 8-fold increase in *MDHAR* mRNA levels within 6 hours. Wounding causes increased oxidative damage to membranes and there could be an increased demand for ascorbate or dehydroascorbate in the processing of hydroxyproline-rich cell wall glycoproteins and in subsequent cross-linking reactions in the cell wall matrix. Thus, the possible upregulation of *MDHAR* gene expression by low tissue levels of ascorbate would be consistent with a rapid rise in its mRNA levels after wounding (Grantz *et al.*, 1995).

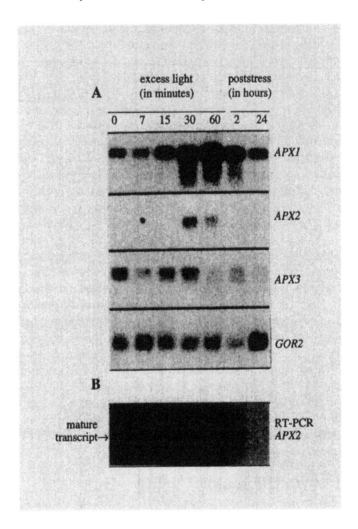

Figure 6. A. RNA gel blot analysis of mRNA levels for *APX1*, *APX2*, *APX3*, and cytosolic GR (*GOR1*). B. DNA blot analysis of *APX2* cDNA obtained by reverse trancription PCR showing *de novo* transcription of *APX2* within 7 min of the light stress. Reprinted with permission, from Karpinski *et al.*, Plant Cell 9, 627–640 (1997), copyright American Society of Plant Physiologists.

GENETIC MANIPULATION OF ASCORBATE-GLUTATHIONE CYCLE GENE EXPRESSION IN TRANSGENIC PLANTS

The advent of plant transformation technology allows experiments to be designed in which the levels of enzymes of the ascorbate-glutathione cycle can be up- or downregulated in order to test the contribution any enzyme could make to the ability of a plant to withstand oxidative stress. In the context of this chapter, it is GR that has been the prime focus of this experimental approach with, to date, seven publications in the peer-reviewed literature of overproduction of GR in either tobacco or poplar in various subcellular compartments (Aono *et al.*, 1991, 1993; Foyer *et al.*, 1991, 1995; Broadbent *et al.*, 1995; Creissen *et al.*, 1995; Badenhorst *et al.*, 1997) and one attempted antisense approach to the downregulation of GR activity in transgenic tobacco (Aono *et al.*, 1995). Almost all of these publications (the exception being that of Creissen *et al.*, 1995) report experiments testing the ability of such transgenic plants to withstand adverse environmental conditions likely to promote oxidative stress.

Manipulation of Ascorbate Peroxidase Expression

Overproduction of APX in the cytosol of transgenic tobacco has been achieved by expressing the *APX1*-coding sequence under the control of the ribulose-1,5-bisphosphate carboxylase small subunit promoter region (Saji *et al.*, 1996). For expression in the chloroplast, the *APX1*-coding sequence was fused to the chloroplast transit peptide from the small subunit of ribulose-1,5-bisphosphate carboxylase, under the control of the CaMV 35S promoter and polyadenylation sequences (Torsethaugen *et al.*, 1997). In the former case, a 2- to 7-fold increase in foliar APX activity was reported to have no effect on paraquat sensitivity (Saji *et al.*, 1996). Plants containing the chloroplast-targeted *APX1*, with up to 13 fold the total extractable APX activity, were exposed to two (acute or chronic) ozone fumigation regimes with no apparent differences between control and transgenic plants in their responses to this stress (Torsethaugen *et al.*, 1997).

Antisense *APX1* tobacco lines with an approximately 50% reduction in extractable foliar APX activity showed increased sensitivity to high-level ozone exposure (500 ppb for 8 h on a single day) and a further increase in sensitivity to a slightly less drastic treatment (250 ppb for 4 hours per day over 3 days) (Örvar and Ellis, 1997).

Clearly the differences in target compartments for overexpression or antisense reduction, coupled with the different stress treatments applied in these three examples? leave many unanswered questions concerning the importance of APX in these stress responses, which will only be answered by more systematic approaches to the problem, as has been attempted with glutathione reductase (see below).

Manipulation of Glutathione Reductase Expression

Most of the attempts to manipulate GR activity have employed the *E. coli GOR*-coding sequence (Aono *et al.*, 1991, 1993; Foyer *et al.*, 1991, 1995) although Broadbent *et al.* (1995) used the pea *GOR1* cDNA as the basis for making chimaeric genes that could target the GR to chloroplasts, mitochondria, or both these organelles simultaneously, or retain the enzyme in the cytosol (Creissen *et al.*, 1995; Broadbent *et al.*, 1995). Treatment of transgenic tobacco leaf discs with varying micromolar concentrations of paraquat showed that increased GR activity did lead to enhanced tolerance as determined by increased retention of chlorophyll compared with control leaf material (Aono *et al.*, 1991, 1993; Broadbent *et al.*, 1995). In one study, the tolerance to paraquat cosegregated with the presence of the GR transgene (Broadbent *et al.*, 1995). However, it should be noted that not all *GOR*-overexpressing transgenic tobacco lines can be identified as paraquat tolerant (Broadbent *et al.*, 1995). One study failed to detect any paraquat tolerance (Foyer *et al.*, 1991). Overexpression of *GOR* by up to 10-fold and 1000-fold in the chloroplasts of transgenic tobacco and poplar, respectively, do not necessarily give more tolerance to paraquat but do reveal other stress tolerances (Broadbent *et al.*, 1995; Foyer *et al.*, 1995). GR-transgenic poplar showed a small, but significant, tolerance to photoinhibitory stress (chilling of leaves in the light) as measured by the light saturation of CO_2 fixation. Ozone tolerance was also detected, as analyzed by changes in stress ethylene emission and PSII efficiency in some but not all transgenic lines of tobacco (Broadbent *et al.*, 1995). It should be noted that in at least some instances the interpretation of data regarding the tolerance (or not) of GR-transgenic plants to oxidative stress is not as straightforward as the above description might imply. It is beyond the scope of this chapter to discuss these concerns and for a more critical review of the current situation regarding the engineering of stress tolerance the reader is referred to Mullineaux and Creissen (1997).

A partial GR cDNA from spinach was used to construct a GR antisense gene fused to the promoter of the small subunit of ribulose-1,5-bisphosphate carboxylase, such that the expression of the antisense gene would be limited to photosynthetically active tissue (Aono *et al.*, 1995). This construct was utilized to transform tobacco and in plants from three lines, independent evidence for a degree of enhanced sensitivity to paraquat was presented. However, more recently it has become apparent that the spinach GR partial cDNA employed in these studies most probably encodes a cytosolic isoform of GR and not a plastidial isoform as stated by the authors (Stevens *et al.*, 1997). Therefore, any inhibition of GR activity in these plants may have been in a subcellular compartment inappropriate for the main site of action of paraquat.

Transgenic plants overexpressing GR have also proven useful in elucidating other properties of GR. The ability of the pea GR preprotein to cotarget to chloroplasts and mitochondria was first revealed in GR-overexpressing transgenic tobacco (see above; Creissen *et al.*, 1995). In addition, overexpression of GR in both chloroplasts and cytosol of transgenic tobacco and poplar has resulted in higher levels of foliar glutathione (up to 50%) and in poplar at least, an increase in the redox state of the glutathione pool in favour of the reduced form (Broadbent *et al.*, 1995; Foyer *et al.*, 1995). Precisely how increasing the activity of an enzyme that recycles glutathione has an effect on the total pool size of this antioxidant is not apparent. However, in a *gor*-deletion mutant of *E. coli* containing no detectable GR activity, the glutathione pool is reduced to 12% of its wild-type level (Kunert *et al.*, 1990), strongly suggesting that the prevailing GR activity and the cellular concentration of glutathione are linked in a wide range of organisms.

Of more relevance to the ascorbate-glutathione cycle is the observation in transgenic poplar overexpressing GR in the chloroplast that the foliar ascorbate pools were elevated up to 3-fold (Foyer *et al.*, 1995). There is some evidence that overexpression of GR in the cytosol of paraquat-treated transgenic tobacco protects, to some extent, the redox state of the foliar ascorbate pool (Foyer *et al.*, 1991), although other GR-transgenic tobacco have not revealed any effects on ascorbate levels (Broadbent *et al.*, 1995). Taken together, these data do lend some support to the existence, in poplar and perhaps in tobacco, of that part of the ascorbate-glutathione cycle concerned with the glutathione-dependent regeneration of reduced ascorbate (Foyer *et al.*, 1991, 1995).

THE ASCORBATE-GLUTATHIONE CYCLE: THE FUTURE OF THE HYPOTHESIS

The ascorbate-glutathione cycle has proved to be a remarkably resilient hypothesis, undergoing only slight modifications since it was proposed in its original form (Foyer and Halliwell, 1976; Asada, 1994). Part of this resilience is that this proposal is a very difficult one to test. Experimental determination of the stoichiometry of the reactions of the ascorbate-glutathione pathway were in good agreement with the predicted values and gave further experimental evidence for a role for DHAR in the chloroplast (Jablonski and Anderson, 1981). The best biochemical evidence has centred around the detection of the constituent enzymes in subcellular compartments of various tissues of different plant species and the observation that under many stress situations the activities of several of the component enzymes are coordinately increased. Therefore the proposed pathway has proven to be very valuable in focussing attention on the roles of ascorbate and glutathione in plant tissues. There is one question that this cycle brings to mind in the context of the chloroplast; is the only function of glutathione in this organelle that of maintaining the redox state of the ascorbate pool? With the advent of the detection of MDHAR in chloroplasts and the possibility of the direct regeneration of reduced ascorbate by PSI (Asada, 1994), the maintenance of millimolar amounts of glutathione in the chloroplast (Foyer and Halliwell, 1976; Gillham and Dodge, 1986) solely as a backup for the ascorbate pool cannot be the whole picture. So new roles for glutathione in the chloroplast need to be found and in this context it is in the molecular biological studies of the response of plants to oxidative stress that the first indications of additional and perhaps fundamentally more important roles for glutathione may be identified, such as the identification of a phospholipid-hydroperoxide glutathione peroxidase in the chloroplast stroma (Mullineaux *et al.*, 1998).

Coordinate regulation of the expression of genes encoding enzyme components of the ascorbate-glutathione cycle may also offer support for this hypothesis. Unfortunately, as can be discerned from this chapter, most studies have not revealed any universal coordinate regulation, although these experiments have been hampered by the lack of availability of probes coding for all the appropriate subcellular isoforms of the relevant enzymes. Some encouragement can be taken from the apparent co-regulation of the levels of mRNAs that encode the cytosolic isoforms of APX and GR during the recovery of peas from drought stress (Mittler and Zilinskas, 1994; Stevens *et al.*, 1997). Therefore such experiments will bear repetition in different plant species as many more gene-specific probes become available.

Perhaps the most valuable contribution that molecular-biological techniques can make is in the use of plant transformation technology to engineer the levels of the enzymic and antioxidant components of the ascorbate-glutathione cycle and ask how far does this affect the other perceived components of the pathway. Such an approach has already revealed some connection between the redox states of the glutathione and ascorbate pools in GR transgenic poplar (Foyer *et al.*, 1995). Further work along these lines and the logical extension of such experiments to the introduction of multiple components of the cycle and the use of such modifications to determine the contribution these enzymes make collectively to the plants' abilities to withstand oxidative stress are probably fruitful avenues to explore. A concerted effort employing transformation and antisense and cosuppression techniques should be made to systematically downregulate the components of this cycle and again determine how far the modification of one perceived component of the pathway affects the functioning and expression of the rest.

The generation of *Arabidopsis* mutants with altered functioning of components of the ascorbate-glutathione cycle would be a major asset and would provide a platform to uncover genes that may regulate the expression of the components of this pathway. With some imagination, it ought to be possible to generate appropriate mutants altered in this way in the near future and with the rapid advancement towards the complete sequence of the *Arabidopsis* genome currently being made it ought to be possible to generate targeted disruptions in key genes and thus allow a systematic dissection of the ascorbate-glutathione cycle hypothesis.

CONCLUDING REMARKS

The considerable body of research that has been carried out on the biochemistry, molecular biology, and genetics of the ascorbate-glutathione pathway has revealed a high level of complexity in terms of the subcellular compartmentation of component enzymes, size of the gene families, and in the regulation of expression of different genes. This complexity, coupled with the evidence for a role for glutathione in signalling in response to photooxidative stress suggests that the existence of these enzymes as contributors to the ascorbate-glutathione pathway represents only part of their function within the plant cell.

The proposed role of the ascorbate-glutathione cycle has been, and will continue to provide, a valuable intellectual framework that permits the many laboratories now engaged in this research to compare their data and experiences. This will probably be the most enduring contribution the ascorbate-glutathione cycle makes to stress research in plants.

REFERENCES

Aono, M., Kubo, A., Saji, H., Natori, T., Tanaka, K., and Kondo, N. (1991) Resistance to active oxygen toxicity of transgenic *Nicotiana tabacum* that expresses the gene for glutathione reductase from *Escherichia coli*. *Plant Cell Physiol.*, **32**, 691–697

Aono, M., Kubo, A., Saji, H., Tanaka, K., and Kondo, N. (1993) Enhanced tolerance to Photooxidative stress of transgenic *Nicotiana tabacum* with high chloroplastic glutathione reductase activity. *Plant Cell Physiol.*, **34**, 129–135.

Aono, M., Saji, H., Fujiyama, K., Sugita, M., Kondo, N., and Tanaka, K. (1995) Decrease in activity of glutathione reductase enhances paraquat sensitivity in transgenic *Nicotiana tabacum*. *Plant Physiol.*, **107**, 645–648.

Asada K. (1994) Production and action of active oxygen species in photosynthetic tissues. In C.H. Foyer and P.M. Mullineaux, (eds.), *Causes of Photooxidative Stress and Amelioration of Defense Systems in Plants*. CRC Press, Boca Raton, pp. 77–104.

Asada, K. (1997) The role of ascorbate peroxidase and monodehydroascorbate reductase in H$_2$O$_2$ scavenging in plants. In J. Scandalios, (ed.), *Oxidative Stress and the Molecular Biology of Antioxidant Defenses*, (Monograph Series Vol. 34), Cold Spring Harbor Laboratory Press, Cold Spring Harbor, pp. 715–735.

Badenhorst, P.W., Amory, A.M., and Huckett, B.I. (1997) Light regulation of native and *Escherichia coli* glutathione reductase in transgenic tobacco. *J. Plant. Physiol.*, **152**, 502–509.

Bielawski, V. and Joy, K.W. (1986) Reduced and oxidised glutathione and glutathione-reductase activity in tissues of *Pisum sativum*. *Planta*, **169**, 267–272.

Broadbent, P., Creissen G.P., Kular, B., Wellburn, A.R., and Mullineaux, P.M. (1995) Oxidative stress responses in transgenic tobacco containing altered levels of glutathione reductase activity. *Plant J.*, **8**, 247–255.

Bunkelmann, J.K. and Trelease, R.N. (1996) Ascorbate peroxidase. A prominent membrane protein in oilseed glyoxysomes. *Plant Physiol.*, **110**, 589–598.

Chen, G.X. and Asada, K. (1989) Ascorbate peroxidase in tea leaves: occurrence of two isozymes and the differences in their enzymatic and molecular properties. *Plant Cell Physiol.*, **30**, 987–998.

Conklin, P.L. and Last, R.L. (1995) Differential accumulation of antioxidant mRNAs in *Arabidopsis thaliana* exposed to ozone. *Plant Physiol.*, **109**, 203–312.

Creissen, G.P. and Mullineaux, P.M. (1995) Cloning and characterisation of glutathione reductase cDNAs and identification of two genes encoding the tobacco enzyme. *Planta*, **197**, 422–425.

Creissen, G.P., Edwards, E.A., Enard, C., Wellburn, A.R., and Mullineaux. (1992) Molecular characterization of glutathione reductase cDNAs from pea (*Pisum sativum* L.). *Plant J.*, **2**, 129–131.

Creissen, G.P., Edwards, E.A., and Mullineaux, P.M. (1994) Glutathione reductase and ascorbate peroxidase In C.H. Foyer and P.M. Mullineaux, (eds.), *Causes of Photooxidative Stress and Amelioration of Defense Systems in Plants*, CRC Press, Boca Raton, pp. 343–364.

Creissen. G.P., Reynolds, H., Xue, Y., and Mullineaux, P.M. (1995) Simultaneous targeting of pea glutathione reductase and of a bacterial fusion protein to chloroplasts and mitochondria in transgenic tobacco. *Plant J.*, **8**, 167–175.

Dipierro, S. and Borraccino, G. (1991) Dehydroascorbate reductase from potato-tubers. *Phytochemistry*, **30**, 427–429.

Edwards, E.A., Rawsthorne, S., and Mullineaux, P.M. (1990) Subcellular distribution of multiple forms of glutathione reductase in leaves of pea (*Pisum sativum* L.). *Planta*, **180**, 278–284.

Edwards, E.A., Enard, C., Creissen, G.P., and Mullineaux, P.M. (1994) Synthesis and properties of glutathione reductase in stressed peas. *Planta*, **192**, 137–143.

Escobar, C. (1998) Regulation of the ascorbate peroxidase gene family in Arabidopsis thaliana. PhD Thesis, University of East Anglia.

Foyer, C.H. and Halliwell, B. (1976) The presence of glutathione and glutathione reductase in chloroplasts: A proposed role in ascorbate metabolism. *Planta*, **133**, 21–25.

Foyer, C.H. and Mullineaux, P.M. (1998) The presence of dehydroascorbate and dehydroascorbate reductase in plant tissues. *FEBS Lett.* **425**, 528–529.

Foyer, C., Lelandais, M., Galap, C., and Kunert, K.J. (1991) Effects of elevated cytosolic glutathione reductase activity on the cellular glutathione pool and photosynthesis in leaves under normal and stress conditions. *Plant Physiol.*, **97**, 863–872.

Foyer, C.H., Souriau, N., Perret, S. Lelandais, M., Kunert, K.J., Pruvost, C., and Jouanin, L. (1995) Overexpression of glutathione reductase but not glutathione synthetase leads to increases in antioxidant capacity and resistance to photoinhibition in poplar trees. *Plant Physiol.*, **109**, 1047–1057.

Gillham, D.J. and Dodge, A.D. (1986) Hydrogen-peroxide-scavenging systems within pea chloroplasts. A quantitative study. *Planta*, **167**, 246–251.

Grantz, A.A., Brummell, D.A., and Bennett, A.B. (1995) Ascorbate free radical reductase mRNA levels are induced by wounding. *Plant Physiol.*, **108**, 411–418.

Hausladen, A. and Alscher, R.G. (1994) Purification and characterisation of glutathione reductase isoforms specific for the state of cold hardiness of red spruce. *Plant Physiol.*, **105**, 205–213.

Ishikawa, T., Sakai, K., Takeda, T., and Shigeoka, S. (1995) Cloning and expression of a cDNA encoding a new type of ascorbate peroxidase from spinach. *FEBS Lett.*, **367**, 28–32.

Ishikawa, T., Kohno, H., Takeda, T., and Shigeoka, S. (1996a) Molecular characterisation of *Euglena* ascorbate peroxidase using monoclonal antibodies. *Bioch. Biophys. Acta.*, **1290**, 69–75.

Ishikawa, T., Sakai, K., Yoshimura, K., Takeda, T., and Shigeoka, S. (1996b) cDNAs encoding spinach stromal and thylakoid-bound ascorbate peroxidase, differing in the presence or absence of their 3'-coding regions. *FEBS Lett.*, **384**, 289–293.

Jablonski, P.P. and Anderson, J.W. (1981) Light-dependent reduction of dehydroascorbate by ruptured chloroplasts. *Plant Physiol.*, **67**, 1239–1244.

Jespersen, H.M., Kjærsgård, I.V.H., Østergaard, L., and Welinder, K.G. (1997) From sequence analysis of three novel ascorbate peroxidases from *Arabiodpsis thaliana* to structure, function and evolution of seven types of ascorbate peroxidase. *Biochem. J.*, **326**, 305–310.

Jimenéz, A., Hernandez, J.A., del Río, L.A., and Sevilla, F. (1997) Evidence for the presence of the ascorbate-glutathione cycle in mitochondria and peroxisomes of pea (*Pisum sativum* L.) leaves. *Plant Physiol.*, **114**, 275–284.

Kaminaka, H., Morita, S., Nakajima, M., Masumura, T., and Tanaka, K. (1998). Gene cloning and expression of cytosolic glutathione reductase in rice (*Oryza sativa* L.). *Plant Cell Physiol.*, **39**, 1269–1280.

Karpinski, S., Wingsle, G., Karpinska, B., and Hallgren, J-E. (1993) Molecular responses to photooxidative stress in *Pinus sylvestris* (L.). *Plant Physiol.*, **103**, 1385–1391.

Karpinski, S., Escobar, C., Karpinska, B., Creissen, G., and Mullineaux, P. (1997) Photosynthetic electron transport regulates the expression of cytosolic ascorbate peroxidase genes in *Arabidopsis* during excess light stress. *Plant Cell*, **9**, 627–640.

Klapheck, S., Zimmer, I., and Cosse, H. (1990) Scavenging of hydrogen peroxide in the endosperm of *Ricinus communis* by ascorbate peroxidase. *Plant Cell Physiol.*, **31**, 1005–1013.

Kato, Y., Urano, J., Maki, Y., and Ushimaru, T. (1997) Purification and characterization of dehydroascorbate reductase from rice. *Plant Cell Physiol.*, **38**, 173–178.

Kubo, A., Saji, H., Tanaka, K. Tanaka, K., and Kondo, N. (1992) Cloning and sequencing of a cDNA encoding ascorbate peroxidase from *Arabidopsis thaliana. Plant Cell Physiol.*, **18**, 691–701.

Kubo, A., Sano, H., Saji, K.,Tanaka, K., Kondo, N., and Tanaka, K. (1993) Primary structure and properties of glutathione reductase from *Arabidopsis thaliana. Plant Cell Physiol.*, **34**, 1259–1266.

Kubo, A., Saji, H., Tanaka, K., and Kondo, N. (1995) Expression of *Arabidopsis* cytosolic ascorbate peroxidase gene in response to ozone or sulfur dioxide. *Plant Mol. Biol.*, **29**, 479–489.

Kubo, A., Aono, M., Nakajima, N., Saji, H., Kondo, N., and Tanaka K. (1998) Genomic DNA structure of a gene encoding glutathione reductase from *Arabidopsis* (Accession No. D89620) (PGR 98–131). *Plant Physiol.*, **117**, 1127.

Kunert, K.J., Cresswell, C.F., Schmidt, A., Mullineaux, P.M., and Foyer, C.H. (1990) Variations in the activity of glutathione reductase and the cellular glutathione content in relation to sensitivity to methylviologen in *Escherichia coli. Arch. Biochem. Biophys.*, **282**, 233–238.

Lee, H., Jo, J., and Son, D.(1998) Molecular cloning and characterization of the gene encoding glutathione reductase in *Brassica campestris. Biochim. Biophys. Acta*, **1395**, 309–314.

Levine, A., Tenhaken, R., Dixon, R., and Lamb, C.J. (1994) H_2O_2 from the oxidative burst orchestrates the plant hypersensitive disease resistance response. *Cell*, **79**, 583–593.

Lopez, F., Vansuyt, G., Casse-Delbart, F., and Fourcroy, P. (1996) Ascorbate peroxidase activity, not the mRNA level, is enhanced in salt-stressed *Raphanus sativus* plants. *Physiol. Plant.*, **97**, 13–20.

Madamanchi, N.R., Anderson, J.V., Alscher, R.G., Cramer, C.L., and Hess, J.L (1992) Purification of multiple forms of glutathione reductase from pea (*Pisum sativum* L.) seedlings and enzyme levels in ozone-fumigated pea leaves. *Plant Physiol.*, **100**, 138–145.

Mittler, R. and Zilinskas, B.A. (1991a) Molecular cloning and nucleotide sequence analysis of a cDNA encoding pea cytosolic ascorbate peroxidase. *FEBS Lett.*, **289**, 257–259.

Mittler, R. and Zilinskas, B.A. (1991b) Purification and characterisation of pea cytosolic ascorbate peroxidase. *Plant Physiol.*, **97**, 962–968.

Mittler, R. and Zilinskas, B.A. (1992) Molecular cloning and characterisation of a gene encoding pea cytosolic ascorbate peroxidase. *J. Biol. Chem.*, **267**, 21802–21807.

Mittler, R. and Zilinskas, B.A. (1994) Regulation of pea cytosolic ascorbate peroxidase and other antioxidant enzymes during the progression of drought stress and following recovery from drought. *Plant J.*, **5**, 397–405.

Mittler, R., Shulaev, V., Seskar, M., and Lam, E. (1996) Inhibition of programmed cell death in tobacco plants during a pathogen-induced hypersensitive response at low oxygen pressure. *Plant Cell*, **8**, 1991–2001.

Mittler, R., Feng, X.Q., and Cohen, M. (1998) Post-transcriptional suppression of cytosolic ascorbate peroxidase expression during pathogen-induced programmed cell death in tobacco. *Plant Cell*, **10**, 461–473.

Miyake, C., Cao, W.H., and Asada, K. (1993) Purification and molecular properties of thylakoid-bound ascorbate peroxidase in spinach chloroplasts. *Plant Cell Physiol.*, **34**, 881–889.

Morell, S., Follmann, H., De Tullio, M., and Haberlein, I. (1997) Dehydroascorbate and dehydroascorbate reductase are phantom indicators of oxidative stress in plants *FEBS Lett.*, **414**, 567–570.

Mullen, R.T. and Trelease, R.N. (1996) Biogenesis and membrane properties of peroxisomes—does the boundary membrane serve and protect? *Trends Plant Sci.*, **1**, 389–394.

Mullineaux, P.M. and Creissen, G.P. (1997) Glutathione reductase: regulation and role in oxidative stress. In Scandalios J, (ed.), *Oxidative Stress and the Molecular Biology of Antioxidant Defenses*, (Monograph Series, Vol. 34), Cold Spring Harbor Laboratory Press, Cold Spring Harbor, pp. 667–713.

Mullineaux, P., Enard, C., Hellens, R., and Creissen, G. (1996) Characterisation of a glutathione reductase gene and its genetic locus from pea (Pisum sativum L). *Planta*, **200**, 186–194.

Mullineaux, P.M., Karpinski, S., Jimenéz, A., Cleary, S.P., Robinson, C., and Creissen, G.P. (1998) Identification of cDNAs encoding plastid-targeted glutathione peroxidase. *Plant J.*, **13**, 375–379.

Murthy, S.S. and Zilinskas, B.A. (1994) Molecular cloning and characterisation of a cDNA encoding pea monodehydroascorbate reductase. *J. Biol. Chem.*, **269**, 31129–31133.

Örvar, B.L. and Ellis, B.E. (1997) Transgenic tobacco plants expressing antisense RNA for cytosolic ascorbate peroxidase show increased susceptibility to ozone injury. *Plant J.*, **11**, 1297–1305.

Saji, H., Aono, M., Kubo, A., Tanaka, K., and Kondo, N. (1996) Paraquat sensitivity of transgenic *Nicotiana tabacum* plants that overproduce a cytosolic ascorbate peroxidase. *Environ. Sci.*, **9**, 241–248.

Sano, S. and Asada, K. (1994) cDNA cloning of monodehydroascorbate radical reductase from cucumber: a high degree of homology in terms of amino acid sequence between this enzyme and bacterial flavoenzymes. *Plant Cell Physiol.*, **35**, 425–437.

Santos, M.C.G. (1995) *Characterisation of ascorbate peroxidase genes from plants.* PhD thesis, University of East Anglia.

Santos, M., Gousseau, H., Lister, C., Foyer, C., Creissen, G., and Mullineaux, P. (1996) Cytosolic ascorbate peroxidase from *Arabidopsis thaliana* L. is encoded by a small multigene family. *Planta*, **198**, 64–69.

Sen Gupta, A., Webb, R.P., Holaday, A.S., and Allen, R.D. (1993) Overexpression of superoxide dismutase protects plants from oxidative stress. *Plant Physiol.*, **103**, 1067–1073.

Stevens, R.G., Creissen, G.P., and Mullineaux, P.M. (1997) Cloning and characterisation of a cytosolic glutathione reductase cDNA from pa (*Pisum sativum* L.) And its expression in response to stress. *Plant Mol. Biol.*, **35**, 641–654.

Strid, Å. (1993) Alteration in expression of defence genes in *Pisum sativum* after exposure to supplementary ultraviolet-B radiation. *Plant Cell Physiol.*, **34**, 949–953.

Tang, X. and Webb, M.A. (1994) Soybean root nodule cDNA encoding glutathione reductase. *Plant Physiol.*, **104**, 1081–1082.

Torsethaugen, G., Pitcher, L.H., Zilinskas, B.A., and Pell, E.J. (1997) Overproduction of ascorbate peroxidase in the tobacco chloroplast does not provide protection against ozone. *Plant Physiol.*, **114**, 529–537.

Trumper, S., Follmann, H., and Haberlein, I. (1994) A novel dehydroascorbate reductase from spinach-chloroplasts homologous to plant trypsin-inhibitor *FEBS Lett.*, **352**, 159–162.

Wells, W.W., Xu, D.P., Yang, Y.F., and Rocque, P.A. (1990) Mammalian thioltransferase (glutaredoxin) and protein disulfide isomerase have dehydroascorbate reductase-activity. *J. Biol. Chem.*, **265**, 15361–15364.

Willekens, H., Van Camp, W., Van Montagu, M., Inzé, D., Langebartels, C., and Sandermann, H. Jr (1994) Ozone, sulphur dioxide and ultraviolet B have similar effects on mRNA accumulation of antioxidant genes in *Nicotiana plumbaginifolia* L. *Plant Physiol.*, **106**, 1007–1014.

Wingsle, G. and Karpinski, S. (1996) Differential redox regulation by glutathione of glutathione reductase and CuZn-superoxide dismutase gene expression in *Pinus sylvestris* L. needles. *Planta*, **198**, 151–157.

Yamaguchi, K., Takeuchi, Y., Mori, H., and Nishimura, M. (1995a) Development of microbody membrane proteins during the transformation of glyoxysomes to peroxisomes in pumpkin cotyledons. *Plant Cell Physiol.*, **36**, 455–464.

Yamaguchi K, Mori, H., and Nishimura, M. (1995b) A novel isoenzyme of ascorbate peroxidase localized on glyoxysomal and leaf peroxisomal membranes in pumpkin. *Plant Cell Physiol.*, **36**, 1157–1162.

Yamaguchi, K., Hayashi, M., and Nishimura, M. (1996) cDNA cloning of thylakoid-bound ascorbate peroxidase in pumpkin and its characterisation. *Plant Cell Physiol.*, **37**, 405–409.

Yoshimura, K., Ishikawa, T., Nakamura, Y., Tamoi, M., Takeda, T., Tada, T., Nishimura, K., and Shigeoka, S. (1998) Comparative study on recombinant chloroplastic and cytosolic ascorbate peroxidase isozymes of spinach. *Arch. Biochem. Biophys.*, **353**, 55–63.?

11 Ascorbate Metabolism and Stress

Mark W. Davey, Marc Van Montagu and Dirk Inzé

INTRODUCTION

L-ascorbic acid is the most abundant hydrophilic antioxidant of higher plant cells, reaching levels of up to 10% of the soluble carbohydrate content of cells (Smirnoff and Pallanca, 1995). Much of our understanding of the biological functions of L-AA is based upon the well-known antioxidant character of this molecule, and L-AA occupies a central role in the protection of plant cells against abiotic and biotic stresses. However, L-AA also serves as a substrate for oxalate and tartrate biosynthesis. An indication of the importance of L-AA to cells is provided by its high intracellular concentrations and the recent isolation of an L-AA-deficient *Arabidopsis thaliana* mutant that is hypersensitive to ozone and to SO_2 (Conklin *et al.*, 1996, 1997). Yet, despite the central role of L-AA in the modulation of processes as diverse as (photo)oxidative defense, senescence, and cell growth, the biosynthetic route for L-AA has only recently been defined in plant tissues (Wheeler *et al.*, 1998). This chapter will therefore look at the current understanding of plant L-AA biosynthesis and metabolism, as well as the biological functions of L-AA in relation to oxidative stress.

CHEMISTRY

Structure and Derivatives of Ascorbic Acid

L-AA is the aldono-1,4-lactone of a hexonic acid (either L-galactonic or L-gulonic acid) (Figure 1). It contains a C_2-C_3 enediol group and delocalization of these π electrons stabilizes the molecule, causing the hydrogen of the C_3 hydroxyl to become acidic. This C_3 hydroxyl dissociates with a pK_a of 4.13, so that at physiological pHs, L-AA exists as a monovalent anion (L-ascorbate). Dissociation of the second hydroxyl only takes place at pH 11.6. Interestingly, yeasts and certain fungi preferentially synthesize D-erythroascorbic acid, a C_5 analogue of L-AA (Figure 1) (Nick *et al.*, 1986; Kim *et al.*, 1996, 1998).

A number of natural derivatives of L-AA are known (Tolbert *et al.*, 1975), including ascorbate-2-sulphate (Dabrowski *et al.*, 1993), ascorbic acid-2-O-β-glucuronide, and 2-O-α-glucoside (Yamamoto *et al.*, 1990; Gallice *et al.*, 1994). In mushroom, several C_5-linked glycosides of 6-deoxyascorbate have been identified (Okamura, 1994) and ascorbinogen, which is a β-substituted indole derivative, has been identified in cabbage (Jaffe, 1984). The possible existence and function of such derivatives in plant tissues has received little attention however.

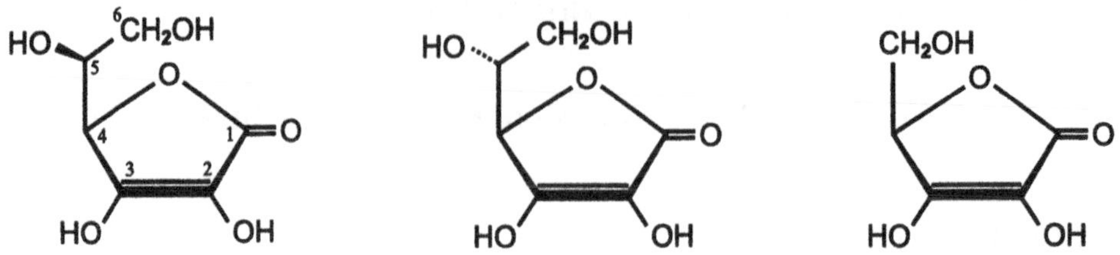

L-ascorbic acid D-isoascorbic acid (D-araboascorbic acid, D-erythorbic acid) D-erythroascorbic acid

Figure 1. Chemical structures of L-AA, D-isoascorbic acid, and D-erythroascorbic acid.

Oxidation

Solutions of L-AA readily oxidize, especially in the presence of trace amounts of transition metal ions and alkali. The first product of this oxidation is the radical monodehydroascorbate (MDA), also known as semidehydroascorbate, or ascorbate-free radical (for a discussion of reaction mechanisms, see Cadenas, 1995; Larson, 1995). MDA is unusually stable for a free radical with a decay constant of $2.8 \times 10^5\ M^{-1}\ s^{-1}$ at pH 7 (Foyer *et al.*, 1991). *In vivo* MDA is reduced back to L-AA by the activity of the NAD(P)-dependent enzyme monodehydroascorbate reductase (MDAR, EC 1.6.5.4), or by electron transfer reactions. However, two molecules of MDA will also spontaneously disproportionate to L-AA and DHA. DHA itself is unstable and undergoes irreversible hydrolytic ring cleavage to 2,3-diketogulonic acid in aqueous solution (Washko *et al.*, 1992; Deutsch 1998) (Figure 2). Clearly though, the rates of L-AA oxidation and DHA hydrolysis will be influenced by their concentrations, the temperature, light, pH, dissolved oxygen, solvent, ionic strength, and the presence of divalent cations.

OCCURRENCE AND COMPARTMENTATION

L-AA is present in all tissues with the exception of dormant seeds (Loewus 1980). High L-AA concentrations are typically associated with tissues undergoing rapid growth and development. L-AA has also been demonstrated to be present in all subcellular compartments including the chloroplast, mitochondria, vacuole, and apoplast (Anderson *et al.*, 1983; Rautenkrantz *et al.*, 1994). Concentrations as high as 50 mM have been measured in spinach chloroplasts, although generally the highest concentrations are found in the cytosol (Foyer *et al.*, 1983; Noctor and Foyer 1998). In certain species with particularly high L-AA levels, L-AA has been reported to be sequestered in the vacuole, where the low pH presumably stabilizes the molecule. The presence of L-AA in the apoplast is now widely recognized to be non-artefactual, and due to the role of apoplastic L-AA in the protection of cells against atmospheric pollutants, such as ozone, pathogen

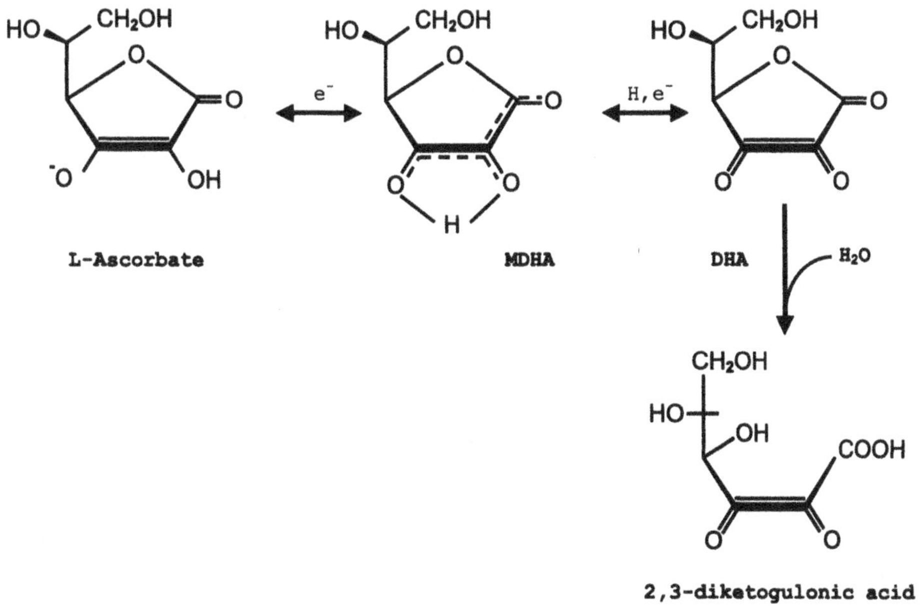

Figure 2. Oxidation of the L-ascorbate anion.

invasion (Vanacker *et al.*, 1998) as well as functions in cell wall metabolism (Smirnoff, 1996). In *Picea* needles, approximately 25% of the total cellular L-AA was recovered from the apoplast (Foyer *et al.*, 1991), with concentrations in the range of 0.15–2 mM (Luwe *et al.*, 1993; Takahama 1993).

BIOSYNTHESIS

L-AA Biosynthesis in Plants

Much of the conflicting data from the previous decades of research have now been resolved by the proposal of a new pathway, involving the conversion of hexose phosphates to L-AA via GDP-D-mannose and GDP-L-galactose (Wheeler *et al.*, 1998) (Equation 1).

$$\text{D-glucose-6-P} \longleftrightarrow \text{D-mannose-1-P} \rightarrow \text{GDP-D-Mannose} \longleftrightarrow$$
$$\text{GDP-L-galactose} \rightarrow \text{(L-galactose-1-P)} \rightarrow \text{L-galactose} \rightarrow$$
$$\text{L-galactono-1,4-lactone} \rightarrow \text{L-AA} \qquad \text{(Eq. 1)}$$

The enzymes catalyzing the interconversions of D-glucose-6-P to GDP-D-mannose are well known and provide a pool of hexose phosphate precursors used for sucrose and polysaccharide biosynthesis as well as glycolysis (Feingold and Avigad, 1980). The double epimerization of GDP-D-mannose to GDP-L-galactose is catalyzed by GDP-D-mannose 3,5-epimerase (EC 5.1.3.18), an enzyme that was originally identified in

Chlorella pyrenoidosa and flax (Barber 1971, 1979; Feingold and Avigad, 1980), but which remains poorly characterized. The mechanism by which GDP-L-galactose is converted to L-galactose is speculative, but may be catalyzed either by a guanylyltransferase, or by the sequential actions of a hydrolase (generating L-galactose 1-phosphate), and a phosphatase. These and subsequent steps are exclusive to L-AA biosynthesis. L-galactose itself is oxidized to L-galactono-1,4-lactone by a newly identified enzyme, L-galactose dehydrogenase, that has been partially purified from pea seedlings and *Arabidopsis thaliana* (Wheeler *et al.*, 1998). L-galactose dehydrogenase catalyzes the NAD-dependent oxidation of the C_1 of L-galactose to form L-galactono-1,4-lactone. It is highly specific for L-galactose as substrate and has an absolute requirement for NAD (Wheeler *et al.*, 1998; Davey *et al.*, 1999a). This substrate specificity distinguishes L-galactone dehydrogenase from the D-arabinose and L-fucose dehydrogenases of yeast and mammals, respectively, which are also able to oxidize L-galactose (Kim *et al.*, 1996). The final step, the oxidation of L-galactono-1,4-lactone to L-AA, is catalyzed by L-galactono-1,4-lactone dehydrogenase (GLDH; EC 1.3.2.3), an enzyme activity that was first identified over 40 years ago (Isherwood *et al.*, 1954; Mapson *et al.*, 1954). GLDH has been purified and characterized from a number of different plant species (Mapson and Breslow 1958; Ôba *et al.*, 1994, 1995; Mutsuda *et al.*, 1995; Østergaard *et al.*, 1997), and has recently been cloned from cauliflower (Østergaard *et al.*, 1997) and sweet potato (Imai *et al.*, 1998). GLDH is able to utilize cytochrome *c* as an electron acceptor (Mapson and Breslow 1958; Ôba *et al.*, 1995; Østergaard *et al.*, 1997) and its activity is commonly assayed by the reduction of cytochrome *c*. It is not clear if cytochrome *c* is the electron acceptor *in vivo* however, and it has recently been shown that GLDH activity in crude cell extracts can be quantitated in the absence of exogenous cytochrome *c* (Davey *et al.*, 1999a). However, this *in vitro* activity does require the presence of active mitochondria (Davey *et al.*, 1999a). The substrate for GLDH, L-galactono-1,4-lactone has also been reported to be a natural component of plant extracts (Østergaard *et al.*, 1997; Wheeler *et al.*, 1998).

The validity of this new biosynthetic route is supported by feeding studies obtained with radiolabelled D-glucose and D-mannose (Wheeler *et al.*, 1998) and by genetic evidence from the study of the L-AA-deficient *Arabidopsis* mutant *vtc-1* (originally called *soz-1*). The *vtc-1* mutant contains only 30% wild-type levels of L-AA and was originally isolated on the basis of increased ozone sensitivity (Conklin *et al.*, 1996, 1997). Recently, the locus of *vtc-1* was identified as being that of D-mannose-1-phosphate-guanylyl transferase (see Eq. 1; Figure 3) (Conklin *et al.*, 1999). In addition to L-AA biosynthesis, GDP-D-mannose is used for the biosynthesis of non-cellulosic structural carbohydrates of the plant cell wall, for protein *N*- and *O*-linked glycosylation as well as the biosynthesis of L-fucose and L-galactose, which are also utilized in cell wall biosynthesis.

The significance of this new pathway therefore, is that it integrates L-AA biosynthesis within the pathways of central hexose phosphate metabolism, providing links to polysaccharide biosynthesis and protein glycosylation. As a result, however, manipulation of L-AA biosynthesis is also likely to influence the flow of carbon into these other processes. In particular, one might expect the partitioning and size of the GDP-D-mannose pool to affect the relative rates of L-AA biosynthesis and cell wall polysaccharide and glycoprotein biosynthesis. Indeed, the *vtc-1* mutant has been shown to have an aberrant protein glycosylation pattern, although plants transformed with an antisense construct of D-mannose-1-phosphate-guanylyl transferase were reported not to have an altered glycoprotein profile (Keller *et al.*, 1999).

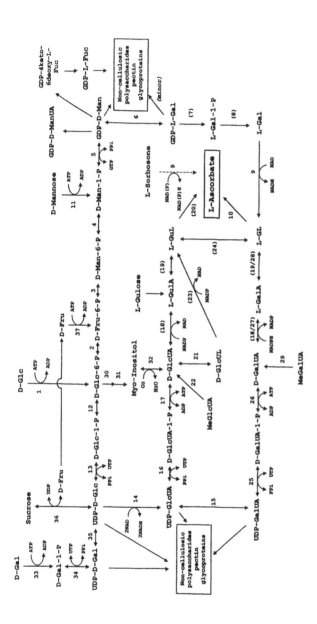

Figure 3. General scheme showing integration of the L-galactose-, uronic acid-, and L-sorbosone-based pathways of L-AA biosynthesis into central plant hexose phosphate metabolism (adapted from Davey *et al.*, 1999b). Abbreviations: D-Fru, D-fructose; L-Fuc, L-fucose; D/L-Gal, D/L-galactose; L-GalA, L-galactonic acid; D-GalUL, D-galacturonolactone; D-Glc, D-glucose; D-GlcUL, D-glucuronolactone; L-GulA, L-gulonic acid; L-GulUL, L-gulonolactone; D-Man, D-mannose. Numbered reactions in brackets have yet to be conclusively demonstrated in the same tissue.

Enzymes catalyzing the individual numbered reactions are: 1, Hexokinase (EC 2.7.1.1) also catalyzes reaction 11; 2, Glucose-6-phosphate isomerase (EC 5.3.1.9); 3, Mannose-6-phosphate isomerase (EC 5.3.1.8); 4, Phosphomannomutase (EC 5.4.2.8); 5, Mannose-1-phosphate guanylyltransferase (EC 2.7.7.22); 6, GDP-mannose 3,5-epimerase (E.C. 5.1.3.18); 7, GDP-L-galactose hydrolase; 8, Sugar phosphatase (EC 3.1.3.23); 9, L-galactose 1-dehydrogenase; 10, L-galactono-1,4-lactone dehydrogenase (EC 1.3.2.3); 11, D-mannose kinase/hexokinase (EC 2.7.1.1); 12, Phosphoglucomutase (EC 5.4.2.2); 13, UTP-glucose-1-phosphate uridylyl transferase (EC 2.7.7.9); 14, UDP-D-glucose dehydrogenase (EC 1.1.1.22); 15, UDP-glucuronate 4-epimerase (EC 5.1.3.6); 16, Glucuronate-1-phosphate uridylyltransferase, (EC 2.7.7.44); 17, D-glucuronokinase (EC 2.7.1.43); 18, D-glucuronate (hexuronate) reductase (EC 1.1.1.19) may also catalyze reaction 27; 19/28, Aldono-lactonase (3.1.1.17)/spontaneous; 20, L-gulono-1,4-lactone oxidase/dehydrogenase; 21, Spontaneous lactonization, or uronolactonase activity (EC 3.1.1.19); 22, (spontaneous) Methylesterase activity; 23, Glucuronolactone reductase activity (EC 1.1.1.20) possibly catalyzes reaction 18, and 27; 24, L-galactono-1,4-lactone 3-epimerase; 25, Galacturonate-1-phosphate uridylyltransferase; 26, Galacturonokinase (EC 2.7.1.44); 27, Hexuronate (D-galacturonate) reductase; 28/19, Aldonolactonase, or spontaneous; 29, (spontaneous) Methylesterase activity; 30, *myo*-inositol 1-phosphate synthase (EC 5.5.1.4); 31, *myo*-inositol 1-phosphate monophosphatase (EC 3.1.3.25); 32, *myo*-inositol oxygenase (EC 1.13.99.1); 33, D-glucose 4-epimerase; 34, D-galactokinase (EC 2.7.1.6); 35, UTP-hexose 1-phosphate uridylyltransferase (EC 2.7.7.10); 36, Sucrose synthase (EC 2.4.1.13); 37, Fructokinase (EC 2.7.1.4).

Alternative Routes of L-AA Biosynthesis in Plants?

Extensive and detailed radiotracer studies (Loewus 1963, 1980, 1988; Saito *et al.*, 1990) demonstrated that in plants there are three critical features regarding the conversion of ^{14}C-D-glucose into L-AA. Firstly, allowing for randomization through the hexose phosphate pool, the majority of radiolabel from D-glucose is incorporated into L-AA without inversion of configuration, i.e. C_1 of D-glucose forms the C_1 of L-AA (a so-called "non-inversion" pathway). Secondly, the C_6 hydroxymethyl group of D-glucose is conserved throughout, and finally there is an epimerization at the C_5 position (Loewus 1963, 1988). The route proposed by Wheeler *et al.* (1998) (Eq. 1) is entirely consistant with these results. By comparison, L-AA biosynthesis in animals involves an "inversion" of label, and radiolabel from 1–^{14}C-D-glucose is incorporated into the C_6 position of L-AA.

L-*sorbosone ("non-inversion" pathway)*

This pathway first proposed in 1990 (Loewus *et al.*, 1990; Saito *et al.*, 1990) was put forward to accomodate earlier radiolabelling data accumulated since the 1950's (see above), which showed no inversion during the formation of L-AA from 1-^{14}C-D-glucose. Here, it was proposed that D-glucose is first oxidized by a pyranose 2-oxidase activity to form the osone D-glucosone, followed by epimerization of C_5 to give L-sorbosone. Finally, C_1 oxidation of L-sorbosone yields L-AA (Loewus *et al.*, 1990; Saito *et al.*, 1990) (Eq. 2).

$$\text{D-glucose} \rightarrow \text{D-glucosone} \rightarrow \text{L-sorbosone} \rightarrow \text{L-AA} \qquad \text{(Eq. 2)}$$

When bean leaves were fed radiolabelled D-glucosone and L-sorbosone, radiolabel was incorporated into L-AA without inversion of configuration, i.e. via a "non-inversion" route, and unlabelled D-glucosone and L-sorbosone competed with ^{14}C-D-glucose for incorporation into L-AA. An enzyme catalyzing the NADP-dependent oxidation of L-sorbosone to L-AA was also partially purified, although substrate affinity was low (Loewus *et al.*, 1990). However, there has been no demonstration to date that plants possess the enzymes capable of converting D-glucose to D-glucosone or of D-glucosone to L-sorbosone. Conklin *et al.* (1997) reported that neither D-glucosone nor L-sorbosone had any effect on L-AA accumulation in their *Arabidopsis* L-AA-deficient mutant, *vtc-1*, and 15–30 mM L-sorbosone had no significant effect on endogenous L-AA levels in *Arabidopsis* cell suspension cultures (Davey *et al.*, 1999b). Wheeler *et al.* (1998) reported that L-galactose dehydrogenase could also oxidize L-sorbosone with low affinity, possibly accounting for the earlier published results (Loewus *et al.*, 1990; Saito *et al.*, 1990). A recent re-examination of this work has also concluded that this pathway is probably not physiologically relevant (Pallanca and Smirnoff, 1999).

Biosynthesis from uronic acids ("inversion" pathway)

Early studies demonstrated that certain plant species are able to efficiently use the uronic acids D-galacturonic acid and D-glucuronic acid, and their derivatives (lactones and esters), as substrates for L-AA biosynthesis (Isherwood *et al.*, 1954; Mapson and Isherwood 1956; Loewus *et al.*, 1958; Loewus 1963; Baig *et al.*, 1970; Leung and Loewus 1985). These conversions to L-AA occur without disruption of the carbon skeleton, but result in an "inversion" of configuration (Loewus, 1963, 1988), and thus in this respect resemble the biosynthesis of L-AA in animals. In animals, L-AA biosynthesis proceeds via D-glucuronic acid and L-gulono-1,4-lactone and also involves an "inversion" of configuration (Nishikimi and Yagi, 1996; Bánhegyi *et al.*, 1997) (Eq. 3).

$$\text{UDP-glucose} \rightarrow \text{UDP-glucuronic acid} \rightarrow \text{D-glucuronic acid 1-phosphate}$$
$$\rightarrow \text{D-glucuronic acid} \rightarrow \text{L-gulono-1, 4-lactone} \rightarrow \text{L-AA} \qquad \text{(Eq. 3)}$$

We have recently shown in *Arabidopsis thaliana* cell suspension culture that D-glucuronolactone, D-glucuronic acid methyl ester, L-gulono-1,4-lactone, D-galacturonic acid methyl ester, L-galactono-1,4-lactone, and L-galactose are the only substrates examined able to increase intracellular L-AA levels *in vivo*. On average, the rates of L-AA biosynthesis over a period of 24 h with these substrates are 3-, 9-, 10-, 35-, 32-, and 68-fold higher than control cultures (sucrose), respectively (Davey *et al.*, 1999b). Clearly therefore, in addition to the substrates of the newly proposed pathway (Eq. 1), *A. thaliana* suspension culture can also utilize the substrates of the animal pathway and D-galacturonic acid methyl ester.

The conversion of D-galacturonic acid methyl ester could conceivably involve a non-specific aldo-reductase-type activity, and this would generate L-galactono-1,4-lactone (via L-galactonic acid), the substrate for GLDH. The same activity would also reduce D-glucuronic acid methyl ester. However, GLDH has been shown to be absolutely specific for L-galactono-1,4-lactone in cauliflower (Mapson and Breslow, 1958; Østergaard *et al.*, 1997), *Arabidopsis thaliana* (unpublished data), and in spinach (*Spinacea oleracea*) leaf (Mutsuda *et al.*, 1995). As a consequence, for the substrates of the animal pathway (D-glucuronolactone, L-gulono-1,4-lactone) to be converted to L-AA, one has to invoke the existence of either a 3'-epimerase interconverting L-gulonolactone and L-galactono-1,4-lactone, isozymes of GLDH with differing substrate specificities, or a separate L-gulono-1,4-lactone oxidase/dehydrogenase activity. In support of the latter hypothesis, we have recently demonstrated the existence L-gulono-1,4-lactone-dependent L-AA formation in crude cellular extracts of *Arabidopsis* cell suspension culture (Davey *et al.*, 1999a) and shown that this activity is primarily localized in the soluble fraction of cell extracts, whereas GLDH activity is mitochondrial (Davey *et al.*, 1999b). We have also identified L-gulono-1,4-lactone to be a minor, natural component of *Arabidopsis* cell suspension extracts (unpublished data). We have further purified and characterized an enzyme catalyzing the NADPH-dependent reduction of D-glucuronolactone.

The physiological relevance of these plant "inversion" conversions must however remain questioned until more data on the enzymes catalyzing these reactions become available, and until identification of the physiological substrates. It may be that L-AA biosynthesis from these compounds is only significant under certain circumstances or in specific tissue types. For example, L-AA biosynthesis from D-galacturonic acid and D-glucuronic acid may take place as part of a salvage pathway for the reutilization of carbon derived from non-cellulosic cell wall polysaccharide breakdown. If this is the case, then labelling with D-glucose will still result in a predominantly "non-inversion" pattern of incorporation via *de novo* GDP-L-galactose synthesis, because uronic acids would be derived from the breakdown of "old" (non-labelled) cell wall polysaccharide. Pectin disassembly occurs during abscission, fruit ripening and softening, pollen grain maturation, and cell expansion as well as during pathogen invasion (Carpita and Gibeaut, 1993; Huysamer et al., 1997; Hadfield and Bennett, 1998; Hadfield et al., 1998; Rose et al., 1998). The resolution of these questions awaits the identification and characterization of the enzymes responsible for these reactions.

The possible interrelationships between the L-galactose, L-sorbosone (non-inversion), and uronic acid (inversion) pathways of L-AA biosynthesis, and hexose phosphate metabolism are summarized in Figure 3 (adapted from Davey et al., 1999b).

Regulation of L-AA Biosynthesis

Little is known about the regulation of L-AA biosynthesis in plants. Nonetheless, it seems unlikely that the terminal enzyme GLDH is subject to end product feedback control, because in the presence of exogenous L-galactono-1,4-lactone, L-AA biosynthesis occurs unabated and can lead to wilting and even death of the plant (Baig et al., 1970; unpublished data). Nonetheless, reports indicate that tissue L-AA concentrations increase in response to wounding (Loewus, 1980), and GLDH activity has been reported to increase in response to wounding in potato tubers (Fukuda et al., 1995; Ôba et al., 1994) and in green pepper (Imahori et al., 1998). Further, the L-AA-deficient vtc-1 mutant appears to have a higher initial capacity to convert L-galactono-1,4-lactone to L-AA than wild-type plants (Conklin et al., 1997). These data suggest that L-AA biosynthesis, and possibly GLDH, is upregulated in response to increased L-AA demand.

Incubations with high concentrations of D-mannose have been reported not to affect endogenous L-AA pool sizes in Arabidopsis and pea (Wheeler et al., 1998; Davey et al., 1999a), indicating that another possible point of control might lie between the conversion of D-mannose to L-galactose (Wheeler et al., 1998), controlling carbon flux between the processes of L-AA biosynthesis, polysaccharide biosynthesis, and protein glycosylation.

L-AA and Light

The biosynthesis of L-AA from hexose phosphates and its involvement in protection against photooxidative stress has long suggested links between photosynthesis, light, and L-AA pool size. It has been shown that the L-AA content of leaves is related to their age, position on the plant, and the light intensity at the leaf surface in maize (Foyer 1993) and in Nicotiana tabacum (unpublished data). Although irradiance has little effect on the L-AA content and redox status of leaves or chloroplasts in the short term (Foyer, 1993), plants grown at high light intensities have higher L-AA concentrations (Foyer, 1993; Smirnoff and Pallanca, 1995). This increase in L-AA during plant adaptation to high light is a process requiring several days (Mishra et al., 1995; Grace and Logan, 1996; Logan et al., 1996). Seasonal and diurnal variations in L-AA content have also been observed in beech (Fagus sylvatica L.) (Luwe, 1996) and pine needles (Polle et al., 1992; Schmieden and Wild 1994). However, the relation between light intensity and L-AA concentration does not appear to be true of all plant species, and it has long been speculated that the availability of soluble carbohydrate may be a more important regulatory factor than light intensity (Isherwood et al., 1954). Both barley and Arabidopsis leaves accumulated significantly more L-AA under high-light compared to low-light conditions (Smirnoff and Pallanca, 1995; Conklin et al., 1997). In barley, this difference was correlated with the soluble carbohydrate content and a linear relationship was found between the L-AA pool size and the soluble carbohydrate content of leaves (Smirnoff and Pallanca, 1995). Significantly, L-AA levels could be maintained in the dark by feeding sugars, demonstrating that L-AA biosynthesis is not strictly light dependent. However, L-AA biosynthesis is much less sensitive to carbohydrate supply in non-photosynthetic tissue (Pallanca and Smirnoff, 1999).

Thus, work to date has hinted at some intriguing relationships between light, photosynthesis, the supply of hexose, and intracellular L-AA levels. However, adaptations of intracellular L-AA to changing environmental conditions will clearly also depend on the balance between the rates and capacity of the tissue for L-AA biosynthesis and L-AA turnover, related to antioxidant demand. Currently, no data are available on the possible role of glycolysis as a source of hexose for L-AA biosynthesis (or even cell wall polysaccharide breakdown under certain circumstances) and whether interorganellar transport can serve to mobilize L-AA, or L-AA precursors under appropriate circumstances. Furthermore, one might reasonably expect the concentrations of other antioxidants, in particular the pool of glutathione, to influence the plant antioxidant capacities and the steady-state levels and turnover of L-AA.

L-AA CATABOLISM

Cleavage of the L-AA carbon skeleton can give rise to oxalate and tartrate by one of two species-specific routes (Figure 4). Few other details on the pathway of L-AA catabolism in plants are currently available. It is not even clear whether L-AA or DHA is the actual substrate for these biosyntheses, although 2,3-diketogulonate (2,3-DKG) is not (Loewus and Loewus, 1987). The only enzyme known to specifically catalyze the oxidation of L-AA is ascorbate oxidase (AO), but there is currently little evidence to indicate that AO is involved in the catabolism of L-AA. Indeed, the cell wall localization of AO might actually argue against such a function. The possible role of AO in relation to cell wall metabolism is discussed below.

Synthesis of Oxalate, Threonate, and Tartrate

Oxalic acid is a normal component of higher plant tissues, possibly involved in osmoregulation and the control of calcium concentrations (Franceschi. 1987). Recently, Smirnoff (1996) has suggested that the formation of oxalate from apoplastic L-AA together with oxalate oxidase activity could also play a role in cell wall expansion, by complexing calcium ions that are involved in the cross-linking of pectin chains, and by the formation of H_2O_2 resulting from by oxalate oxidase activity.

In certain plant species, oxalate and tartrate reach such high levels that these species may be referred to as either oxalate or tartrate accumulators. In tartrate accumulators such as *Pelargonium crispum* (a geranium species), early radiotracer studies with $1\text{-}^{14}C$-ascorbic acid showed that oxalate is produced from the cleavage of L-AA at the C_2/C_3 position and the other fragment, L-threonate, is further oxidized to L-tartrate (Loewus, 1980). Surprisingly oxalate-accumulating species do not always show a concomitant accumulation of L-tartrate. This observation was eventually shown to be due to the decarboxylation of L-threonate to a C_3 product that is recycled into central hexose metabolism (Helsper and Loewus, 1982).

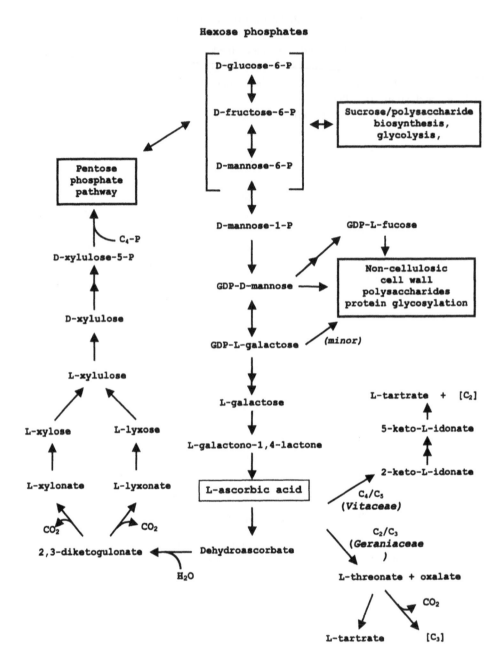

Figure 4. General scheme showing the relationships between L-AA biosynthesis, oxidation, catabolism to oxalate, threonate and tartrate, and possible recycling of 2,3-diketogulonate via the pentose phosphate pathway.

In vitaceous species (such as grape and Virginia creeper), the formation of tartrate arises from cleavage at the C_4/C_5 bond of L-AA (or DHA) (Loewus, 1980) and generates L-tartrate and a C_2 fragment that is recycled into carbohydrate metabolism, possibly as glycoaldehyde. This conversion of L-AA to tartrate in grape was shown to proceed via 5-keto-L-idonic acid (Saito and Kasai, 1984). In species showing C_4/C_5 cleavage and tartrate accumulation, L-AA conversion probably represents the major biosynthetic pathway. However, in other species the biosynthesis of oxalate from L-AA is probably not the major route and oxalate is also formed from the oxidation of glycolic acid, the cleavage of isocitric acid by isocitrate lyase, and the cleavage of oxalacetic acid by oxalacetase (Saito and Kasai 1984; Saito and Loewus 1992; Saito *et al.*, 1997).

Although the pathway for plant oxalate formation has been established by radiotracer studies, none of the enzymes involved have been identified and it is unclear what role this biosynthetic route might play in L-AA turnover. With the high turnover rates reported for L-AA in some species (Smirnoff and Pallanca, 1995), conversion to oxalate would seem to represent an energetic drain on the organism because only 50% of the carbon can be recycled into central metabolism (from L-threonate). Furthermore, oxidation of oxalate by oxalate oxidase generates H_2O_2, which only serves to increase the cellular oxidative stress, and enhance L-AA turnover. Possibly, L-AA cleavage in oxalate accumulators serves another specific, unrelated function. An energetically more favourable, but hypothetical scenario is discussed in the following section.

Recycling of DHA via the Pentose Phosphate Pathway?

DHA is unstable at physiological pHs and spontaneously and enzymatically delactonizes to 2,3-DKG. In animal cells, it has been suggested that DHA can be recycled via 2,3-DKG through the non-oxidative pentose phosphate pathway (Tolbert *et al.*, 1975). This cycle takes place on both an intracellular and an interorganellar level and generates triose phosphates, D-xylulose-5-phosphate, and fructose-6-phosphate (Braun *et al.*, 1996, 1997; Bánhegyi *et al.*, 1997). The idea that L-AA can be recycled in this manner has not been specifically tested in plant tissues, but all the enzymatic conversions necessary are known to be present. Further, under conditions of oxidative stress when DHA accumulates, this cycle would provide a means of regenerating L-AA from gluconeogenic precursors that can again be used for L-AA biosynthesis, with the loss of only one carbon. If this pathway proceeds via the same route as in animal cells, 2,3-DKG is first decarboxylated to L-lyxonate and L-xylonate, before entering the pentose phosphate pathway as D-xylulose 5-phosphate. Together with erythrose 4-phosphate, D-xylulose 5-phosphate can be converted to glyceraldehyde 3-phosphate and the hexose fructose 6-phosphate, by transketolase activity. Fructose 6-phosphate then enters the hexose phosphate pool and can be used for L-AA biosynthesis. In plants, the enzymes of the non-oxidative pentose phosphate pathway are located both in the stroma of the chloroplasts and the cytosol, although it has been questioned whether the cytosol contains the full complement of enzymes necessary for cyclic operation (Schnarrenberger *et al.*, 1995); otherwise, D-xylulose 5-phosphate would have to be transported into the chloroplasts. Operation of such a cycle in plant cells could well explain why specific enzymes associated with the breakdown or catabolism of L-AA have not been detected to date. The conversion of L-AA to oxalate, tartrate, and threonate as well as the recycling of 2,3-diketogulonic acid via the pentose phosphate pathway is summarized in Figure 4.

TRANSPORT OF L-AA

The terminal enzyme of plant L-AA biosynthesis, GLDH, is a mitochondrial enzyme and has consistently been purified from mitochondrial fractions of different plant species (Mapson and Breslow, 1958; Ôba *et al.*, 1994, 1995; Mutsuda *et al.*, 1995; Østergaard *et al.*, 1997; Imai *et al.*, 1998). GLDH activity has also been localized to the inner mitochondrial membranes in bean hypocotyls (Siendones *et al.*, 1999). It is currently unclear whether L-AA is synthesized and released into the mitochondrial matrix, subsequently requiring transport to the cytosol, or whether it is released into the intramembrane space where it can freely diffuse into the cytosol. At physiological pHs, L-AA exists as a monovalent anion and is unable to permeate membranes. It must therefore be transported across the membranes of all other subcellular compartments where it is found.

Chloroplastic Transport

The chloroplastic transport of L-AA serves to maintain stromal L-AA concentrations by replenishing L-AA that is oxidized/degraded during detoxification of photosynthetically generated oxygen radicals and H_2O_2. The uptake of L-AA into chloroplasts was found to be a carrier-mediated process, with transport across the membrane occurring by facilitated diffusion (Anderson *et al.*, 1983; Beck *et al.*, 1983). The surprisingly low affinity of the chloroplastic carrier for its substrate however (K_m of approximately 20 mM) suggests that a rapid influx of extra-chloroplastic L-AA to support peroxide and radical scavenging is not required, presumably because of the highly efficient enzymatic mechanisms for the regeneration of L-AA.

Approximately 10–20% of the chloroplastic L-AA is found bound to the inside of the thylakoid membranes, but entry into the thylakoid lumen is slow and occurs by diffusion alone (Foyer, 1993). Since L-AA is a substrate for the enzyme violaxanthin de-epoxidase, which is also attached to the inside of the thylakoid membrane, this suggests that L-AA availability could modulate this violaxanthin de-epoxidase activity and the operation of the xanthophyll cycle (Foyer *et al.*, 1991).

Plasma Membrane Transport

A significant proportion of the cellular L-AA pool is extracellular and evidence suggests that apoplastic L-AA plays a role in the defense of cells against ozone-induced oxidative stress (Polle *et al.*, 1992; Luwe *et al.*, 1993; Ranieri *et al.*, 1996), pathogen invasion (Vanacker *et al.*, 1998), and wounding (Takahama, 1993). There is also increasing evidence that apoplastic L-AA plays a role in regulating cell wall expansion (for review, see Smirnoff, 1996). Again, the presence of L-AA in the apoplast necessitates the existence of a *trans*-plasma membrane L-AA transport system to maintain the apoplastic L-AA levels. Early work in protoplasts from barley (Rautenkrantz *et al.*, 1994) and pea leaves (Foyer and Lelandais, 1996) showed that this transport is carrier mediated. Recently, it has been clearly demonstrated that DHA is the preferred uptake form of L-AA in purified plasma membrane vesicles of bean, and that DHA is taken up by a facilitated diffusion mechanism (Horemans *et al.*, 1996, 1997, 1998a, 1998b). *In vitro*, the uptake of external DHA results in an exchange with internal (cytosolic) L-AA and preloading vesicles with

L-AA stimulated DHA approximately 3-fold, by a so-called *trans*-stimulation effect. The pH optimum for this high affinity transport was pH 6–7.5, with a K_m for DHA of 24 μM (Horemans *et al.,* 1998a, 1998b). It is presently unclear whether additional L-AA/DHA transporters are also present in the plant plasma membrane, but there is some preliminary evidence that the plasma membrane is also able to transport L-AA via the glucose transporters, similar to the situation in mammalian cell membranes (Vera *et al.,* 1993; Washko *et al.,* 1993).

Trans-Plasma Membrane Electron Transport to MDA

Despite extensive research, the full complement of enzymes associated with the regeneration of the oxidized forms of L-AA or glutathione have not been found outside of the plasma membrane. Therefore, alternative methods to maintain the L-AA/DHA redox status must exist. Apart from the exchange DHA carrier just described, there is evidence for *trans*-plasma membrane electron transport to apoplastic MDA, via the plasma membrane-associated cytochrome *b* (Asard *et al.,* 1992, 1995; Horemans *et al.,* 1994). Cytochrome *b* is a relatively abundant protein that may represent 0.1–0.5% of the total membrane protein and that appears to be present in all tissue types and species examined to date (H. Asard, personal communication). Electron transport from cytosolic L-AA to apoplastic MDA produces cytosolic MDA, which is re-reduced to L-AA by the activity of a NADH-dependent MDAR localized on the inside of the plasma membrane (Bérczi and Møller, 1998). Cytosolic MDA that escapes reduction, disproportionates to L-AA and DHA, and this DHA, together with apoplastic DHA is taken up (by the exchange carrier), and reduced to L-AA via the cytosolic L-AA-GSH cycle. This series of reactions thus provides a mechanism by which the apoplastic L-AA pool can be maintained in a largely reduced state.

Intercellular L-AA Transport

In addition to intracellular transport, L-AA can presumably be transported throughout the plant either through diffusion via the apoplastic space, or by phloem transport, because growth on L-AA stimulates increases in foliar L-AA levels, and radiolabelled L-AA fed to the roots is recovered in other parts of the plant (Mozafar and Oertli 1993). Currently, no information is available on the long-distance transport of L-AA in the xylem or phloem, although the high pH of the phloem would suggest that L-AA is unstable under these conditions.

BIOLOGICAL FUNCTIONS OF L-AA IN PLANTS

Three main types of biological activity can be defined for L-AA: (i) as an enzyme cofactor, (ii) as an antioxidant, and (iii) as a donor/acceptor in electron transport either at the plasma membrane or in the chloroplasts. All three functions are implicated in oxidative stress resistance. Additionally L-AA is a substrate for oxalate/tartrate biosynthesis, at least in certain species. For reviews covering these areas, see Loewus (1988), Foyer (1993), Foyer *et al.* (1991, 1994), Arrigoni (1994), Smirnoff and Pallanca (1995), Smirnoff (1996), Noctor and Foyer (1998), and Foyer (this volume).

Enzymatic Cofactor

In both plant and animal metabolism, L-AA is a cofactor for a number of important enzymatic reactions. These enzymes are characteristically mono- or di-oxygenases that contain iron or copper ions at the active site, such as the propyl lysyl hydroxylases. A study of the reaction mechanism of prolyl hydroxylase indicates that the function of L-AA in these reactions is to maintain the transition metal ion in a reduced state (Padh, 1990). In plants, the hydroxylation of proline and lysine residues is catalyzed by prolyl hydroxylases (EC 1.14.11.2, 1.14.11.7) and lysyl hydroxylases (EC 1.14.11.4), which are required for the post-translational modification of a number of proteins including the extensins. These abundant cell wall glycoproteins are believed to be involved in strengthening the cell wall in response to injury, and extensin genes are induced in response to wounding and pathogen attack (Carpita and Gibeaut, 1993; Reiter, 1998). These hydroxylases require both 2-oxoglutarate and L-AA for activity. L-AA also specifically activates myrosinase (thioglucoside glucohydrolase; EC 3.2.3.1), an enzyme which catalyzes the hydrolysis of glucosinolates to D-glucose and an aglycone fragment. The final step in the biosynthesis of the plant hormone ethylene is catalyzed by 1-aminocyclopropane-1-carboxylate oxidase (ACC oxidase), a member of the Fe(II) oxygenase family of enzymes, which, in addition to CO_2, requires L-AA as an essential cosubstrate. Gibberellin 20-oxidase (gibberellin, 2-oxoglutarate:oxygen oxidoreductase; EC 1.14.11), is another Fe-containing dioxygenase involved in the biosynthesis of the plant hormone gibberelllin. Again, L-AA and 2-oxoglutarate are essential for activity.

Antioxidant Functions

In plant as well as in animal systems, L-AA interacts enzymically and non-enzymically with damaging oxygen radicals and their derivatives, so-called reactive oxygen species (ROS). These detoxification reactions can be considered to be an integral part of the housekeeping duties required of an aerobic existence in eukaryotic cells and the high intracellular concentrations of L-AA reflect the importance of these functions to eukaryotic organisms. The significance of L-AA is that, unlike other low-molecular weight antioxidants (α-tocopherol, glutathione, carotenoids, flavonoids, etc.), L-AA is able to terminate radical chain reactions by disproportionation to non-toxic, non-radical products, *i.e.* to DHA and 2,3-diketogulonic acid. Further, because L-AA is only mildly electronegative, it can accept electrons from a wide range of substrates.

ROS and oxidative stress

ROS include such compounds as superoxide ($O_2^{\bullet-}$), singlet oxygen (O_2^-), hydrogen peroxide (H_2O_2), and the highly reactive hydroxyl radical ($OH^{\bullet}$) (Dalton *et al.*, 1993; Halliwell, 1996). Chloroplasts, as well as mitochondria and peroxisomes all produce ROS as byproducts of normal cellular metabolism, but this production is enhanced by a variety of environmental stresses, including drought, starvation, wounding, high salt, high light, and exposure to pollutants (ozone). The toxicity of ROS arises from their ability to initiate radical cascade reactions that lead to protein damage, lipid peroxidation, DNA damage, and finally cell death. Therefore, aerobic organisms have developed a range of efficient mechanisms to detoxify these species by both enzymatic and non-enzymatic means. Among the enzymatic mechanisms, superoxide dismutases (SOD, E.C. 1.15.1.1) catalyze the dismutation of superoxide to H_2O_2 (Bowler *et al.*, 1992) and ascorbate peroxidase (APx, E.C. 1.11.1.7), glutathione peroxidases (GPX, E.C. 1.1.1.11, 1.11.1.9), and catalases (CAT, E.C. 1.11.1.6) detoxify H_2O_2. Non-enzymatically, low-molecular weight antioxidants such as glutathione (γ-glutamyl-L-cysteinyl glycine, GSH), and L-AA (see above), the lipophilic α-tocopherol (vitamin E), as well as carotenoids and phenolics are able to interact directly with ROS (Dalton, 1995; Inzé and Van Montagu, 1995; Foyer *et al.*, 1997). Saturation of these defense systems leads to the condition of "oxidative stress" and the accumulation of toxic levels of ROS.

Paradoxically, the controlled production of ROS can also be beneficial to the organism and during incompatible plant-pathogen interactions, recognition of an invading pathogen stimulates an "oxidative burst", and a coordinated defense response that is mediated by ROS (for reviews, see Baker and Orlandi, 1995; Bolwell *et al.*, 1995; Mehdy, 1996; Alvarez *et al.*, 1998; Noctor and Foyer, 1998). The production of H_2O_2 during this response can have a direct cytotoxic effect on the invading pathogen in addition to stimulating cell wall peroxidase activity and increasing the cross-linking of cell wall proteins and lignin to hinder pathogen penetration (Schopfer, 1996; Thordal-Christensen *et al.*, 1997; Alvarez *et al.*, 1998). ROS can also function as messengers to induce an entire array of defense response genes. Finally, massive ROS accumulation can trigger a localized hypersensitive response, resulting in cell death, again limiting pathogen development (Levine *et al.*, 1994). In addition to the primary oxidative burst, ROS also induce secondary oxidative bursts in small collections of cells in distant tissues, leading to systemic acquired resistance (Alvarez *et al.*, 1998). Clearly, the size and redox status of the apoplastic and cytoplasmic L-AA pools will be an important determinant in helping to define the strength and duration of these responses.

L-AA and (photo)oxidative stress

In the light, the chloroplasts of higher plants produce ROS as a consequence of the transfer of high-energy electrons from reduced ferredoxin of the photosynthetic electron transport chain to oxygen (instead of NADP). This photoreduction of oxygen in photosystem I (PSI) is termed the Mehler reaction, and the overall transfer of electrons from water to molecular oxygen is called pseudocyclic electron flow. Pseudocyclic electron flow provides a mechanism by which the plant is able to dissipate excess reducing power (and generate ATP), under conditions when carbon fixation is limited (Foyer 1993; Foyer *et al.*, 1994; Alscher *et al.*, 1997). The reduction of O_2 results in the formation of the superoxide radical ($O_2^{\bullet-}$), which disproportionates to H_2O_2, in a reaction catalyzed by SOD (Eq. 5).

$$2O_2^{\bullet-} + 2H^+ \rightarrow H_2O_2 + O_2 \qquad \text{(Eq. 5)}$$

The further detoxification of H_2O_2 is vital to the normal functioning of the chloroplast, because the thiol-modulated enzymes of the Benson-Calvin cycle (pentose phosphate pathway) are highly sensitive to inhibition by low concentrations of H_2O_2. Chloroplasts however lack catalase (which is located in the peroxisomes/glyoxysomes), and H_2O_2 formed in the thylakoids is scavenged by the activity of the APx (Eq. 6). APx isozymes are present in the chloroplasts as a thylakoid and a stromal form, as well as in the cytosol, mitochondria, and peroxisomes (Jiménez *et al.*, 1997, 1998).

$$H_2O_2 + 2\,L\text{-}AA \rightarrow 2\,H_2O + 2\,MDA \qquad\qquad (Eq.\ 6)$$

L-AA can be regenerated from MDA by the action of the stromal, NAD(P)H-dependent, MDAR. Any MDA escaping this reaction then disproportionates to L-AA and DHA. The regeneration of L-AA from this DHA is catalyzed by DHA reductase (DHAR, E.C. 1.8.5.1) in a process involving the oxidation of the hydrophilic thiol GSH to glutathione disulphide (GSSG). GSH in turn is regenerated from GSSG by the NADPH-dependent enzyme, glutathione reductase (E.C. 1.6.4.2). This coupled series of enzymatic reactions, first proposed by Foyer and Halliwell (1976), links APx-generated H_2O_2 to the oxidation of light-generated NADPH, and is termed the ascorbate-glutathione cycle (AA-GSH cycle) (Noctor and Foyer, 1998) or the Halliwell-Asada pathway. In the chloroplasts, MDA that is generated in the reaction catalyzed by APx (Eq. 6) can also act as a direct electron acceptor to PSI by interacting with reduced ferredoxin on the outside of the thylakoid membrane (Miyake and Asada, 1994). In doing so it is re-reduced to L-AA.

The functioning of this pathway depends ultimately on reducing power derived from the light-dependent electron transport reactions of the chloroplast, or on secondary activities, such as glucose-6-phosphate dehydrogenase and malate dehydrogenase, for the generation of NADPH. A high proportion of the enzymes of the L-AA–GSH cycle are localized in the chloroplasts, but sufficient activity is found in other cellular compartments (cytosol, peroxisomes, mitochondria), to drive the H_2O_2-scavenging there as well (e.g. Boraccino et al., 1986; Koshiba, 1993; Dalton et al., 1993; De Leonardis et al., 1995; Yamaguchi et al., 1995; Jiménez et al., 1997, 1998; del Río et al., 1998).

Photorespiration

Under CO_2-limiting conditions, ribulose-1,5-bisphosphate carboxylase, the main enzyme responsible for CO_2 fixation also functions as an oxygenase (Eq. 7):

$$\text{ribulose-1,5-bisphosphate} + O_2 \rightarrow \text{3-phosphoglycerate} +$$
$$\text{2-phosphoglycollate} \qquad\qquad (Eq.\ 7)$$

3-phosphoglycerate is a normal product of photosynthetic CO_2 fixation in the chloroplasts and enters the Benson-Calvin cycle (pentose phosphate pathway). However, in most higher plants 2-phosphoglycollate is transported as glycolate to adjacent peroxisomes, where it is oxidized to glyoxylate by glycolate oxidase and with the concomitant production of H_2O_2. This H_2O_2 is scavenged by catalase activity and the peroxisomal L-AA–GSH cycle (Jiménez et al., 1997, 1998; del Río et al., 1998).

Therefore, L-AA fulfills several functions in photosynthesis: it is a cofactor for enzymes in the biosynthesis of zeaxanthin (a photoprotectant), it is a substrate for APx in the detoxification of chloroplastic and peroxisomal H_2O_2, and an electron acceptor for reduced ferredoxin in the photosynthetic electron transport chain.

Apoplastic L-AA and Cell Wall Metabolism

A significant proportion of the cellular L-AA pool is located outside of the plasma membrane. In addition to providing protection against atmospheric pollutants, there is increasing evidence that L-AA is involved in the modulation of cell wall expansion and lignification.

Apoplastic L-AA and ozone

Changes in the levels and redox status of apoplastic L-AA in response to environmental stresses has led to the realization that apoplastic L-AA could be involved in the protection of cell membranes against oxidative damage, particularly as a result of exposure to ozone and other atmospheric pollutants (Castillo and Greppin, 1988; Peters *et al.*, 1989; Polle *et al.*, 1992; Luwe *et al.*, 1993; Luwe, 1996; Dietz, 1997), although Stokes *et al.* (1998), reported that in velvet beans (*Mucuna pruriens*), the effects of ozone, propene, and isoprene on L-AA were only observed in young leaves (for reviews on ozone effects, see Pell *et al.*, 1997; Schraudner *et al.*, 1997). However, while MDAR activity has been located on the inside of the *Phaseolus* plasma membrane (Asard *et al.*, 1994; Bérczi and Møller, 1998), and an ozone-responsive peroxidase activity with high affinity for L-AA is present in the extracellular matrix of several species (Ranieri *et al.*, 1996), no evidence for the presence in the apoplast of the other enzymes of the GSH-L-AA cycle or significant concentrations of GSH itself in a variety of plant species has been found (Castillo and Greppin, 1988; Polle *et al.*, 1992; Luwe 1996; Ranieri *et al.*, 1996). Therefore, as discussed above, the plant cell must possess alternative mechanisms to maintain the concentration and redox status of apoplastic L-AA, including *trans*-plasma membrane transport of L-AA, DHA, and the cytochrome *b*-mediated, electron transfer reduction of apoplastic MDA.

L-AA and cell wall expansion

There is increasing evidence that L-AA is directly involved in the processes regulating cell wall expansion, and a number of different mechanisms have been proposed for these effects (Córdoba and González-Reyes, 1994; Smirnoff, 1996). Firstly, L-AA is required as a cofactor for the post-translational hydroxylation in cell wall protein biosynthesis (Fry, 1986; Arrigoni, 1994; Arrigoni *et al.*, 1997). L-AA has also been suggested to maintain the hydroxyproline residues of cell wall carbohydrates and proteins in a reduced form, allowing scission or loosening (Lin and Varner, 1991). *Trans*-plasma membrane electron transport to extracellular MDA via the plasma membrane cytochome *b* (Horemans *et al.*, 1994; Asard *et al.*, 1995), and activation of the plasma membrane MDAR activity (Córdoba and González-Reyes, 1994) has been proposed to stimulate the plasma membrane H^+-ATPase, leading to increased solute uptake and vacuolisation (Hidalgo *et al.*, 1989; González-Reyes *et al.*, 1994; de Cabo *et al.*, 1996). However, L-AA also reversibly inhibits the cell wall/ apoplastic peroxidases responsible for the formation of monolignol radicals in *Allium* (Takahama, 1993; Schopfer, 1996), and possibly even the turnover and secretion of peroxidases into the apoplast/cell wall compartments (del Carmen Córdoba-Pedregosa *et al.*, 1996). L-AA may also inhibit lignification by non-enzymically scavenging monolignol radicals produced by peroxidase activity (Takahama, 1993; Otter and Polle, 1994). Recently, it was shown that at least *in vitro*, apoplastic concentrations of L-AA in the presence of Cu^{2+} lead to the production of $OH^{\bullet}$ from H_2O_2 (Fry, 1998) and that this is extremely effective in causing the non-enzymatic cleavage of cell wall polysaccharides. Finally, L-AA can also serve as a substrate for oxalic acid biosynthesis (Loewus, 1988). The degradation of apoplastic L-AA (or DHA) to oxalate could be responsible for the sequestering of calcium ions, which are important for the cross-linking of pectins and cell wall strengthening (Smirnoff, 1996). Germin, an abundant cell wall protein with oxalate oxidase activity (E.C. 1.2.3.4) could then catalyze the breakdown of oxalate to CO_2 and H_2O_2, leading to H_2O_2-mediated cell wall stiffening.

At present the relative contributions of these various mechanisms to cell wall expansion under physiological conditions remains unclear. Undoubtedly, the redox status of apoplastic L-AA, and the balance between L-AA, regeneration and H_2O_2 formation will influence the degree of lignification and cross-linking of cell wall components either directly by radical scavenging, or indirectly by inhibiting peroxidase activities. This is, in turn, dependent on the relative activities of AO, cell wall peroxidases, plasma membrane-associated NAD(P)H oxidases, oxalate oxidase, and the L-AA-regenerating systems. In particular, the role of AO deserves attention, and high AO activities are known to be associated with periods of rapid cell expansion during plant growth and development. Presumably, high AO activity will increase the concentration of apoplastic MDA and DHA. The difficulty in obtaining clear evidence to link apoplastic L-AA/MDA concentrations with different stages of cell wall expansion is in part related to the technical problems involved in the accurate quantitation of this (unstable) antioxidant in extracellular compartments.

Ascorbate Oxidase

Ascorbate oxidase (AO) is a cell wall glycoprotein that catalyzes the oxidation of L-AA to DHA via the free radical MDA, with the concomitant reduction of molecular oxygen to water (Loewus, 1980). In melon, AO is encoded by a small family of at least four genes (Diallinas et al., 1997). This observation is consistent with the presence of the 12 to 14 AO isoforms that have been identified in this species. Although no clear biological function has been assigned to AO activity in plants (Arrigoni, 1994; Smirnoff and Pallanca, 1995; del Carmen Córdoba-Pedrogosa et al., 1996), increasing evidence points to an involvement of AO in plant cell wall expansion and growth. AO is found localized mainly in the cell wall (Chichiricco et al., 1989; Diallinas et al., 1997), and in fast-growing tissues (Kato and Esaka, 1996), suggesting a possible involvement in cell wall loosening during cell wall expansion (Lin and Varner, 1991). AO production has also been correlated with periods of rapid cell expansion following fertilisation in melon. AO function has also been indirectly linked to cell division through control of the levels of L-AA or possibly MDA (Kerk and Feldman, 1995).

L-AA, AO, and Cell Division

Circumstantial data indicate that L-AA is able to influence plant cell division and differentiation (Edgar, 1970; Wahal et al., 1973; Liso et al., 1984, 1988; Arrigoni, 1994). Exogenous L-AA has been found to accelerate the onset of cell proliferation in root primordia of *Allium*, *Pisum*, and *Lupinus* (Arrigoni, 1994; Citterio et al., 1994; de Cabo et al., 1996) because of an increased proportion of cells progressing through the G1-to-S transition (Liso et al., 1984, 1988; Arrigoni, 1994; Citterio et al., 1994). In maize roots, high levels of AO mRNA and protein activity are associated with the slow-cycling cells of the quiescent centre. This is correlated with low or undetectable levels of L-AA, compared to the more rapidly dividing surrounding cells (Kerk and Feldman, 1995). Addition of exogenous L-AA stimulates the cells of the quiescent centre to start dividing (Kerk and

Feldman, 1995), although in the presence of AO the active agent may well be MDA. More recently, it has been shown that in *Arabidopsis* roots, glutathione and non-specific reducing agents such as dithiothreitol also stimulate cell division (Sánchez-Fernández *et al.*, 1997). This observation suggests that the redox status of the L-AA/DHA and/or the GSH/GSSG redox couples could influence the cell cycle, by acting as a transducer of environmental stimuli to key regulatory proteins, for example during stress-induced cell cycle arrest. This would then prevent the replication of damaged DNA under conditions of oxidative stress.

L-AA ANALYSIS

The great interest in L-AA metabolism over the years has given rise to a bewildering array of procedures for the quantification of this antioxidant. Unfortunately these methods have often been developed for a specific tissue type and cannot always readily be transferred. In plant tissues in particular, there is potential interference from many different compounds (for reviews, see Loewus, 1980; Dawes *et al.*, 1991; Washko *et al.*, 1992).

Prior to about 1980, most assays for L-AA content were based on spectrophotometric determinations. Such methods include the reduction of 2,6-dichlorophenol-indophenol (DCIP) by L-AA, and the reduction of ferric iron by L-AA followed by the formation of a chromogenic ferrous iron complex, using chelators such as 2,2-dipyridyl and ferrozine. Another widely used procedure involves the oxidation of L-AA to DHA and 2,3-diketogulonic acid, followed by reaction with 2,4-dinitrophenylhydrazine. In all cases these spectrophotometric procedures suffer from a lack of sensitivity and specificity of reaction (Washko *et al.*, 1992). The use of the enzyme AO avoids the problem of reaction specificity, and the oxidation of L-AA to DHA by AO is followed either by the decrease in L-AA absorbance at 260 nm or by complexation with ferric iron. However, this procedure cannot be used to measure DHA.

Subsequently, HPLC techniques have achieved popularity because of the high specificity and sensitivity of detection. Until recently however, DHA could not be determined directly, because of its low UV absorption coefficient, lack of electrochemical activity, and low concentrations in biological extracts. Rather, extracts are reduced chemically and reanalyzed, after which DHA values are calculated by subtracting the L-AA values from the "total" L-AA. Recently, HPLC methods for the simultaneous analysis of L-AA and DHA have appeared (Tausz *et al.*, 1996; Pappa-Louisi and Pascalidou, 1998), but it remains to be seen how robust these procedures are in the analysis of L-AA in complex biological matrices.

In recent years, the high resolution and high efficiencies of high-performance capillary electrophoresis (HPCE) have increasingly been adapted for L-AA analysis (Chiari *et al.*, 1993; Koh *et al.*, 1993; Davey *et al.*, 1996, 1997; Procházková *et al.*, 1998). In comparison to HPLC separations, HPCE analyses are faster, require only nl of sample, have much higher peak capacities, but the concentration limits of sensitivity are generally higher than in HPLC.

Analytical methods are also available for the analysis of various L-AA derivatives by HPLC (Khaled 1996; Sakai *et al.*, 1996) and HPCE (Pauli and Schuep, 1998).

CONCLUSIONS

Since its initial identification some 60 years ago as the compound able to alleviate the symptoms of scurvy (Svirbely and Szent-Györgyi, 1932), L-AA has received continued and intense attention from the scientific community. The reasons for this are, in part, due to the protective functions assigned to L-AA in a wide variety of common human ailments and diseases. For example, in recent years clear evidence has emerged that elevated dietary intakes of L-AA lower the incidence of a number of cancers, cardiovascular disease, and other oxidative stress-related disorders (for reviews, see Barja, 1996; Diplock *et al.*, 1998). These interests have to some extent obscured research into L-AA metabolism in plants, the importance of which remained largely unappreciated until the identification of L-AA-specific peroxidases (Groden and Beck, 1979). Now, however, with the identification of the L-galactose-based pathway of plant L-AA biosynthesis as well as the recent cloning of some of the enzymes of this pathway and the identification of L-AA-deficient mutants, we can expect to see rapid progress in the understanding of may aspects of L-AA metabolism in plants, eventually leading to controlled manipulation of plant L-AA levels for improved nutritional quality and stress resistance.

Acknowledgments

The authors would like to thank Martine De Cock for help in preparing this manuscript.

REFERENCES

Alscher, R.G., Donahue, J.L., and Cramer, C.L. (1997) Reactive oxygen species and antioxidants: relationships in green cells. *Physiol. Plant.*, **100**, 224–233.

Alvarez, M.E., Pennell, R.I., Meijer, P.-J., Ishikawa, A., Dixon, R.A., and Lamb, C. (1998) Reactive oxygen intermediates mediate a systemic signal network in the establishment of plant immunity. *Cell*, **92**, 773–784.

Anderson, J.W., Foyer, C.H. and Walker, D.A. (1983) Light-dependent reduction of dehydroascorbate and uptake of exogenous ascorbate by spinach chloroplasts. *Planta*, **158**, 442–450.

Arrigoni, O. (1994) Ascorbate system in plant development. *J. Bioenerg. Biomembr.*, **26**, 407–419.

Arrigoni, O., Calabrese, G., De Gara, L., Bitonti, M.B., and Liso, R. (1997) Correlation between changes in cell ascorbate and growth of *Lupinus albus* seedlings. *J. Plant Physiol.*, **150**, 302–308.

Asard, H., Horemans, N., and Caubergs, R.J. (1992) Transmembrane electron transport in ascorbate-loaded plasma membrane vesicles from higher plants involves a *b*-type cytochrome. *FEBS Lett.*, **306**, 143–146.

Asard, H., Horemans, N., and Caubergs, R.J. (1995) Involvement of ascorbic acid and a b-type cytochrome in plant plasma membrane redox reactions. *Protoplasma*, **184**, 36–41.

Baig, M.M., Kelly, S., and Loewus, F. (1970) L-ascorbic acid biosynthesis in higher plants from L-gulono-1, 4-lactone and L-galactono-1,4-lactone. *Plant Physiol.*, **46**, 277–280.

Baker, C.J. and Orlandi, E.W. (1995) Active oxygen in plant pathogenesis. *Annu. Rev. Phytopathol.*, **33**, 299–321.

Bánhegyi, G., Braun, L., Csala, M., Puskás, F., and Mandl, J. (1997) Ascorbate metabolism and its regulation in animals. *Free Rad. Biol. Med.*, **23**, 793–803.

Barber, G.A. (1971) The synthesis of L-glucose by plant enzyme systems. *Arch. Biochem. Biophys.*, **147**, 619–623.

Barber, G.A. (1979) Observations on the mechanism of the reversible epimerization of GDP-D-mannose to GDP-L-galactose by an enzyme from *Chlorella pyrenoidosa. J. Biol. Chem.*, **254**, 7600–7603.

Barja, G. (1996) Ascorbic acid and aging. In J.R. Harris, (ed.), *Ascorbic Acid: Biochemistry and Biomedical Cell Biology*, (Subcellular Biochemistry, Vol. 25), Plenum Press, New York, pp. 157–188.

Beck, E., Burkert, A., and Hofmann, M. (1983) Uptake of L-ascorbate by intact spinach chloroplasts. *Plant Physiol.*, **73**, 41–45.

Bérczi, A. and Møller, I.M. (1998) NADH-monodehydroascorbate oxidoreductase is one of the redox enzymes in spinach leaf plasma membranes. *Plant Physiol.*, **116**, 1029–1036.

Bolwell, G.P., Butt, V.S., Davies, D.R., and Zimmerlin, A. (1995) The origin of the oxidative burst in plants. *Free Rad. Res.*, **23**, 517–532.

Borraccino, G., Dipierro, S., and Arrigoni, O. (1986) Purification and properties of ascorbate free-radical reductase from potato tubers. *Planta*, **167**, 521–526

Bowler, C., Van Montagu, M., and Inzé, D. (1992) Superoxide dismutase and stress tolerance. *Annu. Rev. Plant Physiol. Plant Mol. Biol.*, **43**, 83–116.

Braun, L., Puskás, F., Csala, M., Györffy, E., Garzó, T., Mandl, J., and Bánhegyi, G. (1996) Gluconeogenesis from ascorbic acid: ascorbate recycling in isolated murine hepatocytes. *FEBS Lett.*, **390**, 183–186.

Braun, L., Puskás, F., Csala, M., Mészáros, G., Mandl, J., and Bánhegyi, G. (1997) Ascorbate as a substrate for glycolysis or gluconeogenesis: evidence for an interorgan ascorbate cycle. *Free Rad. Biol. Med.*, **23**, 804–808.

Cadenas, E. (1995) Mechanisms of oxygen activation and reactive oxygen species detoxification. In S. Ahmad, (ed.), *Oxidative Stress and Antioxidant Defenses in Biology*, Chapman and Hall, New York, pp. 1–61.

Carpita, N.C. and Gibeaut, D.M. (1993) Structural models of primary cell walls in flowering plants: consistency of molecular structure with the physical properties of the walls during growth. *Plant J.*, **3**, 1–30.

Castillo, F.J. and Greppin, H. (1988) Extracellular ascorbic acid and enzyme activities related to ascorbic acid metabolism in *Sedum album* L. leaves after ozone exposure. *Environ. Exp. Bot.*, **28**, 231–238.

Chiari, M., Nesi, M., Carrea, G., and Righetti, P.G. (1993) Determination of total vitamin C in fruits by capillary zone electrophoresis. *J. Chromatogr.*, **645**, 197–200.

Chichiricco, G., Ceru, M.P., D'Alessandro, A., Oratore, A., and Avigliano, L. (1989) Immunohistochemical localization of ascorbate oxidase in *Cucurbita pepo* medullosa. *Plant Sci.*, **64**, 61–66.

Citterio, S., Sgorbati, S., Scippa, S., and Sparvoli, E. (1994) Ascorbic acid effect on the onset of cell proliferation in pea root. *Physiol. Plant.*, **92**, 601–607.

Conklin, P.L., Williams, E.H., and Last, R.L. (1996) Environmental stress sensitivity of an ascorbic acid-deficient Arabidopsis mutant. *Proc. Natl. Acad. Sci. USA*, **93**, 9970–9974.

Conklin, P.L., Pallanca, J.E., Last, R.L., and Smirnoff, N. (1997) L-ascorbic acid metabolism in the ascorbate-deficient Arabidopsis mutant *vtc1*. *Plant Physiol.*, **115**, 1277–1285.

Conklin, P.L., Norris, S.R., Wheeler, G.L., Williams, E.H., Smirnoff, N., and Last, R.L. (1999) Genetic evidence for the role of GDP-mannose in plant ascorbic acid (vitamin C) biosynthesis. *Proc. Natl. Acad. Sci. USA*, **96**, 4198–4203.

Córdoba, F., and González-Reyes, J.A. (1994) Ascorbate and plant cell growth. *J. Bioenerget. Biomembr.*, **26**, 399–405.

Dabrowski, K., Lackner, R., and Doblander, C. (1993) Ascorbate-2-sulfate sulfohydrolase in fish and mammal: comparative characterization and possible involvement in ascorbate metabolism. *Comp. Biochem. Physiol. B Comp. Biochem.*, **104**, 717–722.

Dalton, D.A. (1995) Antioxidant defenses of plants and fungi. In S. Ahmad, (ed.), *Oxidative Stress and Antioxidant Defenses in Biology*, Chapman and Hall, New York, pp. 298–355.

Dalton, D.A., Baird, L.M., Langeberg, L., Taugher, C.Y., Anyan, W.R., Vance, C.P., and Sarath, G. (1993) Subcellular localization of oxygen defense enzymes in soybean (*Glycine max* [L.] Merr.) root nodules. *Plant Physiol.*, **102**, 481–489.

Davey, M.W., Bauw, G., and Van Montagu, M. (1996) Analysis of ascorbate in plant tissues by high-performance capillary zone electrophoresis. *Anal. Biochem.*, **239**, 8–19.

Davey, M.W., Bauw, G., and Van Montagu, M. (1997) Simultaneous high-performance capillary electrophoresis analysis of the reduced and oxidised forms of ascorbate and glutathione. *J. Chromatogr. B.*, **697**, 269–276.

Davey, M.W., Bauw, G., and Van Montagu, M. (1999a) Simultaneous analysis of the oxidised and reduced forms of ascorbate and glutathione. *J. Chromatogr. A*, **853**, 381–389.

Davey, M.W., Gilot, C., Persiau, G., Østergaard, J., Han, Y., Bauw, G.C., and Van Montagu, M.C. (1999b) Ascorbate biosynthesis in Arabidopsis cell suspension culture. *Plant Physiol.*, **121**, 535–544.

Dawes, M.B., Austin, J., and Partridge, D.A. (1991) *Vitamin C: its Chemistry and Biochemistry*. Cambridge, Royal Society of Chemistry, pp. 115–146.

de Cabo, R.C., González-Reyes, J.A., Córdoba, F., and Navas, P. (1996) Rooting hastened in onions by ascorbate and ascorbate free radical. *J. Plant Growth Regul.*, **15**, 53–56.

De Leonardis, S., De Lorenzo, G., Borraccino, G., and Dipierro, S. (1995) A specific ascorbate free radical reductase isozyme participates in the regeneration of ascorbate for scavenging toxic oxygen species in potato tuber mitochondria. *Plant Physiol.*, **109**, 847–851.

del Carmen Córdoba-Pedregosa, M., González-Reyes, J.A., del Sagrario Cañadillas, M., Navas, P., and Córdoba, F. (1996) Role of apoplastic and cell-wall peroxidases on the stimulation of root elongation by ascorbate. *Plant Physiol.*, **112**, 1119–1125.

del Río, L.A., Pastori, G.M., Palma, J.M., Sandalio, L.M., Sevilla, F., Corpas, F.J., Jiménez, A., López-Huertas, E., and Hernández, J.A. (1998) The activated oxygen role of peroxisomes in senescence. *Plant Physiol.*, **116**, 1195–1200.

Deutsch, J.C. (1998) Spontaneous hydrolysis and dehydration of dehydroascorbic acid in aqueous solution. *Anal. Biochem.*, **260**, 223–229.

Diallinas, G., Pateraki, I., Sanmartin, M., Scossa, A., Stilianou, E., Panopoulos, N.J., and Kanellis, A.K. (1997) Melon ascorbate oxidase: cloning of a multigene family, induction during fruit development and repression by wounding. *Plant Mol. Biol.*, **34**, 759–770.

Dietz, K.-J. (1997) Functions and responses of the leaf apoplast under stress. *Prog. Bot.*, **58**, 221–254.

Diplock, A.T., Charleux, J.-L., Crozier-Willil, G., Kok, F.J., Rice-Evans, C., Roberfroid, M., Stahl, W., and Viña-Ribes, J. (1998) Functional food science and defence against reactive oxidative species. *Brit. J. Nutrit.*, **80**, Suppl. **1**, S77–S112.

Edgar, J.A. (1970) Dehydroascorbic acid and cell division. *Nature*, **227**, 24–26.

Feingold, D.S. and Avigad, G. (1980) Sugar nucleotide transformation in plants. In J. Preiss, (ed.), *Carbohydrates: Structure and Function* (The Biochemistry of Plants, Vol. 3), Academic Press, New York, pp. 101–170.

Foyer, C. (1993) Ascorbic acid. In R.G. Alscher and J.L. Hess, (eds.), *Antioxidants in Higher Plants*, CRC Press, Boca Raton, pp. 31–58.

Foyer, C.H. and Halliwell, B. (1976) The presence of glutathione and glutathione reductase in chloroplasts: a proposed role in ascorbic acid metabolism. *Planta*, **133**, 21–25.

Foyer, C.H. and Lelandais, M. (1996) A comparison of the relative rates of transport of ascorbate and glucose across the thylakoid, chloroplast and plasmalemma membranes of pea leaf mesophyll cells. *J. Plant Physiol.*, **148**, 391–398.

Foyer, C., Rowell, J., and Walker, D. (1983) Measurements of the ascorbate content of spinach leaf protoplasts and chloroplasts during illumination. *Planta*, **157**, 239–244.

Foyer, C., Lelandais, M., Edwards, E.A., and Mullineaux, P.M. (1991) The role of ascorbate in plants, interactions with photosynthesis, and regulatory significance. In E. Pell and K. Steffen, (eds.), *Active Oxygen/Oxidative Stress and Plant Metabolism*, (Proceedings 6th Annual Penn State Symposium on Plant Physiology), American Society of Plant Physiologists, Rockville, pp. 131–144.

Foyer, C.H., Lelandais, M., and Kunert, K.J. (1994) Photooxidative stress in plants. *Physiol. Plant.*, **92**, 696–717.

Foyer, C.H., Lopez-Delgado, H., Dat, J.F., and Scott, I.M. (1997) Hydrogen peroxide- and glutathione-associated mechanisms of acclimatory stress tolerance and signalling. *Physiol. Plant.*, **100**, 241–254.

Fukuda, M., Kunisada, Y., Noda, H., Tagaya, S., Yamamoto, Y., and Kiday, Y. (1995) Effect of storage time of potatoes after harvest on increase in the ascorbic acid content by wounding. *J. Jpn. Soc. Food Sci. Technol.*, **42**, 1031–1043.

Franceschi, V.R. (1987) Oxalic acid metabolism and calcium oxalate formation in *Lemna minor* L. *Plant Cell Environ.*, **10**, 397–406.

Fry, S.C. (1986) Cross-linking of matrix polymers in the growing cell walls of angiosperms. *Annu. Rev. Plant Physiol.*, **37**, 165–186.

Fry, S.C. (1998) Oxidative scission of plant cell wall polysaccharides by ascorbate-induced hydroxyl radicals. *Biochem. J.*, **332**, 507–515.

Gallice, P., Sarrazin, F., Polverelli, M., Cadet, J., Berland, Y., and Crevat, A. (1994) Ascorbic acid-2-0-β-glucuronide, a new metabolite of vitamin C identified in human urine and uremic plasma. *Biochim. Biophys. Acta*, **1199**, 305–310.

González-Reyes, J.A., Alcaín, F.J., Caler, J.A., Serrano, A., Córdoba, F., and Navas, P. (1994) Relationship between apoplastic ascorbate regeneration and the stimulation of root growth in *Allium cepa* L. *Plant Sci.*, **100**, 23–29.

Grace, S.C. and Logan, B.A. (1996) Acclimation of foliar antioxidant systems to growth irradiance in three broad-leaved evergreen species. *Plant Physiol.*, **112**, 1631–1640.

Groden, D. and Beck, E. (1979) H_2O_2 destruction by ascorbate-dependent systems from chloroplasts. *Biochim. Biophys. Acta.*, **546**, 426–435.

Hadfield, K.A. and Bennett, A.B. (1998) Polygalacturonases: many genes in search of a function. *Plant Physiol.*, **117**, 337–343.

Hadfield, K.A., Rose, J.K.C., Yaver, D.S., Berka, R.M., and Bennett, A.B. (1998) Polygalacturonase gene expression in ripe melon fruit supports a role for polygalacturonase in ripening-associated pectin disassembly. *Plant Physiol.*, **117**, 363–373.

Halliwell, B. (1996) Free radicals, proteins and DNA: oxidative damage versus redox regulation. *Biochem. Soc. Trans.*, **24**, 1023–1027.

Helsper, J.P. and Loewus, F.A. (1982) Metabolism of L-threonic acid in *Rumex* × *acutus* L. and *Pelargonium crispum* (L.) L'Hér. *Plant Physiol.*, **69**, 1365–1368.

Hidalgo, A., González-Reyes, J.A., and Navas, P. (1989) Ascorbate free radical enhances vacuolization in onion root meristems. *Plant Cell Environ.*, **12**, 455–460.

Horemans, N., Asard, H., and Caubergs, R.J. (1994) The role of ascorbate free radical as an electron acceptor to cytochrome *b*-mediated trans-plasma membrane electron transport in higher plants. *Plant Physiol.*, **104**, 1455–1458.

Horemans, N., Asard, H., and Caubergs, R.J. (1996) Transport of ascorbate into plasma membrane vesicles of *Phaseolus vulgaris* L. *Protoplasma*, **194**, 177–185.

Horemans, N., Asard, H., and Caubergs, R.J. (1997) The ascorbate carrier of higher plant plasma membranes preferentially translocates the fully oxidized (dehydroascorbate) molecule. *Plant Physiol.*, **114**, 1247–1253.

Horemans, N., Asard, H., and Caubergs, R.J. (1998a) Carrier mediated uptake of dehydroascorbate into higher plant plasma membrane vesicles shows trans-stimulation. *FEBS Lett.*, **421**, 41–44.

Horemans, N., Asard, H., Van Gestelen, P., and Caubergs, R.J. (1998b) Facilitated diffusion drives transport of oxidised ascorbate molecules into purified plasma membrane vesicles of *Phaseolus vulgaris*. *Physiol. Plant.*, **104**, 783–789.

Huysamer, M., Greve, L.C., and Labavitch, J.M. (1997) Cell wall metabolism in ripening fruit. IX. Synthesis of pectic and hemicellulosic cell wall polymers in the outer pericarp of mature green tomatoes (cv XMT-22). *Plant Physiol.*, **114**, 1523–1531.

Imai, T., Karita, S., Shiratori, G.-i., Hattori, M., Nunome, T., Ôba, K., and Hirai, M. (1998) L-Galactono-γ-lactone dehydrogenase from sweet potato: purification and cDNA sequence analysis. *Plant Cell Physiol.*, **39**, 1350–1358.

Imahori, Y., Zhou, Y.F., Ueda, Y., and Chachin, K. (1998) Ascorbate metabolism during maturation of sweet pepper (*Capsicum annuum* L.). *J. Jpn. Soc. Hortic. Sci.*, **67**, 798–804.

Inzé, D. and Van Montagu, M. (1995) Oxidative stress in plants. *Curr. Opin. Biotechnol.*, **6**, 153–158.

Isherwood, F.A., Chen, Y.T., and Mapson, L.W. (1954) Synthesis of L-ascorbic acid in plants and animals. *Biochem. J.*, **56**, 1–15.

Jaffe, G.M. (1984) Ascorbic acid. In R.E. Kirk and D.F. Othmer, (eds.), *Encyclopedia of Chemical Technology*, 3rd ed., Vol. 24, John Wiley & Sons, New York, pp. 8–40.

Jiménez, A., Hernández, J.A., del Río, L.A., and Sevilla, F. (1997) Evidence for the presence of the ascorbate-glutathione cycle in mitochondria and peroxisomes of pea leaves. *Plant Physiol.*, **114**, 275–284.

Jiménez, A., Hernández, J.A., Ros Barceló, A., Sandalio, L.M., del Río, L.A., and Sevilla, F. (1998) Mitochondrial and peroxisomal ascorbate peroxidase of pea leaves. *Physiol. Plant.*, **104**, 687–692.

Kato, N. and Esaka, M. (1996) cDNA cloning and gene expression of ascorbate oxidase in tobacco. *Plant Mol. Biol.*, **30**, 833–837.

Keller, R., Springer, F., Renz, A., and Kossmann, J. (1999) Antisense inhibition of the GDP-mannose pyrophosphorylase reduces the ascorbate content in transgenic plants leading to developmental changes during senescence. *Plant J.*, **19**, 131–141.

Kerk, N.M. and Feldman, L.J. (1995) A biochemical model for the initiation and maintenance of the quiescent center: implications for organization of root meristems. *Development*, **121**, 2825–2833.

Khaled, M.Y. (1996) Simultaneous HPLC analysis of L-ascorbic acid, L-ascorbyl-2-sulfate and L-ascorbyl-2-polyphosphate. *J. Liq. Chromatogr. Relat. Technol.*, **19**, 3105–3118.

Kim, S.-T., Huh, W.-K., Kim, J.-Y., Hwang, S.-W., and Kang, S.-O. (1996) D-arabinose dehydrogenase and biosynthesis of erythroascorbic acid in *Candida albicans*. *Biochim. Biophys. Acta*, **1297**, 1–8.

Kim, S.-T., Huh, W.-K., Lee, B.-H., and Kang, S.-O. (1998) D-arabinose dehydrogenase and its gene from *Saccharomyces cerevisiae*. *Biochim. Biophys. Acta*, **1429**, 29–39.

Koh, E.V., Bissell, M.G., and Ito, R.K. (1993) Measurement of vitamin C by capillary electrophoresis in biological fluids and fruit beverages using a stereoisomer as an internal standard. *J. Chromatogr.*, **633**, 245–250.

Koshiba, T. (1993) Cytosolic ascorbate peroxidase in seedlings and leaves of maize (*Zea mays*). *Plant Cell Physiol.*, **34**, 713–721.

Larson, R.A. (1995) Antioxidant mechanisms of secondary natural products. In S. Ahmad, (ed.), *Oxidative Stress and Antioxidant Defenses in Biology*, Chapman and Hall, New York, pp. 210–237.

Leung, C.T. and Loewus, F.A. (1985) Ascorbic acid in pollen: conversion of L-galactono-1,4-lactone to L-ascorbic acid by *Lilium longiflorum*. *Plant Sci.*, **39**, 45–48.

Levine, A., Tenhaken, R., Dixon, R., and Lamb, C. (1994) H_2O_2 from the oxidative burst orchestrates the plant hypersensitive disease resistance response. *Cell*, **79**, 583–593.

Lin, L.-S. and Varner, J.E. (1991) Expression of ascorbic acid oxidase in zucchini squash (*Cucurbita pepo* L.). *Plant Physiol.*, **96**, 159–165.

Liso, R., Calabrese, G., Bitonti, M.B., and Arrigoni, O. (1984) Relationships between ascorbic acid and cell division. *Exp. Cell Res.*, **150**, 314–320.

Liso, R., Innocenti, A.M., Bitonti, M.B., and Arrigoni, O. (1988) Ascorbic acid-induced progression of quiescent centre cells from G_1 to S phase. *New Phytol.*, **110**, 469–471.

Loewus, F.A. (1963) Tracer studies on ascorbic acid formation in plants. *Phytochemistry*, **2**, 109–128.

Loewus, F.A. (1980) L-Ascorbic acid: metabolism, biosynthesis, function. In J. Preiss, (ed.), *Carbohydrates, Structure and Function*, (The Biochemistry of Plants, Vol. 3), Academic Press, New York, pp. 77–99.

Loewus, F.A. (1988) Ascorbic acid and its metabolic products. In J. Preiss, (ed.), *Carbohydrates*, (The Biochemistry of Plants: A comprehensive Treatise, Vol. 14), Academic Press, San Diego, pp. 85–107.

Loewus, F.A. and Loewus, M.W. (1987) Biosynthesis and metabolism of ascorbic acid in plants. *CRC Crit. Rev. Plant Sci.*, **5**, 101–119.

Loewus, F.A., Jang, R., and Seegmiller, C.G. (1958) The conversion of C^{14}-labeled sugars to L-ascorbic acid in ripening strawberries. IV. A comparative study of D-galacturonic acid and L-ascorbic acid formation. *J. Biol. Chem.*, **232**, 533–541.

Loewus, M.W., Bedgar, D.L., Saito, K., and Loewus, F.A. (1990) Conversion of L-sorbosone to L-ascorbic acid by a NADP-dependent dehydrogenase in bean and spinach leaf. *Plant Physiol.*, **94**, 1492–1495.

Logan, B.A., Barker, D.H., Demmig-Adams, B., and Adams III, W.W. (1996) Acclimation of leaf carotenoid composition and ascorbate levels to gradients in the light environment within an Australian rainforest. *Plant Cell Environ.*, **19**, 1083–1090.

Luwe, M. (1996) Antioxidants in the apoplast and symplast of beech (*Fagus sylvatica* L.) leaves: seasonal variations and responses to changing ozone concentrations in air. *Plant Cell Environ.*, **19**, 321–328.

Luwe, M.W.F., Takahama, U., and Heber, U. (1993) Role of ascorbate in detoxifying ozone in the apoplast of spinach (*Spinacia oleracea* L.) leaves. *Plant Physiol.*, **101**, 969–976.

Mapson, L.W. and Breslow, E. (1958) Biological synthesis of ascorbic acid: L-galactono-γ-lactone dehydrogenase. *Biochem. J.*, **68**, 395–406.

Mapson, L.W. and Isherwood, F.A. (1956) Biological synthesis of ascorbic acid: the conversion of derivatives of D-galacturonic acid into L-ascorbic acid by plant extracts. *Biochem. J.*, **64**, 13–22.

Mapson, L.W., Isherwood, F.A., and Chen, Y.T. (1954) Biological synthesis of L-ascorbic acid: the conversion of L-galactono-γ-lactone into L-ascorbic acid by plant mitochondria. *Biochem. J.*, **56**, 21–28.

Mehdy, M.C., Sharma, Y.K., Sathasivan, K., and Bays, N.W. (1996) The role of activated oxygen species in plant disease resistance. *Physiol. Plant.*, **98**, 365–374.

Mishra, N.P., Fatma, T., and Singhal, G.S. (1995) Development of antioxidative defense system of wheat seedlings in response to high light. *Physiol. Plant.*, **95**, 77–82.

Miyake, C. and Asada, K. (1994) Ferredoxin-dependent photoreduction of the monodehydroascorbate radical in spinach thylakoids. *Plant Cell Physiol.*, **35**, 539–549.

Mozafar, A. and Oertli, J.J. (1993) Vitamin C (ascorbic acid): uptake and metabolism by soybean. *J. Plant Physiol.*, **141**, 316–321.

Mutsuda, M., Ishikawa, T., Takeda, T., and Shigeoka, S. (1995) Subcellular localization and properties of L-galactono-γ-lactone dehydrogenase in spinach leaves. *Biosci. Biotech. Biochem.*, **59**, 1983–1984.

Nick, J.A., Leung, C.T., and Loewus, F.A. (1986) Isolation and identification of erythroascorbic acid in *Saccharomyces cerevisiae* and *Lypomyces starkeyi*. *Plant Sci.*, **46**, 181–187.

Nishikimi, M. and Yagi, K. (1996) Biochemistry and molecular biology of ascorbic acid biosynthesis. In J.R. Harris, (ed.), *Ascorbic Acid: Biochemistry and Biomedical Cell Biology* (Subcellular Biochemistry, Vol. 25), Plenum Press, New York, pp. 17–39.

Noctor, G. and Foyer, C.H. (1998) Ascorbate and glutathione: keeping active oxygen under control. *Annu. Rev. Plant Physiol. Plant Mol. Biol.*, **49**, 249–279.

Ôba, K., Fukui, M., Imai, Y., Iriyama, S., and Nogami, K. (1994) L-Galactono-γ-lactone dehydrogenase: partial characterization, induction of activity and role in the synthesis of ascorbic acid in wounded white potato tuber tissue. *Plant Cell Physiol.*, **35**, 473–478.

Ôba, K., Ishikawa, S., Nishikawa, M., Mizuno, H., and Yamamoto, T. (1995) Purification and properties of L-galactono-γ-lactone dehydrogenase, a key enzyme for ascorbic acid biosynthesis, from sweet potato roots. *J. Biochem.*, **117**, 120–124.

Okamura, M. (1994) Distribution of ascorbic acid analogs and associated glycosides in mushrooms. *J. Nutr. Sci. Vitaminol.*, **40**, 81–94.

Østergaard, J., Persiau, G., Davey, M., Bauw, G., and Van Montagu, M. (1997) Isolation of a cDNA coding for L-galactono-γ-lactone dehydrogenase, an enzyme involved in the biosynthesis of ascorbic acid in plants. Purification, characterization, cDNA cloning, and expression in yeast. *J. Biol. Chem.*, **272**, 30009–30016.

Otter, T. and Polle, A. (1994) The influence of apoplastic ascorbate on the activities of cell wall-associated peroxidase and NADH oxidase in needles of Norway spruce (*Picea abies* L.). *Plant Cell Physiol.*, **35**, 1231–1238.

Padh, H. (1990) Cellular functions of ascorbic acid. *Biochem. Cell Biol.*, **68**, 1166–1173.

Pallanca, J.E. and Smirnoff, N. (1999) Ascorbic acid metabolism in pea seedlings. A comparison of D-glucosone, L-sorbosone, and L-galactono-1,4-lactone as ascorbate precursors. *Plant Physiol.*, **120**, 453–461.

Pappa-Louisi, A. and Pascalidou, S. (1998) Optimal conditions for the simultaneous ion-pairing HPLC determination of L-ascorbic acid, dehydro-L-ascorbic, D-ascorbic, and uric acids with on-line ultraviolet absorbance and electrochemical detection. *Anal. Biochem.*, **263**, 176–182.

Pauli, N.M. and Schuep, W. (1998) Capillary zone electrophoretic determination of the four vitamin C esters L-ascorbyl-2-phosphate, L-ascorbyl-2-sulfate, L-ascorbyl-2-diphosphate and L-ascorbyl-2-triphosphate in fish feed, plasma and tissue. *J. Chromatogr. B*, **715**, 369–377.

Pell, E.J., Schlagnhaufer, C.D., and Arteca, R.N. (1997) Ozone-induced oxidative stress: mechanisms of action and reaction. *Physiol. Plant.*, **100**, 264–273.

Peters, J.L., Castillo, F.J., and Heath, R.L. (1989) Alteration of extracellular enzymes in pinto bean leaves upon exposure to air pollutants, ozone and sulfur dioxide. *Plant Physiol.*, **89**, 159–164.

Polle, A., Chakrabarti, K., Chakrabarti, S., Seifert, F., Schramel, P., and Rennenberg, H. (1992) Antioxidants and manganese deficiency in needles of Norway spruce (*Picea abies* L.) trees. *Plant Physiol.*, **99**, 1084–1089.

Procházková, A., Křivánková, L., and Boček, P. (1998) Quantitative trace analysis of L-ascorbic acid in human body fluids by on-line combination of capillary isotachophoresis and zone electrophoresis. *Electrophoresis*, **19**, 300–304.

Ranieri, A., D'Urso, G., Nali, C., Lorenzini, G., and Soldatini, G.F. (1996) Ozone stimulates apoplastic antioxidant systems in pumpkin leaves. *Physiol. Plant.*, **97**, 381–387.

Rautenkranz, A.A.F., Li, L., Mächler, F., Märinoia, E., and Oertli, J.J. (1994) Transport of ascorbic and dehydroascorbic acids across protoplast and vacuole membranes isolated from barley (*Hordeum vulgare* L. cv Gerbel) leaves. *Plant Physiol.*, **106**, 187–193.

Reiter, W.-D. (1998) The molecular analysis of cell wall components. *Trends Plant Sci.*, **3**, 27–32.

Rose, J.K.C., Hadfield, K.A., Labavitch, J.M., and Bennett, A.B. (1998) Temporal sequence of cell wall disassembly in rapidly ripening melon fruit. *Plant Physiol.*, **117**, 345–361.

Saito, K. and Kasai, Z. (1984) Synthesis of L-(+)-tartaric acid from L-ascorbic acid via 5-keto-D-gluconic acid in grapes. *Plant Physiol.*, **76**, 170–174.

Saito, K. and Loewus, F.A. (1992) Conversion of D-glucosone to oxalic acid and L-(+)-tartaric acid in detached leaves of *Pelargonium*. *Phytochemistry*, **31**, 3341–3344.

Saito, K., Nick, J.A., and Loewus, F.A. (1990) D-Glucosone and L-sorbosone, putative intermediates of L-ascorbic acid biosynthesis in detached bean and spinach leaves. *Plant Physiol.*, **94**, 1496–1500.

Saito, K., Ohmoto, J., and Kuriha, N. (1997) Incorporation of ^{18}O into oxalic, L-threonic and L-tartaric acids during cleavage of L-ascorbic and 5-keto-D-gluconic acids in plants. *Phytochemistry*, **44**, 805–809.

Sakai, T., Murata, H., and Ito, T. (1996) High-performance liquid-chromatographic analysis of ascorbyl-2-phosphate in fish tissues. *J. Chromatogr. B*, **685**, 196–198.

Sánchez-Fernández, R., Fricker, M., Corben, L.B., White, N.S., Sheard, N., Leaver, C.J., Van Montagu, M., Inzé, D., and May, M.J. (1997) Cell proliferation and hair tip growth in the *Arabidopsis* root are under mechanistically different forms of redox control. *Proc. Natl. Acad. Sci. USA*, **94**, 2745–2750.

Schmieden, U. and Wild, A. (1994) Changes in levels of α-tocopherol and ascorbate in spruce needles at three low mountain sites exposed to Mg^{2+}-deficiency and ozone. *Z. Naturforsch.*, **49c**, 171–180.

Schnarrenberger, C., Flechner, A., and Martin, W. (1995) Enzymatic evidence for a complete oxidative pentose phosphate pathway in chloroplasts and an incomplete pathway in the cytosol of spinach leaves. *Plant Physiol.*, **108**, 609–614.

Schopfer, P. (1996) Hydrogen peroxide-mediated cell wall stiffening in vitro in maize coleoptiles. *Planta*, **199**, 43–49.

Schraudner, M., Langebartels, C., and Sandermann, H. (1997) Changes in the biochemical status of plant cells induced by the environmental pollutant ozone. *Physiol. Plant.*, **100**, 274–280.

Siendones, E., González-Reyes, J.A., Santos-Ocaña, C., Navas, P., and Córdoba, F. (1999) Biosynthesis of ascorbic acid in kidney bean. L-galactono-γ-lactone dehydrogenase is an intrinsic protein located at the mitochondrial inner membrane. *Plant Physiol.*, **120**, 907–912.

Smirnoff, N. (1996) The function and metabolism of ascorbic acid in plants. *Ann. Bot.*, **78**, 661–669.

Smirnoff, N. and Pallanca, J.E. (1995) Ascorbate metabolism in relation to oxidative stress. *Biochem. Soc. Trans.*, **24**, 472–478.

Stokes, N.J., Terry, G.M., and Hewitt, C.N. (1998) The impact of ozone, isoprene and propene on antioxidant levels in two leaf classes of velvet bean (*Mucuna pruriens* L.). *J. Exp. Bot.*, **49**, 115–123.

Svirbely, J.L. and Szent-Györgyi, A. (1932) CV. The chemical structure of vitamin C. *Biochem. J.*, **26**, 865–870.

Takahama, U. (1993) Redox state of ascorbic acid in the apoplast of stems of *Kalanchoë daigremontiana*. *Physiol. Plant.*, **89**, 791–798.

Tausz, M., Kranner, I., and Grill, D. (1996) Simultaneous determination of ascorbic acid and dehydroascorbic acid in plant metarials by high performance liquid chromatography. *Phytochem. Anal.*, **7**, 69–72.

Thordal-Christensen, H., Zhang, Z., Wei, Y., and Collinge, D.B. (1997) Subcellular localization of H_2O_2 in plants. H_2O_2 accumulation in papillae and hypersensitive response during the barley—powdery mildew interaction. *Plant J.*, **11**, 1187–1194.

Tolbert, B.M., Downing, M., Carlson, R.W., Knight, M.K., and Baker, E.M. (1975) Chemistry and metabolism of ascorbic acid and ascorbate sulfate. *Annals N.Y. Acad. Sci.*, **258**, 48–69.

Vanacker, H., Harbinson, J., Ruisch, J., Carver, T.L.W., and Foyer, C.H. (1998) Antioxidant defences of the apoplast. *Protoplasma*, **205**, 129–140.

Vera, J.C., Rivas, C.I., Fischbarg, J., and Golde, D.W. (1993) Mammalian facilitative hexose transporters mediate the transport of dehydroascorbic acid. *Nature*, **364**, 79–82.

Wahal, C.K., Bhattacharya, N.C., and Talpasayi, E.R.S. (1973) Ascorbic acid and heterocyst development in the blue-green alga *Anabaena ambigua*. *Physiol. Plant.*, **28**, 424–429.

Washko, P.W., Welch, R.W., Dhariwal, K.R., Wang, Y. and Levine, M. (1992) Ascorbic acid and dehydroacorbic acid analyses in biological samples. *Anal. Biochem.*, **204**, 1–14.

Washko, P.W., Wang, Y., and Levine, M. (1993) Ascorbic acid recycling in human neutrophils. *J. Biol. Chem.*, **268**, 15531–15535.

Wheeler, G.L., Jones, M.A., and Smirnoff, N. (1998) The biosynthetic pathway of vitamin C in higher plants. *Nature*, **393**, 365–369.

Yamaguchi, K., Mori, H., and Nishimura, M. (1995) A novel isoenzyme of ascorbate peroxidase localized on glyoxysomal and leaf peroxisomal membranes in pumpkin. *Plant Cell Physiol.*, **36**, 1157–1162.

Yamamoto, I., Muto, N., Nagata, E., Nakamura, T., and Suzuki, Y. (1990) Formation of stable L-ascorbic acid α-glucoside by mammalian α-glucosidase-catalyzed transglucosylation. *Biochim. Biophys. Acta*, **1035**, 44–50.

12 Glutathione Biosynthesis in Plants

Teva Vernoux, Rocío Sánchez-Fernández and Mike May

INTRODUCTION

A plant's tolerance to environmental stresses is one of the parameters that influences the variety of conditions where it can grow. Thus, factors that contribute to environmental stress tolerance will determine the distribution of a given plant species in the natural habitat and will also affect the practice of commercial agriculture. Amongst these, the tripeptide thiol glutathione (GSH; γ-glutamylcysteinyl glycine) plays a pivotal role in the mechanisms plants have evolved to minimize the potential for cellular dysfunction, which arises through stress-induced redox perturbation. This interest stems from the fact that most cellular components tolerate only minor perturbations in the cellular redox potential, and in most organisms, GSH is the principle redox buffer. The notable stability of GSH, which derives from the γ-glutamyl linkage and the strong nucleophilic nature of the central cysteine, contributes to its function as the major cellular store of reduced sulfur, its function as a scavenger of peroxides, and its role in the redox regulation of many cellular processes. It is not surprising, therefore, that GSH has been the focus of considerable attention from all fields of biotechnological research, effort being largely targeted towards strategies for the engineering of stress tolerance (May et al., 1998a). Interest in the potential benefits of genetically engineered cellular GSH concentrations in plants was prompted by the observations that elevated GSH levels often correlate with stress tolerance: activation of GSH biosynthesis at the onset of stress may be an intrinsic part of stress tolerance, and GSH appears to play a key role in the modulation of defense gene expression (Alscher, 1989).

It is our aim to assess current understanding of how GSH is synthesized and the contribution of biosynthesis to cellular GSH concentrations. We also consider current opinion of how alterations in the concentration and redox status of GSH impinge on the integration of cellular processes with adaptations to environmental stress. Our final aim is to identify future experimental priorities that will enrich our understanding of the function of GSH in plants and will provide the foundations for the rational design of strategies for the improvement of plant stress tolerance.

THE CONTROL OF GSH HOMEOSTASIS IN PLANTS

General Considerations

The ability of reduced GSH to function efficiently as a redox buffer in the regulation of cellular processes is dependent upon its concentration. Thus, identification of the factors that modify the concentration, redox status, and spatial distribution of GSH in plants will be essential to understand the function of GSH. Cellular GSH concentrations in plants will be determined by a complex interplay between a number of metabolic factors (cysteine availability, biosynthetic activity, degradation, oxidation, and chelation), superimposed on which, environmental constraints (photoperiod, xenobiotics, extremes of temperature, and oxidative stress) will exert critical influences.

The list of clones needed for the genetic engineering of GSH levels in plants is now almost complete and the current state of the field is summarized in Figure 1. Glu, derived from the N assimilation pathway, Cys, derived from the S assimilation pathway, and Gly, to a large extent derived from photorespiration, are the building blocks for the synthesis of GSH. In some species, β-Ala or Ser may substitute for Gly (Price, 1957; Klapheck *et al.*, 1992). Synthesis of GSH occurs exclusively through the activity of two biosynthetic enzymes and both have been cloned (Figure 1; see below). The reduced GSH pool may be oxidized, to form a glutathione dimer (GSHG) for example, during the destruction of peroxides. Molecular aspects of this process are described by Creissen and Mullineaux (Chapter 10; the Asada-Halliwell pathway). GSHG is reduced to GSH through the activity of the NADPH-dependent enzyme glutathione reductase (GR; Figure 1). Degradation of GSH in plants is poorly characterized, although recently a member of a γ-glutamyl transpeptidase (γ-GT; Figure 1) gene family was isolated from *Arabidopsis* (Kushnir *et al.*, 1995). The identification of additional members of this and other pathways for the degradation of GSH is awaited (Figure 1). Molecular characterization of glutathione *S*-transferases (GST; Figure 1), which catalyze the formation of GSH conjugates (GS-X; Figure 1) during the detoxification of xenobiotics or in the processing of endogenous compounds, is extensive (Kreuz *et al.*, 1996; Marrs, 1996). Ultimately, GS-X conjugates are transported into the vacuole by a class of transporters with biochemical properties and sequence homologies that assign them to the family of multidrug resistance proteins (MRP; Figure 1). The cloning of members of this family from *Arabidopsis* has recently been reported (Lu *et al.*, 1997; Sánchez-Fernández *et al.*, 1998). Most of these GS-X pumps are probably located in the vacuolar membrane, albeit at this stage the possiblity of GS-X pumps in the plasmalemma cannot be ruled out (Figure 1). Although mutants of *Arabidopsis* that are defective in the formation of phytochelatins, which are heavy metal-binding peptides, have been described (Howden *et al.*, 1995) as well as mutants affected in phytochelatin synthetase activity (PCS; Figure 1), the corresponding cDNAs or genes have not been reported. Map-based cloning approaches should shortly resolve this issue. Similarly, it is not known whether members of the GS-X pump family so far cloned from *Arabidopsis* can transport phytochelatin-heavy metal complexes or indeed GSH-heavy metal complexes as has been described in yeast (Li *et al.*, 1997).

Clearly, rapid progress in recent years has significantly advanced the molecular characterization of GSH metabolism. It is reasonable to predict that in coming years, transgenic plants that express combinations of recently isolated genes will be made. A rational evaluation of the functions of GSH in plants and the potential for engineering many GSH-related aspects of plant metabolism is clearly now possible (May *et al.*, 1998a).

GSH Biosynthesis

In this chapter we will focus primarily on progress towards understanding the biosynthesis of GSH. Biosynthesis of GSH in all organisms studied to date occurs in two steps. In the first, catalyzed by γ-glutamylcysteine synthetase (γECS; EC 6.3.2.2), γ-glutamylcysteine is synthesized from L-Glu and L-Cys (Eq. 1). In the second step, catalyzed by glutathione synthetase (GSHS; EC 6.3.2.3), Gly is added to the C-terminal site of γ-glutamylcysteine to yield GSH (Eq. 2) (Hell and Bergmann, 1988, 1990).

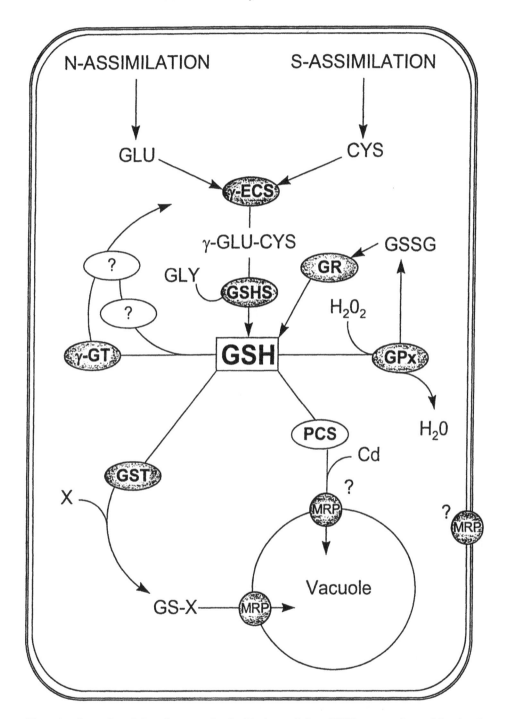

Figure 1. Current knowledge of processes involved in the regulation of GSH concentrations and functions in plants. Enzymes responsible for the catalysis of the different steps of GSH metabolism are represented as ovals. Enzymes for which cDNAs or genes have been isolated are represented by shaded ovals. Steps that must exist, but for which biochemical or molecular data are lacking are represented by ovals containing a question mark.

$$\text{L-Glu} + \text{L-Cys} + \text{ATP} \rightarrow \text{L-}\gamma\text{-Glu-L-Cys} + \text{ADP} + \text{P}_i \qquad \text{(Eq. 1)}$$

$$\text{L-}\gamma\text{-Glu-L-Cys} + \text{Gly} + \text{ATP} \rightarrow \text{L-}\gamma\text{-Glu-L-Cys-Gly} + \text{ADP} + \text{P}_i \qquad \text{(Eq. 2)}$$

The GSH-Biosynthetic Pathway is Highly Responsive to Environmental Stress

Interest in the enzymes of the GSH-biosynthetic pathway and the factors that regulate their activity comes from observations that considerable increases in the cellular concentration of GSH occur in response to chilling, heat shock, pathogen attack, oxidative stress, air pollution, or drought (Guri, 1983; Nieto-Sotelo and Ho, 1986; Dhindsa, 1991; Edwards et al., 1991; Madamanchi and Alscher, 1991; Sen Gupta et al., 1991; May and Leaver, 1993; Kocsy et al., 1996; May et al., 1996a, 1996b). In many instances, increases in the level of GSH are associated with an activation of the GSH-biosynthetic pathway, most notably γECS, and in some cases, increased γECS activity correlates with stress tolerance. Elevated GSH levels arising specifically through increased γECS activities have been measured in extracts of pea, tobacco, maize, and tomato treated with cadmium (Rüegsegger et al., 1990; Rüegsegger and Brunold, 1992; Chen and Goldsbrough, 1994) and cadmium tolerance in tomato cell lines relates to increased γECS activity (Chen and Goldsbrough, 1994). Similarly, increased γECS activity and GSH levels correlates with chilling tolerance in maize (Kocsy et al., 1996) and herbicide safener treatment increases GSH levels through enhanced γECS activity in maize and tobacco (Rennenberg et al., 1982) and in Arabidopsis (May et al., 1998b). The latter observation is of particular interest because we have shown previously that stress activation of GSH biosynthesis in Arabidopsis is an essential component of stress tolerance (May and Leaver, 1993).

γECS is the Rate-Limiting Step of GSH Biosynthesis

For the design of strategies for the manipulation of GSH concentrations, it is important to have a grasp of the kinetic constraints on the biosynthetic pathway. In mammals and yeast it is generally accepted that γECS is the rate-limiting enzyme in GSH biosynthesis (Richman and Meister, 1975; Grant and Dawes, 1996), and recent evidence has demonstrated this to be the case in plants too (Cobbett et al., 1998).

It is generally thought that a major limitation on flux through the GSH-biosynthetic pathway is exerted through feedback inhibition of γECS by GSH. Although feedback inhibition of γECS by non-covalent binding of reduced GSH can be readily demonstrated in vitro and K_i values have been calculated (Richman and Meister, 1975; Hell and Bergmann, 1990), it is now clear that in vivo synthesis can continue or be stimulated even when the levels of GSH considerably exceed the K_i of γECS. Data from transgenic plants rather imply that in vivo cysteine availability plays an important role in determining cellular GSH concentration through kinetic restriction of γECS. Evidence for the rate-limiting nature of γECS comes from the analysis of transgenic poplars in which bacterial genes that encode γECS or GSHS are expressed (Strohm et al., 1995). In these plants, the in vivo capacity of GSHS does not appear to kinetically limit the synthesis of GSH, because strong overproduction of GSHS in poplar cytosol or chloroplasts does not affect GSH levels. Furthermore, addition of γEC to leaf discs with a 50-fold higher GSHS activity was shown to give only a 3-fold higher rate of GSH accumulation than in control leaf discs (Strohm *et al.*, 1995). In contrast, overproduction of *Escherichia coli* γECS brought about marked increases in foliar GSH demonstrating that increase of this activity removes a major limitation over the rate of GSH synthesis and implies that the rate of synthesis of γEC is held out of equilibrium in untransformed poplars (Strohm *et al.*, 1995; Noctor *et al.*, 1996). Direct evidence for the rate-limiting nature of γECS came recently from analysis of the γECS gene of a cadmium-sensitive mutant of *Arabidopsis*, *cad2-1*, in which the GSH content of leaves is decreased to 30% of that in the wild-type parental accession (Cobbett and May, 1997). GSH depletion in this mutant results from a 6-base pair deletion in the γECS-encoding gene and leads to an in-frame, two-amino acid deletion in the γECS protein, which causes a 60% reduction in the extractable γECS activity compared to wild-type *Arabidopsis*. In this mutant, the GSHS activity is unaltered. Whereas transformation of *cad2-1* with the wild-type γECS gene restored cadmium tolerance, GSH levels, and γECS activity to, or above, wild-type levels, the GSHS activity did not change, unequivocally demonstrating that γECS *in planta* is the limiting enzyme of GSH biosynthesis (Cobbett and May, 1997; Cobbett *et al.*, 1998). Similarly, treatment of *Arabidopsis* suspensions with a number of stressors, which at sublethal doses raise GSH levels, do so uniquely through stimulation of γECS activity; GSHS activities remain unaltered by any of these treatments (May *et al.*, 1998b).

Taken together, these data demonstrate that the supply of γEC restricts the production of GSH, that γECS is the rate-limiting step in GSH biosynthesis in plants, and that *in vivo* feedback inhibition by GSH is not a major control point. If this is indeed the case, then alternative mechanisms for the control of γECS must be envisaged. These data also indicate that plants overproducing γECS may have increased tolerance to a number of environmental stresses through removal of the limitation on GSH synthesis imposed by endogenously low levels of γECS activity.

γECS and *GSHS* cDNAs

The potential to enhance GSH levels by genetic engineering of its own biosynthetic pathway is clearly an attractive proposition. Because of their low activity and lability, however, neither of the two biosynthetic enzymes have been purified to homogeneity from any plant species. Furthermore, with the exception of tobacco (Hell and Bergmann, 1988, 1990), our understanding of the biochemistry of GSH biosynthesis in plants was, until recently, rudimentary. Although the lack of structural data prevented conventional cloning procedures, this problem has been circumvented recently through complementation of *E. coli* and *Saccharomyces cerevisiae* mutants deficient in these enzymes. cDNAs encoding γECS (May and Leaver, 1994; May *et al.*, 1998b) and GSHS (Rawlins *et al.*, 1995; Ullmann *et al.*, 1996) have been isolated from *Arabidopsis thaliana*. By using these cDNAs as probes, clones that encode both enzymes from tomato and *Brassica juncea* have been identified and display very high sequence homology (Kovari *et al.*, 1997; Schäfer *et al.*, 1997). In addition, homologous expressed sequence tags (ESTs) from rice are available in the databases. Whereas both genes are represented by a single copy in the *Arabidopsis* genome, gene families exist in *Brassica juncea* and tomato, possibly reflecting organelle or tissue-specific forms of the gene products. The availability of these cDNAs will allow the construction of transgenic plants as well as rigorous evaluation of the role of GSH in plant development and stress responses.

*At*γECS Posesses Unique Structural Properties

Although expression of the *Arabidopsis* GSHS sequence in the mutant *E. coli* background raised GSH levels to those of wild type (Rawlins *et al.*, 1995), expression of the γECS cDNA did not (May and Leaver, 1994). Furthermore, whereas the derived amino acid sequence of *AtGSHS* shares extensive homology with that of other eukaryotes, the *Arabidopsis* γECS sequence is highly divergent with only apparently critical regions, particularly at the active site, being conserved between plants and other organisms (May and Leaver, 1994; Lueder and Phillips, 1996).

The puzzling structural and functional peculiarities of the *Arabidopsis* γECS clone, the failure of the cDNA product to restore GSH levels in the heterologous mutant host, and the potential ambiguities of the assay used to measure the activity of the gene product raised questions as to its true identity (Brunold and Rennenberg, 1996). These questions have been resolved through isolation of the same cDNA by functional complementation of a γECS-deficient yeast mutant and measurement of γECS activity in yeast extracts, which expressed these cDNAs, by post-reaction derivatization of reaction products with monobromobimane and HPLC analysis (May *et al.*, 1998a). Independent proof of the identity of this clone came through mapping and cloning of γECS from *cad2-1*, a cadmium-sensitive mutant of *Arabidopsis* with only 15–30% wild type GSH. The identification of a 6-base pair deletion in the γECS gene of *cad2-1* and complementation of the mutant with a wild-type *At*γECS genomic clone that restored cadmium resistance, GSH levels, and γECS activity in 11 independent transformants proved beyond doubt the identity of this clone (Cobbett and May, 1997). More recently, a root meristemless mutant (*rml1*) was found to be allelic to *cad2-1* and is in fact a GSH null mutant (Vernoux *et al.*, 2000). The use of such mutants in future analyses will be central to unravel the function of GSH in plants.

These data raise the question of why *At*γECS has evolved to be so structurally distinct. The unique structural properties of *At*γECS may be a reflection of how the gene has evolved to tune γECS activity to environmental demands and permit survival under changeable and often hostile conditions. Genetic approaches will allow the identification of additional mutant alleles in *Arabidopsis* and biochemical analysis of the mutant gene products will unravel structure-function relationships.

Activation of γECS Biosynthesis is Complex

Measurements of activation of γECS in response to stress and correlations of increased γECS activity with stress tolerance imply the existence of stress sensors and signal transduction cascades that elicit the response of the biosynthetic pathway of GSH and also show that GSH biosynthesis in plants is finely tuned to the prevailing environmental conditions. If we are to understand and ultimately engineer the synthesis of GSH in plants, the characterization of these sensory systems and the identification of the factors controlling γECS activity are clearly major priorities. The integration of molecular and genetic approaches with biochemical techniques is now providing indications of how γECS activity is regulated in plants. For example, in *Brassica juncea*, a heavy metal-tolerant plant, stimulation of γECS transcript accumulation by heavy metal treatment has been reported (Schäfer *et al.*, 1997). In *Arabidopsis*, however, the situation seems more complex. A number of lines of evidence indicate that *Arabidopsis* γECS is regulated by post-transcriptional mechanisms. In yeast and *E. coli* mutants deficient in γECS, which express AtγECS, the level of GSH correlates with the extractable γECS activity, yet in both never exceeds 10% of the wild-type concentration, indicating that features intrinsic to the protein may be responsible for the poor GSH synthesis in heterologous hosts (May and Leaver, 1994; May *et al.*, 1998a). In support of this proposition, expression of a genomic clone corresponding to *At*ECS in the GSH-deficient mutant of *Arabidopsis*, *cad2-1*, raised the cellular concentration of GSH up to 2.3-fold that of wild type (Cobbett and May, 1997; Cobbett *et al.*, 1998). The regulation of plant γECS thus appears to be rather complex, and plant-specific factors, absent in heterologous hosts, may be necessary for full catalytic activity. Evidence for the existence of such factors came from analysis of GSH synthesis in *Arabidopsis* cell cultures treated with a number of stress agents. In response to oxidative stress, safener or heavy metals, substantial increases in the levels of GSH correlated uniquely with increases in the activity of γECS; changes in the activity of GSHS or in the steady state levels of γECS or *GSHS* transcripts were undetectable (May *et al.*, 1998b). This observation demonstrated the existence of post-transcriptional mechanisms that specifically and rapidly activate γECS. Most importantly, we demonstrated that the kinetics of changes in the activity of γECS appear to be adapted to the type of stress that elicits this response. Thus, discrete signal transduction pathways exist that allow discrimination between stresses of different nature and that activate appropriately adapted responses to specific stimuli. These results imply the existence of one or more proteins that act as stress sensors and that relay this information to molecules, which interact directly with γECS. In theory, a single sensor protein could discriminate between different stresses and evidence for such mechanisms has recently been demonstrated in the responses of yeast to oxidative stress arising from different stimuli (Wemmie *et al.*, 1997). As an adaptive response, such levels of complexity will provide rapid responses of the GSH biosynthetic pathway to a variety of signals and restoration of cellular GSH homeostasis under stress conditions. Obvious candidates that could directly and rapidly activate γECS are protein kinases or phosphatases as is known for animal γECS (Sun *et al.*, 1996). Indeed, sequence analysis with the PROSITE program identified amino acid motifs that could potentially be phosphorylated by protein kinases A and C, calmodulin-dependent kinases, and cyclin-dependent kinases (May *et al.*, 1998a). Strategies for the identification of such proteins will be of fundamental importance for our understanding of GSH biosynthesis in plants. Given that GSH biosynthesis will be regulated by interactions of these proteins with γECS, introduction of genes encoding these proteins into plants may provide a more effective means of modulating plant GSH concentrations than simply increasing the copy number of γECS.

Gearing Demand to Supply; Activation of S Assimilation

Recent studies strongly support the view that cysteine availability in plants, as in animals (Richman and Meister, 1975), plays an important role in determining cellular GSH concentrations through kinetic restrictions of the reaction catalyzed by γECS (Farago *et al.*, 1994; Strohm *et al.*, 1995; Noctor *et al.*, 1996, 1997). Environmental conditions that stimulate synthesis of GSH will therefore increase demand upon S assimilation into cysteine. Available evidence suggests that environmental conditions that place increased demand on GSH synthesis are not only met by activation of γECS, but also by parallel increases in the activity of key enzymes in S assimilation (Rüegsegger *et al.*, 1990; Rüegsegger and Brunold, 1992; Farago *et al.*, 1994). Whereas activation of γECS appears to be regulated by post-transcriptional mechanisms in *Arabidopsis* at least (May *et al.*, 1998b), the upregulation of S assimilation in *Arabidopsis* results seemingly from transcriptional activation (Hell, 1997; K. Saito, personal communication). As an environmental adaptation, the ability to coordinate increases in the flux through both the pathways of sulfur assimilation and GSH synthesis is of obvious value because the kinetic restriction of γECS activity by cysteine would be alleviated, thereby assuring efficient synthesis of GSH during stress responses. Clearly, mechanisms that function to sense the demand for increased GSH synthesis communicate with those for the supply of cysteine. Alternatively, the sensor that orchestrates these two responses may be the same. Indications that the latter possibility may indeed be the case come from recent studies that the sensor of root sulfur status, which regulates the expression of genes in the S assimilation pathway in *Brassica* and *Arabidopsis*, is GSH (Lappartient and Touraine, 1996, 1997). A future priority will be the isolation of (the) redox-regulated transcriptional activator(s) and components of the intercommunicating signal transduction pathways that regulate the flux of sulfur into GSH.

THE IMPACT OF CHANGES IN GSH CONCENTRATION ON CELLULAR PROCESSES

Given progress in recent years in gene cloning strategies, we are in a position where we can envisage efficient strategies for the engineering of cellular GSH concentrations. Indications that plants with altered cellular GSH concentrations or altered capacity for GSH synthesis will be of agronomic importance are now forthcoming.

Responses to biotic and abiotic stresses essentially require the activation of the appropriate defenses and the activation of mechanisms to ensure the correct partitioning and supply of energy. In addition, the selective inactivation of certain cellular processes, for example, the cell cycle, may be critical for the preservation of limited energy reserves and, most importantly, the limitation of heritable damage. It is clear that plants possess highly sensitive sensory circuits for monitoring environmental change and coordinated transduction pathways that convey this information to elicit the appropriate responses. Here, we assess recent evidence which demonstrates that the cellular concentration of GSH plays a determinative role in many cellular processes and in the coordination of plant stress responses.

Whereas it has been known for some time that GSH plays a key role in the regulation of photosynthetic enzymes (Wolosiuk and Buchannan, 1977), new data suggest that GSH may have many more functions in the regulation of chloroplast function. Emerging evidence indicates that redox-based sensing and signalling mechanisms collectively serve to maintain chloroplast redox poise, and promote the prevention or repair of oxidative damage (Danon and Mayfield, 1994; Allen, 1995; Huner *et al.*, 1996; Link, 1996; Karpinski *et al.*, 1997; Liere and Link, 1997; Link *et al.*, 1997), and, in addition, provide the nucleus with the cues for the activation of cytosolic defenses (Karpinski *et al.*, 1997). Whether similar redox-based signalling mechanisms exist in the mitochondria is unknown, but could be predicted because the initial effects of environmental changes that affect redox chemistry are most clearly seen in the chloroplasts and mitochondria, being the bioenergetic and assimilatory interface of the cell with the environment. The responses described highlight the sophistication of mechanisms that have evolved to link the consumption of energy with supply and information exchange between organelles under potentially debilitating conditions. With the battery of clones now at our disposition, it is now possible to design transgenic plants in which the role of GSH in chloroplast function can be assessed, and to predict whether such plants would have more efficient energy conversion and improved quantum yield factors critical for crop productivity.

Another field of intense activity in which functions for GSH feature heavily is in the responses of plants to oxidative stress. Regulatory or signalling functions have been ascribed to redox changes in the responses of plants to high light (Karpinski *et al.*, 1997), drought (Dhindsa, 1991; Smirnoff, 1993), heavy metals (Lobréaux *et al.*, 1995), extreme temperatures (Prasad *et al.*, 1994), UV irradiation and atmospheric pollutants (Yalpini *et al.*, 1994), mechanical stress (Legendre *et al.*, 1993), and pathogen attack (Apostol *et al.*, 1989; Chen *et al.*, 1993). There is no doubt that among the battery of responses which have evolved, reactive oxygen species (ROS) serve a multitude of functions in defense signalling and the advent of large-scale mutant screening programs will allow the dissection of the pathways in which ROS mediate plant defenses. Such approaches are essential because a number of crucial issues remain obscure and must be addressed. Firstly, how do ROS function in the establishment of an effective defense response? Secondly, what are the events downstream of ROS accumulation that regulate changes in gene expression? Thirdly, how do components of redox-regulated signalling interact with other signal transduction pathways? At present, it is difficult to form a clear picture of events in plant stress signalling and the information to be gleaned from the literature to a large extent adds to the confusion, making rational speculation necessary for experimental design almost impossible. For example, initial interest in the role of redox chemistry in plant defense signalling was fuelled by the demonstration that GSH could elicit a massive and selective activation of defense gene expression, and a role as a regulator of phytoalexin accumulation was proposed (Wingate *et al.*, 1988). In direct contrast, depletion of GSH or increases in its oxidation state were shown to prompt accumulation of phytochelatins (Stossel, 1984; Gustine, 1987; Guo *et al.*, 1993). Despite the identification of GSH-responsive elements on the promoters of genes involved in the synthesis of phytoalexins (Dron *et al.*, 1988) and demonstrations of net increases in the level of GSH in plant cells treated with fungal elicitors (Edwards *et al.*, 1991; Nakagawara *et al.*, 1993), increases in phytoalexins after elicitor treatment precede and do not follow increases in the GSH level. In addition, further complexity comes from experiments in which the effects of ROS can be mimicked by GSH (Bradley *et al.*, 1992; Hérouart *et al.*, 1993; Levine *et al.*, 1994; Wingsle and Karpinski, 1996). A puzzling question raised by these data is how does an oxidant and an antioxidant elicit the same response both in terms of kinetics and magnitude? It is clear that transgenic plants and mutants with altered GSH metabolism will help resolve these issues and substantially improve experimental design. Indeed, the complexity of responses elicited in animal cells to changes in cellular redox chemistry and the pleiotropic consequences of ROS on downstream signalling serve as important indicators of the complexity to be expected in plant ROS signalling pathways. Perhaps a more rational way to view ROS signalling is to consider responses in terms of alterations in cellular redox status because the redox potential of electron and hydrogen carriers is concentration dependent and thus responses to ROS will not only be dependent on the initial concentration of ROS, but also on spatial and temporal variations (cellular and organellar) in the levels and redox state of antioxidants. This view is particularly relevant because physiological signalling through a redox mechanism is probably channelled within the confines of specific cellular compartments. Clearly advances in cellular imaging techniques will play a central role in the future (see below).

Fortunately, approaches for rationalizing our understanding of redox in stress signalling are now being applied. For example, the use of the GSH biosynthesis inhibitor, L-buthionine [S,R] sulfoximine (BSO) is a powerful tool for understanding how the concentration of GSH regulates stress responses. Indeed, much of what has been learnt about the antioxidant function of GSH, the function of GSH in mitochondrial integrity, and the function of GSH in redox signalling and in compensatory interactions between GSH and ascorbate in animals has come from the use of this powerful and highly selective drug (Griffith and Meister, 1979; Jain *et al.*, 1992; Martensson and Meister, 1992; Esposito *et al.*, 1995; Mihm *et al.*, 1995). BSO is now being used to modify GSH concentrations in plants too and its use has allowed analysis of the functions of GSH in stress responses (May and Leaver, 1993), developmental processes (Sánchez-Fernández *et al.*, 1997), cell cycle progression (Vernoux *et al.*, 2000), plant pathogen interactions (Guo *et al.*, 1993), and redox signalling (Price *et al.*, 1994; Lobréaux *et al.*, 1995). Such approaches allow discrimination between the effects of oxidative stress *per se* or indirect effects through modifications in the redox poise and size of the antioxidant pool during development and environmental responses. The widespread adoption of such approaches will not only broaden our understanding of stress signal transduction, but will also give an indication of its dynamics.

NEW TECHNOLOGIES FOR THE ANALYSIS OF GSH IN PLANTS

Spatial and temporal variations in the cellular concentration of GSH and subcellular compartmentation will underly the role of GSH in signal transduction and will be central to the specificity of redox changes in a given organ or even within specific cell types. It is therefore necessary to have an idea of the spatial distribution of GSH in intact plants because many of the pathways in which GSH acts are spatially, temporally, and developmentally regulated. Current methods for the determination of cellular GSH concentration, albeit specific and sensitive, require the destruction of the material, are not easily applicable to large numbers of small samples, and, most importantly, fail to detect populational heterogeneity. The application of the non-toxic, cell-permeant, non-fluorescent dye, monochlorobimane (mBCl; [1H, 7H-pyrazolo(1,2-α)pyrazole-1,7-dione,3-(chloromethyl)-2,5,6-trime thyl-), which reacts specifically with GSH to yield a strong blue-fluorescent bimane-GSH (GS-B) conjugate, and analysis of the fluorescent conjugate directly *in vivo* in animal systems has addressed all of these limitations (Rice *et al.*, 1986; Fernández-Checa and Kaplowitz, 1990). Confocal laser scanning and conventional fluorescence microscopy of intact tissues stained with mBCl have allowed clear demonstrations of marked heterogeneity in the concentration of GSH between different cell organelles and between cells in a population or organ (Bellomo *et al.*, 1992; Briviba *et al.*, 1993). Correlations between high GSH concentrations and the proliferative capacity of cells (Poot *et al.*, 1995; Kavanagh *et al.*, 1990) or correlations with drug resistance in subpopulations of tumour cells (Shrieve *et al.*, 1988) have also been demonstrated.

In order to unravel the molecular details of GSH biosynthesis in plants and the processes in which GSH acts, mBCl is now being used to spatially map GSH in intact plants and to investigate GSH concentrations in *Arabidopsis* mutants (Figure 1; Sánchez-Fernández et al., 1997; May et al., in preparation; Meyer et al., in preparation). mBCl is also being utilized as a tool to study xenobiotic detoxification mechanisms, because it is conjugated to GSH through the activity of GSTs and is loaded into the vacuole by GS-X pumps in a fashion analagous to xenobiotics, such as herbicides, and endogenous compounds, such as anthocyanins (Coleman et al., 1997). The development of suitable computer software for confocal imaging data will in theory allow approaches to address physiological aspects of GSH metabolism *in vivo* and for the estimation of GSH biosynthesis rates, rates of conjugation, or the kinetics of transport into the vacuole (Meyer et al., in preparation). Such developments would allow the rigorous analysis of mutants or transgenics with modified GSH metabolism and will lead to the identification of tissue- or cell-specific differences in the activities of different enzymes.

FUTURE PERSPECTIVES

The molecular nature of mechanisms by which plants undertake massive reorganization of their growth and metabolism in a commitment to the maintenance of redox poise during growth under stress is now becoming obvious. In principle, glutathione could clearly act both as a direct link between environmental stress and a number of the key adaptive responses plants have evolved. We are still far from understanding the details of these interactions but progress within the last years has at least put us in a position to begin to unravel this complexity. It is expected that with the battery of molecular, genetic, and biochemical tools now available, we will see rapid progress towards understanding how the pool size of GSH is regulated in plants. In addition, new techniques for detection and imaging of GSH in living cells (Coleman et al., 1996, 1997; Sánchez-Fernández et al., 1997) will allow us to get a better understanding of how changes in spatial distribution of GSH in whole plants contribute to its role in physiological processes. The identification of the factors that sense alterations in the GSH pool will clearly be a valuable addition to the tools available for the manipulation of plant stress responses.

REFERENCES

Allen, J.F. (1995) Thylakoid protein phosphorylation, state I-state 2 transitions and photosystem stoichiometry adjustment: Redox control at multiple levels. *Physiol. Plant.*, **93**, 196–205.

Alscher, R.G. (1989) Biosynthesis and antioxidant function of glutathione in plants. *Physiol. Plant.*, **77**, 457–464.

Apostol, I., Heinstein, P.F., and Low, P.S. (1989) Rapid stimulation of an oxidative burst during elicitation of cultured plant cells. *Plant Physiol.*, **90**, 109–116.

Bellomo, A., Viretti, M., Stivala, L., Mirabelli, F., Richemi, P., and Orrenius, S. (1992) Demonstration of nuclear compartmentalization of glutathione in hepatocytes. *Proc. Natl. Acad. Sci. USA*, **89**, 4412–4416.

Bradley, D.J., Kjellbom, P., and Lamb, C.J. (1992) Elicitor- and wound-induced oxidative cross-linking of a proline-rich plant cell-wall protein: A novel, rapid defense response. *Cell*, **70**, 21–30.

Briviba, K., Fraser, G., Sies, H., and Ketterer, B. (1993) Distribution of the monochlorobimane-glutathione conjugate between nucleus and cytosol in isolated hepatocytes. *Biochem. J.*, **294**, 631–633.

Brunold, C. and Rennenberg, H. (1996) Regulation of sulfur metabolism in plants: first molecular approaches. *Progr. Bot.*, **58**, 164–186.

Chen, J. and Goldsbrough, P.B. (1994) Increased activity of γ-glutamylcysteine synthetase in tomato cells selected for cadmium tolerance. *Plant Physiol.*, **106**, 233–239.

Chen, Z., Silva, H., and Klessig, D.F. (1993) Active oxygen species in the induction of plant systemic acquired resistance by salicylic acid. *Science*, **262**, 1883–1886.

Cobbett, C. and May, M. (1997) The cadmium sensitive, glutathione-deficient mutant *cad2-1*, is deficient in γ-glutamylcysteine synthetase. *Plant Physiol.* (Suppl.), **114**, 123.

Cobbett, C.S., May, M.J., Howden, R., and Rolls, B. (1998) The glutathione-deficient, cadmium-sensitive mutant, *cad2-1*, of *Arabidopsis thaliana* is deficient in γ-glutamylcysteine synthetase. *Plant J.*, **16**, 73–78.

Coleman, J.O.D., Blake-Kalff, M.M.A., and Davies, T.G.E. (1996) Detoxification of xenobiotics by plants: chemical modification and vacuolar compartmentation. *Trends Plant Sci.*, **2**, 144–151.

Coleman, J.O.D., Randall, R., and Blake-Kalff, M.M.A. (1997) Detoxification of xenobiotics in plant cells by glutathione conjugation and vacuolar compartmentalization: a fluorescent assay using monochlorobimane. *Plant Cell Environ.*, **20**, 449–460.

Danon, A. and Mayfield, S.P. (1994) Light regulated translation of chloroplast messenger RNAs through redox potential. *Science*, **266**, 1717–1719.

Dhindsa, R.S. (1991) Drought stress, enzymes of glutathione metabolism, oxidation injury and protein synthesis in *Tortula ruralis*. *Plant Physiol.*, **95**, 648–651.

Dron, M., Clouse, S.D., Dixon, R.A., Lawton, M.A., and Lamb, C.J. (1988) Glutathione and fungal elicitor regulation of a plant defense gene promoter in electroporated protoplasts. *Proc. Natl. Acad. Sci. USA*, **85**, 6738–6742.

Edwards, R., Blount, J.W., and Dixon, R.A. (1991) Glutathione and elicitation of the phytoalexin response in legume cell cultures. *Planta*, **184**, 403–409.

Esposito, F., Cuccovillo, F., Morra, F., Russo, T., and Cimino, F. (1995) DNA binding activity of the glucocorticoid receptor is sensitive to redox changes in intact cells. *Biochem. Biophys. Acta*, **1260**, 308–314.

Farago, S., Brunold, C., and Kreuz, K. (1994) Herbicide safeners and glutathione metabolism. *Physiol. Plant.*, **91**, 537–542.

Fernández-Checa, J.C. and Kaplowitz, N. (1990) The use of monochlorobimane to determine hepatic GSH levels and synthesis. *Anal. Biochem.*, **190**, 212–219.

Grant, C.M. and Dawes, I.W. (1996) Synthesis and role of glutathione in protection against oxidative stress in yeast. *Redox Rep.*, **2**, 223–229.

Griffith, O.W. and Meister, A. (1979) Potent and specific inhibition of glutathione synthesis by buthionine sulfoximine (S-N butyl homocysteine sulfoximine). *J. Biol. Chem.*, **254**, 7558–7560.

Guo, Z.-J., Nakagawara, S., Sumitani, K., and Ohta, Y. (1993) Effect of intracellular glutathione level on the production of 6-methoxymellein in cultured carrot (*Daucus carota*) cells. *Plant Physiol.*, **102**, 45–51.

Guri, A. (1983) Variation in glutathione and ascorbic acid content among selected cultivars of *Phaseolus vulgaris* prior to and after exposure to ozone. *Can. J. Plant Sci.*, **63**, 733–737.

Gustine, D.L. (1987) Induction of medicarpin biosynthesis in ladino clover by p-chloromercuribenzoic acid is reversed by dithiothreitol. *Plant Physiol.*, **84**, 3–6.

Hell, R. (1997) Molecular physiology of plant sulfur metabolism. *Planta*, **202**, 138–148.

Hell, R. and Bergmann, L. (1988) Glutathione synthetase in tobacco suspension cultures: catalytic properties and localisation. *Physiol. Plant.*, **72**, 70–76.

Hell, R. and Bergmann, L. (1990) γ-Glutamylcysteine synthetase in higher plants: catalytic properties and subcellular location. *Planta*, **180**, 603–612.

Hérouart, D., Inzé, D., and Van Montagu, M. (1993) Redox-activated expression of the cytosolic copper/zinc superoxide dismutase gene in *Nicotiana*. *Proc. Natl. Acad. Sci. USA*, **90**, 3108–3112.

Howden, R., Andersen, C.R., Goldsbrough, P.B., and Cobbett, C.S. (1995) Cadmium-sensitive, glutathione-deficient mutant of *Arabidopsis thaliana*. *Plant Physiol.*, **107**, 1067–1073.

Huner, N.P.A., Maxwell, D.P., Gray, G.R., Savitch, L.V., Krol, M., Ivanov, A.G., and Falk, S. (1996) Sensing environmental temperature change through imbalances between energy supply and energy consumption: Redox state of photosystem II. *Physiol. Plant.*, **98**, 358–364.

Jain, A., Martensson, J., Mehta, T., Krauss, A.N., Auld, P.A.M., and Meister, A. (1992) Ascorbic acid prevents oxidative stress in glutathione-deficient mice: Effects on lung type 2 cell lamellar bodies, lung surfactant, and skeletal muscle. *Proc. Natl. Acad. Sci. USA*, **89**, 5093–5097.

Kapheck, S., Chrost, B., Shake, J., and Zimmermann, H. (1992) γ-glutamylcysteinylserine–a new homologue of glutathione in plants of the family Poaceae. *Bot. Acta*, **105**, 174–179.

Karpinski, S., Escobar, C., Karpinska, B., Creissen, G., and Mullineaux, P.M. (1997) Photosynthetic electron transport regulates the expression of cytosolic ascorbate peroxidase genes in excess light stress. *Plant Cell*, **9**, 627–640.

Kavanagh, T.J., Grossman, A., Jacks, E.P., Jinneman, J.C., Eaton, D.L., Martin, G.M., and Rabinovitch, P.S. (1990) Proliferative capacity of human peripheral blood lymphocytes sorted on the basis of glutathione content. *J. Cell Physiol.*, **145**, 472–480.

Kocsy, G., Brunner, M., Rüegsegger, A., Stamp, P., and Brunold, C. (1996) Glutathione synthesis in maize genotypes with different sensitivity to chilling. *Planta*, **198**, 365–370.

Kovari, I.A., Cobbett, C., and Goldsbrough, P.B. (1997) Expression of tomato gamma-Glu-Cys synthetase in the Arabidopsis *cad2* mutant restores cadmium tolerance. *Plant Physiol.* (Suppl.), **114**, 126.

Kreuz, K., Tommasini, R., and Martinoia, E. (1996) Old enzymes for a new job. *Plant Physiol.*, **111**, 349–353.

Kushnir, S., Babiychuk, E., Kampfenkel, K., Belles-Boix, E., Van Montagu, M., and Inzé, D. (1995) Characterization of *Arabidopsis thaliana* cDNAs that render yeasts tolerant to the thiol-oxidizing drug diamide. *Proc. Natl. Acad. Sci. USA*, **92**, 10580–10584.

Lappartient, A.G., and Touraine, B. (1996) Demand driven control of root ATP sulfurylase activity and SO_4^{2-} uptake in intact canola. *Plant Physiol.*, **111**, 147–157.

Lappartient, A.G., and Touraine, B. (1997) Glutathione-mediated regulation of ATP sulfurylase activity, SO_4^{2-} uptake and oxidative stress response in intact canola roots. *Plant Physiol.*, **114**, 177–183.

Legendre, L., Rueter, S., Heinstein, P.F., and Low, P.S. (1993) Characterization of the oligogalacturonide-induced oxidative burst in cultured soybean (*Glycine max*) cells. *Plant Physiol.*, **102**, 233–240.

Levine, A., Tenhaken, R., Dixon, R., and Lamb, C. (1994) H_2O_2 from the oxidative burst orchestrates the plant hypersensitive disease resistance response. *Cell*, **79**, 1–20.

Li, Z.-P., Lu, Y.-P., Zhen, R.-G., Szczypka, M., Thiele, D., and Rea, P.A. (1997) A new pathway for vacuolar cadmium sequestration in *Saccharomyces cerevisiae*: YCF1-catalyzed transport of bis(glutathionato)cadmium. *Proc. Natl. Acad. Sci. USA*, **94**, 42–47.

Liere, K. and Link, G. (1997) Chloroplast endoribonuclease p54 involved in RNA 3'-end processing is regulated by phosphorylation and redox state. *Nucleic Acids Res.*, **25**, 2403–2408.

Link, G. (1996) Green life: control of chloroplast gene transcription. *BioEssays*, **18**, 465–471.

Link, G., Tiller, K., and Baginsky, S. (1997) Glutathione, a regulator of chloroplast transcription. In K.K. Hatzios, (ed.), *Regulation of Enzymatic Systems Detoxifying Xenobiotics in Plants*, (NATO Sciences Partnership Sub-Series, Vol. 37), Kluwer Academic Publishers, Dordrecht, pp. 125–138.

Lobréaux, S., Thoiron, S., and Briat, J.-F. (1995) Induction of ferritin synthesis in maize leaves by an iron-mediated oxidative stress. *Plant J.*, **8**, 443–449.

Lu, Y.-P., Li, Z.-S., and Rea, P.A. (1997) *AtMRP1* gene of Arabidopsis encodes a glutathione *S*-conjugate pump: isolation and functional definition of a plant ATP-binding cassette transporter. *Proc. Natl. Acad. Sci. USA*, **94**, 8243–8248.

Lueder, D.V. and Phillips, M.A. (1996) Characterization of *Trypanosoma brucei* γ-glutamylcysteine synthetase, an essential enzyme in the biosynthesis of trypanothione (diglutathionylspermidine). *J. Biol. Chem.*, **271**, 17485–17490.

Madamanchi, N.R. and Alscher, R.G. (1991) Metabolic bases for differences in sensitivity of two pea cultivars to sulfur dioxide. *Plant Physiol.*, **97**, 88–93.

Marrs, K.A. (1996) The functions and regulation of glutathione *S*-transferases in plants. *Ann. Rev. Plant Physiol. Plant Mol. Biol.*, **47**, 127–158.

Martensson, J. and Meister, A. (1992) Glutathione deficiency increases hepatic ascorbic acid synthesis in adult mice. *Proc. Natl. Acad. Sci. USA*, **89**, 11566–11568.

May, M.J. and Leaver, C.J. (1993) Oxidative stimulation of glutathione synthesis in *Arabidopsis thaliana* suspension cultures. *Plant Physiol.*, **103**, 621–627.

May, M.J. and Leaver, C.J. (1994) *Arabidopsis thaliana* γ-glutamylcysteine synthetase is structurally unrelated to mammalian, yeast and *E. coli* homologs. *Proc. Natl. Acad. Sci. USA*, **91**, 10059–10063.

May, M.J., Hammond-Kosack, K., and Jones, J.D.G. (1996a) Involvement of reactive oxygen species, glutathione metabolism, and lipid peroxidation in the *Cf*-gene-dependent defense response of tomato cotyledons induced by race-specific elicitors of *Cladosporium fluvum*. *Plant Physiol.*, **110**, 1367–1379.

May, M.J., Parker, J.E., Daniels, M.J., Leaver, C.J., and Cobbett, C.S. (1996b) An *Arabidopsis* mutant depleted in glutathione shows unaltered responses to fungal and bacterial pathogens. *Mol. Plant-Microbe Interact.*, **9**, 349–356.

May, M., Vernoux T., Leaver, C., Van Montagu, M., and Inzé, D. (1998a) Glutathione homeostasis in plants: implications for environmental sensing and plant development. *J. Exp. Bot.*, **49**, 649.

May, M.J., Vernoux, T., Sánchez-Fernández, R., Van Montagu, M., and Inzé, D. (1998b) Evidence for posttranscriptional activation of γ-glutamylcysteine synthetase during plant stress responses. *Proc. Natl. Acad. Sci. USA*, **95**, 12049

Mihm, S., Galter, D., and Dröge, W. (1995) Modulation of transcription factor NFκB activity by intracellular glutathione levels and by variations of the extracellular cysteine supply. *FASEB J.*, **9**, 246–252.

Nakagawara, S., Nakamura, N., Guo, Z.-J., Sumitani, K., Katoh, K., and Ohta, Y. (1993) Enhanced formation of a constitutive sesquiterpenoid in cultured cells of a liverwort, *Calypogeia granulata* Inoue during elicitation; effects of vanadate. *Plant Cell Physiol.*, **34**, 421–429.

Nieto-Sotelo, J. and Ho, T.-H.D. (1986) Effect of heat shock on the metabolism of glutathione in maize roots. *Plant Physiol.*, **82**, 1031–1035.

Noctor, G., Strohm, M., Jouanin, L., Kunert, K.-J., Foyer, C.H., and Rennenberg, H. (1996) Synthesis of glutathione in leaves of transgenic poplar overexpressing γ-glutamylcysteine synthetase. *Plant Physiol.*, **112**, 1071–1078.

Noctor, G., Arisi, A.-C.M., Jouanin, L., Valadier, M.-H., Roux, Y., and Foyer, C.H. (1997) The role of glycine in determining the rate of glutathione synthesis in poplar: Possible implications for glutathione production during stress. *Physiol. Plant.*, **100**, 255–263.

Poot, M., Teubert, H., Rabinovitch, P.S., and Kavanagh, T.J. (1995) De novo synthesis of glutathione is required for both entry into and progression through the cell cycle. *J. Cell Physiol.*, **163**, 555–560.

Prasad, T.K., Anderson, M.D., Martin, B.A., and Stewart, C.R. (1994) Evidence for chilling-induced oxidative stress in maize seedlings and a regulatory role for hydrogen peroxide. *Plant Cell*, **6**, 65–74.

Price, C.A. (1957) A new thiol in legumes. *Nature*, **180**, 148–149.

Price, A.H., Taylor, A., Ripley, S.J., Griffiths, A., Trewavas, A.J., and Knight, M.R. (1994) Oxidative signals in tobacco increase cytosolic calcium. *Plant Cell*, **6**, 1301–1310.

Rawlins, M.R., Leaver, C.J., and May, M.J. (1995) Characterisation of a cDNA encoding *Arabidopsis* glutathione synthetase. *FEBS Lett.*, **376**, 81–86.

Rennemberg, H., Birk, C., and Schaer, B. (1982) Effect of N,-N-diallyl-2,2-dichloroacetamide (R-25788) on efflux and synthesis of glutathione in tobacco suspension cultures. *Phytochemistry*, **21**, 2778–2781.

Rice, G.C., Bump, E.D., Shrieve, D.C., Lee, W., and Kovacs, M. (1986) Quantitative analysis of cellular glutathione by flow cytometry utilizing monochlorobimane: some applications to radiation and drug resistance *in vitro* and *in vivo*. *Cancer Res.*, **46**, 6105–6110.

Richman, P.G. and Meister, A. (1975) Regulation of γ-glutamylcysteine synthetase by non-allosteric feedback inhibition by glutathione. *J. Biol. Chem.*, **250**, 1422–1426.

Rüegsegger, A. and Brunold, C. (1992) Effect of cadmium on γ-glutamylcysteine synthesis in maize seedlings. *Plant Physiol.*, **99**, 428–433.

Rüegsegger, A. and Brunold, C. (1993) Localization of γ-glutamylcysteine synthetase and glutathione synthetase activity in maize seedlings. *Plant Physiol.*, **101**, 561–566.

Rüegsegger, A., Schmutz, D., and Brunold, C. (1990) Regulation of glutathione synthesis by cadmium in *Pisum sativum* L. *Plant Physiol.*, **93**, 1579–1584.

Sánchez-Fernández, R., Fricker, M., Corben, L.B., White, N.S., Sheard, N., Leaver, C.J., Van Montagu, M., Inzé, D., and May, M.J. (1997) Cell proliferation and hair tip growth in the *Arabidopsis* root are under mechanistically different forms of redox control. *Proc. Natl. Acad. Sci. USA*, **94**, 2745–2750.

Sánchez-Fernández, R., Ardiles-Díaz, W., Van Montagu, M., Inzé, D., and May, M.J. (1998) Cloning and expression analysis of AtMRP4, a novel *Arabidopsis thaliana* MRP-like gene. *Mol. Gen. Genet.*, **258**, 655–662.

Schäfer, H.J., Greiner, S., Rausch, T., and Haag-Kerwer, A. (1997) In seedlings of the heavy metal accumulator *Brassica juncea* Cu^{2+} differentially affects transcript amounts for γ-glutamylcysteine synthetase (γECS) and metallothionein (MT2). *FEBS Lett.*, **404**, 216–220.

Sen Gupta, A., Alscher, R.G., and McCune, D. (1991) Response of photosynthesis and cellular antioxidants to ozone in *Populus* leaves. *Plant Physiol.*, **96**, 650–655.

Shrieve, D.C., Bump, E.A., and Rice, G.C. (1988) Heterogeneity of cellular glutathione among cells derived from a murine fibrosarcoma or a human renal cell carcinoma detected by flow cytometric analysis. *J. Biol. Chem.*, **263**, 14107–14114.

Smirnoff, N. (1993) The role of active oxygen in the response of plants to water deficit and dessication. *New Phytol.*, **125**, 27–58.

Stossel, P. (1984) Regulation by sulfhydryl groups of glyceollin accumulation in soybean hypocotyls. *Planta*, **160**, 314–319.

Strohm, M., Jouanin, L., Kunert, K.-J., Pruvost, C., Polle, A., Foyer, C.H., and Rennenberg, H. (1995) Regulation of glutathione synthesis in leaves of transformed poplar (*Populus tremula* x *P. alba*) overexpressing glutathione synthetase. *Plant J.*, **7**, 141–145.

Sun, W.-M., Huang, Z.-Z., and Lu, S.C. (1996) Regulation of γ-glutamylcysteine synthetase by protein phosphorylation. *Biochem. J.*, **320**, 321–328.

Ullmann, P., Gondet, L., Potier, S., and Bach, T.J. (1996) Cloning of the *Arabidosis thaliana* glutathione synthetase (*GSH2*) by functional complementation of a yeast *gsh2* mutant. *Eur. J. Biochem.*, **236**, 662–669.

Vernoux, T., Wilson, R.C., Seeley, K.A., Reichheld, J.-P., Muroy, S., Maughan, S.C., Cobbett, C.S., Van Montagu, M., Inzé, D., May, M.J., and Sung, Z.R. (2000) The *ROOT MERISTEMLESS1/CADMIUM SENSITIVE2* gene defines a glutathione-dependent pathway involved in initiation and maintenance of cell division during postembryonic root development. *Plant Cell*, **12**, 97–110.

Wemmie, J.A., Steggerda, S.M., and Moye-Rowley, W.S. (1997) The *Saccharomyces cerevisiae* AP-1 protein discriminates between oxidative stress elicited by the oxidants H_2O_2 and diamide. *J. Biol. Chem.*, **272**, 7908–7914.

Wingate, V.P.M., Lawton, M.A., and Lamb, C.J. (1988) Glutathione causes a massive and selective induction of plant defense genes. *Plant Physiol.*, **87**, 206–210.

Wingsle, G. and Karpinski, S. (1996) Differential redox regulation by glutathione of glutathione reductase and CuZn-superoxide dismutase gene expression in *Pinus sylvestris* L. needles. *Planta*, **198**, 151–157.

Wolosiuk, R.A. and Buchanan, B.B. (1977) Thioredoxin and glutathione regulate photosynthesis in chloroplasts. *Nature*, **266**, 565–567.

Yalpini, N., Endeyi, AJ., Leon, J., and Raskin, I. (1994) Ultraviolet ligth and ozone stimulate accumulation of salicylic acid, pathogen-related proteins and virus resistance in tobacco. *Planta*, **193**, 372–376.

Index